Renewable Energy Integration and Application in Buildings for Carbon Neutrality 2nd Edition

Renewable Energy Integration and Application in Buildings for Carbon Neutrality 2nd Edition

Guest Editors

Lin Lu
Jianheng Chen

MDPI

Basel • Beijing • Wuhan • Barcelona • Belgrade • Novi Sad • Cluj • Manchester

Guest Editors

Lin Lu
Department of Building
Environment and Energy
Engineering
The Hong Kong Polytechnic
University
Hong Kong
China

Jianheng Chen
College of Pipeline and Civil
Engineering
China University of Petroleum
(East China)
Qingdao
China

Editorial Office
MDPI AG
Grosspeteranlage 5
4052 Basel, Switzerland

This is a reprint of the Special Issue, published open access by the journal *Sustainability* (ISSN 2071-1050), freely accessible at: https://www.mdpi.com/journal/sustainability/special_issues/A3846NAAUP.

For citation purposes, cite each article independently as indicated on the article page online and as indicated below:

Lastname, A.A.; Lastname, B.B. Article Title. *Journal Name* **Year**, *Volume Number*, Page Range.

ISBN 978-3-7258-7841-3 (Hbk)
ISBN 978-3-7258-7842-0 (PDF)
https://doi.org/10.3390/books978-3-7258-7842-0

Contents

About the Editors

Lin Lu

Lin Lu is a professor at The Hong Kong Polytechnic University, where she has worked since 2006. Her research interests include solar and offshore wind energy, radiative cooling to deep space, smart nanomaterial building envelopes, and heat and mass transfer for building energy systems. She has led numerous projects funded by the Hong Kong Research Grant Council, Innovation and Technology Commission, Environmental Protection Department, and industry partners, including Sinopec. A Highly Cited Researcher by Clarivate Analytics and a World's Top 2% Scientist by Stanford University, she has published over 350 SCI papers and six books, received multiple prestigious awards, serves on several government committees, and holds editorial roles including Associate Editor of *Renewable and Sustainable Energy Reviews* and Editor-in-Chief of *Energy Engineering*.

Jianheng Chen

Jianheng Chen is a professor at China University of Petroleum (East China), where his research expertise spans radiative cooling and thermal management, cold energy conservation in energy storage and transportation systems, and low-carbon energy technologies. He earned his Ph.D. in renewable energy from The Hong Kong Polytechnic University, followed by postdoctoral research at City University of Hong Kong. He has secured major research funding, including projects on efficient cold energy utilization. His achievements include numerous publications in high-impact journals, inclusion in the World's Top 2% Scientists list, service on the editorial boards of several international journals, and delivery of numerous keynote and invited talks at international conferences.

sustainability

MDPI

Editorial

Renewable Energy Integration and Application in Buildings for Carbon Neutrality

Jianheng Chen [1,*] and Lin Lu [2]

1 College of Pipeline and Civil Engineering, China University of Petroleum (East China), Qingdao 266580, China
2 Department of Building Environment and Energy Engineering, The Hong Kong Polytechnic University, Hong Kong, China; vivien.lu@polyu.edu.hk
* Correspondence: jianheng.chen@upc.edu.cn

The global pursuit of carbon neutrality has indisputably placed the building sector at the heart of climate change mitigation strategies [1]. As a major contributor to global energy consumption and greenhouse gas emissions, the built environment is not merely a collection of static structures but a dynamic system deeply influenced by thermodynamic necessities, complex occupant behaviors, and evolving technological constraints [2]. While early green building efforts focused primarily on improving insulation or applying singular renewable technologies (such as rooftop photovoltaics), the current global climate paradigm demands a profound and comprehensive transformation [3]. This shift from simply pursuing "energy efficiency" to achieving "carbon neutrality" necessitates a systemic approach: the deep integration of multi-carrier energy supply networks, high-performance physical envelopes constructed with advanced materials, and the implementation of data-driven intelligent operational controls. As the urgency of the climate crisis intensifies and international commitments like the Paris Agreement are agreed [4], policymakers, engineers, and urban planners face unprecedented pressure to find comprehensive solutions that are scalable, reliable, and economically viable to address the growing challenges of urbanization and energy.

However, the path to a net-zero future remains fraught with challenges. The inherent intermittency of renewable energy sources (such as solar and wind) leads to significant temporal mismatches between peak energy production and peak building demand, placing immense strain on grid stability [5]. Simultaneously, the physical limitations of existing, vast building infrastructure pose formidable barriers to energy-saving retrofits. Furthermore, the increasing technological complexity of modern building systems, ranging from HVAC and lighting to distributed generation, often exceeds the capabilities of traditional management models [6]. This Special Issue, "Renewable Energy Integration and Application in Buildings toward Carbon Neutrality", aims to address these critical knowledge and technical gaps. Moving beyond isolated analyses of single technologies (such as simply installing heat pumps or batteries), it explores the synergistic effects among three fundamental pillars: the diversified supply and efficient storage of clean energy (Source); the optimization of physical building carriers, thermal performance, and structural integrity (Flow); and intelligent decision-making in operating systems and user interaction (Control). Through this holistic perspective, this collection seeks to provide theoretical foundations and practical pathways for building a climate-resilient human habitat of the future.

The first pillar of this transformation addresses the comprehensive decarbonization of energy supply, the advancement of storage capabilities, and the modernization of associated grid infrastructure. The shift toward distributed energy generation requires robust systems capable of flexibly handling multiple energy carriers (electricity, heat, cooling, and gas). In this domain, Hybrid Renewable Energy Systems (HRESs) that synthesize wind, solar, and green hydrogen demonstrate immense transformative potential. Gabbar et al. (Contribution 1) reveal that the integration of HRESs not only increases the renewable fraction of the energy mix but also can effectively cover up to 63% of key electrical loads, such as lighting and equipment. By coupling HRESs with sector-coupling strategies, including blending green hydrogen for heating and transportation, these systems can reduce annual greenhouse gas emissions by over 200 tons of CO_2 per building while delivering substantial cost benefits. However, clean generation is only half the equation: efficient energy storage serves as the indispensable temporal bridge connecting intermittent supply with continuous demand. Innovative approaches are now unlocking the latent potential of the ground itself. Zdankus et al. (Contribution 2) demonstrate that utilizing the soil beneath buildings as a seasonal thermal accumulator can fundamentally alter a building's thermodynamic profile. By deploying separated sand storage structures, this method can reduce heat flux density through the floor by up to 21%, reaching a new thermodynamic equilibrium within approximately one week and improving annual energy efficiency by more than 10%. Similarly, addressing the legacy of the fossil fuel era, Liu et al. (Contribution 3) explore the creative repurposing of abandoned industrial infrastructure by changing oil wells into deep coaxial geothermal heat exchangers. This approach not only remediates environmental liabilities but also converts them into assets for the renewable transition, achieving heat extraction rates exceeding 400 kW while significantly reducing drilling costs. Crucially, the sustainability of these building-level innovations relies on macro-level grid support. Che et al. (Contribution 4) highlight that the operational phase of electrical substations, the critical interface between buildings and the grid, accounts for up to 70% of their life-cycle carbon footprint. Their review confirms that transitioning to high-efficiency equipment, combined with on-site photovoltaic–battery hybrid systems and ecological landscaping, provides a necessary pathway to address the "last-mile" challenges of green power distribution.

The second pillar focuses on the building's physical fabric and thermal management systems, specifically encompassing advanced envelope materials and cooling technologies. No amount of renewable generation can compensate for a structure that is thermally inefficient. The research presented here emphasizes a return to bio-based materials. Brien et al. (Contribution 5) validate the hygrothermal performance of Cross-Laminated Timber (CLT) in cold climates. Through field monitoring (2 years) and simulation (5 years) of a CLT–wood fiber insulation (WFI) building, they demonstrate stable temperature and humidity conditions with no risk of mold. Crucially, the study reveals that vapor-open assemblies allow internal moisture to dissipate efficiently, confirming that low embodied carbon does not come at the expense of durability. Complementing these structural innovations is the integration of dynamic thermal regulation. Jiao et al. (Contribution 6) review phase change materials (PCMs) and illustrate how embedding latent heat storage directly into envelopes buffers thermal fluctuations. They highlight that micro/nanoencapsulated PCMs significantly enhance thermal inertia, while hybrid applications incorporating radiant cooling panels and thermochromic smart windows synergistically improve thermal comfort. Beyond passive regulation, innovative cooling strategies are evolving to meet the demands of a warming world. Passive radiative cooling offers a zero-energy method of transferring heat to outer space. Liang et al. (Contribution 7) demonstrate that radiative cooling materials, including multilayer films, inorganic pigments, and polymer-based coatings, can achieve significant sub-ambient temperature reductions. In real-world conditions, these materials have achieved cooling powers frequently exceeding 100 W/m^2 and daytime temperature drops of 5 to 16 °C. Additionally, advancements have focused on combining

 https://doi.org/10.3390/su1809431

high solar reflectance with strong mid-infrared emissivity, while also improving durability and self-cleaning properties. Nanofluid-enhanced HVAC systems promise to breach the thermal limits of conventional fluids. Myat et al. (Contribution 8), synthesizing experimental and AI-driven studies, highlight that nanofluids, colloidal suspensions of nanoparticles in base fluids, demonstrate up to 45% enhancement in heat transfer coefficients and notable increases in the system coefficient of performance (COP). Despite remarkable laboratory gains, the scalable application of nanofluid-based thermal enhancement still requires advances in long-term stability, cost optimization, and full-system integration with smart energy management technologies. High-load environments have also received special attention. The feasibility of indirect evaporative cooling for data centers has emerged as a key focus, given the sector's growing energy demand. Shi et al. (Contribution 9) indicate that multi-stage indirect evaporative cooling systems, when assisted by liquid desiccant, can reduce dependence on traditional mechanical cooling in hot–humid climates. This hybrid approach effectively improves cooling performance and minimizes the operational hours of energy-intensive supplemental systems, demonstrating strong potential for achieving low-carbon data center operations even in challenging environments. However, as buildings transform into power plants, thermal optimization must be matched by structural reliability. Scrinzi et al. (Contribution 10) present wind tunnel analyses of rooftop photovoltaic arrays. Their findings reveal that local wind loads often exceed standard code estimates, particularly under oblique wind conditions, highlighting the necessity of comprehensive, site-specific assessments to ensure that the physical integration of renewable assets does not compromise structural safety.

The third and final pillar is the digital infrastructure connecting supply and demand: sensing and intelligent operation. As buildings evolve into active energy nodes, traditional static management approaches become obsolete. The foundation of this intelligence is established through advanced sensing and data processing technologies, which empower the Internet of Things (IoT) to monitor complex service interactions. As demonstrated by Elkhoukhi et al. (Contribution 11), scalable IoT architectures supported by big data analytics frameworks such as Apache Kafka and Storm facilitate real-time collection, fusion, and analysis of environmental and occupancy data within buildings. These technologies enable low-latency, fault-tolerant data pipelines, forming the backbone for adaptive control, predictive maintenance, and the seamless integration of smart services across diverse building platforms. Such comprehensive data streams enable the creation of digital twins, which allow for the visualization of thermo-hygrometric conditions within Building Information Modeling (BIM) environments. De Rubeis et al. (Contribution 12) highlight that BIM-based digital twin workflows now incorporate real-time sensor data with three-dimensional semantic models, providing facility managers with intuitive visual tools to monitor temperature and humidity at the room level. These digital twins facilitate proactive maintenance, anomaly detection, and occupant engagement, while serving as an expandable foundation for future predictive analytics and energy optimization. To close the loop between data and action, adaptive grey-box models provide a practical approach for predicting thermal dynamics and optimizing retrofit measures, thereby ensuring that energy interventions are grounded in empirical observations rather than mere theoretical assumptions. Yang et al. (Contribution 13) demonstrate that these flexible grey-box models, validated through real-world case studies, utilize readily accessible measurements including temperature, solar radiation, and heating system variables to capture building dynamics at the room scale. When integrated with digital twins, these models enable multi-objective optimization of retrofit strategies and control schemes, effectively balancing energy savings with occupant comfort while remaining adaptable as new data streams emerge.

https://doi.org/10.3390/su18094310

Synthesizing the contributions collected in this Special Issue, a clear, interdependent pathway to carbon neutrality emerges. Rather than viewing renewable integration as a mere add-on, the research confirms that "Source", "Flow", and "Control" must operate as a unified system. Current studies underscore that achieving "Zero Carbon" is economically unviable if relying on active generation alone; the load must first be minimized. Innovative envelope materials and advanced cooling technologies (Pillar 2), including bio-based solutions, phase change materials, and radiative cooling, embody a "Passive First" philosophy that renders the Hybrid Renewable Energy Systems (Pillar 1) more effective and affordable. Furthermore, the complexity of managing these coupled systems, where a building is simultaneously a consumer, a generator, and a storage unit, necessitates a shift from manual operation to the digital twin-based controls identified in Pillar 3. This digital layer acts as the brain, managing the temporal mismatch between renewable availability and occupant demand. Therefore, the articles in this Special Issue collectively argue that future progress lies not in the breakthrough of a single technology, but in the convergence of energy systems engineering, material science, and digital technologies. By dissolving the boundaries between these disciplines, future research can foster a seamless symbiosis where buildings adapt to both their occupants and the broader ecological context, ultimately transforming the built environment from a passive energy consumer into an active steward of environmental sustainability.

Acknowledgments: The authors would like to acknowledge the partial support from the National Natural Science Foundation of China (52504069), Natural Science Foundation of Shandong Province (2025HWYQ-039), China Post-doctoral Science Foundation (2025M770581 and GZB20250679) and Shandong Provincial Postdoctoral Innovation Project (SDCX-ZG-202501025).

Conflicts of Interest: The authors declare no conflicts of interest.

List of Contributions:

1. Gabbar, H.A.; Ramadan, A. Integrated Renewable Energy Systems for Buildings: An Assessment of the Environmental and Socio-Economic Sustainability. *Sustainability* **2025**, *17*, 656.
2. Zdankus, T.; Bandarwadkar, S.; Vaiciunas, J.; Stelmokaitis, G.; Vaicaitis, A. Research on Increasing the Building's Energy Efficiency by Using the Ground Beneath It for Thermo-Accumulation. *Sustainability* **2025**, *17*, 262.
3. Liu, R.-J.; Jia, L.-R.; Zhang, W.-S.; Yu, M.-Z.; Zhao, X.-D.; Cui, P. Study of Heat Transfer Characteristics and Economic Analysis of a Closed Deep Coaxial Geothermal Heat Exchanger Retrofitted from an Abandoned Oil Well. *Sustainability* **2024**, *16*, 1603.
4. Che, W.; Wang, Y.; Zhu, W.; Hong, L.; Fang, C.; Wu, H.; Liu, J. A Review of Carbon Emission Reduction During the Operation Stage of Substations. *Sustainability* **2024**, *16*, 10017.
5. O'Brien, L.; Li, L.; Herzog, B.; Snow, J.; Friess, W.A. Field Monitoring and Modeling of the Hygrothermal Performance of a Cross-Laminated Timber and Wood Fiber-Insulated Building Located in a Cold Climate. *Sustainability* **2025**, *17*, 7879.
6. Jiao, K.; Lu, L.; Zhao, L.; Wang, G. Towards Passive Building Thermal Regulation: A State-of-the-Art Review on Recent Progress of PCM-Integrated Building Envelopes. *Sustainability* **2024**, *16*, 6482.
7. Liang, L.; Bai, S.; Lin, K.; Kwok, C.T.; Chen, S.; Zhu, Y.; Tso, C.Y. Advancing Sustainable Development: Broad Applications of Passive Radiative Cooling. *Sustainability* **2024**, *16*, 2346.
8. Myat, A.; Rahman, M.M.; Akbar, M. Nanofluid-Enhanced HVAC&R Systems (2015–2025): Experimental, Numerical, and AI-Driven Insights with a Strategic Roadmap. *Sustainability* **2025**, *17*, 7371.
9. Shi, W.; Ma, X.; Min, Y.; Yang, H. Feasibility Analysis of Indirect Evaporative Cooling System Assisted by Liquid Desiccant for Data Centers in Hot-Humid Regions. *Sustainability* **2024**, *16*, 2011.

https://doi.org/10.3390/su1809431

10. Scrinzi, G.; Mazzucchelli, E.S.; Muggiasca, S. Photovoltaic Strings on Large, Flat Roofs: Experimental Wind Loads on Representative Configurations. *Sustainability* **2025**, *17*, 5914.
11. Elkhoukhi, H.; Elmouatamid, A.; Haibi, A.; Bakhouya, M.; El Ouadghiri, D. Overview of Sensing and Data Processing Technologies for Smart Building Services and Applications. *Sustainability* **2025**, *17*, 4029.
12. De Rubeis, T.; Ciccozzi, A.; Ragnoli, M.; Stornelli, V.; Brusaporci, S.; Tata, A.; Ambrosini, D. A Workflow for a Building Information Modeling-Based Thermo-Hygrometric Digital Twin: An Experimentation in an Existing Building. *Sustainability* **2024**, *16*, 10281.
13. Yang, Y.; Jradi, M. Adaptive Grey-Box Modelling for Energy-Efficient Building Retrofits: Case Studies in Denmark. *Sustainability* **2025**, *17*, 1702.

eferences

Kashif, U.; Abbas, S.; Kousar, S.; Lu, H. Linking of bio-energy and carbon neutrality: Navigating economic policy uncertainty and climate change policy in the USA. *Energy* **2025**, *324*, 136012. [CrossRef]

Kim, J.; Yeom, S.; Hong, T. Analyzing the cooling effect, thermal comfort, and energy consumption of integrated arrangement of high-rise buildings and green spaces on urban heat island. *Sustain. Cities Soc.* **2025**, *119*, 106105. [CrossRef]

Houchmand, L.J.; Martí, M.M.; Gassó-Domingo, S. Photovoltaics and green roofs: Holistic analysis in built environments. *Renew. Sustain. Energy Rev.* **2025**, *207*, 114987. [CrossRef]

Agreement, P. Paris agreement. In *Proceedings of the Report of the Conference of the Parties to the United Nations Framework Convention on Climate Change (21st Session, 2015: Paris). Retrived December*; No. 2017: HeinOnline; UNFCCC: Bonn, Germany, 2015; Volume 4, p. 2.

Mousavi, R.; Mousavi, A.; Mousavi, Y.; Tavasoli, M.; Arab, A.; Kucukdemiral, I.B.; Alfi, A.; Fekih, A. Revolutionizing solar energy resources: The central role of generative AI in elevating system sustainability and efficiency. *Appl. Energy* **2025**, *382*, 125296. [CrossRef]

Sha, X.; Ma, Z.; Sethuvenkatraman, S.; Li, W. Online learning-enhanced data-driven model predictive control for optimizing HVAC energy consumption, indoor air quality and thermal comfort. *Appl. Energy* **2025**, *383*, 125341. [CrossRef]

https://doi.org/10.3390/su18094310

sustainability

MDPI

Article

Integrated Renewable Energy Systems for Buildings: An Assessment of the Environmental and Socio-Economic Sustainability

Hossam A. Gabbar [1,*] and A. Ramadan [1,2]

1 Faculty of Engineering and Applied Science, Ontario Tech University, Oshawa, ON L1G0C5, Canada; ahmed.mohamed@ontariotechu.ca

2 College of Engineering (Cairo Branch), Arab Academy for Science Technology and Maritime Transport (AASTMT), Elhorria, Cairo P.O. Box 2033, Egypt

* Correspondence: hossam.gaber@ontariotechu.ca; Tel.: +1-9057218668 (ext. 5497)

Abstract: Developing a green energy strategy for municipalities requires creating a framework to support the local production, storage, and use of renewable energy and green hydrogen. This framework should cover essential components for small-scale applications, including energy sources, infrastructure, potential uses, policy backing, and collaborative partnerships. It is deployed as a small-scale renewable and green hydrogen unit in a municipality or building demands meticulous planning and considering multiple elements. Municipality can promote renewable energy and green hydrogen by adopting policies such as providing financial incentives like property tax reductions, grants, and subsidies for solar, wind, and hydrogen initiatives. They can also streamline approval processes for renewable energy installations, invest in hydrogen refueling stations and community energy projects, and collaborate with provinces and neighboring municipalities to develop hydrogen corridors and large-scale renewable projects. Renewable energy and clean hydrogen have significant potential to enhance sustainability in the transportation, building, and mining sectors by replacing fossil fuels. In Canada, where heating accounts for 80% of building energy use, blending hydrogen with LPG can reduce emissions. This study proposes a comprehensive approach integrating renewable energy and green hydrogen to support small-scale applications. The study examines many scenarios in a building as a case study, focusing on economic and greenhouse gas (GHG) emission impacts. The optimum scenario uses a hybrid renewable energy system to meet two distinct electrical needs, with 53% designated for lighting and 10% for equipment with annual saving CAD$ 87,026.33. The second scenario explores utilizing a hydrogen-LPG blend as fuel for thermal loads, covering 40% and 60% of the total demand, respectively. This approach reduces greenhouse gas emissions from 540 to 324 tCO_2/year, resulting in an annual savings of CAD$ 251,406. This innovative approach demonstrates the transformative potential of renewable energy and green hydrogen in enhancing energy efficiency and sustainability across sectors, including transportation, buildings, and mining.

Keywords: wind energy; solar energy; energy transition; hybrid renewable energy; green hydrogen; GHG

1. Introduction

Renewable energy from natural elements is more environmentally friendly than fossil fuels and offers an alternative to conventional energy sources. In the context of buildings, renewable energy involves incorporating sustainable energy sources like solar, wind, hydrogen, and biomass throughout the entire building lifecycle spanning design, construction, operation, and maintenance to lessen reliance on fossil fuels and traditional energy, fostering environmental sustainability and helping combat climate change [1]. The use of renewable energy in buildings primarily depends on the energy requirements of the building and the specific types of energy sources available. Among the various renewable energy options, solar, wind, hydro, tidal, geothermal, and hydrogen are widely acknowledged as leading and well-established technologies in the renewable energy industry. However, solar, wind, geothermal, and hydrogen energy hold the greatest potential for meeting the energy demands of buildings [2]. Figure 1 visually summarizes the progression of energy policies and the growing integration of renewable energy and energy efficiency measures in building design and operation. It reflects the increasing emphasis on sustainability and global commitments towards reducing carbon footprints in the construction and operation of buildings.

The International Energy Agency (IEA) reports that buildings are responsible for approximately 40% of global energy use and contribute 30% of worldwide carbon dioxide emissions [3]. Therefore, reducing energy consumption in buildings becomes essential for lowering greenhouse gas emissions and minimizing environmental impact. Implementing renewable hybrid energy systems in buildings can significantly contribute to this objective. The demand for renewable hybrid energy systems in buildings is anticipated to increase consistently over the coming decade. The building sector currently accounts for 20–40% of total energy consumption in developed countries [4,5]. With two-thirds of greenhouse gas (GHG) emissions stemming from energy consumption, transitioning from fossil fuels to low-carbon alternatives is essential [6]. Renewable energy and energy efficiency offer effective solutions for reducing greenhouse gas emissions while meeting the energy needs of billions, steering the world toward sustainable economic and social development [7]. Furthermore, renewable energy sources offer viable alternatives to fossil fuels, helping to address the building industry's energy challenges and reduce environmental pollution. Renewable options such as solar, hydrogen, and wind energy have seen considerable growth in recent years [8,9]. Clean hydrogen offers significant potential to enhance the environmental sustainability of the transportation and mining sectors as an alternative to fossil fuels. Heating currently constitutes 80% of energy use in Canadian homes, and blending hydrogen with natural gas offers an alternative for reducing emissions. The combustion process results in 541 Mt CO_2e annually, encompassing emissions from electricity and fuel production (192 Mt CO_2e) and the end use of carbon-based fuels (349 Mt CO_2e) across transportation, buildings, and industry. Table 1 provides a breakdown of carbon emissions by application.

Table 1. Carbon emission yearly due to application.

No.	Application	Mt CO_2e/yr
1	Fugitive emissions from fuel and electricity production	55
2	Process emissions from non-energy industries such as cement	56
3	Agricultural emissions, primarily from nitrous oxide and methane	59
4	Waste management emissions, primarily from methane	18

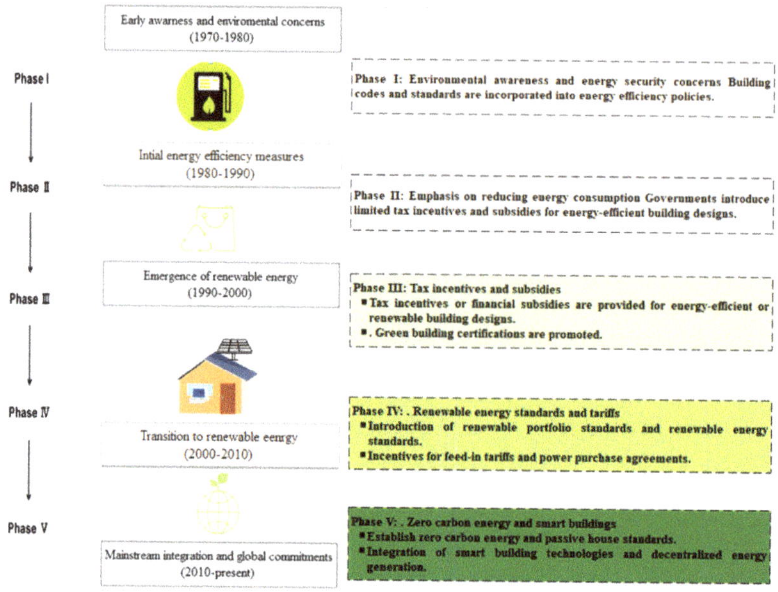

Figure 1. Five-phase diagram illustrating the progression of energy efficiency and renewable energy integration in buildings [4].

Utilizing low-emission hydrogen and hydrogen-based fuels is projected to yield a moderate reduction in CO_2 emissions by 2030, especially when compared to other significant mitigation strategies such as expanding renewable energy, direct electrification, and promoting behavior change. Figure 2 illustrates the global share of renewable energy from 2018 to 2020, highlighting variations across different renewable sources. Canada's pathways to achieving net zero carbon rely on a mix of economic, political, cultural, and technological factors that will shape tangible results and interact dynamically throughout the process. While Canada has full control over some of these factors, others can influence only partially, and some remain entirely beyond its control. Technological innovation, particularly in negative emission solutions like direct air capture, hydrogen generation, and carbon-sequestering bioenergy, is expected to play a pivotal role in reaching the net zero target. These choices will shape Canada's net zero journey and existing pathways.

The shift towards reducing GHG emissions should focus on moving from a heavy reliance on fossil fuels to an increased use of renewable energy sources and hydrogen. Key strategic points for transitioning from fossil fuel energy to net zero emissions are summarized in Table 2.

Using renewable energy and hydrogen (H_2) for building loads is gaining traction as residents, industries, and governments focus on sustainability and carbon reduction. The use of renewable energy and hydrogen (H_2) for building loads is indeed gaining momentum, driven by the dual goals of sustainability and carbon footprint reduction. Governments and industries worldwide are increasingly prioritizing these technologies, as they not only reduce dependence on fossil fuels but also support grid stability and energy resilience.

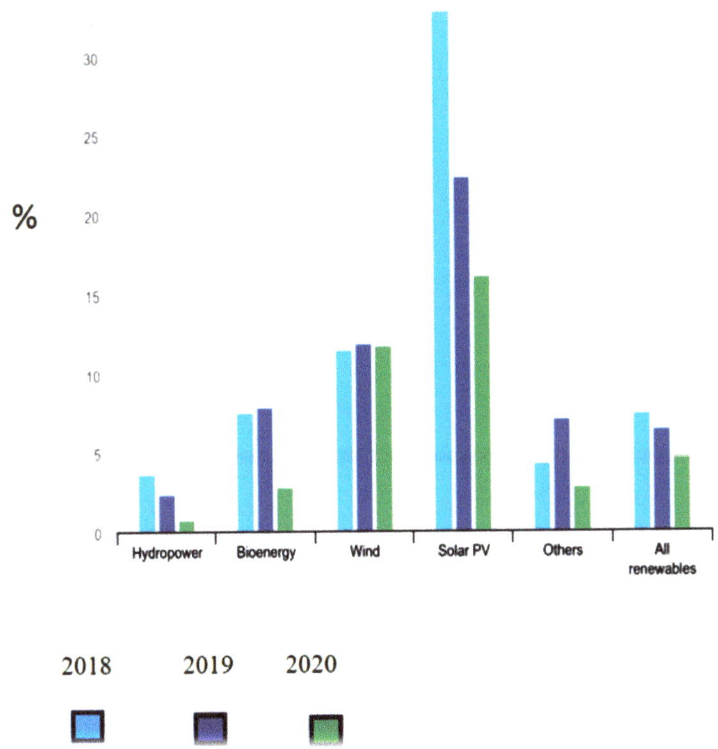

Figure 2. Annual growth for renewable electricity generation by source, 2018–2020 [10].

Table 2. The strategies sequences needed to achieve net zero emissions.

Strategy	Methodology
Decarbonizing Electricity	• Renewable Energy Expansion • Nuclear Power • Energy Storage
Electrification of End-Use Sectors	• Promoting electric vehicles (EVs), expanding • EV charging infrastructure, and supporting public transit electrification. • Enhancing energy efficiency in new and existing buildings, • Transitioning to electric heating systems, such as heat pumps. • Electrifying industrial processes • Adopting advanced low-carbon technologies.
Energy Efficiency Improvements	• Implementing energy-saving technologies • Retrofitting buildings and infrastructure to reduce energy consumption.
Carbon Capture, Utilization, and Storage	• Capturing CO_2 emissions from industrial processes and power generation. • Utilizing captured CO_2 in products • Storing CO_2 underground.

Table 2. *Cont.*

Strategy	Methodology
Hydrogen Economy	• Developing green hydrogen production capacity using renewable energy. • Utilizing hydrogen in transportation, industry, and as an energy storage medium.

Figure 3 illustrates the different applications of Hydrogen (H_2) as a central energy source, categorized into three main areas: Fuel, Heat, and Feedstock:

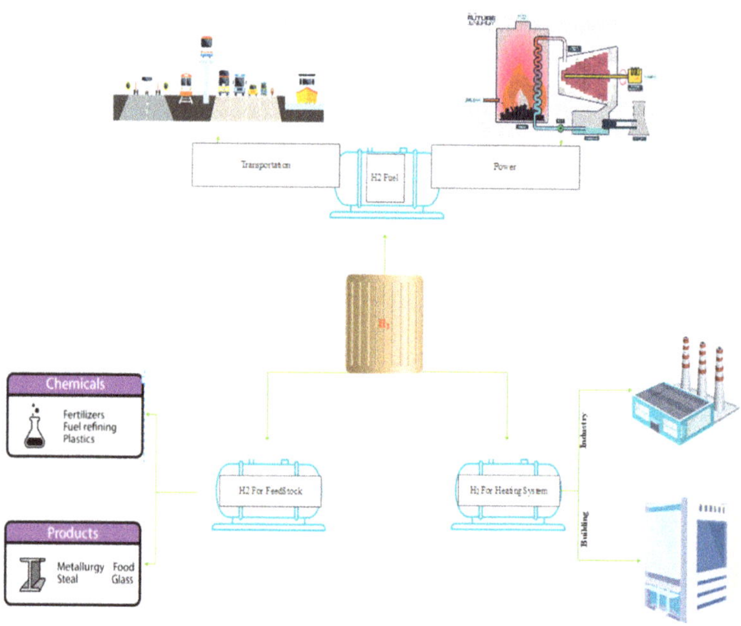

Figure 3. Hydrogen (H_2) with its uses for fuel, heat, and feedstock.

1.1. Fuel for

- Transportation: Hydrogen is used as a clean fuel for various transportation modes, including cars, trucks, buses, and even airplanes. Hydrogen fuel cells generate electricity through a chemical reaction, powering electric motors without producing harmful emissions.
- Power: Hydrogen is also used in electricity generation, particularly in electricity peaking plants. These plants rely on hydrogen to generate power during periods of high electricity demand, providing a clean and efficient solution to meet peak energy needs.

1.2. Heat for

- Industry: Hydrogen provides heat for industrial processes, playing a crucial role in high-temperature applications such as steel, cement, and aluminum production, as well as industries like food and paper manufacturing. It offers a clean alternative to fossil fuels traditionally used in these energy-intensive sectors.
- Buildings: Hydrogen can provide heating for residential and commercial buildings, offering an environmentally friendly way to maintain comfortable indoor temperatures and meet energy needs in urban areas.

1.3. Feedstock for

- Chemicals: Hydrogen is a key raw material in the production of chemicals such as fertilizers, fuel refining, and plastics. It plays an essential role in processes like ammonia synthesis for fertilizers and in refining crude oil into useful fuels.
- Products: In addition to chemicals, hydrogen is used in various products, including metallurgy, where it is involved in processes like reducing metal ores, as well as in food and glass production.

The municipality can effectively collaborate with local businesses and communities to enhance the adoption of small-scale renewable energy systems through some approaches as:

1. Municipality can partner with local businesses to co-fund and implement small-scale renewable projects, such as rooftop solar installations or microgrids. These partnerships can reduce financial burdens and encourage private-sector participation.
2. Support programs as offering tax breaks, subsidies, or low-interest loans to small businesses and residents can make small-scale renewable energy systems more accessible and affordable.
3. Municipality can host workshops, and stakeholder meetings to raise awareness about the benefits of renewable energy and gather input on local energy needs.
4. Municipality can support the establishment of community energy co-operatives that allow residents and businesses to collectively invest in and benefit from renewable energy systems.
5. Simplifying regulations and approval processes for small-scale renewable installations reduces barriers for businesses and homeowners.
6. Municipality can collaborate with local businesses to implement pilot projects that showcase the feasibility and benefits of renewable energy, inspiring wider adoption.
7. Facilitate resource-sharing initiatives, such as solar panel leasing programs or community solar gardens, to make renewable energy systems accessible to those who cannot install them independently.
8. Provide technical expertise, data, and tools to help businesses and communities make informed decisions about renewable energy adoption.

2. Literature Survey

Wind energy, for instance, is utilized at both large and small scales, with wind turbines often installed in remote or offshore locations [11]. Solar energy is another promising power source, particularly suitable for integration with existing structures in urban areas where installation space is limited, vibration control is essential, and wind conditions are unfavorable. However, energy storage technologies are necessary to ensure a reliable power supply, as both solar and wind energy are typically intermittent and unpredictable [12], leading to inconsistency with buildings' energy demands. In addition to cutting greenhouse gas emissions and boosting energy efficiency, hybrid energy systems in buildings can also result in long-term cost savings. A report by the Rocky Mountain Institute indicates that hybrid systems can offer a cost-effective alternative to conventional energy sources, potentially reducing energy bills by 10% to 60%, depending on specific conditions. These systems integrate renewable sources like solar, hydrogen, and wind with traditional sources like fossil fuels to create a sustainable and efficient energy solution. This combination enhances energy efficiency, lowers dependency on traditional sources, reduces emissions, and cuts operating costs for building owners. As more recognize these advantages, hybrid systems are increasingly adopted in commercial and residential buildings to lower energy expenses and support sustainability [13]. Following a similar approach to ecosystem development for building demandsAbdin, Z.; and Mérida, W conducted a techno-economic analysis across various locations for an off-grid renewable hybrid energy system designed to generate power and hydrogen. Using HOMER, nine different renewable energy system configurations were simulated to identify the most cost-effective setup to meet load demand. The results indicate that improved solar radiation and wind speed profiles can help achieve a lower cost of energy (COE) for renewable hybrid systems. Additionally, the analysis suggests that hydrogen technologies, such as fuel cells, electrolyzers, and hydrogen tanks, could potentially replace battery banks. However, the current high capital costs of these components lead to an overall higher system cost [14]. However, R. Al Afif et al. assessed the techno-economic feasibility of hybrid renewable energy systems, both off-grid and on-grid, for Jordan's Al-Karak governorate. Using HOMER Pro software, the study evaluates configurations to maximize renewable energy integration by exploring grid-connected and stand-alone systems comprising wind turbines, biogas plants, photovoltaic (PV) panels, flywheels, and batteries. The goal is to minimize net present costs, the leveled cost of energy, and CO_2 emissions. Results indicate that the PV/Wind system, grid-connected with battery storage, offers the optimal setup for sustainable electrification in Al-Karak, combining environmental benefits with dependable, uninterrupted energy access. This configuration achieves the lowest net present cost (USD 298,359) and Levelized cost (USD 0.024/kWh), with a renewable energy share of 71.8% (wind-generated), while 28.2% is purchased from the grid. This setup reduces CO_2 emissions to 220 tons annually, marking a 53% decrease compared to a grid-only system, contributing to energy independence and environmental quality improvements [15]. Moreover, P. Malik et al. studied the technical, economic, and environmental feasibility of a grid-connected hybrid microgrid system in the western Himalayan region, utilizing local renewable resources such as solar, wind, and biomass (pine needles). Five unique system configurations are evaluated, considering both the presence and absence of a storage unit. Comparative analysis indicates that the grid-connected PV/BG hybrid system is the most cost-effective, achieving the lowest levelized cost of energy (LCOE) at US$0.099 per kWh. Key findings reveal that the BG system contributes the largest portion (62%) of total power generation, followed by PV (20%) and grid supply (18%). Environmental assessments show that CO_2 emissions from the optimal hybrid system are reduced by approximately 85% compared to a DG-only system [16]. In 2021, B.K. Das et al. studied and explored the optimal sizing of a Hybrid Renewable Energy System (HRES) that integrates photovoltaic (PV), wind turbine, battery storage, and a converter to meet the electricity demands of a remote community in Broome, WA. Additionally, any surplus energy produced by the system is channeled through Thermal Load Control (TLC) to address the community's thermal needs, specifically for hot water. The optimized HRES, which supports both electrical and thermal demands, is evaluated against systems that fulfill only electrical needs. Findings reveal that the PV/Wind/Batt/TLC-based HRES achieves a lower Cost of Energy (COE) at 0.255 US$/kWh and Net Present Cost (NPC) of US$1,416,590, outperforming both the PV/Batt/TLC (COE: 0.282 US$/kWh, NPC: US$1,567,361) and the Wind/Batt/TLC (COE: 0.406 US$/kWh, NPC: US$2,241,687) systems under conditions optimized for combined electric and thermal load fulfillment [17]. On the other hand, K. Muruga Perumal and P. Ajay D Vimal Raj studied primarily aims to identify a technically practical and economically viable hybrid energy solution for off-grid electrification in Korkadu village, Puducherry. The goal is to determine the least-cost configuration of solar PV, wind turbines, a bio-generator, and battery storage to reliably meet the projected energy demand at an estimated cost of Rs.10.18 per kWh. This design and analysis focus on a hybrid renewable system that leverages solar energy as a primary source, supplemented by wind power and the village's biomass resources as alternative energy sources [18]. Wang et al. developed an environmentally efficient system for power and cooling, achieving a total efficiency of 77.5%, primarily due to the high hydrogen-to-electricity conversion rate and waste heat recovery from AFCs [19]. However, Behzadi et al.

introduced and rigorously evaluated an innovative approach to integrate solar and wind energy sources through hydrogen storage, aiming to improve grid stability and manage peak loads. Using the Transient System Simulation (TRNSYS) tool and the Engineering Equation Solver, the study comprehensively analyzed the technical, economic, and environmental aspects of a residential building in Sweden. A four-objective optimization was performed using MATLAB, combining the grey wolf optimization algorithm with an artificial neural network to balance key indicators. Results showed optimal conditions could yield an 80.6% improvement in primary energy savings, a 219% reduction in CO_2 emissions, a cost of $14.8/h and purchased energy of 24.9 MWh. Scatter distribution analysis highlighted the importance of minimizing fuel cell voltage and collector length, while electrode area had a limited impact. The smart renewable system proposed could meet 70% of the building's annual energy needs, with surplus energy sellable back to the local grid, establishing it as a practical and sustainable solution [20].

Li et al. proposed a novel solar-based system featuring an Alkaline Fuel Cell (AFC) paired with an electrolyzer. The study found that, along with significant electricity generation and low exergy loss, the AFC's waste heat could be effectively utilized to co-generate power via a Stirling engine and provide cooling with an absorption chiller [21]. A case study for different buildings was performed by Chen et al. by reviewing and examining new approaches to incorporating renewable energy within the construction industry, concentrating on energy types, policies, advancements, and future perspectives. The energy sources discussed include solar, wind, geothermal, and biomass. Case studies from Seattle, USA, and Manama, Bahrain, illustrate these practices. Key perspectives cover self-sufficiency, microgrid systems, carbon neutrality, intelligent building design, cost reduction, energy storage solutions, policy backing, and market visibility. Integrating wind power into buildings can meet approximately 15% of energy needs, while solar can boost renewable energy use by up to 83%. Additionally, financial incentives, such as a 30% subsidy for renewable technology adoption, enhance the appeal of these innovations [3]. T. Ahmad and D. Zhang's study integrates renewable energy sources and conducts a techno-economic feasibility analysis in both grid-connected and islanded modes to address community and commercial energy demands. Four sensitivity cases were developed, each further divided into four loops with varying combinations of photovoltaics, wind turbines, battery storage, power grids, converters, and diesel generators. The analysis utilized parameters such as capital, replacement, and operational costs, along with maintenance, salvage, fuel costs, real-time load demands, and climate data. The optimal results from sensitivity case 2 were applied to evaluate the remaining cases. In sensitivity case-2, the net present cost and annualized cost for the islanded mode in loop-1 were found to be $1,381,080.42 and US$106,832.62, respectively, whereas, in the grid-connected mode, they were noted at US$512,792.1 and US$39,666.71, respectively. The study demonstrates that the proposed utility grid-connected mode incurs costs that are half those of the islanded mode, highlighting its cost-effectiveness in meeting energy demands [22]. Patil, A et al. conducted a techno-economic and environmental analysis on grid-connected Hybrid Renewable Energy Systems (HRES) with and without a battery bank to determine the optimal solution for a remote area in the Bhorha Village, India., Among the six feasible configurations analyzed, the grid-connected PV/BG (biogas) hybrid setup was identified as the most cost-effective, with a Levelized Cost of Energy (LCOE) of $0.309 per kWh. Results from the optimal configuration show that 62% of energy production is from the PV system, followed by 20% from wind turbine and 18% from the biogas. Environmentally, this configuration resulted in an approximately 85% reduction in CO_2 emissions compared to a system relying solely on diesel generators (DG) [23]. Furthermore, H. Karunathilake et al. study aims to introduce a structured framework for integrating renewable energy into community development. A scenario-based analysis was carried out for a planned neighborhood in Okanagan, British Columbia (BC), Canada, with each scenario evaluated based on life cycle cost (LCC) and greenhouse gas (GHG) emission reduction. Scenarios were then compared by weighing their costs and benefits to key stakeholders. The study's findings showed that increasing the share of renewable sources in the energy mix does not always reduce LCC or GHG emissions. The advantages of adding renewables to a region's energy mix vary depending on the existing energy sources. Consequently, the composition of the regional power grid affects the feasibility and acceptance of renewable energy integration in any given area. The suitability of renewable options depends on local conditions, resource availability, and the current energy mix. To accurately assess the potential for emissions reduction and the benefits of integrating renewable energy, an examination of the regional energy mix specific to the community's location is essential. The study indicates that the extent of environmental impact reduction through renewable integration varies significantly by province within Canada. Developers should consider the optimal investment level that can be wholly or partially recovered through premium home pricing [24]. Moreover, they stated that a comprehensive framework that includes decision-making, design, and occupancy considerations is essential for realizing the full potential of energy demand reduction initiatives in residential communities. This framework also serves as an initial step toward planning energy-sustainable communities. Long-term energy modeling provides valuable insights into the effectiveness and the costs and benefits of demand reduction measures. The life cycle costs and impacts should play a key role in selecting suitable interventions for a residential community, as it's essential to assess whether the investment and effort justify the anticipated results. Given the diverse climatic conditions across Canada, demand reduction interventions must be tailored for regional applicability to achieve maximum impact. Additionally, information dissemination, along with regulatory and policy-level measures, is crucial for effectively managing residential energy consumption [25]. K. Singh and C. Hachem-Vermette developed a new optimization approach for cost-effective energy resource planning in neighborhoods, designed to meet energy load demands, achieve a targeted return on investment, and ensure an optimal payback period. This methodology incorporates various tariff structures alongside renewable energy sources (RES), alternative energy sources (AES), borehole thermal energy storage (BTES), and other cost models. The proposed approach is demonstrated through a case study of a mixed-use neighborhood, developing multiple energy system scenarios and waste-to-energy (WtE) operational strategies. In an all-electric scenario (SC1), a combination of PV panels and a continuously operated WtE-CHP plant offers the shortest payback period. For the scenario where domestic hot water (DHW) is supplied by non-electrical systems (SC2), it is not recommended to install a WtE-CHP plant or BTES. In the scenario where DHW and space heating are provided by non-electrical systems (SC3), the use of natural gas, WtE-CHP, and BTES is advised, achieving a payback period of 15.2 years [26]. F.D. Minuto, P. Lazzeroni, R. Borchiellini et al. developed multiple retrofit scenarios for buildings to identify optimal solutions. Data-driven retrofit models were simulated for an 87-unit condominium in Northwestern Italy, using various technology combinations—such as rooftop photovoltaic (PV) systems, air-source heat pumps, battery energy storage, and electric vehicle chargers—to meet both electric and heating demands. Each scenario's techno-economic feasibility was assessed through multi-criteria analysis, evaluating compliance with EU climate goals. The PV system retrofit consistently yields positive environmental and economic benefits, achieving an internal rate of return of 18.2%. The air-to-water heat pump also reduces primary energy demand by 26% and CO_2 emissions by 30%. However, it introduces electric load volatility on the grid, posing challenges potentially greater than those presented by PV intermittency [27].

Hydrogen complements direct electrification by enhancing renewable energy integration through its time-shifting and energy storage capabilities. While Canada's primary renewable energy source is hydropower, which inherently provides energy storage, wind energy capacity has grown steadily over the last decade. Wind power is among the fastest-growing electricity sources globally, accounting for 4% of Canada's national electricity generation, with Ontario and Quebec leading in capacity. Notably, Prince Edward Island generates 98% of its local electricity from wind but relies on imported power from New Brunswick's grid. Hydrogen offers a dispatchable energy solution to increase energy independence and can serve as a hybrid system for winter heating, reducing peak electric heating demand. Its versatility as an energy carrier allows for tailored applications across Canada [28]. While solar technologies have matured significantly, challenges such as high initial costs and intermittent energy production remain. Wind power, in contrast, is abundant, sustainable, and cost-effective for domestic energy applications. Combining solar and wind energy sources offers a strategic approach to producing reliable, clean, and cost-effective energy. Recent studies have explored integrating solar and wind systems for multi-generation purposes, including hydrogen generation, electricity generation, and heating/cooling, with a focus on techno-economic and environmental benefits [29]. Parra et al. developed a hydrogen energy storage system incorporating a polymer electrolyte membrane electrolyzer, a metal hydride tank, and a proton-exchange membrane fuel cell, achieving a round-trip efficiency of 52% for mid- to long-term energy storage. Izadi et al. simulated hybrid systems integrating solar panels, wind turbines, and hydrogen storage across four locations, creating zero-energy buildings where renewables and hydrogen systems provided 70–88% of required electricity. Wei et al. conducted similar simulations in Saskatoon, Canada, finding solar panels to have greater energy generation potential than wind turbines [30]. Further research by Mansir et al. analyzed residential systems using TRNSYS software, comparing hydrogen fuel cells and conventional batteries for energy storage. While hydrogen systems had double the capital cost, their larger capacities improved overall performance [31]. Behzadi et al. explored combining solar and wind energy with hydrogen storage to enhance grid stability and reduce peak loads in Sweden. Their optimized system demonstrated an 80.6% improvement in primary energy savings, a 219% reduction in carbon emissions, and the ability to meet 70% of annual energy needs while selling excess energy back to the grid [32].

Moreover, Ishaq et al. developed a system combining a wind turbine with high-temperature solar collector, and a $CuCl_2$ thermochemical hydrogen cycle, achieving a 50% performance efficiency. Hasan and Genç analyzed a hybrid solar-wind system with electrolyzers, showing that selling surplus power to the grid could reduce hydrogen production costs by US\$0.03/m^3 [33]. Similarly, Al-Buraiki and Al-Sharafi designed a wind-solar system for hydrogen production in Dhahran, Saudi Arabia, capable of independently meeting electricity demands while cutting carbon emissions by 9.6 tons annually [34].

Table 3 summarizes the findings from the previous literature by categorizing the studies based on their focus, key results, technologies utilized, and locations, offering a clear and comprehensive overview.

Table 3. Literature survey summery.

Authors	Objective	Key Findings	Technologies	Location
Z. Abdin & W. Mérida [14]	Techno-economic analysis of off-grid hybrid energy systems for power and hydrogen	Improved solar radiation and wind speed reduce the cost of energy; hydrogen storage can replace batteries but increases costs.	Solar, wind, hydrogen storage	Various
R. Al Afif et al. [15]	Techno-economic feasibility of hybrid systems for Al-Karak, Jordan	PV/wind system with battery storage achieved the lowest cost and emissions; 53% CO_2 reduction compared to grid-only systems.	PV, wind, battery, biogas	Al-Karak, Jordan
P. Malik et al. [16]	Grid-connected hybrid microgrid for the Western Himalayas	PV/biogas hybrid system achieved 85% CO_2 emission reduction and the lowest Levelized Cost of Energy (US$0.099/kWh).	PV, biogas, grid	Western Himalayas, India
B.K. Das et al. [17]	Optimized hybrid system for remote community in Broome, WA	PV/wind/battery/TLC system achieved the lowest cost and net present value compared to alternatives.	PV, wind, battery, TLC	Broome, WA, Australia
K. Muruga Perumal & P. Ajay [18]	Off-grid hybrid system for Korkadu, Puducherry	Least-cost configuration combines solar, wind, and bio-generator with battery storage for Rs. 10.18/kWh.	Solar PV, wind, bio-generator, battery	Korkadu, India
Wang et al. [19]	Power and cooling system integration	Achieved 77.5% efficiency with hydrogen-to-electricity conversion and waste heat recovery.	Hydrogen (AFCs), waste heat recovery	Not specified
Behzadi et al. [20]	Solar-wind-hydrogen system for grid stability	80.6% energy savings, 219% CO_2 reduction, and 70% annual energy met through renewables; excess energy sold back to the grid.	Solar, wind, hydrogen storage	Sweden
Chen et al. [3]	Renewable energy integration in buildings	Up to 83% renewable energy integration using wind and solar; financial incentives enhance adoption.	Solar, wind, policy frameworks	Seattle, USA; Manama, Bahrain
K. Singh & C. Hachem-Vermette [26]	Optimization of energy resource planning for neighborhoods	Mixed-use scenarios with PV and waste-to-energy systems offered shortest payback periods and highest returns.	PV, waste-to-energy (WtE), thermal storage	Not specified
F.D. Minuto et al. [27]	Retrofit analysis for energy-efficient buildings	Retrofit models with PV and heat pumps reduced CO_2 emissions by 30% and increased economic benefits.	PV, heat pumps, energy storage	Northwestern Italy
H. Karunathilake et al. [25]	Renewable integration in a Canadian neighborhood	Renewable integration's benefits vary by region; energy mix affects feasibility and emission reductions.	Renewable sources (varied by scenario)	Okanagan, BC, Canada

Table 3. *Cont.*

Authors	Objective	Key Findings	Technologies	Location
Ishaq et al. [33]	Hybrid renewable system with $CuCl_2$ thermochemical cycle	Combined wind, solar, and thermochemical hydrogen cycle achieved 50% efficiency.	Wind, solar, hydrogen (thermochemical cycle)	Not specified
Al-Buraiki & Al-Sharafi [34]	Wind-solar hydrogen system for Dhahran, Saudi Arabia	System met electricity demands and cut carbon emissions by 9.6 tons annually.	Wind, solar, hydrogen	Dhahran, Saudi Arabia

The previous survey demonstrates that all research aligns with the objectives outlined in the paper. Consequently, this paper aims to address various scenarios, detailed in the following sections, with the primary goal of conducting a comparative analysis of renewable energy and hydrogen as power sources for buildings. This analysis includes power calculations and cost assessments, focusing specifically on the impact of wind turbines, hydrogen as fuel mixture with LPG, and photovoltaic (PV) panels. The comparison seeks to evaluate and contrast the efficiency, performance, and economic factors related to wind, hydrogen–LPG fuel mixture, and solar energy. This study depending on the adopting of a hybrid renewable energy system combined with a hydrogen-LPG mix for building energy in Canada entails upfront expenses for renewable energy infrastructure such as solar panels (CAD$2.5–$3/Watt installed), wind turbines (CAD$1500–CAD$2500/kW), electrolyzers, and storage systems, as well as retrofitting LPG setups and ongoing maintenance (solar and wind maintenance: ~1–2% of system cost annually). Operational costs include electricity for hydrogen production and system upkeep, with hydrogen priced at ~CAD$3–CAD$5/kg and LPG at ~CAD$0.60/L. Financial support is available through programs like the Canada Greener Homes Grant (up to CAD$5000), the Clean Energy Innovation Program, provincial incentives (e.g., Alberta Hydrogen Roadmap, Quebec's ÉcoPerformance), carbon tax savings, and tax advantages such as accelerated depreciation for clean energy systems. These measures lower long-term energy expenses, boost resilience, and enhance property value [35].

The paper introduces a transformative approach to addressing critical gaps in energy sustainability for buildings by leveraging Hybrid Renewable Energy Systems (HRES). Unlike previous studies that often focus on isolated renewable energy solutions, this research emphasizes the integration of solar panels, wind turbines, and green hydrogen production to achieve environmental and socio-economic sustainability.

The innovation of this study lies in its comprehensive analysis and dynamic simulation of HRES using TRNSYS 16 software under diverse climatic conditions. By addressing key challenges such as optimizing energy generation, reducing greenhouse gas (GHG) emissions, and enhancing energy reliability, the manuscript provides actionable solutions for transitioning to sustainable energy systems.

Through this assessment, the study aims to provide valuable insights into the most efficient and cost-effective approach to meeting building energy demands.

3. Methodology

To effectively implement a Hybrid Energy System (HES), a comprehensive approach is required, encompassing:

3.1. Energy Available Analysis

This step evaluates energy available patterns to understand the optimum energy system needed dynamically. It includes two primary functions:

- Weather Data: This is done by analyzing historical data, weather conditions, and usage patterns. This helps ensure that the system can meet expected energy needs.
- Demand Profiling: Demand profiling focuses on analyzing daily, weekly, and seasonal energy consumption trends. It identifies peak and off-peak periods, enabling the optimization of energy supply and system operations [36].

3.2. Energy Supply Analysis

This step evaluates the availability and performance of different energy sources within the system. It involves two main functions:

- Renewable Energy Assessment: This function assesses the potential of renewable energy sources, such as solar, wind, and others, based on geographic and environmental data. It determines the viability of these sources in meeting energy demands.
- Supply Profiling: Supply profiling develops detailed profiles of energy supply from various sources, aligning them with demand profiles. This ensures a balance between energy generation and consumption, optimizing the system's efficiency [7].

The methodology adopted in this paper is depicted in Figure 4 as a flowchart, offering a clear visual representation of the workflow and the steps involved in obtaining and calculating the results.

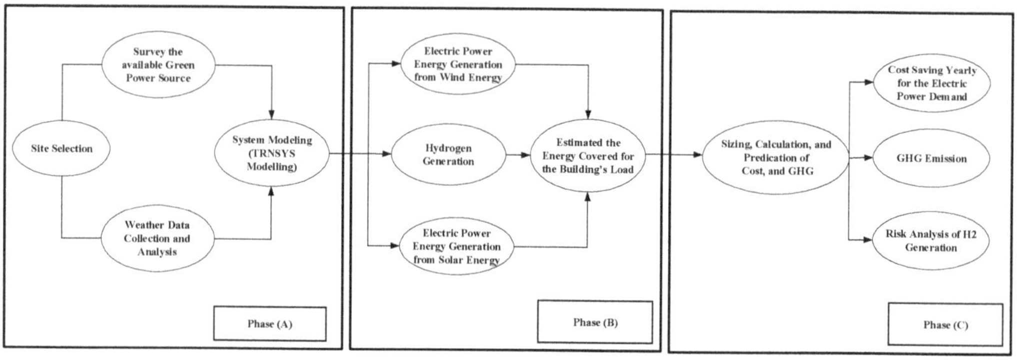

Figure 4. The Block Diagram for The Workflow.

- The implementation of hybrid renewable energy system: Launch a hybrid power system in a select residential area or community to install and monitor power and hydrogen generation for both electric demand and heating systems.
- Data Collection and Analysis: Collect data on electric power demand, heating efficiency, energy consumption, cost, and user satisfaction.
- Comparative Study: Contrast the collected data with similar metrics from utility as grid and natural gas for electric power demand and heating systems in comparable settings.

Developing and implementing hybrid renewable energy systems and a green hydrogen strategy for buildings necessitates the creation of an algorithm, as depicted in Figure 5, to facilitate the generation and utilization of green energy and hydrogen within local communities. This algorithm should address multiple aspects of small-scale applications, including energy sources and their specific uses. The focus of this paper is to investigate the deployment of a small-scale wind turbine and photovoltaic (PV) system integrated with a green hydrogen production unit via electrolysis for use in a building. Moreover, to integrate wind turbines and solar panels as power sources for buildings and an H_2-LPG mixture for heating systems without compromising reliability and safety can be summarized into points as:

I. Use batteries to manage intermittent energy from wind and solar sources, ensuring consistent power supply.

II. Implement advanced controllers to balance power generation and heating demands efficiently.

III. Adhere to strict safety standards for handling and using the H_2-LPG mixture to prevent leaks or combustion risks.

IV. Include backup systems, such as grid connections or additional energy sources, to ensure reliability.

V. Develop a real-time monitoring system for power generation and heating components, coupled with regular maintenance to address potential issues promptly.

VI. Ensure all systems comply with building codes and energy standards for renewable integration and fuel safety.

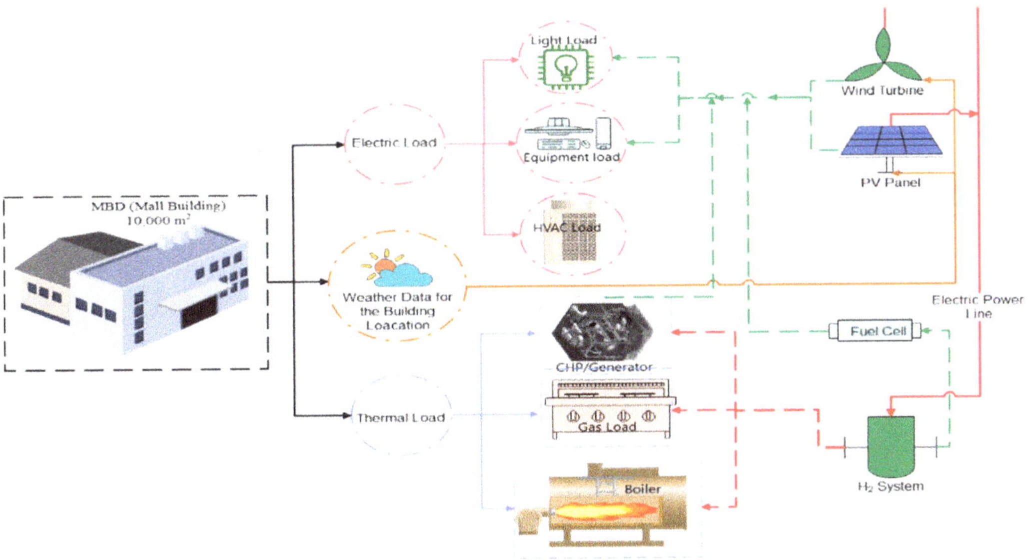

Figure 5. Algorithm for Electric Power and Heat Energy in Buildings.

The paper highlights several scenarios as examples to demonstrate the potential outcomes that can be achieved.

MBD is specified according to some parameters inside the algorism and scenarios in Figures 5 and 6 and definition of each one in Table 4, these parameters are:

1. MBD physical and dimension specifications.
2. MBD loads (Electrical-Thermal-Gas).
3. Location & weather specification at the building location.
4. Building facility.

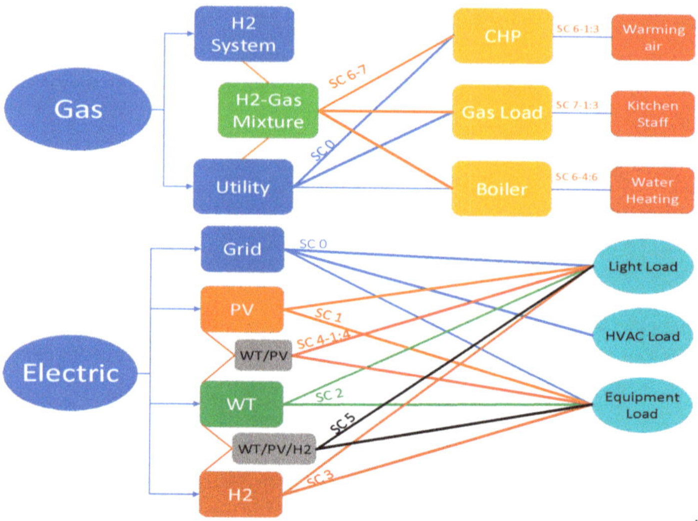

Figure 6. Algorithm of the Building Electric Power and Heat Energy.

Table 4. The symbols definition for the selected building.

Symbol	Definition
MBD	Mall Building Description
Eq—E1:E3	Electrical Equipment one to three E1: Light system–E2: Equipment system
Gl—K1: K2	Kitchen staff working with gas
Tl—T1: T2	Thermal loads for heating system T1: Air heating system–T2: Water heating system
SC	The proposed scenario

MBD encompasses numerous scenarios that together address all demand loads, starting with SC0, where the utility grid supplies 100% of the energy requirements. The scenarios are grouped based on load type:

- Scenario 0 (SC0): All electric, thermal, and gas loads are fully dependent on the utility grid (Electricity and Gas).
- Scenario 1 (SC1): A portion of the electric loads (lighting and equipment) is supplied by photovoltaic (PV) panels.
- Scenario 2 (SC2): A portion of the electric loads (lighting and equipment) is covered by a wind turbine.
- Scenario 3 (SC3): A portion of the electric loads (lighting and equipment) is supported by hydrogen (H_2) and a fuel cell.
- Scenario 4 (SC4): A portion of the electric loads (lighting and equipment) is supplied by both wind turbines and PV panels.
- Scenario 5 (SC5): A portion of the electric loads (lighting and equipment) is supported by wind turbines, fuel cells, and PV panels.
- Scenario 6 (SC6): The thermal loads (air and water heating) are partially covered by the hydrogen system.
- Scenario 7 (SC7): The gas loads (e.g., kitchen use) are partially supplied by the hydrogen system.

The specification of the building that is used as a case study can be shown in Table 5.

Table 5. Building Characteristics.

Area: sq. m	10,000
Volume: cu. M	40,000
Insulation (R-value) for walls: ($m^2 \cdot K/W$)	15
Outdoor design temperature: °C	−30
Indoor design temperature: °C	20
Window area: sq. m with a U-value of $W/(m^2 \cdot K)$	500
Ventilation rate: m^3/h	10,000
Assume walls make up sq. m of the building envelope.	8000
Location	Oshawa/ON/Canada

4. Modelling

The TRNSYS is a versatile simulation software widely used for analyzing the performance of dynamic systems, including solar energy applications. In this study, TRNSYS version 16 was utilized to simulate system performance over an entire year, with hourly simulations covering a total of 8760 h. The system under analysis includes solar panels, wind turbines, an electric controller, and an electrolyzer. Each component is detailed in this section. As shown in Figure 7, the TRNSYS software provides a schematic representation of two renewable energy simulations, with offering a clear visualization of the hydrogen system and illustrating the power output of the renewable energy and hydrogen production.

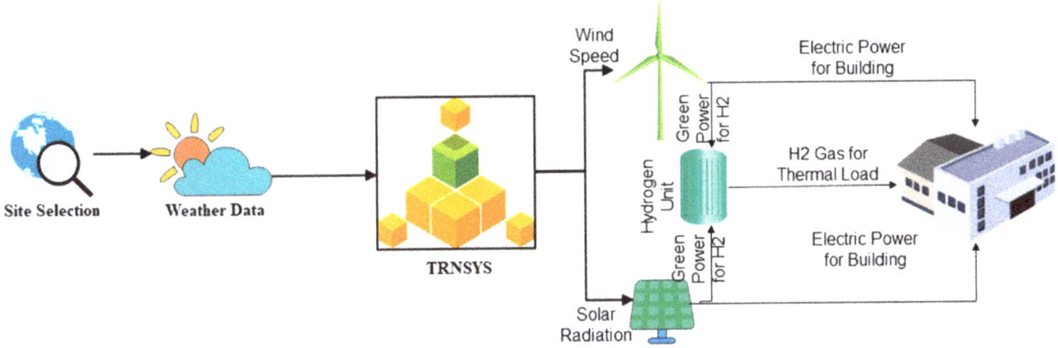

Figure 7. Schematic of the renewable energy and the hydrogen system in TRNSYS software.

The following components of the TRNSYS simulation and their interactions are thoroughly outlined:

- Data Reader:

This component reads data from external files at regular intervals, converts it into the required units, and makes it available for use by other components in the TRNSYS simulation as time-dependent forcing functions. It is highly flexible, capable of handling various data file formats, provided the data is organized at consistent time intervals. It simplifies the management of weather data and solar radiation inputs within the TRNSYS environment.

- Wind Turbine:

This mathematical model represents a Wind Energy Conversion System (WECS) and calculates its power output based on wind speed and the turbine's power characteristics, typically stored in a table within an external file. The model considers variations in air density and wind speed at different heights, enabling accurate simulation of WECS performance. This approach is crucial for evaluating wind energy systems and optimizing their efficiency under changing conditions. The selected wind turbine has a nominal maximum power output of 29 kW. TRNSYS employs a power curve-based model, where the turbine's power output (P_w) is determined by the Equation (1) [37]:

$$\mathbf{P_w} = \frac{1}{2}\boldsymbol{\rho}\mathbf{A}\mathbf{v}^3\mathbf{C_p} \tag{1}$$

ρ: Air density (kg/m^3).
A: Swept area of the turbine blades (m^2).
v: Wind speed (m/s).
C_p: Power coefficient, representing the efficiency of the turbine in extracting power from the wind.

Solar Panels:

PV panels are integral to the system, converting solar irradiation into direct electric current. A portion of the incident solar energy is transformed into electricity, while the rest is dissipated as heat. The selected PV panels, installed on the laboratory rooftop, have a nominal maximum power rating of 550 W. TRNSYS uses the Type 94a model to simulate the electrical output of the PV array, with the power output calculated using Equation (2):

$$P = \eta G A \tag{2}$$

η: Conversion efficiency of the PV module.
G: Incident solar radiation on the module (W/m^2).
A: Area of the PV module (m^2).

- Electrolyzer:

The electrolyzer facilitates water electrolysis, splitting water into hydrogen and oxygen, thereby producing green hydrogen. This study uses an alkaline electrolyzer simulated with the TRNSYS Type160a model. The electrolysis process adheres to Faraday's law, which relates the amount of chemical reaction to the electrical charge passed through the system. The reaction is:

Faraday's Law

$$2H_2O \rightarrow 2H_2 + O_2 \tag{3}$$

The Faraday's law equation is given by Equation (4):

$$Q = nF \tag{4}$$

Q: Charge (Coulombs)
n: No. of moles for electrons exchanged in the reaction.
F: Faraday's constant (approximately 96,485 C/mol)
Cell Voltage:

The cell voltage in an electrolysis system can be related to the standard cell potential and overpotentials:

$$V_{cell} = E_{cell} - \eta_{activation} - \eta_{Ohmic} - \eta_{Concentration} \tag{5}$$

E_{cell}: Standard cell potential
$\eta_{activation}$: Activation overpotential
η_{Ohmic}: (electrical resistance) overpotential
$\eta_{Concentration}$: Concentration overpotential

- Control Unit:

This module governs the electrolyzer's operation within an integrated mini-grid that includes a wind turbine, and electrolyzer. The electrolyzer operates in variable power mode, adjusting its power setpoint based on available excess renewable energy. If excess power exceeds the idling power, the electrolyzer's setpoint is adjusted to utilize it efficiently. Otherwise, the setpoint remains at the idling power level.

- Power Conditioning:

This component models power conditioning units, including electrical converters (DC/DC) and inverters (DC/AC or AC/DC). Using empirical efficiency curves, it calculates output power based on known input power, enabling system optimization.

These components are seamlessly integrated within TRNSYS to simulate the annual performance of wind turbines and solar panels, as well as hydrogen production. This integration accounts for the dynamic nature of energy generation and consumption.

Together, these components are integrated into the TRNSYS software to simulate the annual performance of the hydrogen energy system, accounting for the dynamic variability in energy generation and consumption.

5. Results and Discussion

A dynamic simulation of the system is performed using TRNSYS software, covering various hours and months throughout the year. The simulation operates with a time step of one hour and spans a full year, running from 1 January to 31 December 2023. This comprehensive simulation enables a detailed analysis of the system's behavior across different seasons and under varying conditions. The weather data used corresponds to Oshawa City, located within the Durham region of Ontario, Canada. This data is essential for capturing the specific climatic and environmental conditions of the region, playing a critical role in energy system simulations and environmental modeling. The simulation incorporates two primary power sources: a wind turbine and a photovoltaic (PV) panel. The characteristics, input data, and output performance of these power sources will be thoroughly detailed in the next section to provide a complete understanding of their roles and contributions to the energy system.

5.1. Turbine Simulation Parameters

The simulation is conducted with TRNSYS package for the power output of the wind turbine, which is subsequently used in the building power demand and the electrolysis process to produce green hydrogen. The primary input for this simulation is weather data, specifically the correlation between wind speed and hours. This data is sourced from the official Government of Canada website [38], ensuring it is accurate and reliable due to its authoritative origin. The simulation focuses on the power output for building loads with the conventional green hydrogen generation system. The simulation aims to assess and evaluate the performance of this stage within the overall energy system.

Table 6 outlines the key specifications and parameters used to model the wind turbine in the simulation.

Table 6. Turbine specification.

No.	Specification	Value	Description
1	$Rotor_{\_Ht}$	14.5	Rotor center height above ground level, which determines the height at which wind speeds are measured and utilized by the turbine.
2	$Rotor_{\_Di}$	5	Rotor diameter, representing the span of the turbine blades. This dimension influences the swept area and the amount of wind energy captured.
3	$Sensr_{\_Ht}$	14.5	Sensor height, often aligned with the rotor centre height, where wind data (speed and direction) is recorded for accurate turbine performance analysis.
4	$Sher_{\_Exp}$	0.14	Shear exponent, indicating the power-law relationship for vertical wind speed variation, dependent on terrain and atmospheric conditions.
5	$Turb_{\_Int}$	0.10	Turbulence intensity, representing the variability in wind speed as a fraction of the mean wind speed.
6	Air_Dens	1.225	Air density under standard conditions (15 °C and 1 atm), crucial for calculating wind power and turbine performance. kg/m^3
7	$Pwr_{\_Ratd}$	12.158	Rated power output of the turbine, achieved at the rated wind speed, reflecting the maximum power the turbine is designed to produce, kW
8	$Spd_{\_Ratd}$	19	Rated wind speed at which the turbine achieves its rated power output, indicating its performance threshold, m/s
9	Nominal turbine Efficiency	0.35	Coefficient of performance (Cp), representing the efficiency of the turbine in converting wind energy into electrical power

Figure 8 depicts the turbine's performance across different wind speeds, with its cut-off speed of 12 m/s with 29.63 kW. It demonstrates how the turbine's power output fluctuates with varying wind speeds and identifies the point where power generation is restricted or ceases. This visualization is crucial for understanding the turbine's operational behavior and efficiency under various wind conditions.

5.2. PV Simulation Parameter

The simulation employs TRNSYS package for modelling and simulate the PV panel's power output. It is subsequently utilized to meet the building's electrical load and to support the electrolysis process for hydrogen production. The primary input for this simulation is solar irradiance data, provided as hourly irradiance values. This data is sourced from the official Government of Canada website [38], ensuring accuracy and reliability for the simulation.

Table 7 likely presents the specifications of the photovoltaic (PV) panel employed in the TRNSYS simulation. These specifications generally include essential details such as the PV panel's electrical characteristics, physical dimensions, efficiency, temperature coefficients, and performance under standard test conditions (STC). Table 8 shows the electrolysis specification used in the simulation for H_2 generation.

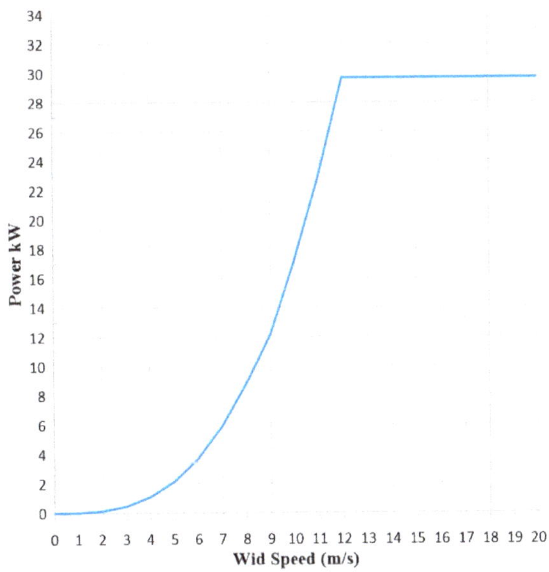

Figure 8. The power Curve for the Wind Turbine with cut-off power.

Table 7. Electrical specification of PV panel at STC.

Rated Maximum Power (P_{max}) [W]	550
Open Circuit Voltage (V_{oc}) [V]	49.9
Maximum Power Voltage (V_{mp}) [V]	41.7
Short Circuit Current (I_{sc}) [A]	14
Maximum Power Current (I_{mp}) [A]	13.2
Module Efficiency [%]	20.3
Power Tolerance	0~+5 W
Temperature Coefficient of I_{sc} (α_I_{sc})	0.05%/°C
Temperature Coefficient of V_{oc} (β_V_{oc})	−0.26%/°C
Temperature Coefficient of Pmax (γ_P_{mp})	−0.351%/°C
STC	Irradiance 1000 W/m^2, cell temperature 25 °C, AM1.5 G
Cell Maximum Efficiency	Mono Type with efficiency 21.3%

Table 8. The Electrolysis Specification.

No.	Specification	Value	Description
1	Unit Area	0.066	Total area of the unit (m^2)
2	N cells	2	
3	I_{ELY}	4	Current through single electrolyzer stack (A)
4	P_{ELY}	1	Electrolyzer pressure (atm)
5	T_{ROOM}	20	Room Temperature (°C)
6	T_{ELY}	50	Temperature of electrolyzer (°C).

5.3. Electrolysis Characteristics

In the following section, various scenarios are generated for both the electric and thermal loads of the building specified in Table 5. A selection of a few scenarios from the hundreds available is analyzed as follows:

5.4. Electric Load Scenarios

5.4.1. SC 0

This scenario represents a power demand for the building mall that is fully met (100% coverage) by the utility supply. The statistical analysis of this scenario is presented in the following Table 9.

Table 9. Electric Load.

Load	kW	kWh/Day	Cost CAD$/Day
Light Load	300.00	2400.00	395.20
HVAC Load	3000.00	24,000.00	3952.00
Equipment load	200.00	1600.00	263.47
Miscellaneous	100.00	800.00	131.73
Subtotal	3600.00	28,800.00	4742.40
Diversity factor 0.7	2520.00	20,160.00	3319.68
Safety Margin 10%	2772.00	22,176.00	3651.65

The cost analysis of the grid utility is considering in Table 9 due the price of kWh in Table 10 with considering the total annual cost in Table 11.

Table 10. Price list for Non-Residential.

Ultra-low Overnight	¢2.8/kWh
Weekend Off-peak	¢7.6/kWh
Mid-peak	¢12.2/kWh
On-peak	¢28.4/kWh

Table 11. Utility Cost Analysis.

Utility Grid Power Consumed in the building kWh/year	7,761,600.00
Utility Cost CAD $/year	1,278,076.80

5.4.2. SC 4-1

The proposed hybrid renewable energy system efficiently covers the mall's light load demand while utilizing wind and solar energy resources. It emphasizes the environmental benefits of reducing reliance on conventional energy sources and showcases an effective design for renewable energy integration in urban infrastructure.

This scenario offers a power rate of green power due to the available wind and solar energy at the building mall location. In this scenario, wind turbines and solar panels (PV) are used as a power source to cover a percentage of the building's light load (Eq-E1) (300 kW), which runs for 8 h. Wind turbines work 18 h, and PV with 550 W (1 kW needs 10 m^2) works for 5 h per day. To cover this load, it is simulated the power output from 6 wind turbines with 29 kW for one, and PV 15 in series and 15 in parallel.

Figures 9 and 10 present the annual solar irradiation and wind speed data for the selected zone, based on the available weather data for this location. This data serves as input to the TRNSYS simulation, enabling the calculation and modeling of power generation from both PV panels and wind turbines.

Figures 11 and 12 depict the simulated power output from the PV panels and wind turbines. This generated power is utilized to partially meet the building's load requirements, as indicated in Scenario 4-1, detailed in Table 12.

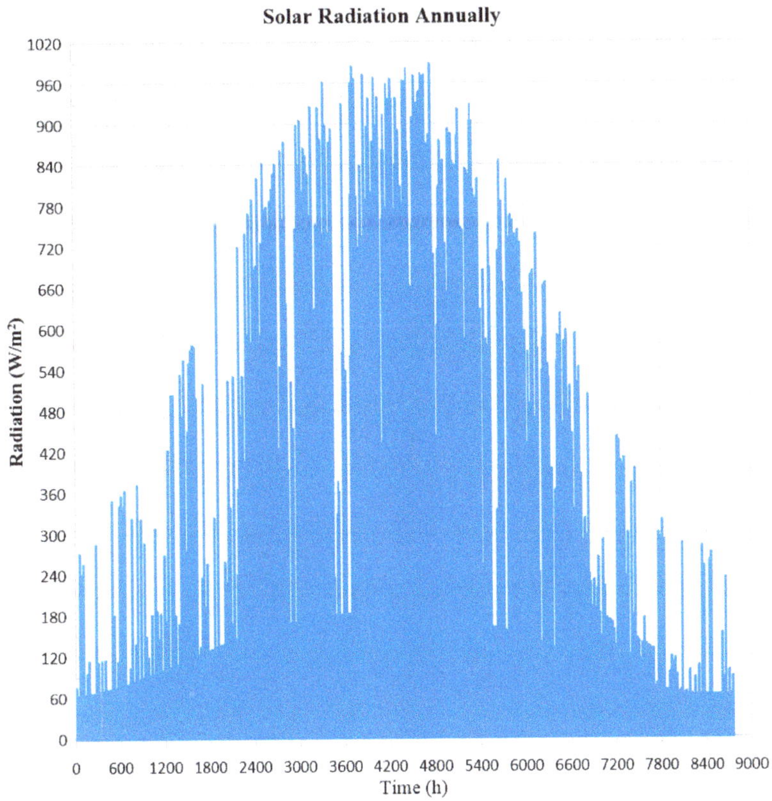

Figure 9. Annually hour solar irradiance data for Oshawa City, located within the Durham zone.

Table 12. Scenario SC 4-1 analysis.

	Wind Turbine 70 kW (Working 18 h/Day)	PV 50 kW (Working 5 h/Day)	Total
Power kW/day	1260.00	250.00	1510.00
Covering Percent	53%	10%	63%
SC 4-1 Power kWh/year	441,000.00	87,500.00	528,500.00
Cost saving SC 4-1 CAD $/year	72,618.00	14,408.33	87,026.33

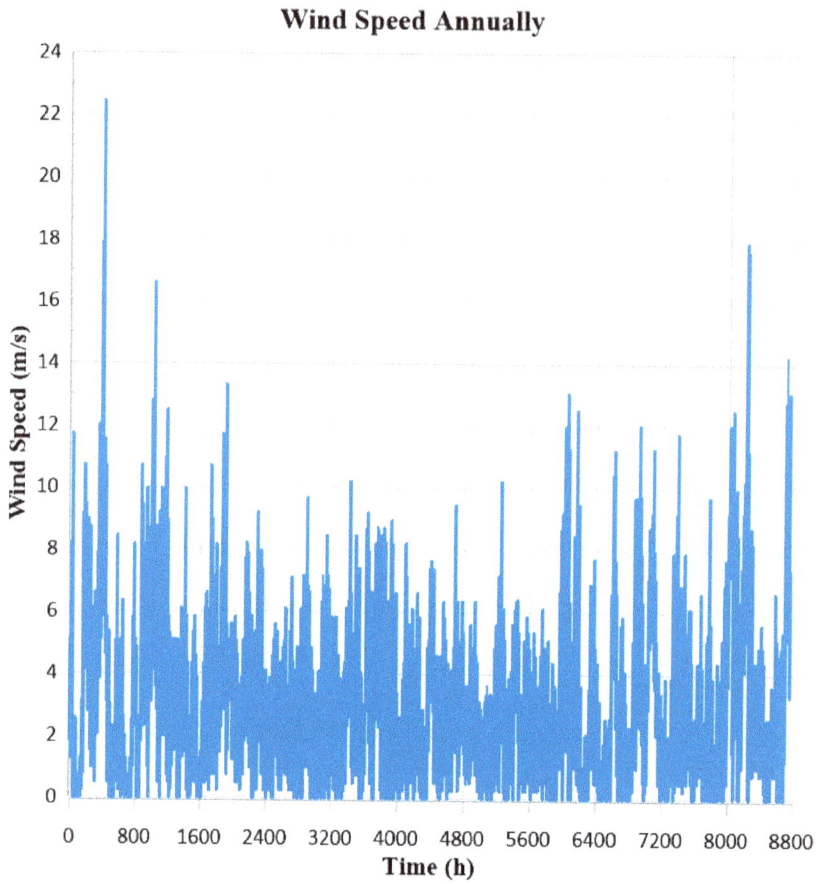

Figure 10. Annual wind speed data for Oshawa City, located within the Durham zone.

5.4.3. Sc 4-2

This scenario provides a green power supply based on the available wind and solar energy at the building mall location. In this setup, both wind turbines and photovoltaic (PV) solar panels are used as power sources to cover a portion of the equipment load (Eq-E2) of 200 kW, operating for 8 h daily. The wind turbines operate for 18 h per day, while the PV panels function for 5 h per day. The statistical analysis of this scenario is presented in Table 13.

Table 13. Scenario SC 4-2 analysis.

	Wind Turbines 70 kW (Working 18 h/Day)	PV 50 kW (Working 5 h/Day)	Total
Power kW/day	1260.00	250.00	1510.00
Covering Percent	79%	16%	95%
SC 4-2 Power kWh/year	441,000.00	87,500.00	528,500.00
Cost saving SC 4-2 CAD $/year	72,618.00	14,408.33	87,026.33

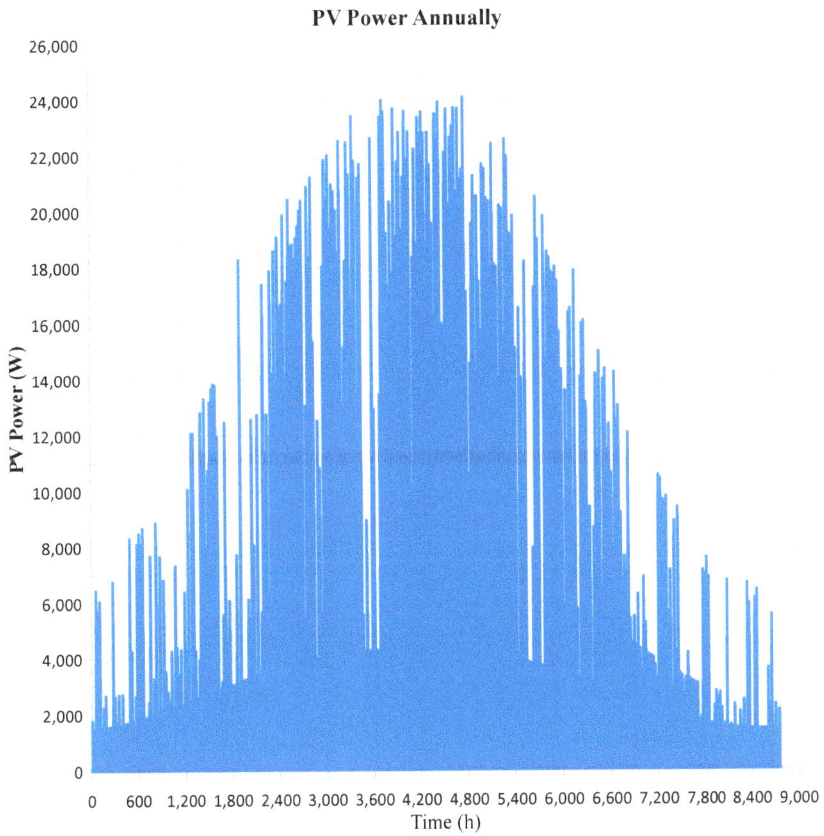

Figure 11. Annually PV Power output data for Oshawa City, located within the Durham zone Sc 4-1 and Sc 4-2.

5.4.4. Sc 4-3

This scenario utilizes green power derived from wind and solar energy resources available at the building mall location. In this setup, both wind turbines and photovoltaic (PV) solar panels are employed as power sources to partially meet the building's light load (Eq-E1), which has a capacity of 300 kW and operates for 8 h daily. Wind turbines work 18 h, and PV with 550 W works for 5 h per day. To cover this load, it is simulated the power output from 4 wind turbines with 29 kW for one, and PV 9 in series and 7 in parallel as can be seen in Figures 13 and 14.

5.4.5. Sc 4-4

This scenario offers a power rate of green power due to the available wind and solar energy at the building mall location. In this scenario, both wind turbines and solar panels (PV) are used as a power source to cover a percentage of the equipment load (Eq-E2) (200 kW) of the building running for 8 h. Wind turbines work for 18 h, and PV works for 5 h per day. To cover this load, it is simulated the power output from 4 wind turbines with 29 kW for one, and PV 9 in series and 7 in parallel.

The statistical analysis can be shown in the following Tables 14 and 15 for Sc 4-3 and Sc 4-4.

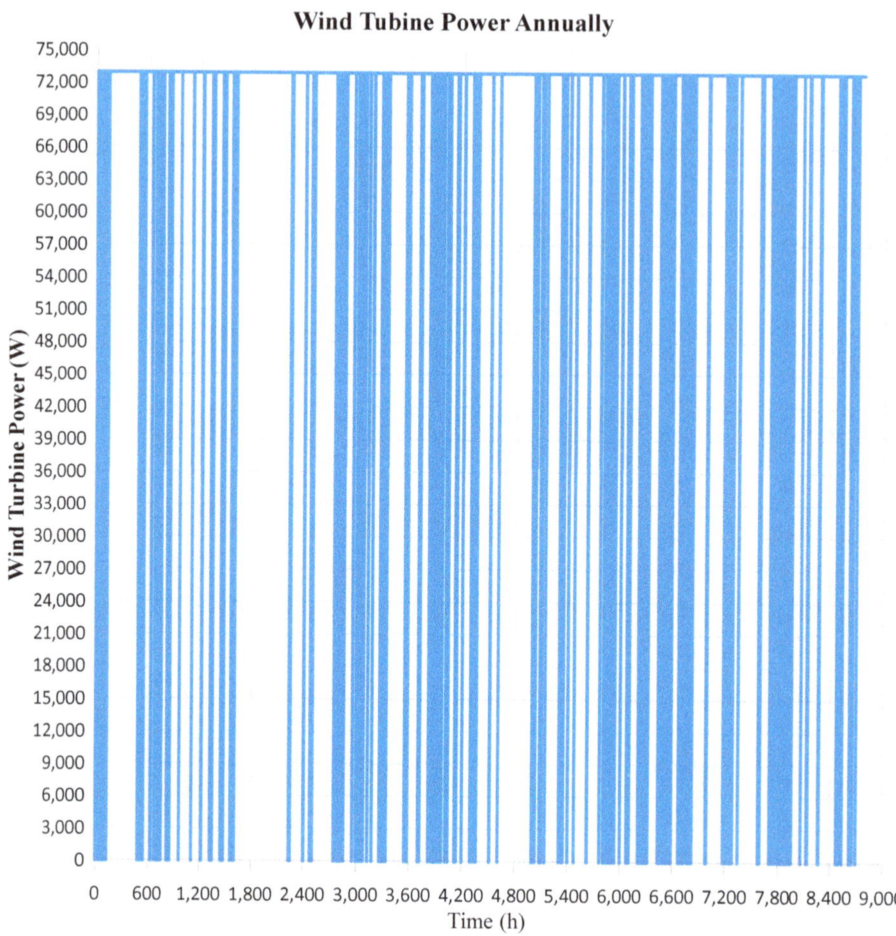

Figure 12. Annually Wind Turbines Power output data for Oshawa City, located within the Durham zone Sc 4-1 and Sc 4-2.

Table 14. Scenario SC 4-3 analysis.

	Wind Turbine 40 kW (Working 18 h/Day)	PV 30 kW (Working 5 h/Day)	Total
Power kW/day	720.00	150.00	870
Covering Percent	30%	6%	36%
SC 4-3 Power kWh/year	252,000.00	52,500.00	304,500.00
Cost saving SC 4-3 CAD $/year	41,496.00	8645.00	50,141.00

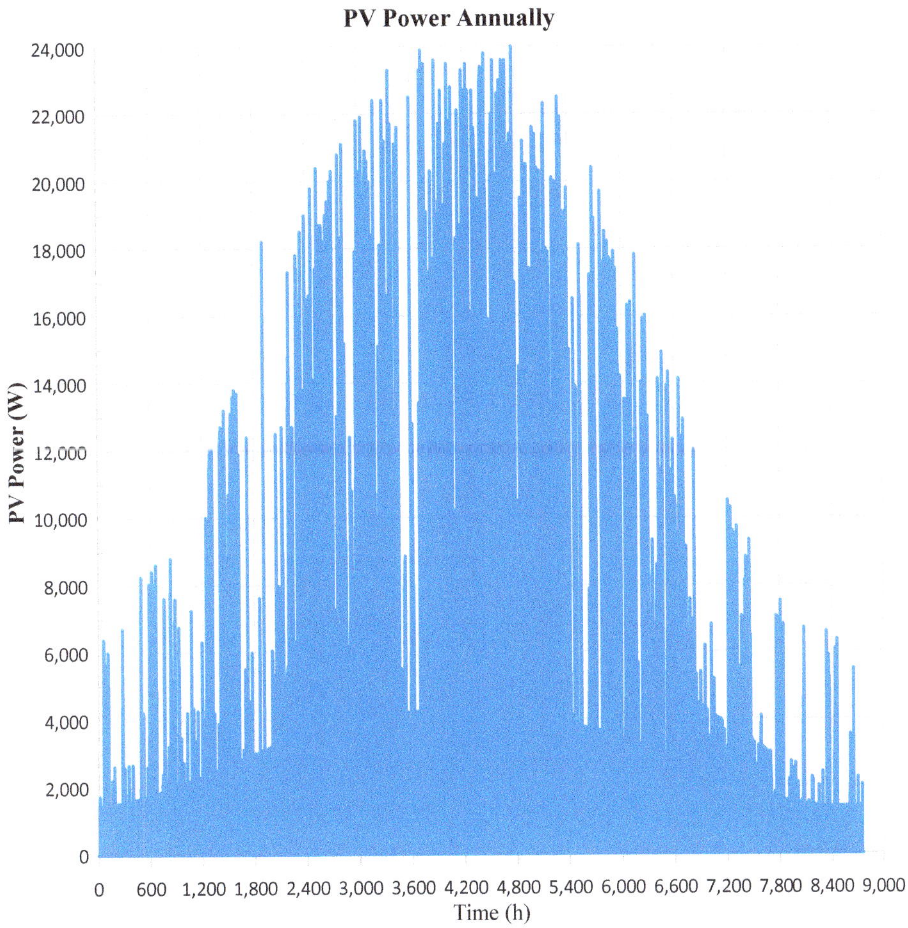

Figure 13. Annually PV Power output data for Oshawa City, located within the Durham zone Sc 4-3 and Sc 4-4.

Table 15. Scenario SC 4-4 analysis.

	Wind Turbine 40 kW (Working 18 h/Day)	PV 30 kW (Working 5 h/Day)	Total
Power (kW/day)	720	150	870.00
Covering Percent	45%	9%	54%
SC 4-4 Power kWh/year	252,000.00	52,500.00	304,500.00
Cost saving SC 4-4 CAD $/year	41,496.00	8645.00	50,141.00

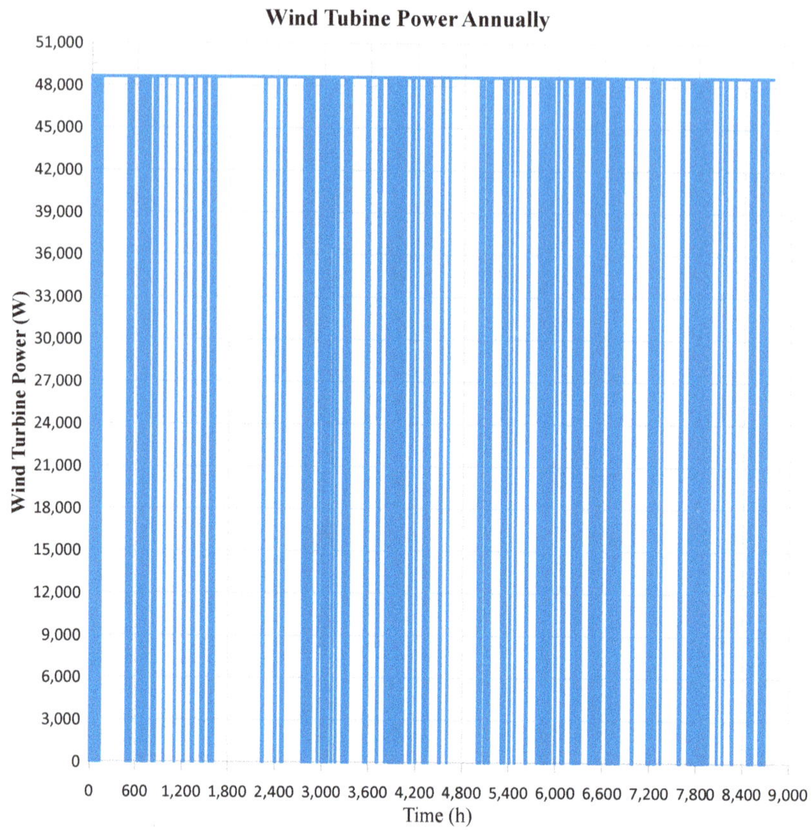

Figure 14. Annually Wind Turbines Power output data for Oshawa City, located within the Durham zone Sc 4-3 and Sc 4-4.

Figures 15–17 identify Scenario 4-1 (SC 4-1) as the most efficient configuration for addressing the electric light load demand in the specified building. This scenario optimally combines renewable energy sources, utilizing:

- Wind Turbine Capacity: 70 kW
- Photovoltaic (PV) System Capacity: 50 kW

Key Details: SC 4-1 and 4-2 integrate both wind turbines and solar panels to ensure a balanced and reliable power supply. This combination maximizes the use of renewable energy resources available at the location, such as solar irradiance and wind speed, to reduce dependence on utility-supplied electricity. The system Performance of the wind turbine generates power for extended periods, leveraging the consistent wind speeds in the area, while the PV panels contribute additional power during peak sunlight hours, complementing the wind turbine's output. This led to specifically targets the electric light and equipment loads demand, ensuring that the renewable energy sources provide sufficient power while minimizing gaps in coverage. From the point of sustainability, it reduces carbon footprint by minimizing reliance on traditional utility power. And for meanwhile, the cost-effectiveness decreases for the operational costs by harnessing renewable energy, particularly during peak production hours.

Furthermore, Figure 18 depicts the proposed arrangement of PV panels and turbines on the building's rooftop.

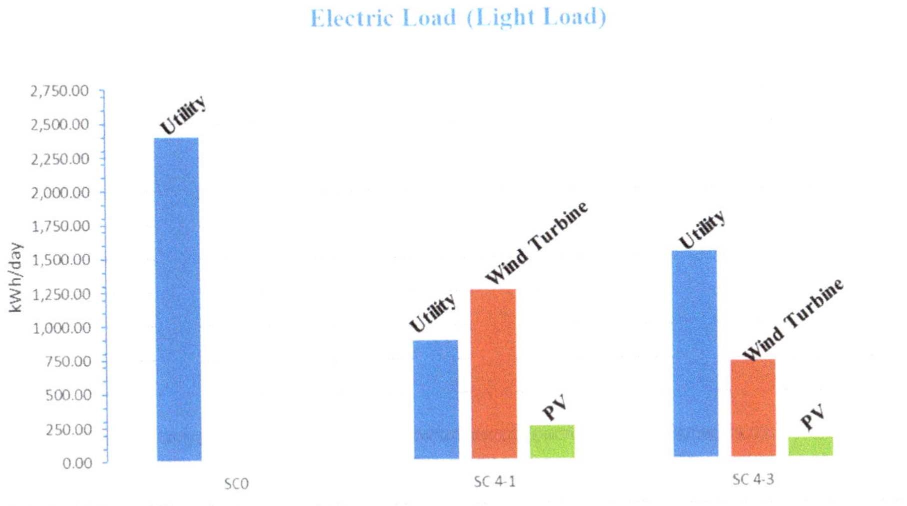

Figure 15. Electric Loads (Light) Summary.

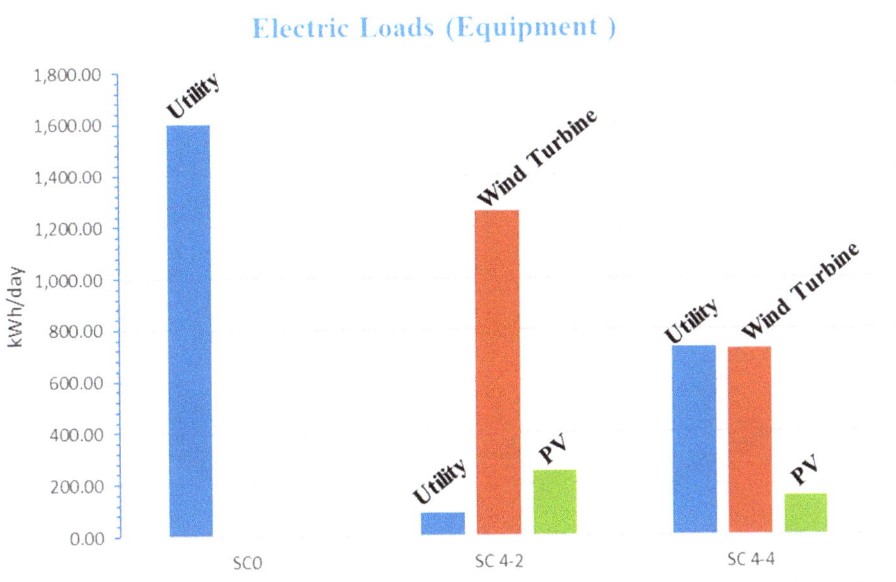

Figure 16. Electric Loads (Equipment) Summary.

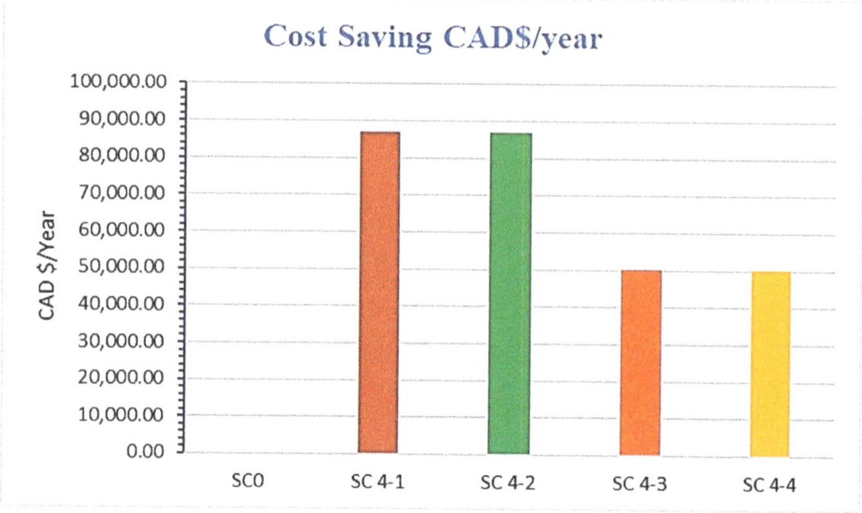

Figure 17. Cost Saving Analysis Yearly Summary.

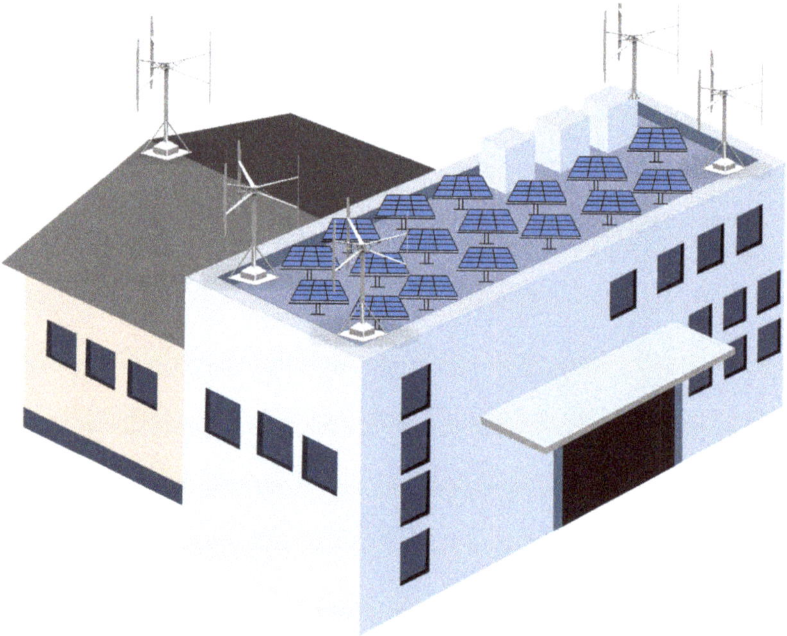

Figure 18. Building Roof Layout with PV and Wind turbines Installed.

Thermal Loads (TL—T1).

This scenario explores the use of a blended fuel approach, combining hydrogen (H_2) and propane, to meet the energy demands of the building's heating air system. By integrating hydrogen, a clean and sustainable energy source, with propane, the system aims to enhance energy efficiency while reducing environmental impact. This innovative approach leverages the advantages of both fuels, contributing to a more sustainable and reliable heating solution.

SC 0

In this scenario, the thermal load demands in Table 16 are entirely reliant on the utility supply, with 100% of the required energy being provided by the utility. The detailed statistical analysis of this scenario is presented in the following Tables 16–18.

Table 16. Thermal Load.

Load	kW
Heat Loss through Walls	26.67
Heat Loss through Windows	50.00
Ventilation Heat Loss	174.48
Total Heat Load Per Hour	251.15

Table 17. Fuel used for the thermal load.

Effective Heat Required kJ	295.47
Propane heating value kJ/L	1,063,676.47
Amount of Propane needed L/h	42.04
Total Fuel Consumption (L)	1345.36
Running Cost (CAD $/year)	628,552.91
GHG Emission (tCO_2/year)	540.29

Table 18. Cost Analysis.

Utility Gas Running Cost in the building (CAD $/year)	28,552.91
GHG Emission from the building fuel consumption (tCO_2/year)	540.29

A TRNSYS simulation is conducted in Figure 19 for the green hydrogen generation system, emphasizing a partial power supply from wind turbines and photovoltaic (PV) panels. This simulation focuses on the alkaline electrolysis system, with a vital component of the hydrogen production process. The goal is to analyze and evaluate its performance, particularly in terms of the annual hydrogen output generated.

Sc 6-1

This scenario utilizes a fuel mixture of hydrogen (H_2) and propane to supply the building mall's heating air system (TL-T1). The mixture consists of 40% hydrogen and 60% propane, providing an efficient and balanced fuel source to meet the thermal load demands with analysis of cost reduction which can be seen in Table 19.

Table 19. H_2 Propane mixture analysis for the thermal load (Air heating).

Amount of H_2 as 40% mixture needed L/h	17
Amount of Propane 60% mixture needed L/h	25
Total Propane Consumption (L)	807
H_2—Propane Cost (CAD $/year)	377,109
GHG Emission (tCO_2/year)	324
Saving (CAD $/year)	251,406

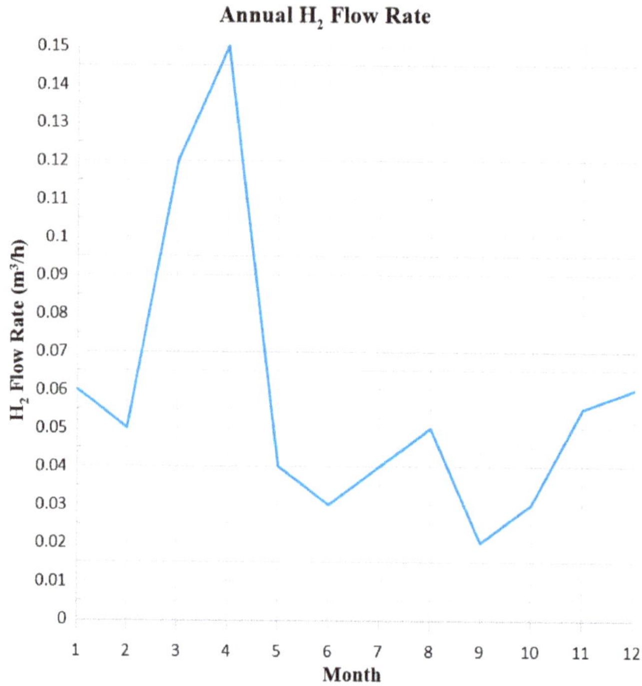

Figure 19. Annually Hydrogen Production.

Sc 6-2

This scenario offers a fuel mixture of H_2 and Propane at the building mall. In this scenario, H_2 gas and propane are used as a fuel mixture with 25% H_2 and 75% propane to feed the building's air heating system (TL-T1) with analysis of cost reduction which can be seen in Table 20.

Table 20. H2 Propane mixture analysis for the thermal load (Air heating).

Amount of H_2 as 25% mixture needed L/h	11
Amount of Propane 75% mixture needed L/h	32
Total Propane Consumption (L)	1009
H_2—Propane Cost (CAD $/year)	471,415
GHG Emission (tCO_2/year)	405
Saving (CAD $/year)	157,129

SC 6-3

This scenario offers a fuel mixture of H_2 and Propane at the building mall. In this scenario, both H_2 gas and propane are used as a fuel mixture with 15% H_2 and 85% propane to feed the building's air heating system (TL-T1) with analysis of cost reduction which can be seen in Table 21.

Table 21. H2 Propane mixture analysis for the thermal load (Air heating).

Amount of H_2 as 15% mixture needed L/h	6
Amount of Propane 85% mixture needed L/h	35
Total Propane Consumption (L)	1009
H_2—Propane Cost (CAD $/year)	534,238
GHG Emission (tCO_2/year)	450
Saving (CAD $/year)	94,277

The amount of fuel consumption for a thermal load of air heating of the building can be shown in Figure 20. SC 6-1 is the optimum one for the Propane-H_2 mixture. This scenario has many privileges as compared to others as can be seen in Figure 21 as saving costs yearly, and in Figure 22 for GHG emissions.

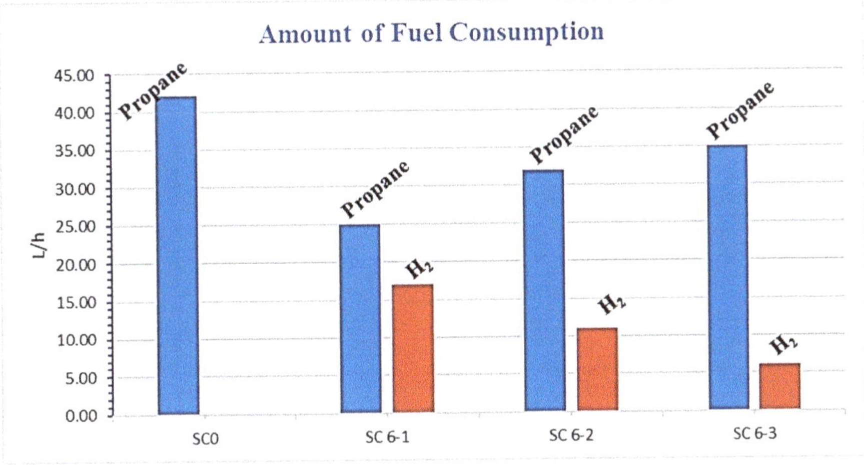

Figure 20. Fuel Consumption for Different Scenarios with and Without Propane -H_2 mixture.

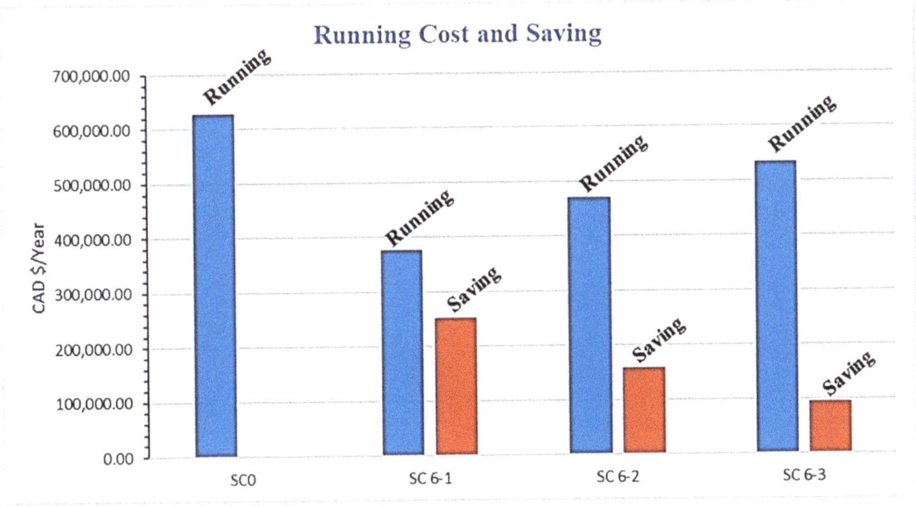

Figure 21. Running Cost and Saving for Different Scenarios.

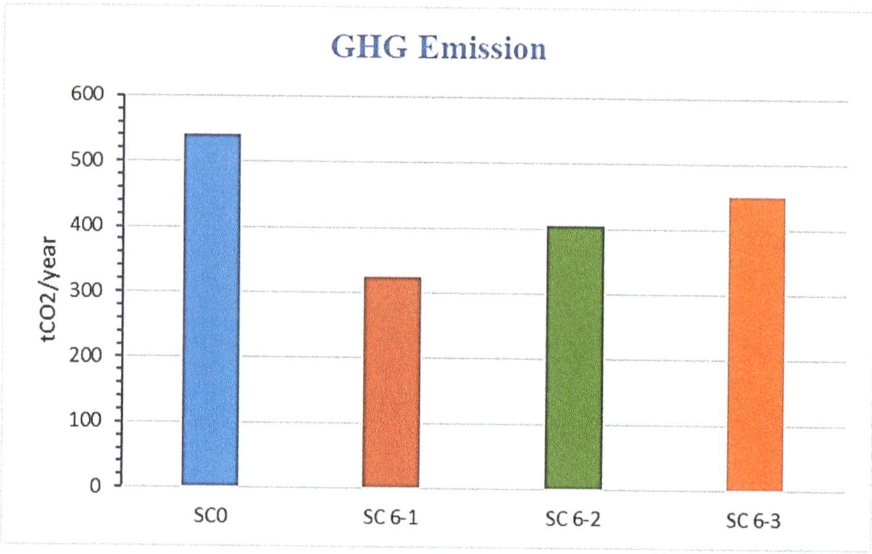

Figure 22. GHG Emissions for Different Scenarios.

Risk Analysis for H_2 Generation Unit

Conducting a risk analysis for an alkaline electrolysis system designed to generate hydrogen involves evaluating potential hazards associated with each component and the system as a whole with some assumptions built out as in Table 22.

Table 22. Assumed Probabilities for Basic Events.

Basic Event	Probability
Sensor Malfunction	0.02
Electrical Short Circuit	0.03
Control System Failure	0.04
Hydrogen Leak	0.01
Electrolyte Leak	0.02
Power Supply Interruption	0.03

Figures 23 and 24 outlines a Risk Assessment and Scenarios Framework designed for evaluating and managing risks in a hydrogen (H_2) system. The framework is divided into three interconnected components:

1. Risk Assessment of H_2 Unit: This component focuses on identifying and analyzing potential risks specific to the hydrogen unit, including operational, safety, and environmental hazards.
2. Risk Scenarios: Based on the assessment, various potential risk scenarios are developed. These scenarios help in understanding how different risk factors could materialize under specific conditions.
3. Risk Level Calculation: This step involves quantifying the level of risk associated with each scenario. It calculates risk severity and probability, providing a clear understanding of the overall risk profile.

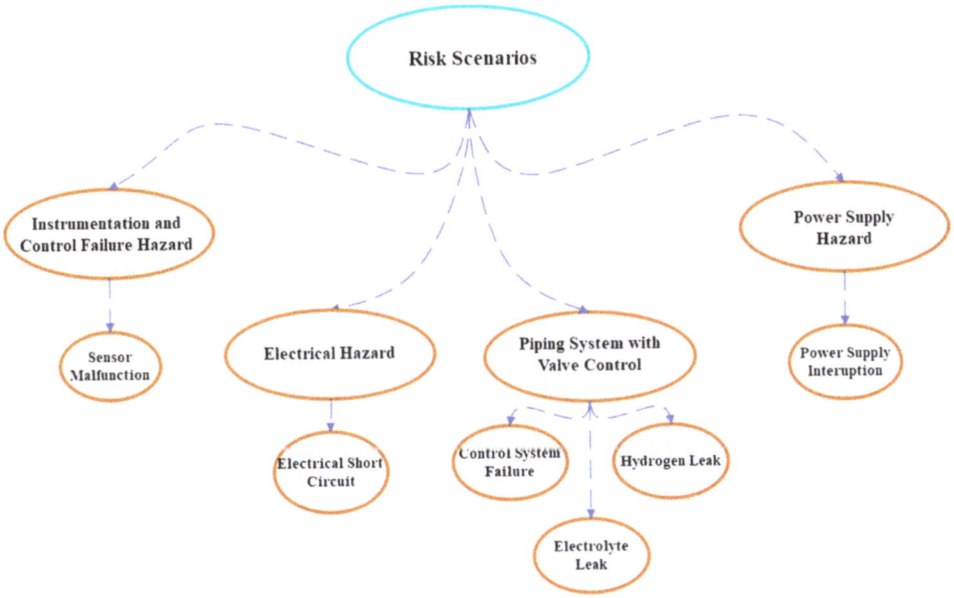

Figure 23. The Definition of the Risk Scenarios.

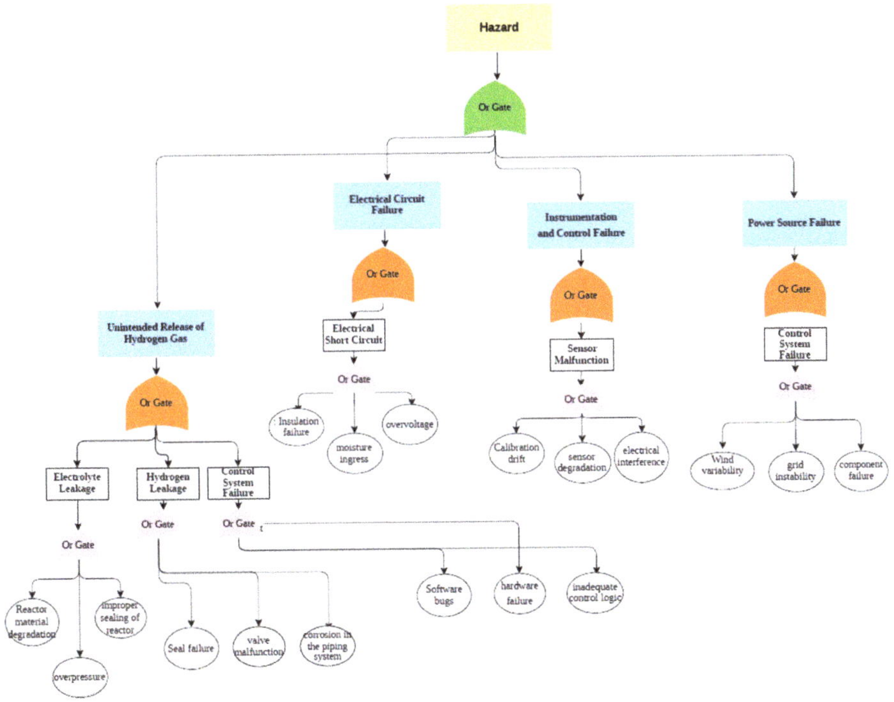

Figure 24. Fault tree Analysis of hydrogen hazard.

Intermediate Events and Top-Level Calculations

1. Instrumentation and Control Failure Hazard

This is caused by a Sensor Malfunction (single basic event):

$$I_{nstrumentation} = P_{\text{Sensor Malfunction}} = 0.02 \tag{6}$$

2. Electrical Hazard (OR Gate)

Combines:

- Electrical Short Circuit (P1): 0.03
- Control System Failure (P2): 0.04

$$\begin{aligned} P_{\text{Electrical Hazard}} &= 1 - (1 - P1) \times (1 - P2) \\ &= 1 - (1 - 0.03) \times (1 - 0.04) \\ &= 1 - (0.97) \times (0.96) = 1 - 0.9312 = 0.0688 \end{aligned} \tag{7}$$

Piping System with Valve Control (OR Gate)
Combines:

- Control System Failure (P3): 0.04
- Hydrogen Leak (P4): 0.01
- Electrolyte Leak (P5): 0.02

$$\begin{aligned} P_{\text{Piping System}} &= 1 - (1 - P3) \times (1 - P4) \times (1 - P5) \\ &= 1 - (1 - 0.04) \times (1 - 0.01) \times (1 - 0.02) \\ &= 1 - (0.96) \times (0.99) \times (0.98) \\ &= 1 - 0.931 \\ &= 0.069 \end{aligned} \tag{8}$$

Power Supply Hazard (Single Event)
This is caused by a Power Supply Interruption:

$$P_{\text{Power Supply Hazard}} = 0.03$$

Top-Level Event (Risk Scenarios)
The top event Risk Scenarios combines:

- Instrumentation and Control Failure Hazard (P6): 0.02
- Electrical Hazard (P7): 0.0688
- Piping System with Valve Control (P8): 0.069
- Power Supply Hazard (P9): 0.03

OR Gate Calculation:

$$\begin{aligned} P\ \text{Risk Scenarios} &= 1 - (1 - P6) \times (1 - P7) \times (1 - P8) \times (1 - P9) \\ &= 1 - (1 - 0.02) \times (1 - 0.0688) \times (1 - 0.069) \times (1 - 0.03) \\ &= 1 - (0.98) \times (0.9312) \times (0.931) \times (0.97) \\ &= 1 - (0.98) \times (0.9312) \times (0.931) \times (0.97) \\ &= 0.186 \end{aligned} \tag{9}$$

The overall probability of the Risk Scenarios is 18.6%. This value reflects the combined likelihood of the hazards associated with instrumentation, electrical systems, piping systems, and power supply interruptions. If actual probabilities are available, these calculations can be further refined.

6. Conclusions

This study fills a significant void in the literature by presenting a clear pathway to replacing fossil fuels in buildings with hybrid renewable systems. The integration of green hydrogen amplifies the systems' environmental impact, offering a versatile energy carrier for diverse applications. It highlights the transformative capabilities of hybrid renewable energy systems (HRES) in meeting the energy needs of buildings, emphasizing their pivotal role in achieving environmental and socio-economic sustainability. By integrating solar panels, wind turbines, and green hydrogen production, the research highlights a pathway to reducing greenhouse gas (GHG) emissions, lowering operational costs, and enhancing energy reliability. Through dynamic simulations using TRNSYS software, the paper validates the effectiveness of hybrid systems in optimizing energy generation and consumption under varying climatic conditions, as demonstrated through carefully analyzed scenarios.

Key findings include:

1. Hybrid systems, such as Scenario SC 4-1, effectively meet up to 63% of electrical light loads, combining solar and wind energy to maximize renewable energy utilization and reduce reliance on utility grids, achieving annual savings of CAD$ 87,026.33.
2. The introduction of hydrogen-propane fuel blends for thermal loads, exemplified by SC 6-1, reduces greenhouse gas emissions by 216 tCO_2/year and generates annual cost savings of CAD$ 251,406 achieves significant reductions in GHG emissions and operational costs while enhancing energy efficiency.
3. Risk assessments and fault tree analyses provide actionable insights into mitigating hazards associated with hydrogen generation units, ensuring the safety and reliability of these advanced systems.

The results clearly demonstrate that the hybrid renewable energy systems are not only practical but have extra advantageous in meeting energy demands sustainably. They present a compelling case for replacing fossil fuels in buildings with renewable energy sources, driving progress toward global net-zero targets. The integration of green hydrogen into these systems further amplifies their environmental benefits, offering a clean, versatile energy carrier for diverse applications.

Future work should focus on scaling these systems, optimizing their cost-effectiveness, and integrating emerging technologies to address the ever-evolving challenges of energy sustainability. By embracing these innovative solutions, municipalities, industries, and governments can accelerate the transition to a cleaner, more sustainable energy future.

Author Contributions: A.R.: Conceptualization and Methodology, Visualization, Investigation, Writing—Original Draft Preparation. H.A.G.: Funding Acquisition, Principal Investigator, Methodology, Writing—Reviewing and Editing. All authors have read and agreed to the published version of the manuscript.

Funding: The research is supported by the Government of Canada and NSERC.

Institutional Review Board Statement: Not Applicable.

Informed Consent Statement: Not Applicable.

Data Availability Statement: The authors will provide the raw data that supports the conclusions of this article upon request.

Conflicts of Interest: The authors declare no conflicts of interest.

Abbreviations

PV	Photovoltaic panel.
WT	Wind turbine
STC	Standard Test Conditions
DG	Diesel Generator
BG	Biogas
LCE	Levelized Cost of Energy
LCC	Life Cycle Cost
GHG	Green House Gas
Nomenclature	
$Rotor_{Ht}$	Turbine rotor center height (m)
$Rotor_{\_Di}$	Turbine rotor diameter (m)
$Sher_{\_Exp}$	Power-law exponent for vertical wind profile
$Turb_{\_Int}$	Turbulence intensity valid for this curve
Air_Dens	air density (kg/m^3)
$Pwr_{\_Ratd}$	Rated power of the turbine (W)
$Spd_{\_Ratd}$	Rated wind speed (m/s)
I_{ELY}	Current through single electrolyzer stack (A)
P_{ELY}	Electrolyzer pressure (Pa)
T_{ROOM}	Room temperature (°C)
T_{ELY}	Temperature of electrolyzer (°C)
P_{max}	Rated Maximum Power (W)
V_{oc}	Open Circuit Voltage (V)
V_{mp}	Maximum Power Voltage (V)
Isc	Short Circuit Current (A)
I_{mp}	Maximum Power Current (A)
Creek Letter	
α	Temperature Coefficient of Isc
β	Temperature Coefficient of V_{oc}
γ	Temperature Coefficient of Pmax
P	Air Density (kg/m^3)
η	Efficiency

References

1. Dey, S.; Sreenivasulu, A.; Veerendra, G.T.N.; Rao, K.V.; Babu, P.S.S.A. Renewable energy present status and future potentials i India: An overview. *Innov. Green Dev.* **2022**, *1*, 100006. [CrossRef]
2. Khan, S.A.; Al-Ghamdi, S.G. Renewable and Integrated Renewable Energy Systems for Buildings and Their Environmental an Socio-Economic Sustainability Assessment. In *Energy Systems Evaluation (Volume 1)*; Ren, J., Ed.; Springer International Publishin Cham, Switzerland, 2021; pp. 127–144.
3. Energy Efficiency 2019—Analysis. Available online: https://www.iea.org/reports/energy-efficiency-2019 (accessed on 1 October 2024).
4. Chen, L.; Hu, Y.; Wang, R.; Li, X.; Chen, Z.; Hua, J.; Osman, A.I.; Farghali, M.; Huang, L.; Li, J.; et al. Green building practices t integrate renewable energy in the construction sector: A review. *Environ. Chem. Lett.* **2024**, *22*, 751–784. [CrossRef]
5. Pérez-Lombard, L.; Ortiz, J.; Pout, C. A review on buildings energy consumption information. *Energy Build.* **2008**, *40*, 394–39 [CrossRef]
6. Shahzad, U.; Gupta, M.; Sharma, G.D.; Rao, A.; Chopra, R. Resolving energy poverty for social change: Research directions an agenda. *Technol. Forecast. Soc. Change* **2022**, *181*, 121777. [CrossRef]
7. Gielen, D.; Boshell, F.; Saygin, D.; Bazilian, M.D.; Wagner, N.; Gorini, R. The role of renewable energy in the global energ transformation. *Energy Strategy Rev.* **2019**, *24*, 38–50. [CrossRef]
8. Ramadan, A.; Gabbar, H.A. Evaluation of Hydrogen Generation with Hybrid Renewable Energy Sources. *Appl. Sci.* **2024**, *14* 6235. [CrossRef]
9. Nazari-Heris, M.; Asadi, S. Reliable energy management of residential buildings with hybrid energy systems. *J. Build. Eng.* **2023** *71*, 106531. [CrossRef]

Canada's Clean Hydrogen Gets Set to Power the World—Financial Times—Partner Content by Invest in Canada. Available online: https://canada-next-best-place-to-home.ft.com/canadas-clean-hydrogen-gets-set-to-power-the-world (accessed on 16 October 2024).

Ma, T.; Yang, H.; Lu, L. Development of hybrid battery–supercapacitor energy storage for remote area renewable energy systems. *Appl. Energy* **2015**, *153*, 56–62. [CrossRef]

Ghahramani, M.; Nazari-Heris, M.; Zare, K.; Mohammadi-Ivatloo, B. A two-point estimate approach for energy management of multi-carrier energy systems incorporating demand response programs. *Energy* **2022**, *249*, 123671. [CrossRef]

The Economics of Load Defection: How Grid-Connected Solar-Plus-Battery Systems Will Compete with Traditional Electric Service, Why It Matters, and Possible Paths Forward. This Report, Published by the Rocky Mountain Institute in May 2015. Available online: https://rmi.org/wp-content/uploads/2017/04/2015-05_RMI-TheEconomicsOfLoadDefection-FullReport-1.pdf (accessed on 16 December 2024).

Abdin, Z.; Mérida, W. Hybrid energy systems for off-grid power supply and hydrogen production based on renewable energy: A techno-economic analysis. *Energy Convers. Manag.* **2019**, *196*, 1068–1079. [CrossRef]

Al Afif, R.; Ayed, Y.; Maaitah, O.N. Feasibility and optimal sizing analysis of hybrid renewable energy systems: A case study of Al-Karak, Jordan. *Renew. Energy* **2023**, *204*, 229–249. [CrossRef]

Malik, P.; Awasthi, M.; Sinha, S. A techno-economic investigation of grid integrated hybrid renewable energy systems. *Sustain. Energy Technol. Assess.* **2022**, *51*, 101976. [CrossRef]

Das, B.K.; Tushar, M.S.H.K.; Hassan, R. Techno-economic optimisation of stand-alone hybrid renewable energy systems for concurrently meeting electric and heating demand. *Sustain. Cities Soc.* **2021**, *68*, 102763. [CrossRef]

Murugaperumal, K.; Ajay D Vimal Raj, P. Feasibility design and techno-economic analysis of hybrid renewable energy system for rural electrification. *Sol. Energy* **2019**, *188*, 1068–1083. [CrossRef]

Wang, J.; Pan, Z.; Ge, H.; Zhao, H.; Xia, T.; Wang, B. Economic Dispatch of Integrated Electricity–Heat–Hydrogen System Considering Hydrogen Production by Water Electrolysis. *Electronics* **2023**, *12*, 4166. [CrossRef]

Behzadi, A.; Arabkoohsar, A.; Perić, V.S. Innovative hybrid solar-waste designs for cogeneration of heat and power, an effort for achieving maximum efficiency and renewable integration. *Appl. Therm. Eng.* **2021**, *190*, 116824. [CrossRef]

Shen, Y.; Li, X.; Wang, N.; Li, L.; Hoseyni, A. Introducing and investigation of a pumped hydro-compressed air storage based on wind turbine and alkaline fuel cell and electrolyzer. *Sustain. Energy Technol. Assess.* **2021**, *47*, 101378. [CrossRef]

Ahmad, T.; Zhang, D. Renewable energy integration/techno-economic feasibility analysis, cost/benefit impact on islanded and grid-connected operations: A case study. *Renew. Energy* **2021**, *180*, 83–108. [CrossRef]

Patil, A.A.; Arora, R.; Sridhara, S. Techno-economic analysis of solar, wind, and biomass hybrid renewable energy systems in Bhorha Village, India. *J. Atmos. Sol. -Terr. Phys.* **2024**, *265*, 106362. [CrossRef]

Karunathilake, H.; Perera, P.; Ruparathna, R.; Hewage, K.; Sadiq, R. Renewable energy integration into community energy systems: A case study of new urban residential development. *J. Clean. Prod.* **2018**, *173*, 292–307. [CrossRef]

Karunathilake, H.; Hewage, K.; Sadiq, R. Opportunities and challenges in energy demand reduction for Canadian residential sector: A review. *Renew. Sustain. Energy Rev.* **2018**, *82*, 2005–2016. [CrossRef]

Singh, K.; Hachem-Vermette, C. Economical energy resource planning to promote sustainable urban design. *Renew. Sustain. Energy Rev.* **2021**, *137*, 110619. [CrossRef]

Minuto, F.D.; Lazzeroni, P.; Borchiellini, R.; Olivero, S.; Bottaccioli, L.; Lanzini, A. Modeling technology retrofit scenarios for the conversion of condominium into an energy community: An Italian case study. *J. Clean. Prod.* **2021**, *282*, 124536. [CrossRef]

Environment and Climate Change Canada (ECCC). Air Health Trends. Available online: https://www.canada.ca/en/environment-climate-change/services/environmental-indicators/air-health-trends.html (accessed on 18 December 2024).

Nematollahi, O.; Alamdari, P.; Jahangiri, M.; Sedaghat, A.; Alemrajabi, A.A. A techno-economical assessment of solar/wind resources and hydrogen production: A case study with GIS maps. *Energy* **2019**, *175*, 914–930. [CrossRef]

Wei, D.; Zhang, L.; Alotaibi, A.A.; Fang, J.; Alshahri, A.H.; Almitani, K.H. Transient simulation and comparative assessment of a hydrogen production and storage system with solar and wind energy using TRNSYS. *Int. J. Hydrogen Energy* **2022**, *47*, 26646–26653. [CrossRef]

Mansir, I.B.; Bani Hani, E.H.; Farouk, N.; AlArjani, A.; Ayed, H.; Nguyen, D.D. Comparative transient simulation of a renewable energy system with hydrogen and battery energy storage for residential applications. *Int. J. Hydrogen Energy* **2022**, *47*, 26198–26208. [CrossRef]

Behzadi, A.; Alirahmi, S.M.; Yu, H.; Sadrizadeh, S. An efficient renewable hybridization based on hydrogen storage for peak demand reduction: A rule-based energy control and optimization using machine learning techniques. *J. Energy Storage* **2023**, *57*, 106168. [CrossRef]

Ishaq, H.; Dincer, I.; Naterer, G.F. Development and assessment of a solar, wind and hydrogen hybrid trigeneration system. *Int. J. Hydrogen Energy* **2018**, *43*, 23148–23160. [CrossRef]

34. Al-Buraiki, A.S.; Al-Sharafi, A. Hydrogen production via using excess electric energy of an off-grid hybrid solar/wind syste based on a novel performance indicator. *Energy Convers. Manag.* **2022**, *254*, 115270. [CrossRef]
35. Aide Financière | Programme Chauffez Vert | Innovation et Transition Énergétiques. Available online: https: transitionenergetique.gouv.qc.ca/residentiel/programmes/chauffez-vert/aide-financiere (accessed on 2 January 2025).
36. Pandey, A.K.; Singh, P.K.; Nawaz, M.; Kushwaha, A.K. Forecasting of non-renewable and renewable energy production in Ind using optimized discrete grey model. *Environ. Sci. Pollut. Res.* **2023**, *30*, 8188–8206. [CrossRef]
37. Solar Energy Laboratory, University of Wisconsin–Madison. *TRNSYS 16: A Transient System Simulation Program—User Manu* University of Wisconsin–Madison: Madison, WI, USA, 2017.
38. Historical Data—Climate—Environment and Climate Change Canada. Available online: https://climate.weather.gc.c historical_data/search_historic_data_e.html (accessed on 19 December 2024).

sustainability

MDPI

rticle

ʀesearch on Increasing the Building's Energy Efficiency by Jsing the Ground Beneath It for Thermo-Accumulation

adas Zdankus *, Sandeep Bandarwadkar, Juozas Vaiciunas, Gediminas Stelmokaitis and Arnas Vaicaitis

Faculty of Civil Engineering and Architecture, Kaunas University of Technology, Studentu Str. 48, LT-51367 Kaunas, Lithuania; sandeep.bandarwadkar@ktu.edu (S.B.); juozas.vaiciunas@ktu.lt (J.V.); gediminas.stelmokaitis@ktu.lt (G.S.); arnas.vaicaitis@ktu.edu (A.V.)
* Correspondence: tadas.zdankus@ktu.lt

Abstract: A whole series of factors influence the temperature of the soil surface and surface layers. The soil surface is heated by solar radiation during the day. It radiates some of the obtained heat at night. The heat exchange between the soil and the atmosphere depends on the air and soil temperatures and the speed of air movement. Precipitation may also affect surface soil layers, but this was not considered in this study. In the mentioned interaction, a specific temperature field of the surface layers of the soil is established. To increase the building's energy efficiency, the aim is to optimize the operation of its heating and cooling systems and to reduce heat loss to the environment as much as possible. Heat loss through the floor of the building or the walls of the recessed part into the ground changes the established temperature field of the ground. The heat spreads in the soil and is given to the atmospheric air. During the research, to validate the numerical model, the heat flow density was analysed to determine how it changes while maintaining a constant temperature of the heating surface at a certain depth of the soil. It was found that the new thermodynamic equilibrium, depending on the seasonality, can be reached in a time interval of up to a week. The temperature change in the artificially limited volume of the ground under the building or next to it can be treated as the work of the ground thermo-accumulator: its charge or discharge by heat. This makes it possible to reduce the annual energy costs of the building by more than ten percent.

Keywords: heat transfer into the ground; heat loss; heat accumulation; thermal equilibrium; building's energy efficiency

1. Introduction

Buildings account for a significant share of global energy consumption. To address this, many countries have implemented policies aimed at improving energy efficiency in buildings. Heat production constitutes a larger portion of global energy use (47%) compared to transport (27%), electricity (17%), and non-energy purposes (9%) [1,2]. Over the past decade, global energy consumption in buildings has increased by 1% annually [3,4]. In northern Europe, nearly 45% of the demand for heating in residential buildings is driven by the need for domestic hot water and space heating and is responsible for approximately 80% of CO_2 emissions from the building sector [5]. However, the growing threat of human-induced climate change, fueled by pollution from fossil fuel-based energy sources, is driving increased interest in small-scale renewable energy technologies. Shallow geothermal systems, a renewable technology used for heating and cooling buildings, have the potential to reduce energy consumption by a factor of 3–4 [6–10]. These systems' efficiency depends on the consistency of ground temperatures with depth throughout the year. One of the

 https://doi.org/10.3390/su17010262

main challenges in decarbonizing heating systems is the seasonal imbalance between demand and heat generation from sustainable sources. Many sustainable heating systems involve high initial investments but have relatively low operational costs. As a result, designing capacity to meet peak demand is not cost-effective. To address this, extending the operation of the system over the entire year can help meet energy needs, reduce costs, and promote decarbonization. The optimal usage of sustainable heat requires long-term storage to account for seasonal fluctuations in supply and demand [4,8].

Soil temperature is an environmental variable that plays a crucial role in energy balance applications. In addition, knowledge of soil thermal dynamics depends on understanding its properties, such as thermal conductivity, thermal diffusivity, density, and volumetric heat capacity. These properties are fundamental for estimating the soil temperature and soil heat flux, which are essential for determining the overall energy balance [11,12]. The nature of the soil, such as its physical structure, organic matter content, texture, and mineral composition, decides the soil's thermal variations. These characteristics are particularly valuable for applications in passive solar heating systems, where the goal is to capture solar energy, store it, and use it when needed [13]. Studies have shown that the soil's moisture content and solar radiation influence the dynamics of the thermal variables and soil properties. The diverse material that composes the soil exerts a substantial impact on the configuration of the thermal diffusivity. The spatial arrangements of these materials, in addition to the presence of air voids, affect thermal diffusion [14–16]. At shallow depths, the temperature of the soil undergoes significant seasonal variations due to direct exposure to surface conditions and solar heating. The ability of soil to store heat, or its timing of accumulation, plays a critical role in this process. For example, the soil near the surface has a shorter accumulation time, which means it heats up and cools down more quickly in response to seasonal changes. As the depth increases, the accumulation time also increases, allowing the soil to store heat more slowly and for longer periods. This longer accumulation time dampens the amplitude of the temperature fluctuations, leading to more stable conditions at the greater depths [17–19]. Several authors have investigated the accumulation of heat in the soil, focusing on different methodologies and implications for energy efficiency and environmental management. Research was carried out that involved laboratory experiments and field conditions to assess heat accumulation under varying atmospheric influences. The study explored using the soil layers for thermal energy accumulation, highlighting soil temperatures during the heating seasons [20]. Another investigation examined the effect of soil heat storage on energy balance in field conditions. It emphasized the importance of measuring the soil heat flux at different depths to improve energy balance closure, noting the significant phase shifts between surface and subsurface heat fluxes. This study developed methods to determine the surface soil heat flux by incorporating heat storage data [21]. Modelling focused on the heat flow to understand the temperature distributions and their ecological impacts. The study discussed the challenges of measuring soil temperatures and the need for predictive models that consider environmental variables and moisture content [22,23]. However, most of the research is focused on short-term data or specific soil depths without accounting for long-term variations in heat accumulation and transfer between different seasons and soil layers.

For an isotropic and homogeneous medium soil, analytical solutions of a one-dimensional conductive equation can be used to determine thermal properties using the amplitude method, arctangent, logarithmic, phase shift, and harmonic methods [24–26]. However, the heat transfer mechanism in the soil is also determined by heat convection within the pores. Therefore, research that describes the soil temperature considers heat transfer by conduction and convection [27–29]. Several researchers have attempted to determine the thermal accumulation properties using the Laplace transform method and

Fourier series using field data despite quiet weather conditions. In 2009, researchers employed harmonic and conduction–convection methods to estimate soil retention using soil temperature data, although they used data from only seven days under clear skies. In 2011, Wang et al. compared six different methods to evaluate how efficiently soil can absorb and transfer heat in the Loess Plateau in China, conducting the study under a single set of weather conditions [30,31]. In 2013, Tadiran and Otunla evaluated the performance of six methods to evaluate the soil heat storage potential during both dry and wet seasons in West Africa [32]. However, their study was conducted under similar conditions. The accuracy of these methods was determined by comparing the error between the observed and simulated temperatures using the estimated values of thermal conductivity, which were derived from soil temperature profiles over the same period. In 2012, Miao attempted to use heat flux to evaluate the amplitude, conduction-convection, phase shift, and harmonic methods to estimate soil heat conservation of the soil [33]. However, their study was based on only five days of data collected under clear skies conditions. As a result, the application of these methods to estimate thermal sky remains insufficiently validated, lacking comprehensive and systematic assessment using long-term field data across various weather conditions.

A detailed understanding of both surface and subsurface thermal dynamics throughout all seasons is crucial to the reliable and consistent operation of soil-based heat storage systems. Seasonal variations in soil temperature, moisture content, and atmospheric conditions significantly influence the capacity of the soil to store and transfer heat, particularly in shallow areas where temperature fluctuations are most pronounced. It is imperative to analyze long-term data to fully capture the seasonal dynamics that affect heat accumulation and retention.

The change in soil surface temperature over the year was analyzed during the research. The experimental part included heat transfer in the soil from the heated plane surface, thus changing the soil temperature distribution and transitioning to a new thermodynamic equilibrium. The results validated a numerical model for the heat transfer simulation through the building floor to the soil. It was determined that it would be appropriate to use the soil under the building for heat accumulation, thus increasing the energy efficiency of the building itself.

2. Materials and Methods

By slowing down the heat dissipation in the ground, it would be possible to reduce heat losses through the building floor and the walls of the deepened part to the ground. Lower heat losses mean lower energy costs for heating and cooling the building. Therefore, the research was focused on slightly deepened buildings with a semi-basement or basement. In this case, the lower surface of the building was located in the surface soil layers at a depth from 0 to 1.5 m.

The research consisted of numerical simulation and experimental parts. The experimental investigation was intended to study heat dissipation in the surface soil layers. For this purpose, a heating device was placed at a depth of one meter (Figure 1). Its lower flat surface was initially heated to the set temperature, and then it was kept constant at the set value until the new thermodynamic balance was reached in the ground. The experiments were provided in the field conditions. There was a clay at a depth of 1 m. Parameters of the clay were such: density: ρ = 1460 kg/m^3, moisture: μ = 15%, thermal conductivity: λ = 0.76 W/(m·K), and specific heat capacity: c_p = 878 J/(kg·K). Polystyrene EPS 70 NEO as thermal insulation was placed above the heated device to simulate the heat loss through the building floor to the ground. The thickness of the thermal insulation layer was equal to δ_p = 0.1 m, and the thermal conductivity of the polystyrene was equal to λ_p = 0.031 W/(m·K). The heating was performed electrically. The heating cable CTAV-18

(24 m) was mounted onto the chipboard with a thickness of δ_{ch} = 0.01 m. The heating cable was separated from the ground by a flat steel surface heated during the experiment and whose thickness was equal to δ_{st} = 0.001 m. The heating surface was square in shape. The edge length was equal to a = 0.5 m.

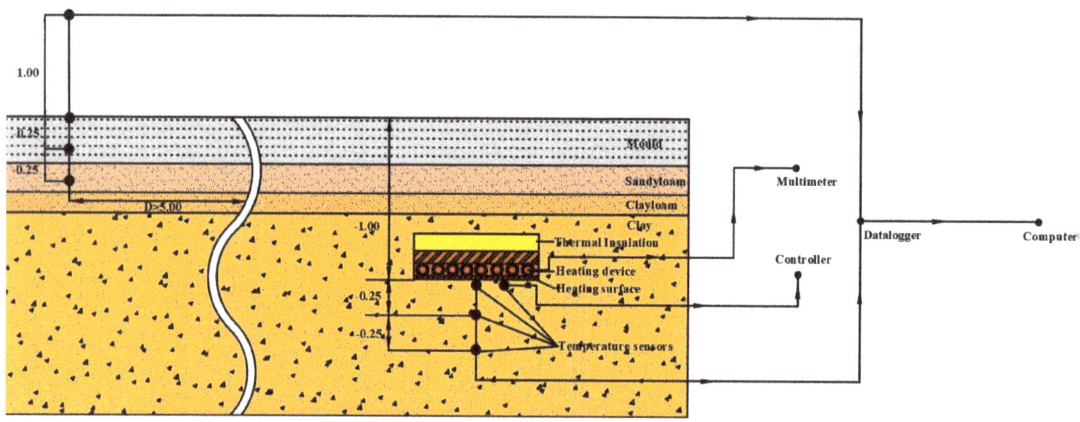

Figure 1. Experimental setup.

When heating the building, a specific constant temperature is maintained inside it, which also means a specific temperature of the external surface of the building's floor in contact with the ground. Since it was difficult to predict the possible surface temperature in advance and, on the other hand, to have a higher temperature difference and a more evident effect on the ground temperature distribution, the temperature of the heated surface was set to a value of 20 °C. The temperature was maintained discretely by connecting and disconnecting the heating device to the power supply. This meant that the controller of the type ATR244-12ABC (manufactured by Pixsys Srl, Pianiga, Italy) worked by the discrete "on-off" method. One sensor of the Pt1000 type (precision class A) placed at the center of the heated surface from the ground side was used as the input for the temperature control for the ATR244-12ABC. The controller turned on the heating device when the soil temperature was below 20 C. When the temperature reached 20 °C, the regulator turned off the heating surface. It was assumed that all the energy from the electricity network was converted into heat. At the beginning of the experiment, part of the energy was used to heat the heating device itself, but this amount of energy was insignificant and was not evaluated. To observe the change in heat flux density, the heating cable's electric current (I) and voltage (V) were measured by a multimeter of the ESCORT 3136A type (the resolution of 100 μA, accuracy of 2% for the range of 5A for AC, and resolution of 10 mV, and accuracy of 0.5% for the range of 500 V for AC, produced by Escort Instruments Corporation in Taiwan). Both multimeter inputs were used for that. The power of the heating device was, on average, equal to $P = VI = 375 \pm 10$ W. An average value of the heat flux density for the set time interval was computed by Equation (1):

$$q_{av,t_{set}} = \frac{\sum_{i=1}^{n} V_{av,i} I_{av,i} t_i}{a^2 t_{set}} \quad (1)$$

where t_{set} is the time interval in seconds for which the average heat flux density was calculated, i is the number indicating the specific turn-on of the heating device, n is the number of turns-on of the heating device per set time interval, and t_i is the duration of a specific turn-on.

The average temperature of the heated surface was determined by the output signal of one more temperature sensor, Pt1000 (h = −1.00 m), which was installed at the center of the heated surface from the soil side. Data Logger PT–104 (Pico Technology, Cambridgeshire, United Kingdom) was applied to transmit the output signal of Pt1000 sensors to the computer. The measured values were read and recorded at an interval of 1 s. Two Pt1000 sensors were installed at 0.25 m (h = −1.25 m) and 0.5 m (h = −1.50 m) from the heating surface's center to monitor the heat transfer in the ground. The formation of a new thermodynamic balance in the ground was expected and observed. It was determined by the same unchangeable value of the heat flux density for several hours in a row.

Temperature measuring sensors Pt1000 (UAB "Auregis", Kaunas, Lithuania) were tested and calibrated in the thermodynamics laboratory in the Faculty of Mechanical Engineering and Design at Kaunas University of Technology before the measurements were made in the field conditions. It was necessary to maintain the same soil structure during all experiments. Therefore, the devices were tested and calibrated a second time only after the experimental research was completed, and they were dismantled from the soil. The coincidence of their output signals under melting ice conditions was checked; it did not exceed ±0.1 °C.

Experimental research on heat dissipation in the soil started in December and was finished in July. One experiment was performed each month. The number of experiments was limited by the fact that a return to the naturally formed soil temperature distribution in the surface layers required a time of about two weeks after the completion of the soil heating part of the study.

Heat transfer in the soil during the operation of the heating device was simulated numerically using COMSOL software, version 6.1. This simulation was named as the "Case 1". The results obtained by simulation and compared with experimental ones were used to validate the numerical model.

The intensity of heat loss through the building's floor to the ground depends on environmental conditions and varies over the year. Therefore, the change in the temperature of the ground surface and soil layers near the surface was observed over the year. The atmospheric air temperature was also measured. It was performed at 1 m (h = 1) above the ground surface, following the requirements for meteorological measurements. The ground surface temperature was measured by covering the sensor from direct sunlight and precipitation, allowing wind flow. Soil temperatures were measured at depths (h) equal to 0.25, 0.50, 1.00 m, 1.25 m, and 1.50 (h = −0.25; −0.50; −1.00, −1.25, and −1.50 m). Platinum resistance thermometers, Pt1000, were used to measure the temperature of the soil at different depths.

Soil temperatures at the mentioned depths and the air temperature were measured before each experiment with the heating device. The duration of the measurements was not less than one day. The average daily values were calculated and analyzed in the graphs or tables. When the naturally formed conditions had been reached after the experiment was finished, the measurements were repeated. Attention was paid to the analysis of the influence of seasonality on the ground temperatures in the surface layers. Since the soil temperature distribution defined the initial conditions of the experiment, the results chapter began with an analysis of the influence of seasonality on the air temperature and the naturally formed soil temperature field.

The numerical simulation was focused on heat transfer through the floor and the deepened part of the walls of the building to the ground. The architecture didn't play a significant role here. Therefore, the simplified building architecture with a semi-basement was chosen for the numerical simulation (Figure 2). The external surface of the building floor was at a depth of 1 m (h = −1 m). This scheme for the simulation was named as the "Case 2". The building comprised several key components: the roof, external walls, partition walls, and floor, all constructed from concrete. The concrete material had a thermal conductivity of λ_{con} = 2 W/(m·K), a density of ρ_{con} = 2300 kg/m^3, and a specific heat capacity of $c_{p,con}$ = 880 J/(kg·K). The structure was surrounded by multiple soil layers, each with distinct thermal properties, which influenced the heat transfer dynamics in the system. The thickness of the floor concrete layer was equal to δ_{fl} = 0.1 m, and the thickness of the thermal insulation layer was equal to δ_{tin} = 0.1 m as well.

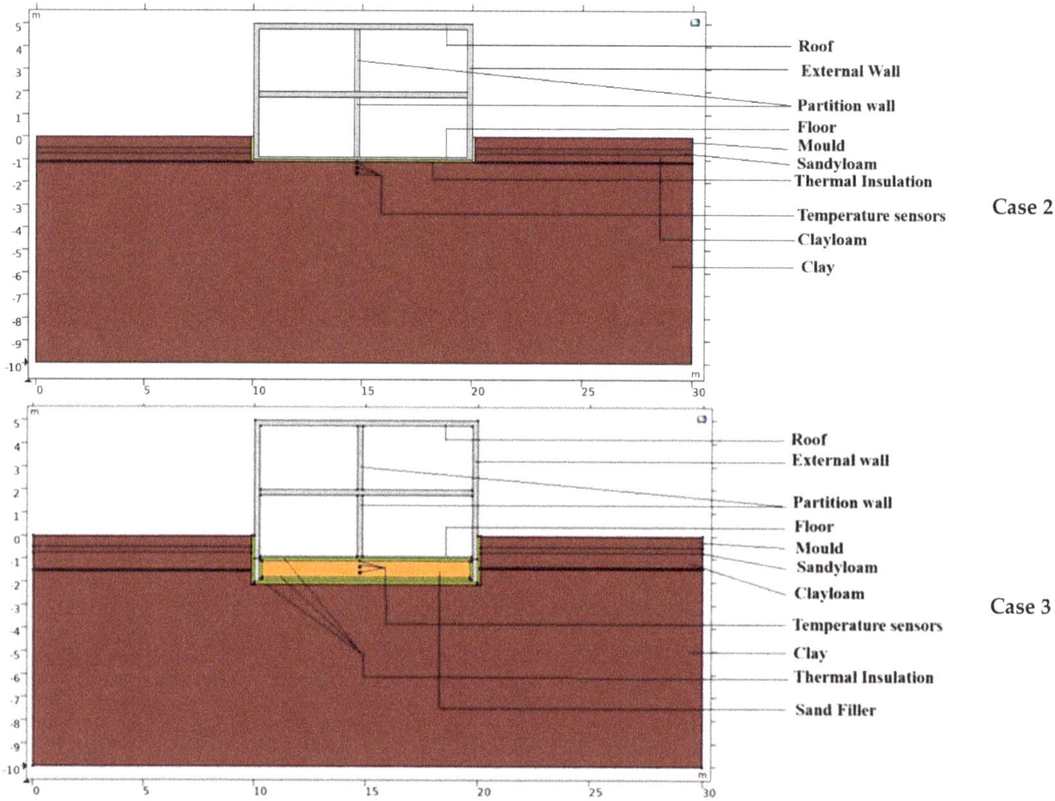

Figure 2. Scheme of the building with a semi-basement for numerical simulation. The external surface of the building floor is at a depth of 1 m—Case 2. Building with sand filler layer separated from the surrounding environment—Case 3.

Directly beneath the building's floor was a mold layer characterized by a thermal conductivity of λ_m = 0.9 W/(m·K), a density of ρ_m = 1570 kg/m^3, and a specific heat capacity of $c_{p,m}$ = 1250 J/(kg·K). Below this, a sandy loam layer extended further down, with a thermal conductivity of λ_{sl} = 1.3 W/(m·K), ρ_{sl} = 1550 kg/m^3 density, and a specific heat capacity of $c_{p,sl}$ = 1000 J/(kg·K). The soil transitioned to clay loam and clay layers at greater depths, with a thermal conductivity of λ_{cl} = 0.6 W/(m·K), ρ_{cl} = 1850 kg/m^3 density, and a specific heat capacity of $c_{p,cl}$ = 1500 J/(kg·K).

The building structure was surrounded by a layer of thermal insulation strategically placed around its sides and base to reduce thermal losses to the surrounding soil. This insulation material had a thermal conductivity of λ_{tin} = 0.035 W/m·K, a ρ_{tin} = 30 kg/m^3 density, and a specific heat capacity of $c_{p,tin}$ = 1200 J/kg·K. Additionally, temperature sensors were simulated at different locations within the soil layers to monitor heat transfer and ground temperature distribution, providing insights into the thermal performance of the structure and its interaction with the surrounding environment.

Beneath the structure, a sand filler layer separated from the surrounding environment had been added to act as a thermal storage medium. The height of the internal chamber was 1 m. The properties of the sand were such: λ_s = 1.45 W/(m·K) thermal conductivity, ρ_s = 1600 kg/m^3 density, and $c_{p,s}$ = 900 J/(kg·K) specific heat capacity. This sand filler enhanced the thermal dynamics of the system, aiding in heat retention and regulation. This scheme (Figure 2) for the numerical simulation was named as "Case 3".

The COMSOL Multiphysics software was used to simulate numerically all mentioned cases.

3. Results

At shallow ground depths, atmospheric factors' influence on the soil temperature distribution is significant. The ground surface is exposed to solar radiation. Due to the motion of the atmospheric air, its temperature can change quite quickly. Due to the temperature difference, heat is transferred from the soil to the atmosphere or in the opposite direction. Water evaporation or condensation processes occur at the ground surface. Precipitation water soaks into the surface soil layers. The absorbed water changes the thermal properties of the soil, and at the same time, heat exchange occurs due to the temperature difference between the water moving through the pore channels and the soil.

Air temperature is one of the most essential factors that needs to be known for analyzing the heat exchange between the ground surface layers and atmospheric air. It defines not only the intensity of thermal processes but also determines the direction of their occurrence. The values of the average daily air temperatures measured and computed during experimental studies, that is, in the period from September 2022 to August 2023, are shown in Figure 3. For comparison, the average monthly air temperature values measured at the Lithuanian Meteorological Station near Kaunas, in Noreikiskes, in the period from 1961 to 1990 [34] and in the period from 1991 to 2020 [35] are also shown in Figure 3. It was noticed that the average annual air temperature value, based on the data for the period from 1961 to 1990, in the mentioned place was equal to T = 6.3 °C [34], and from 1991 to 2020, it was equal to T = 7.5 °C [35]. That means the temperature value increased by ΔT = 1.2 ° in thirty years. That very clearly reflects climate change.

Based on the values of air temperature measurements, the average monthly temperature values were calculated for the period from 2022 to 2023. The change in air temperature during the year was analyzed in a way that was different than usual (Figure 4). Analysis started with the coldest January month, going to the hottest July month, and then returning to January. The difference between the average monthly temperatures of those months that were at the same time interval starting from the coldest month was analyzed (Table 1). If the middle of January was taken as the reference "0", then the time interval from mid-January to mid-February and mid-December was one month—"1". Accordingly, from January to March and November was a two-month time interval ("2"), from April to October was three months ("3"), from May to September was four—"4", and from January to June and August was a five-month time interval—"5". The middle of July was half a year from January ("6").

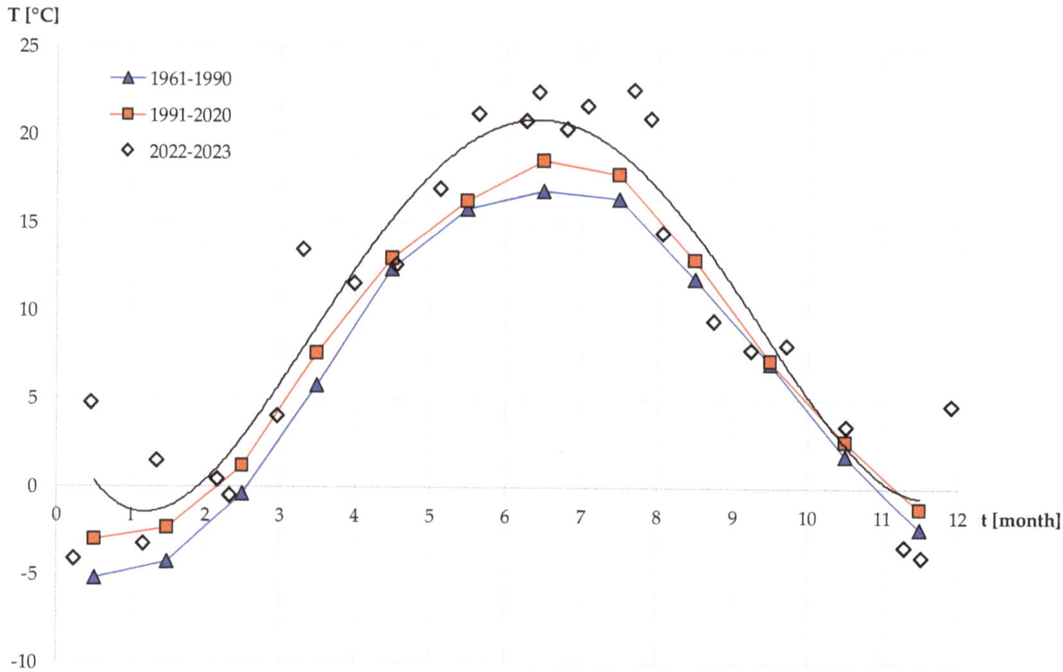

Figure 3. Comparison of the air temperatures measured in the period from 2022 to 2023 with average monthly temperature values given in documents of the construction climatology [34,35].

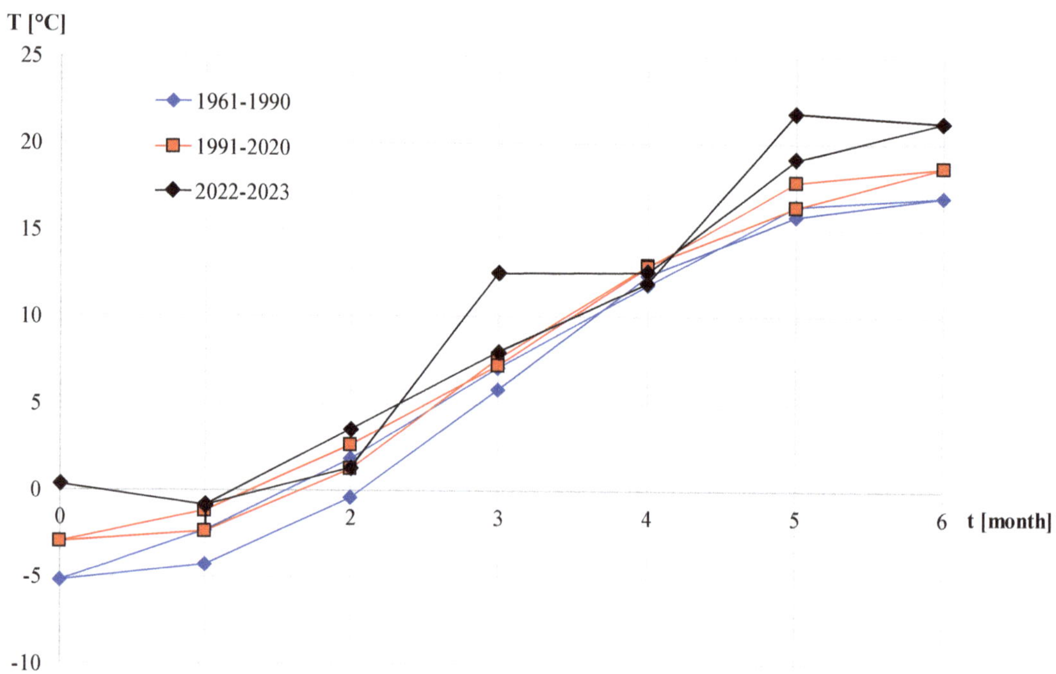

Figure 4. Change in air temperature over the year, starting from January, going to July and returning to January.

Table 1. Change in air temperature over the year, from January to July and back.

Time Interval	Row	January	February	March	April	May	June	July
1961 ÷ 1990	1	−5.2	−4.3	−0.4	5.8	12.4	15.8	16.9
1991 ÷ 2020	2	−3.0	−2.4	1.2	7.6	13.0	16.3	18.6
2022 ÷ 2023	3	0.3	−0.9	1.3	12.5	12.6	19.1	21.2
			December	November	October	September	August	
1961 ÷ 1990	4		−2.3	1.8	7.1	11.9	16.4	
1991 ÷ 2020	5		−1.2	2.6	7.2	12.9	17.8	
2022 ÷ 2023	6		−0.9	3.5	8.0	12.0	21.8	
Δ (1961 ÷ 1990)	1–4		−2.0	−2.2	−1.3	0.5	−0.6	
Δ (1991 ÷ 2020)	2–5		−1.2	−1.4	0.4	0.1	−1.5	
Δ (2022 ÷ 2023)	3–6		0	−2.2	4.5	0.6	−2.7	
interval in months from January		0	1	2	3	4	5	6

The difference between monthly air temperature values of the months, which were at the same time interval from January, was not high (Table 1). As expected, the highest difference in temperature equal to ΔT_{max} = 4.5 °C was noticed in the case of the 2022–2023 period. The maximal temperature difference was lesser and equal to ΔT_{max} = 2.2 °C in the case of the 1961–1990 and 1991–2020 periods [34,35]. In the spring, the temperature was slightly lower than in autumn. However, it can be stated that the inertia was very slight.

The temperature change in the ground surface layers was analyzed for 2022–2023 (Figure 5). During the winter, the higher soil temperature was in the deeper layers ($h = -1.50$ m) and decreased in the surface direction (to $h = 0$ m). Such character of the soil temperature dependence on the depth may be described by the function: $T\downarrow = f(h\uparrow)$. The temperature measurements on March 31st stood out. It can be said that it was like a critical point at which the nature of the soil temperature distribution changed the character: from $T\downarrow = f(h\uparrow)$ to $T\uparrow = f(h\uparrow)$. At that point, the temperature difference within the mentioned depth values (from $h = -1.50$ m to $h = 0$ m) was only $\Delta T = 2.1$ °C. The second critical point was around 10 September (Figure 5), but temperatures were not measured on that day.

The average monthly values of the ground temperature were calculated at different depths, from the surface ($h = 0$ m) to the depth of 1.5 m ($h = -1.5$ m). Results are presented in Figure 6, similar to the character of the atmospheric air temperature change (Figure 4). At the ground surface ($h = 0$ m) and near it ($h = -0.5$ m), the lowest soil temperature T_{min} was in February. The minimal temperature T_{min} was observed in March in the deeper ground layers ($h = -1.0$ m and $h = -1.5$ m). Here, the delay in time appeared while the heat transferred from the deeper soil layers to the surface. In the following months of the spring, the soil temperature increased at all the mentioned depths (from $h = 0$ m to $h = -1.5$ m)—the ground surface layers were warming up. The highest soil temperature was in August (Figure 6). Later in autumn, the ground temperatures decreased. At depths $h = 0$ m and $h = -0.5$ m, the character of soil temperature variation over the year was analogous to that of atmospheric air temperature change (Figure 4).

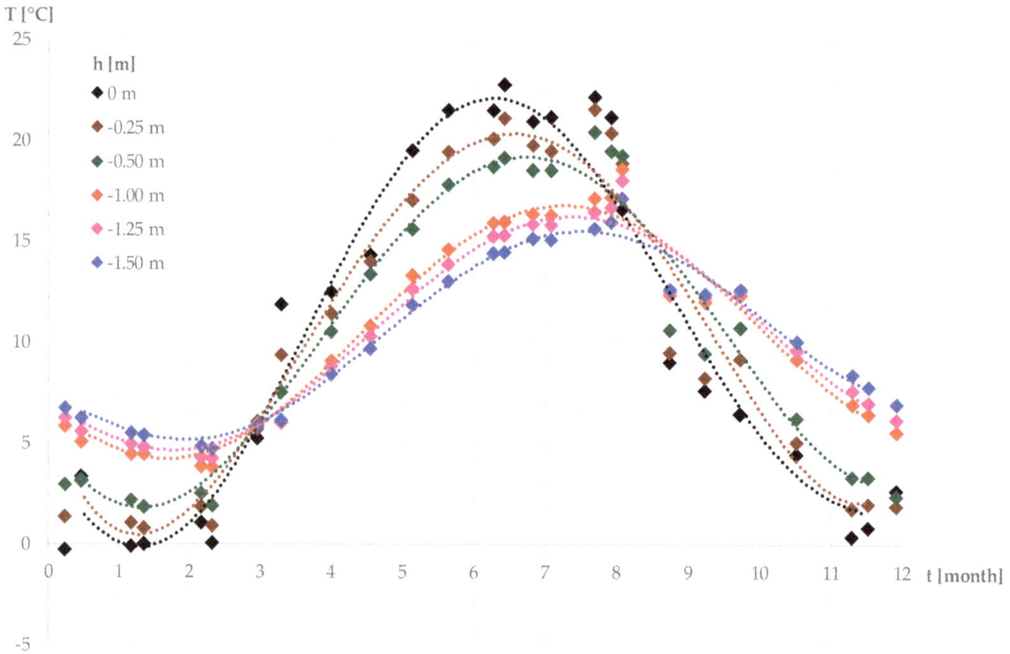

Figure 5. Change in soil temperature at different depths during the year.

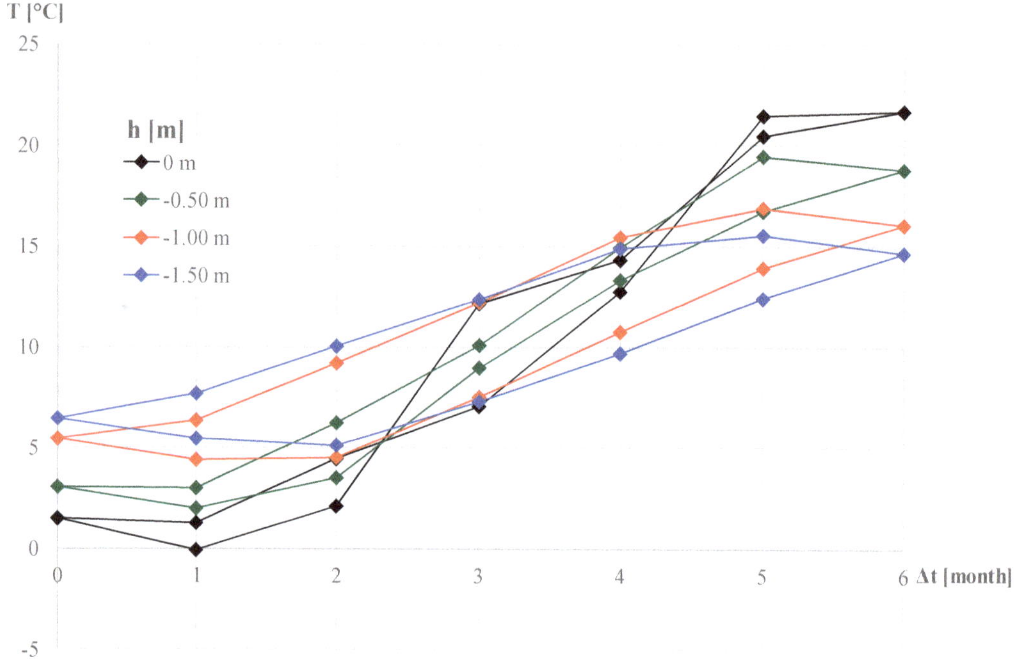

Figure 6. Change in monthly ground temperature at different depths over the year, starting from January to July and returning to January.

February was chosen as the datum month instead of January to analyze the change in soil temperature at depths $h = -1.0$ m and $h = -1.5$ m (Figure 7). In a similar time period from the new datum month, the temperatures at the corresponding depths were almost the same (Figure 7). Such character of the dependence (Figure 7), compared with the previous one (Figure 6), showed that the inertia of the process is about one month, and the speed of heat transfer in the soil and release to the atmospheric air was the same in both directions.

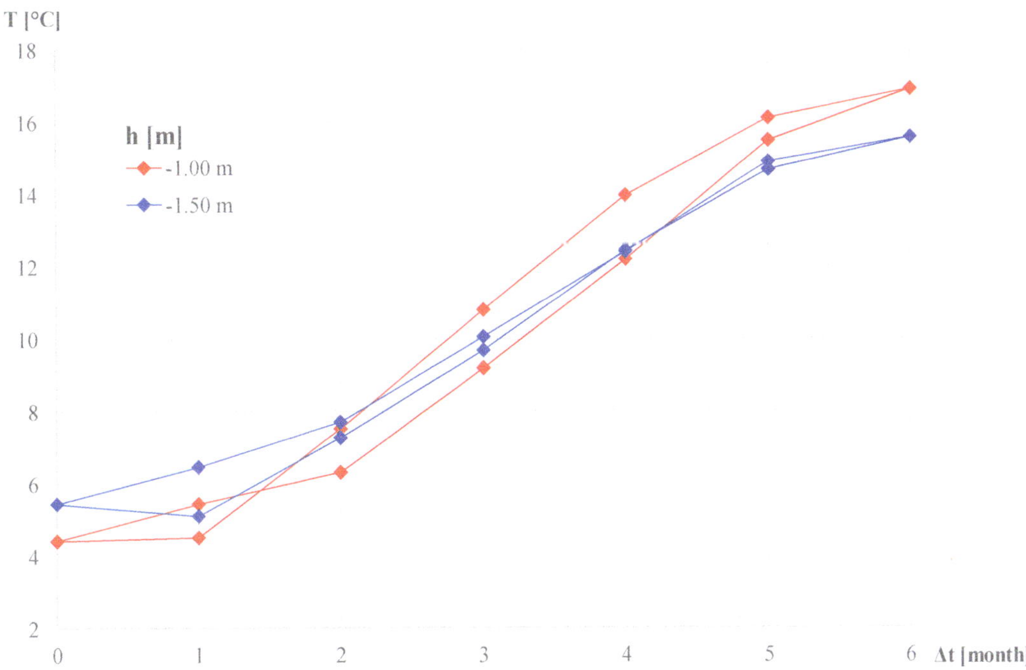

Figure 7. Change in monthly ground temperature at the depths: $h = -1.0$ m and $h = -1.5$ m, over the year, starting from February to August and returning to February.

During experimental research, the heating device was placed at a depth of 1 m ($h = -1.0$ m), and the set temperature was kept constant ($T_{set} = 20$ °C). Since the heated surface was covered with thermal insulation from above, the main part of the heat flux was directed downwards. After about one week, a new thermal balance was reached. During this period, the soil was heated from the initial temperature T_{in} to the new thermal balance T_{bal} temperature. The ground temperature distribution was changed. An average temperature at the heated surface ($h = -1.0$ m) was calculated from measured temperature values. It was equal to $T_{-1.0} = 19.5 \pm 0.1$ °C. The soil temperature was measured at the distances of 0.25 m ($h = -1.25$ m) and 0.50 m ($h = -1.50$ m) from the heated surface. The results are presented in Figure 8.

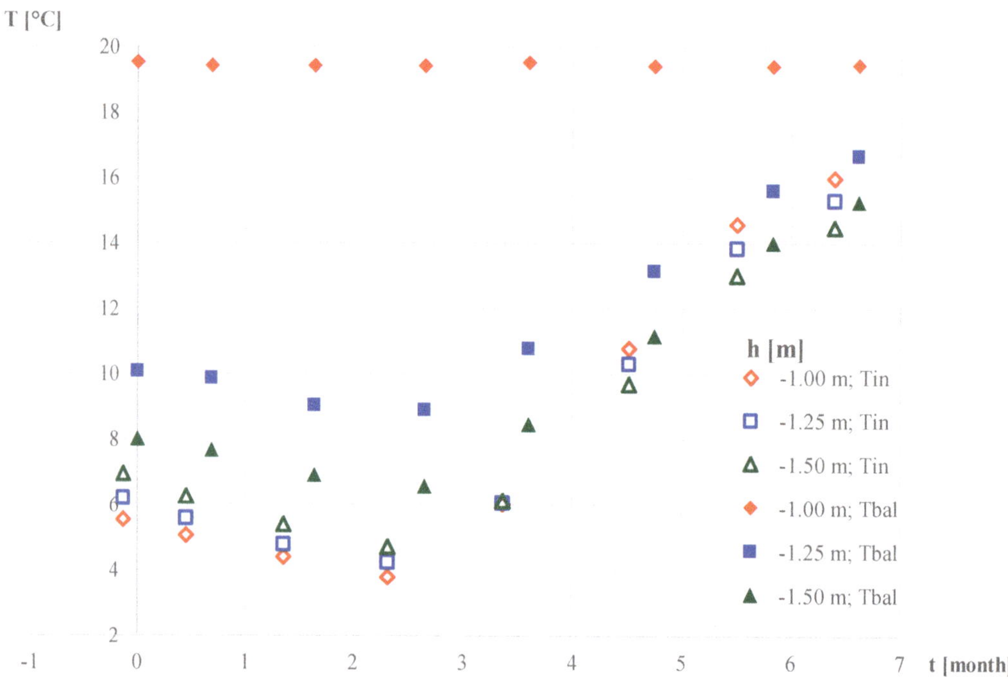

Figure 8. Change in soil temperature distribution at depths $h = -1.0$ m, -1.25 m and -1.5 m one week after the heating device was turned on.

The experimental research started in December and continued until July. The average hourly heat flux density $q_{av,h}$ was determined based on the heating device's electricity consumption for each experiment. The experiment started with the naturally formed soil temperature distribution. After the turn-on of the heating device, heat transfer to the soil began. Initially, the temperature difference between the heated surface and the soil was the largest. Therefore, the intensity of the heat flux density was the highest. The heat from the soil layer, which was in contact with the heated surface, was transferred to other soil layers below. The heat was spreading in the soil volume. After some time, the other soil temperature distribution settled, depending on the depth. The heat flux density obtained a constant value. A new thermodynamic balance was reached while the heating device was operating. It was found that this balance was achieved in approximately one week.

In the first hour, the average hourly heat flux density was maximal and exceeded the value of the second hour a few times. In the first hour, the heat flux density for the corresponding month was equal to $q_{Dec} = 592$ W/m^2, $q_{Jan} = 646$ W/m^2, $q_{Feb} = 615$ W/m^2, $q_{Mar} = 679$ W/m^2, $q_{Apr} = 567$ W/m^2, $q_{May} = 386$ W/m^2, $q_{June} = 226$ W/m^2, $q_{July} = 159$ W/m^2. The research results were presented from the second hour to analyze the subsequent change in the heat flux density in detail (Figure 9). The high change in the heat flux density was observed in the first four hours.

The steady-state values of the heat flux density for different months of the year are presented in Figure 10. From December to March, the heat flux density varied slightly (up to 5%): $q_{av,h} = 117.2 \pm 2.5$ W/m^2. By June, the value of the heat flux density decreased by an average of 25 percent, computed from the average value of q of the cold period. In July, the heat flux density stabilized at $q_{av,h} = 20.2$ W/m^2.

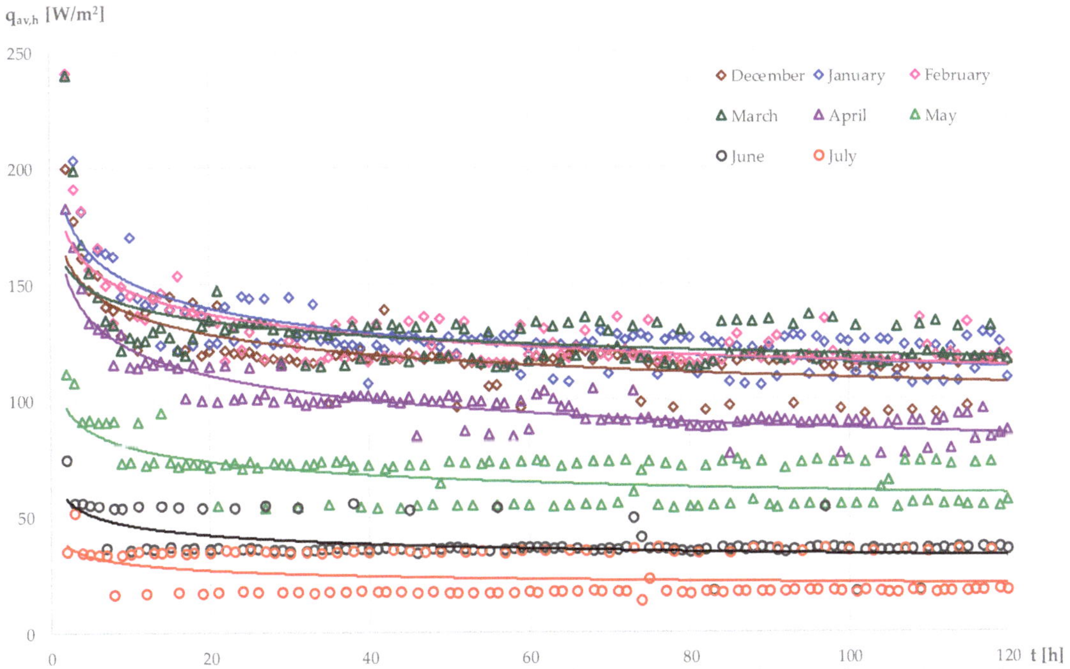

Figure 9. Change in the heat flux density generated by the heated surface over time. Note: data are presented starting from the second hour.

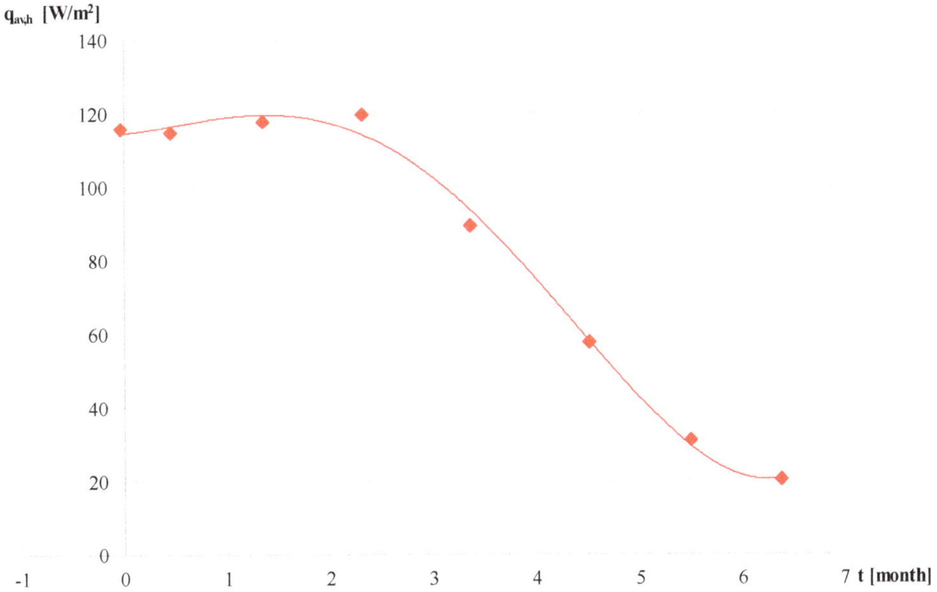

Figure 10. Change in heat flux density from the heated surface of the constant temperature to the ground from December to July.

The operation of the heating device and the heat transfer in the soil were simulated using COMSOL Multiphysics software. January month was chosen, and the initial ground temperatures were set according to the results of the experimental measurements (Figures 3 and 5). The time until the constant ground temperature values were established. The obtained data are presented in Figure 11 and Table 2. The difference was insignificant, and it was concluded that the model was valid.

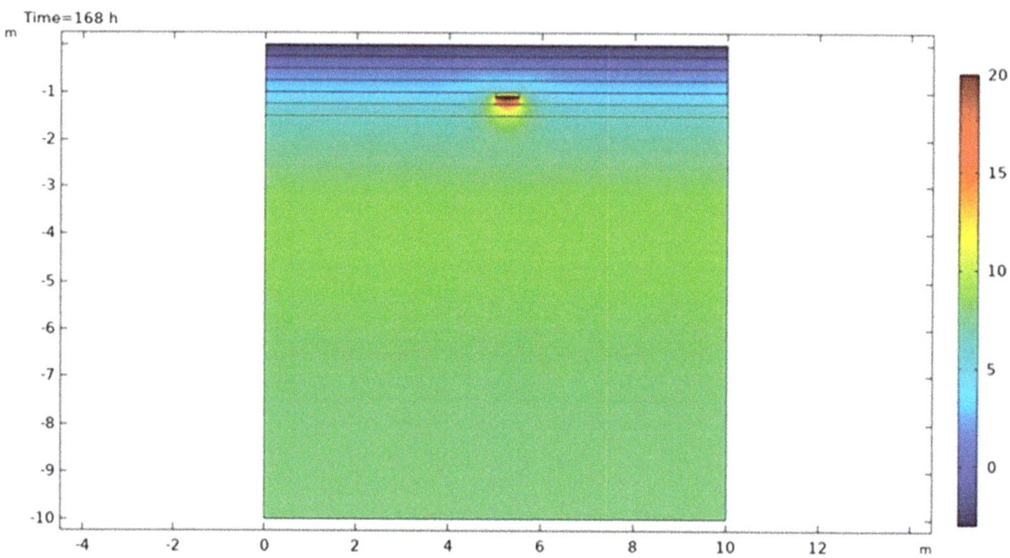

Figure 11. The temperature profile of the ground after 168 h (one week) of operation of the heating device placed at $h = -1$ m.

Table 2. Comparison of the temperatures near the heated surface after 168 h of its operation.

Results Type	Parameter, Units	$h = -1.00$ m	$h = -1.25$ m	$h = -1.50$ m
Experimental	T, °C	19.4	9.8	7.6
Simulated	T, °C	19.4	9.6	8.0
	ΔT, °C	0	0.2	−0.4
	$100 \cdot \Delta T/T$, %	0	2.1	−5.1

Simulation of the heat transfer through the typical building's floor to the ground started from Case 2. After 168 h, an average heat flux density was found, and it was, on average, equal to $q_{av} = 4.7$ W/m^2. The contact surface temperature between the floor and the ground was equal to $T = 6.9$ °C (Figure 12). Taking the inside temperature of 20 °C, the heat flux density value may be calculated theoretically for these temperatures: $q_T = 4.5$ W/m^2. The difference can be explained by the fact that the energy was used to warm the ground surrounding the building basement at the sides. That means that heat dissipated not only vertically downward but also in the horizontal direction at the sides as well.

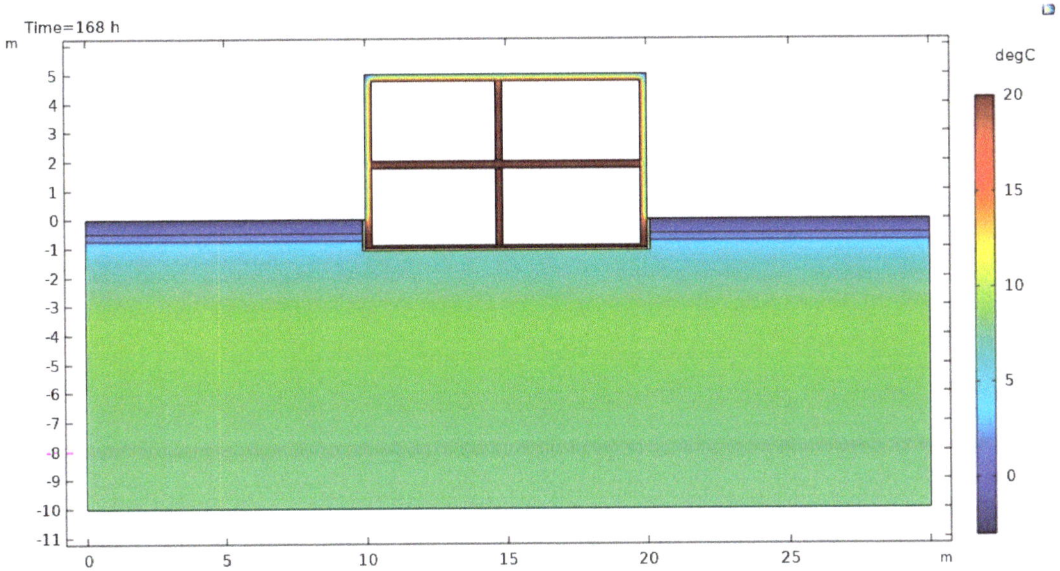

Figure 12. The temperature profile of the ground after one week in Case 2.

At present, when there are real possibilities to install renewable energy sources into the engineering systems of buildings, the issue of network overload becomes increasingly relevant. One of the possible solutions is energy storage. The need for and possibility of thermal energy storage still seem unusual, especially in the case of an individual building when the heat accumulator is installed under the building. It can be stated that a separate study needs to be performed to analyze such an issue. During this research, the numerical simulation was performed in the case of the special chamber filled with sand and installed under the building. It acted as a brake on heat dissipation to the ground. The significant inertia of the process was noticed. The soil in this chamber must be heated to begin to interfere with the transfer of heat through the building floor to the ground. The time period during which the thermodynamic equilibrium was reached was observed. The time was extended so that the change in the parameter per day did not exceed one percent. The numerical simulation of Case 3 was performed (Figure 13). After three weeks, an average heat flux density through the floor to the chamber filled with sand was found, and it was, on average, equal to q_{av} = 3.7 W/m^2. It is 21% less than that in Case 2. That meant that even the usage of the container of the ground-type heat accumulator without its charge by heat increased the energy efficiency of the building.

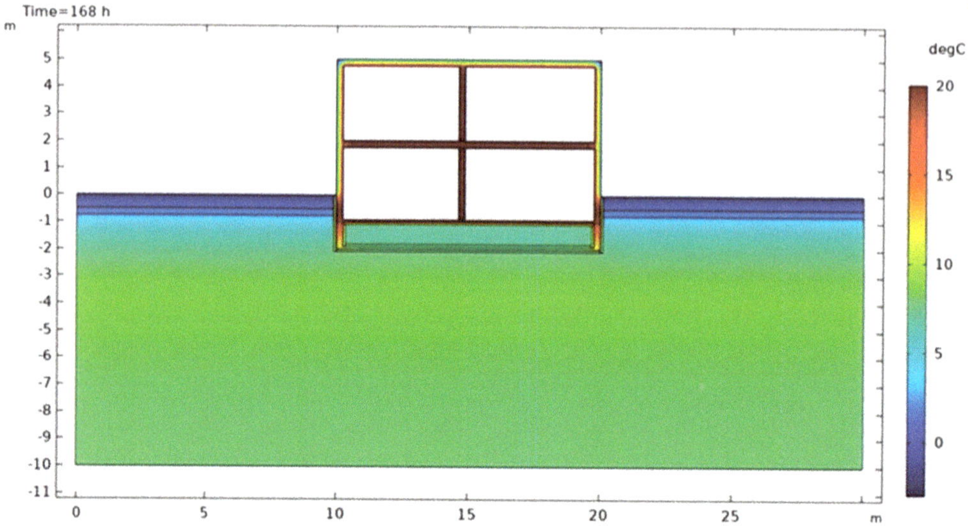

Figure 13. The temperature profile of the ground after three weeks in Case 3.

4. Discussion

During the experimental investigation, the temperature of the ground surface layers was measured to a depth of 1.5 m. The atmospheric air temperature influences the soil surface layers. Therefore, the air temperature was also measured, and the average monthly values were calculated. The experimental data covered one year, from September 2022 to August 2023. That corresponded to the academic but not the calendar year. Official meteorological observations do not include measurements of soil temperatures in the ground surface layers, but the atmospheric air temperature is measured and recorded at all Lithuanian meteorological stations. In this article, the air temperature measured during research was compared with that measured at the nearest meteorological station over a number of years, which is presented in the official document Lithuanian Construction Climatology [34,35]. The difference was not very large. By analogy, the assumptions can be made as to how much the measured soil temperature would differ insignificantly from the average value obtained from multi-year observations.

Precipitation can affect the temperature of the surface soil layers. The temperature of rainwater entering the soil is usually different from the temperature of the ground. Due to the temperature difference, heat exchange begins. The temperature difference is highest near the surface. This difference decreases till water moves downward through the pore channels. The heat transfer slows down, and temperatures equalize. It can be stated that the heat transfer resulting from the penetration of precipitation and the temperature difference depends on the intensity of the downpour itself, its duration, and the temperature of the rainwater. On the other hand, this is also influenced by the soil's permeability to water, the soil temperature, its thermal capacity, and conductivity. Water changes the soil's moisture content and, at the same time, its thermal capacity and conductivity, which can significantly affect the intensity of heat transfer taking place here. This effect can persist for a more extended period.

As mentioned in the paper, the soil temperature at the different depths and the air temperature were measured before each experiment with the heating device. The measurement lasted at least one day. However, no significant rainfall was recorded during these measurements. A similar situation occurred during eight experiments with the heated surface. Snow that fell during the cold period covered the soil surface episodically until it completely melted in some days. No significant influence of the precipitations on the results was observed. Since precipitation is difficult to predict, it would be appropriate to conduct simulations in the laboratory initially. Here, different precipitation temperatures and intensities could be simulated.

An economic assessment of the installation and use of a thermo-accumulator is necessary. Unfortunately, the studies described in the article partially covered only one operating mode of the storage tank, charging with heat. The simulation was performed in the case of a private building. In the case of thermo-accumulation, a critical question is how long and with what heat losses the heat can be kept. There is a need to charge the storage ground volume with heat in summer. In this case, renewable energy sources, such as solar photovoltaic cells, can also be used, and they could operate at maximum capacity, regardless of the energy demand, that is, exceeding it and converting the surplus energy to heat. The problem arises in heat storage with low losses for at least several months. Another issue is heat utilization, that is, discharging the thermo-accumulator. The most effective would be to use heat pumps. However, the above-mentioned issues require a separate study and economic assessment throughout the system's lifecycle.

Soil-type thermo-accumulators could be used in urban districts' heating networks, where cogeneration power plants and heat pumps could be combined. Industrial soil-type heat storage tanks could balance the operation of DH networks and prevent electricity grid overloads. However, in this case, electricity would be irreversibly converted into heat. Determining the optimal operating mode of industrial ground-type thermos-accumulators, integrating them into the district heating system, etc., also requires separate studies.

We focused on how to use the soil under the building for thermal storage. In the case of this study, it was used to suppress heat loss through the building floor or walls of the deepened part. The operating modes of this ground volume, such as heat charging, storage, and discharge, were not examined.

5. Conclusions

Soil temperatures were measured from the ground surface to a depth of 1.5 m, and their change over the year was observed. The delay in time was observed while the heat transferred from the deeper soil layers to the surface or vice versa. The lowest atmospheric air temperature was measured in January. The lowest soil temperatures at the surface layers were measured in February and at a depth of 1.0 m to 1.5 m—in March. The highest atmospheric air temperature was in July, and the highest soil temperature was in August.

The character of the ground temperature distribution varied during the year. The soil temperature increased with an increase in depth from August to March. The end of March was the first critical point. From that point, the soil temperature decreased with the increase in depth. Such distribution continued until the end of August, the second critical point. Soil temperature distribution returned to the previous type.

The heating device was placed in the ground at a depth of one meter. The heat flux was directed downward from the heated surface, the temperature of which was kept constant. The highest amount of energy was needed to heat the surrounding soil at the beginning. Therefore, the average heat flux density was highest in the first hour and was more than five times higher than that after some days. Nearly constant values of the measured ground temperature indicated that a new thermal balance was reached. It was approximately after one week. The steady-state values of the heat flux density from December to March were equal to $q_{av,h}$ = 117.2 ± 2.5 W/m^2. In July, the heat flux density stabilized at $q_{av,h}$ = 20.2 W/m^2.

With the rapid growth of solar panel usage in the private sector, the problem of overloading city energy networks appears. The installation of soil-type thermal accumulators may be the solution to that problem. Installing a soil-filled container under a building would bring benefits even without using it directly for heat accumulation. The numerical simulations allowed us to compare the amount of heat loss through the floor of a building with a semi-basement with that of the same building with an underground-installed, soil-filled chamber. The average value of the heat flux density decreased by twenty-one percent in the case of a building with a container.

Author Contributions: Conceptualization, T.Z.; methodology, T.Z., S.B., J.V. and G.S.; validation, T.Z. (experimental results) and S.B. (numerical simulation results); formal analysis, T.Z., S.B., J.V. and G.S.; investigation, T.Z., S.B., J.V., G.S. and A.V.; resources, T.Z.; data curation, T.Z. and J.V.; writing—original draft preparation, T.Z., S.B., J.V., G.S. and A.V.; writing—review and editing, T.Z., S.B., J.V. and G.S.; visualization, T.Z., S.B., J.V. and G.S.; supervision, T.Z. All authors have read and agreed to the published version of the manuscript.

Funding: This research received no external funding.

Data Availability Statement: Data used in the paper are available in the KTU Faculty of Civil Engineering and Architecture, in the data archive of the experimental research of geothermal properties (tadas.zdankus@ktu.lt).

Conflicts of Interest: The authors declare no conflicts of interest.

References

1. Beerepoot, M.; Marmion, A. Policies for Renewable Heat: An Integrated Approach. Available online: www.iea.org (accessed on 13 September 2024).
2. Energy Efficiency for Buildings. Available online: www.unep.org/energy (accessed on 13 September 2024).
3. Eze, F.; Lee, W.; An, Y.; Joo, H.; Lee, K.; Ogola, J.; Mwabora, J. Development and simulated evaluation of inter-seasonal power-to-heat and power-to-cool with underground thermal storage for self-consumption of surplus solar energy in buildings. *Energy Convers. Manag.* **2024**, *320*, 119013. [CrossRef]
4. Cao, X.; Dai, X.; Liu, J. Building energy-consumption status worldwide and the state-of-the-art technologies for zero-energy buildings during the past decade. *Energy Build.* **2016**, *128*, 198–213. [CrossRef]
5. World Energy Outlook 2023 (International Energy Agency). Available online: https://iea.blob.core.windows.net/assets/86ede3 9e-4436-42d7-ba2a-edf61467e070/WorldEnergyOutlook2023.pdf (accessed on 13 September 2024).
6. Omer, A.M. Ground-source heat pumps systems and applications. *Renew. Sustain. Energy Rev.* **2008**, *12*, 344–371. [CrossRef]
7. Nair, N.K.C.; Garimella, N. Battery energy storage systems: Assessment for small-scale renewable energy integration. *Energy Build.* **2010**, *42*, 2124–2130. [CrossRef]
8. 9-Buildings (Climate Change 2022). pp. 953–1048. Available online: https://www.cambridge.org/core/books/climate-change-2022-mitigation-of-climate-change/buildings/458CA0414DBB715BF532153CB9872CA3 (accessed on 14 October 2024).
9. Small Scale Renewables Solar Water Heating. Available online: http://www.sepa.org.uk/ (accessed on 14 September 2024).
10. REGEND-Small Scale Renewable Energy Technologies with Income Generating Activities in Rural Areas Within a Nexus Framework Addressing the WEF Nexus in the Context of Climate Change and Sustainable Development-22 October 2020. Available online: https://www.unescwa.org/sites/default/files/event/materials/bentouati_regend_escwa.pdf (accessed on 13 September 2024).
11. Farouki, O.T. *Thermal Properties of Soils*; CRREL Monograph: Hanover, NH, USA, 1981; pp. 1–151.
12. Zimmer, T.; Souza, V.A.; Romio, L.C.; Buligon, L.; Veeck, G.P.; Stefanello, M.B.; Roberti, D.R. Estimation of soil thermal properties using conduction and conduction-convection heat transfer equations in the Brazilian Pampa biome. *Agric. For. Meteorol.* **2023**, *338*, 109517. [CrossRef]
13. Mo, W.; Zhang, G.; Yao, X.; Li, Q.; DeBacker, B.J. Assessment of Passive Solar Heating Systems' Energy-Saving Potential across Varied Climatic Conditions: The Development of the Passive Solar Heating Indicator (PSHI). *Buildings* **2024**, *14*, 1364. [CrossRef]
14. Sugathan, N.; Biju, V.; Renuka, G. Influence of soil moisture content on surface albedo and soil thermal parameters at a tropical station. *J. Earth Syst. Sci.* **2014**, *123*, 1115–1128. [CrossRef]

5. Yaowen, C.; Lixing, Z.; Jing, D.; Xueqin, W.; Xiaoxi, L.; Yunhu, X.; Wenbang, G.; Chunxing, H.; Ruiqiang, Z. Effects of Soil Moisture on Surface Radiation Balance and Water-Heat Flux in Desert Steppe Environment of Inner Mongolia. *Pol. J. Environ. Stud.* **2021**, *30*, 1881–1891. [CrossRef] [PubMed]
6. Zhang, T.; Shen, S.; Cheng, C. Impact of radiations on the long-range correlation of soil moisture: A case study of the A'rou superstation in the Heihe River Basin. *J. Geogr. Sci.* **2019**, *29*, 1491–1506. [CrossRef]
7. Park, K.; Kim, Y.; Lee, K.; Kim, D. Development of a shallow-depth soil temperature estimation model based on air temperatures and soil water contents in a permafrost area. *Appl. Sci.* **2020**, *10*, 1053. [CrossRef]
8. Sofyan, S.E.; Hu, E.; Kotousov, A.; Indra Riayatsyah, T.M. A new approach to modelling of seasonal soil temperature fluctuations and their impact on the performance of a shallow borehole heat exchanger. *Case Stud. Therm. Eng.* **2020**, *22*, 100781. [CrossRef]
9. Islam, M.A.; Lubbad, R.; Ghoreishian Amiri, S.A.; Isaev, V.; Shevchuk, Y.; Uvarova, A.V.; Afzal, M.S.; Kumar, A. Modelling the seasonal variations of soil temperatures in the Arctic coasts. *Polar Sci.* **2021**, *30*, 100732. [CrossRef]
0. Zdankus, T.; Jonynas, R.; Vaiciunas, J.; Bandarwadkar, S.; Lenkas, T. Investigation of Thermal Energy Accumulation Using Soil Layer for Buildings' Energy Efficiency. *Sustainability* **2022**, *14*, 5247. [CrossRef]
1. Liu, X.; Yang, S.; Xu, J.; Zhang, J.; Liu, J. Effects of soil heat storage and phase shift correction on energy balance closure of paddy fields. *Atmósfera* **2017**, *30*, 39–52. [CrossRef]
2. Soil Temperature Protocol. Available online: https://www.globe.gov/documents/352961/353769/Soil+Temperature+protocol/87a3491a-25af-4123-9e0c-08f341bfc004 (accessed on 13 September 2024).
3. Yolcubal, I.; Brusseau, M.L.; Artiola, J.F.; Wierenga, P.; Wilson, L.G. Environmental Physical Properties and Processes. In *Environmental Monitoring and Characterization*; Elsevier Inc.: Amsterdam, The Netherlands, 2004; pp. 207–239. [CrossRef]
4. Nerpin, S.V.; Chudnovskii, A. *Physics of the Soil*; IPST: Jerusalem, Israel, 1970; p. 926.
5. Sellers, W.D. *Physical Climatology*; University of Chicago Press: Chicago, IL, USA, 1965; p. 272.
6. Horton, R.; Wierenga, P.J.; Nielsen, D.R. Evaluation of Methods for Determining the Apparent Thermal Diffusivity of Soil Near the Surface. *Soil Sci. Soc. Am. J.* **1983**, *47*, 25–32. [CrossRef]
7. Passerat de Silans, A.M.B.; Monteny, B.A.; Lhomme, J.P. Apparent soil thermal diffusivity, a case study: HAPEX-Sahel experiment. *Agric. For. Meteorol.* **1996**, *81*, 201–216. [CrossRef]
8. Gao, Z. Determination of Soil Heat Flux in a Tibetan Short-Grass Prairie. *Bound. Layer Meteorol.* **2005**, *114*, 165–178. [CrossRef]
9. Zhiqiu, G.; Xingang, F.; Lingen, B. An Analytical Solution to One-Dimensional Thermal Conduction-Convection in Soil. *Soil Sci.* **2003**, *168*, 99–107. Available online: https://journals.lww.com/soilsci/abstract/2003/02000/an_analytical_solution_to_one_dimensional_thermal.4.aspx (accessed on 15 October 2024).
0. An, K.; Wang, W.; Zhao, Y.; Huang, W.; Chen, L.; Zhang, Z.; Wang, Q.; Li, W. Estimation from Soil Temperature of Soil Thermal Diffusivity and Heat Flux in Sub-surface Layers. *Bound. Layer Meteorol.* **2016**, *158*, 473–488. [CrossRef]
1. Gao, Z.; Wang, L.; Horton, R. Comparison of six algorithms to determine the soil thermal diffusivity at a site in the Loess Plateau of China. *Hydrol. Earth Syst. Sci. Discuss.* **2009**, *6*, 2247–2274. [CrossRef]
2. Otunla, T.A.; Oladiran, E.O. Evaluation of soil thermal diffusivity algorithms at two equatorial sites in West Africa. *Ann. Geophys.* **2013**, *56*, R0101. [CrossRef]
3. Miao, Y.; Liu, S.H.; Lu, S.H.; Zhang, Y. A comparative study of computing methods of soil thermal diffusivity, temperature and heat flux (SCI). *Chin. J. Geophys.* **2012**, *55*, 441–451. [CrossRef]
4. *Republican Construction Standards 156-94*; Construction Climatology. Ministry of Civil Engineering and Urban of the Republic of Lithuania: Vilnius, Lithuania, 1995. (In Lithuanian)
5. *Technical Regulations for Construction Str 2.01.12:2024*; Construction Climatology. Ministry of Environment of the Republic of Lithuania: Vilnius, Lithuania, 2024; (In Lithuanian). Available online: https://e-seimas.lrs.lt/portal/legalAct/lt/TAD/60c1acd17f6311ef84ff9693ecd03ff5?jfwid=-4mbsyfsha (accessed on 15 October 2024).

sustainability

MDPI

Article

Study of Heat Transfer Characteristics and Economic Analysis of a Closed Deep Coaxial Geothermal Heat Exchanger Retrofitted from an Abandoned Oil Well

Rui-Jia Liu [1], Lin-Rui Jia [1,2], Wen-Shuo Zhang [1], Ming-Zhi Yu [1], Xu-Dong Zhao [1] and Ping Cui [1,*]

1 Faculty of Thermal Energy Engineering, Shandong Jianzhu University, Jinan 250101, China; ymz@sdjzu.edu.cn (M.-Z.Y.); xudong.zhao@hull.ac.uk (X.-D.Z.)
2 Faculty of Construction, Hong Kong Polytechnic University, Hong Kong SAR 999077, China
* Correspondence: sdcuiping@sdjzu.edu.cn

Abstract: It is economical to transform abandoned oil/geothermal wells into closed deep geothermal heat exchangers with coaxial tubes. A numerical model of a coaxial geothermal heat exchanger (CGHE) with varying borehole diameters is established according to an abandoned well in Northern China. The finite difference method is adopted to solve the temperature distribution, and the accuracy of the model is validated with experimental data. Based on the existing structure of the abandoned well with different depths, the feasibility of its conversion into a deep CGHE is discussed, and this study uses the orthogonal experimental method to analyze the influence of four main factors and their significance level on the average heat extraction rate, with the heat extraction rate up to 422.18 kW in the optimal combination. This study also integrates with actual project considerations and conducts an economic analysis to determine the most appropriate circulation fluid flow rate. The results highlight the key factors on the heat transfer performance of the CGHE, with the inlet water temperature to the CGHE being the most significant, followed by the configuration of the CGHE retrofitted from abandoned. From the economic perspective, given that the CGHE in this study is retrofitted from the abandoned oil Wells, the drilling cost can be reduced by up to CNY 1800 thousand, and the flow rate design of 35 m^3/h is the optimal choice, ensuring a cost-effective system operation while meeting the operational requirements of the deep CGHE.

Keywords: deep coaxial geothermal heat exchanger; transformation of abandoned well; heat extraction simulation; economic analysis

1. Introduction

It is estimated that coal remains the primary energy source for 92% of space heating systems in China [1], precipitating a substantial contribution to greenhouse gas emissions. Consequently, it is necessary to utilize renewable energy for building heating to reduce coal consumption and alleviate air pollution. Geothermal energy has shown significant potential in space heating because of its environmental friendliness and reduced energy consumption [2,3]. The installed capacity of geothermal heat pumps for space heating and cooling exceeds 26,450 MWt in China by the end of 2020 [4].

Deep geothermal space heating systems have recently aroused growing interest in academic and engineering fields because of their enhanced thermal capacities and decreased installation space requirements in contrast to shallow geothermal systems [5,6]. The most crucial component of a deep geothermal heating system is the deep coaxial geothermal heat exchanger (CGHE), also known as coaxial borehole heat exchanger or wellbore heat exchanger. Nevertheless, drilling a borehole incurs the most significant financial investment of any geothermal system, accounting for up to 70% of the total investment costs on average, and therefore makes these systems economically unprofitable [7]. This prompts the desire to explore alternative solutions for efficient and economical acquisition of geothermal energy.

In recent years, retrofitting an abandoned oil well into a CGHE has emerged as a promising method to enhance geothermal energy applications. For instance, Morita et al. [8,9] reported the pioneering CGHE experiment in an abandoned well on the island of Hawaii. Following this, Kohl et al. [10,11] reported the performance of a CGHE at Weissbad and Weggis, Switzerland. Abandoned oil and gas wells (AOGWs) with high bottom-hole temperatures contain abundant geothermal energy. This energy can be harnessed by retrofitting these wells into novel geothermal systems for various uses without the need for costly drilling [12]. Furthermore, Cheng et al. [13,14] successfully employed the organic Rankine cycle to exploit the geothermal energy of abandoned oil wells. They also presented a novel method to enhance geothermal utilization efficiency through the development of thermal reservoirs.

Previous studies have demonstrated the feasibility of harvesting geothermal energy from existing abandoned oil and gas wells. A wide range of wells across different countries have been covered, such as horizontal wells [15], enhanced geothermal systems (EGS) and low-temperature deep borehole heat exchangers (DBHE) in California [16], U-tube heat exchanger located in Southern Iran [17], and double-pipe geothermal and Jachowka K-2 [18] heat exchanger in the Persian Gulf [19]. Some studies also proposed mathematical models to analyze key factors affecting heat outputs from CGHEs. A software package (ANSYS 2019) was developed based on the finite difference method for thermal analysis of a CGHE [20]. An unsteady-state heat transfer model [21] was also established for a deep CGHE and validated with experimental data. A transient 2D finite volume method (FVM) was applied to solve the thermal interference under seasonal effects and dynamic heat flux within a vertical coaxial borehole heat exchangers field [22]. In consideration of well kick occurrences during the drilling process, the models were solved using the fully implicit finite difference method coupled with the clustering method [23].

For the performance optimization of the CGHE, Bu et al. [24] developed mathematical models to investigate the heat exchange between fluid and rocks. Based on the models, they also examined the effects of critical parameters and determined their optimal values. Yekoladio et al. [25] identified the optimal mass flow of geothermal fluid under minimum pumping power and maximum extracted heat energy. Beier et al. [26] characterized and analyzed the vertical temperature profiles of the working fluid of a CGHE. Nian et al. [27] developed a building and CGHE integrated model that comprehensively takes the heat transfer of wellbores, system formation, and building thermal loads into consideration. This model is successfully used to examine critical system performance parameters, including the production of geothermal energy, room temperature, and outlet water temperature.

Currently, the number of AOGWs worldwide has reached approximately 30 million [16,27]. In addition, regions with favorable geothermal parameters are inherently rich in geothermal energy [7]. The conversion of those petroleum wells into geothermal heat exchangers can substantially reduce the investment costs of borehole drilling [16,24] and enhance the utilization of geothermal resources, while also mitigating environmental risks [28]. Therefore, repurposing old petroleum boreholes for geothermal energy extraction is highly desirable in the pursuit of global carbon neutrality.

Despite the extensive research, a comprehensive performance analysis of a CGHE retrofitted from AOGW, particularly in terms of varied borehole diameters, remains lacking. There is also a lack of a computational model for the CGHE with different diameters Most existing studies merely focus on retrofitting processes, ignoring an in-depth examination of factors that affect the heat extraction performance of such retrofitted heat exchangers. Moreover, there is a notable scarcity of insightful research on optimizing critical parameters during the retrofitting process and analyzing the economic performance of these systems. Given these research gaps, this study aims to develop a numerical heat transfer model of the retrofitted CGHE with varied outer pipe diameters based on the finite difference method (FDM) [20]. Based on the developed CGHE model, the influence of four critical factors on the thermal performance of CGHE is evaluated. In addition, the optimal flow rate of CGHE is also investigated from an economic standpoint. The findings of this study can further enhance the adoption of deep CGHE from abandoned oil wells and extend the utilization of deep geothermal energy.

2. Abandoned Oil Well Structure and Working Principles of a CGHE

The oil well usually consists of a steel pipe with varying pipe diameters along the borehole depth. The drilling process and equipment used for oil well engineering are almost identical to those used for CGHE, which makes retrofitting the oil well into CGHE possible. In this study, the abandoned well, which will be retrofitted to CGHE with coaxial tubes, refers to an oil well that no longer has economic potential after several years of oil production. Compared with a conventional CGHE, the main distinction of the retrofitted CGHE derived from an abandoned well lies in its decreasing diameter of the outer pipe along the borehole depth. Thus, developing a CGHE with varied outer pipe diameters for different depth regions is critical to this transformation. There are various types of retrofitted CGHE, and currently, the three-section CGHE and four-section CGHE are the most commonly used types and are more suitable for heat extraction. Figure 1 describes a cross-section structure of the retrofitted CGHE with four sections along the borehole depth. As shown, the diameter of the outer pipe decreases along the depth, while the diameter of the inner pipe remains constant.

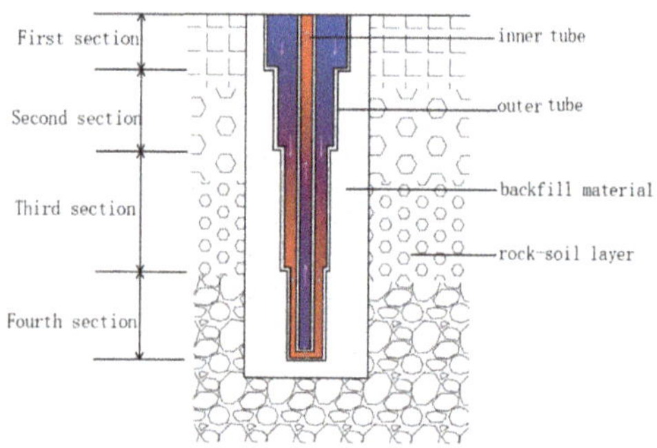

Figure 1. Schematic diagram of CGHE with varied outer pipe diameters. (The colors in the figure represent the water temperature, which is red when it is higher and blue when it is lower).

Due to the high ground temperature, the deep CGHE is only used for heat extraction to provide space heating for buildings. Under heat extraction conditions, low-temperature water flows into the CGHE through the outer pipe. During this process, the circulating water is heated by the surrounding ground. As the water reaches the bottom of the CGHE, it reaches its highest temperature and flows into the inner pipe. One advantage of this system is that it operates as a closed circulation system, meaning that it does not extract groundwater directly during the heat extraction process, which can help protect the groundwater environment to some extent.

3. Development of Numerical Heat Transfer Model

3.1. Model Assumptions

To improve the calculation efficiency while maintaining the prediction accuracy, the following assumptions are adopted:

1. The thermal properties of ground, pipe, water, and backfilling material are dependent on their specific temperatures.
2. The properties of the ground are homogeneous in each ground layer.
3. The penetration of groundwater is neglected.
4. The lateral and bottom boundaries of the borehole model are set to be temperature constant.
5. The geothermal heat flux in the ground is assumed to be uniform and constant.

3.2. Initial and Boundary Conditions

The initial temperature of the ground has a uniform geothermal gradient along the depth direction in different ground layers. As the geothermal heat flux is assumed to be uniform and constant, the initial ground temperature at any depth in the stratum can be calculated by Equation (1) [20]:

$$t(r, z, \tau) = t_a + \frac{q_g}{h_a} + \sum_{j=1}^{m-1} \frac{q_g}{k_j}(H_j - H_{j-1}) + \frac{q_g}{k_m}(z - H_{m-1}) \quad (1)$$

$$\tau = 0, H_{m-1} \leq z \leq H_m, r_b \leq r \leq r_{bnd}$$

where t_a is the ambient air temperature (°C), h_a is the convective heat transfer coefficient at the ground surface, (W/(m^2·K)), H_j is the bottom coordinate of the j-layer, q_g is the geothermal heat flux (W/m^2), k is the ground thermal conductivity (W/(m·K)); $\frac{q_g}{h_a}$ represents the temperature responses induced by q_g at the ground surface, with a convective heat transfer coefficient of h_a; and $\frac{q_g}{k_j}$ represents the temperature responses induced by q_g within the j-layer ground, with a conductive heat transfer coefficient of k_j.

The bottom boundary of the whole calculation region is set far below the bottom of the borehole, with an extra distance of 200 m beneath the borehole bottom. As no thermal interference occurred at this bottom boundary, the temperature remains constant.

The heat convection boundary condition is set at the ground surface, which can be expressed by Equation (2) [20].

$$k\frac{\partial t}{\partial z} = h_a(t - t_a), z = 0, r_b \leq r \leq r_{bnd}, \tau \geq 0 \quad (2)$$

Although the assumption of constant temperature slightly deviates from the actual conditions, which are influenced by variations in ground surface temperature, this impact can be ignored. Considering the drilling depth of more than 2000 m, the model does not need to introduce more complex boundary conditions.

For the two partial differential equations of fluid temperature distribution in the inner and outer tubes, it is considered that the initial temperature of the fluid in the tube is the same as that of the ground at the same depth, which can also be expressed by Equation (1).

The boundary conditions of the partial differential Equation (3) [20] of the flow direction of the tube is

$$\begin{cases} t_{f1} = t_{f2} - \frac{Q}{Mc}, \ z = 0 \\ t_{f1} = t_{f2}, \ z = H \end{cases} \quad (3)$$

where Q (W) is the total heat transfer rate of the CGHE, and c (J/(kg·K) is the specific heat capacity of the circulating fluid.

3.3. Heat Transfer Governing Equation

Based on the above assumptions, the heat transfer governing equations of the CGHE are described as follows. Since each layer of ground is axially symmetric, the heat transfer process of each layer of the ground can be regarded as a two-dimensional unsteady heat transfer in the cylindrical coordinate [20]:

$$\frac{1}{a}\frac{\partial t}{\partial \tau} = \frac{1}{r}\frac{\partial}{\partial r}\left(r\frac{\partial t}{\partial r}\right) + \frac{\partial^2 t}{\partial z^2} \quad (4)$$

In the case of heat extraction, the heat flux exchanged between the CGHE and ground peaks at the borehole wall and decreases radially outward. As a result, the ground temperature gradually decreases and can hardly be detected at the radial boundary. Considering heat transfer characteristics in the ground, the varied space steps in the radial direction are employed to improve the computing efficiency, as shown in Figure 2.

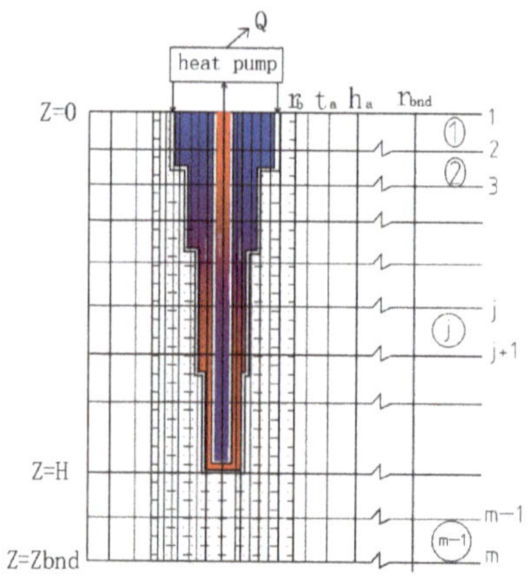

Figure 2. Numerical discretization of the CGHE. (The colors in the figure represent the water temperature, which is red when it is higher and blue when it is lower).

Thus, a new radial coordinate is adopted, which can be described as follows:

$$\sigma = ln(\frac{r}{r_0}), \frac{r_{i+1}}{r_i} = \frac{r_1}{r_0} = exp(\Delta\sigma) = \beta$$

If we apply the new radial coordinate in the control Equation (6), it can be rewritten as follows:

$$\frac{1}{a}\frac{\partial t}{\partial \tau} = \frac{1}{r^2}\frac{\partial^2 t}{\partial \sigma^2} + \frac{\partial^2 t}{\partial z^2} \tag{5}$$

The flow pattern of the CGHE in this study is the fluid flowing down from the sandwich of two tubes and up from the inner tube. The energy differential equation of the fluid in the inner tube is

$$C_2\frac{\partial t_{f2}}{\partial \tau} = \frac{t_{f2} - t_{f1}}{R_2} + C\frac{\partial t_{f2}}{\partial z} \tag{6}$$

The energy differential equation for the fluid in the outer tube is

$$C_1\frac{\partial t_{f1}}{\partial \tau} = \frac{t_{f2} - t_{f1}}{R_2} + \frac{t_b - t_{f1}}{R_1} - C\frac{\partial t_{f1}}{\partial z} \tag{7}$$

where C = Mc is the heat capacity (J/(m·K)) of the circulating fluid; ρ_f and c_f are the density and specific heat capacity of the circulating fluid; ρ_{f1} and c_{f1} are the density and specific heat of the outer pipe wall; and ρ_{f2} and c_{f2} are the density and specific heat of the inner pipe wall.

Considering the inner pipe and circulating water inside the inner pipe, the specific heat capacities should be calculated by

$$C_1 = \frac{\pi}{4}\left(d_{1i}^2 - d_{2o}^2\right)\rho_f c_f + \frac{\pi}{4}\left(d_{1o}^2 - d_{1i}^2\right)\rho_1 c_1 + \frac{\pi}{4}\left(d_b^2 - d_{1o}^2\right)\rho_g c_g \tag{8}$$

$$C_2 = \frac{\pi}{4}d_{2i}^2\rho_f c_f + \frac{\pi}{4}\left(d_{2o}^2 - d_{2i}^2\right)\rho_2 c_2 \tag{9}$$

where C_1 is the per-length heat capacity of the outer channel, including the outer pipe, circulating water, and backfilling material (J/m·K); and C_2 is the per-length heat capacity of the inner channel.

Since the model divides the CGHE into four sections, the thermal resistances of the four sections can be calculated according to different radial boundaries. The thermal resistances between the circulating water in the outer pipe and the borehole wall, R_1, are given as follows:

$$R_1 = \begin{cases} \frac{1}{\pi d_{11i} h_1} + \frac{1}{2\pi k_{p1}} ln \frac{d_{11o}}{d_{11i}} + \frac{1}{2\pi k_g} ln \frac{d_b}{d_{21i}} \; 0 < Z \leq Z_1 \\ \frac{1}{\pi d_{12i} h_1} + \frac{1}{2\pi k_{p1}} ln \frac{d_{12o}}{d_{12i}} + \frac{1}{2\pi k_g} ln \frac{d_b}{d_{22i}} \; Z_1 < Z \leq Z_2 \\ \frac{1}{\pi d_{13i} h_1} + \frac{1}{2\pi k_{p1}} ln \frac{d_{13o}}{d_{13i}} + \frac{1}{2\pi k_g} ln \frac{d_b}{d_{23i}} \; Z_2 < Z \leq Z_3 \\ \frac{1}{\pi d_{14i} h_1} + \frac{1}{2\pi k_{p1}} ln \frac{d_{14o}}{d_{14i}} + \frac{1}{2\pi k_g} ln \frac{d_b}{d_{24i}} \; Z_3 < Z \leq Z_4 \end{cases} \quad (10)$$

R_2 is the thermal resistance between the circulating water in the inner and outer pipes, which can be calculated by

$$R_2 = \begin{cases} \frac{1}{\pi d_{21i} h_2} + \frac{1}{2\pi k_{p2}} ln \frac{d_{21o}}{d_{2i}} + \frac{1}{\pi d_{21o} h_1} \; 0 < Z \leq Z_1 \\ \frac{1}{\pi d_{22i} h_2} + \frac{1}{2\pi k_{p2}} ln \frac{d_{21o}}{d_{2i}} + \frac{1}{\pi d_{22o} h_1} \; Z_1 < Z \leq Z_2 \\ \frac{1}{\pi d_{23i} h_2} + \frac{1}{2\pi k_{p2}} ln \frac{d_{23o}}{d_{2i}} + \frac{1}{\pi d_{23o} h_1} \; Z_2 < Z \leq Z_3 \\ \frac{1}{\pi d_{24i} h_2} + \frac{1}{2\pi k_{p2}} ln \frac{d_{24o}}{d_{2i}} + \frac{1}{\pi d_{24o} h_1} \; Z_3 < Z \leq Z_4 \end{cases} \quad (11)$$

where k_g, k_{p1}, and k_{p2} are the thermal conductivity of backfill material, outer pipe, and inner pipe, respectively; d_o and d_i represent the pipe's outer diameter and inner diameter, respectively; and d_b represents the borehole diameter.

Moreover, h_1 and h_2 are the convective heat transfer coefficients of the annular pipe and inner pipe, which can be calculated by

$$h = \frac{Nu \times k}{d} \quad (12)$$

Nu is the dimensionless criterion number of the reaction on the strength of the heat transfer process, which can be obtained from [29,30]

$$Nu = 0.023Re^{0.8}Pr^{0.333} \; Re \geq 10^4 \quad (13)$$

$$Nu = 1.86Re^{0.8}\left(\frac{Re \times Pr \times d}{L_i}\right)^{0.333} \; Re \leq 2200, \; i = 1,2,3,4 \quad (14)$$

$$Nu = 0.116(Re^{0.667} - 125)Pr^{0.333} \; Re > 2200 \quad (15)$$

where L is the length of the borehole in each section, m; and Pr is the dimensionless number of the effect of reactive fluid property on heat transfer.

Re can be calculated by

$$Re = \frac{u \times d}{v} \quad (16)$$

where u is the average section velocity, m/s; d is the equivalent diameter, m; and v is the kinematic viscosity, m^2/s.

3.4. Numerical Discretization

The heat transfer model for the deep CGHE can be considered a two-dimensional unsteady heat conduction problem with specific convective heat transfer boundary conditions on the borehole wall. This study employs the alternative direction finite difference method (ADFDM) [31]. The ADFDM combines the forward and backward FDM and uses them alternatively in adjacent time steps. This approach ensures that no more than three unknowns need to be solved for each numerical node, which significantly improves the calculation efficiency.

The model assumes that there is no influence from groundwater flow. Based on the geometric characteristics of the CGHE, the FDM is utilized to establish the computational grid division for the efficient simulation model of the heat exchanger in abandoned oil wells. The grid division is shown in Figure 2.

In the case of converting an abandoned well into a CGHE, there is a variation in the diameter of the outer pipe of the ground heat exchanger along the depth of the borehole. The calculation process differs from the conventional medium-depth CGHE, where the outer tube diameter remains constant. To address this, the fluid temperature in each section of the pipe with the same diameter needs to be calculated separately, as illustrated in Figure 3.

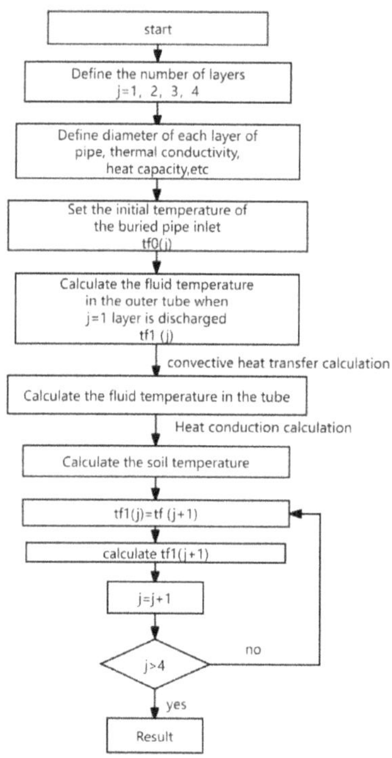

Figure 3. Flowchart to calculate the circulating fluid temperature of the CGHE with varying outer pipe diameters.

Based on the ADFDM, the numerical equations for fluid in inner and outer pipes can be obtained. It should be noted that the fluid in the pipe is assumed to be a one-dimensional model, indicating that the water temperature in the pipe on the horizontal level is uniform. The numerical equations of fluid in the outer pipe are as follows:

$$-B_3 t_{f2,j}^{p+1} + (1 + B_1 + B_3) t_{f1,j}^{p+1} - B_1 t_{0,j}^{p+1} = -B_4 t_{f1,j-1}^{p} + t_{f1,j}^{p} - B_4 t_{f1,j+1}^{p} \quad (17)$$

$$B_4 t_{f1,j-1}^{p+1} + t_{f1,j}^{p+1} - B_4 t_{f1,j+1}^{p+1} = B_3 t_{f2,j}^{p} + (1 - B_1 - B_3) t_{f1,j}^{p} + B_1 t_{0,j}^{p} \quad (18)$$

where $B_1 = \frac{\Delta\tau}{C_1 R_1}$, $B_3 = \frac{\Delta\tau}{C_1 R_2}$, $B_4 = \frac{C\Delta\tau}{2C_1 \Delta z}$.

The numerical equations of fluid in the inner pipe are as follows:

$$-B_2 t_{f1,j}^{p+1} + (1+B_2) t_{f2,j}^{p+1} = B_5\left(t_{f2,j-1}^{p} - t_{f2,j+1}^{p}\right) + t_{f2,j}^{p} \tag{19}$$

$$-B_5 t_{f2,j-1}^{p+1} + t_{f2,j}^{p+1} + B_5 t_{f2,j+1}^{p+1} = B_2 t_{f1,j}^{p} + (1-B_2) t_{f2,j}^{p} \tag{20}$$

where $B_2 = \frac{\Delta\tau}{C_2 R_2}$, $B_5 = \frac{C\Delta\tau}{2C_z\Delta z}$.

The numerical equations of the ground are

$$-B_r t_{i-1,j}^{p+1} + (1+2B_r) t_{i+1,j}^{p+1} - B_r t_{i+1,j}^{p+1} = B_z t_{i,j-1}^{p} + (1-2B_z) t_{i,j}^{p} - B_z t_{i,j+1}^{p} \tag{21}$$

$$-B_z t_{i,j-1}^{p+1} + (1+2B_z) t_{i,j}^{p+1} - B_z t_{i,j+1}^{p+1} = B_r t_{i-1,j}^{p} + (1-2B_r) t_{i,j}^{p} + B_r t_{i+1,j}^{p} \tag{22}$$

where $B_r = \frac{a\Delta\tau}{(r\Delta\sigma)^2}$, $B_z = \frac{a\Delta\tau}{(\Delta z)^2}$.

3.5. Equation Solving Algorithm

The complete difference equation can be broken down into several sets of second-order equations, and each set of algebraic equations has a tridiagonal coefficient matrix. These tridiagonal matrices satisfy the solution rule of the catch-up method. The tridiagonal linear algebraic equations can be expressed as follows [20]:

$$\left.\begin{aligned} b_1t_1 + c_1t_2 &= d_1 \\ a_2t_1 + b_2t_2 + c_2t_3 &= d_2 \\ &\cdots \\ a_it_{i-1} + b_it_i + c_it_{i+1} &= d_i \\ &\cdots \\ a_nt_{n-1} + b_nt_n &= d_n \end{aligned}\right\} \tag{23}$$

Since the first equation contains only two unknowns, it can be written as

$$t_1 = U_1t_2 + V_1 \tag{24}$$

Therefore, when $i = 1$, the $U_1 = -\frac{c_1}{b_1}$. $V_1 = \frac{d_1}{b_1}$.

Using this method, in turn, the general expression of the coefficient is

$$t_i = U_it_{i+1} + V_i,\ i = 2,3,\ldots,n-1 \tag{25}$$

where $U_i = -\frac{c_i}{a_iU_{i-1}+b_i}$; $V_2 = -\frac{d_i-a_iV_{i-1}}{a_iU_{i-1}+b_i}$; and $i = 2, 3, \ldots, n-1$

It can be obtained according to the solution process of the catch-up method.

$$t_n = \frac{d_n - a_nV_{n-1}}{a_nU_{n+1} + b_n} \tag{26}$$

where $U_i = -\frac{c_i}{a_iU_{i-1}+b_i}$; $i = 2, 3, \ldots$; $V_i = -\frac{d_i-a_iV_{i-1}}{a_iU_{i-1}+b_i}$; and $i = 2, 3, \ldots, n-1$.

3.6. Model Validation

This study selected the experimental data from Xi'an's non-interference geothermal heat supply project [32] to validate the proposed numerical model. In this project, the drilling depth is 2500 m, the average geothermal gradient is 0.0283 °C/m, the import flow rate of a single well is 13.97 m^3/h, and the heating period is from 15 November to the 15 March. The inputs include the inlet fluid temperatures, specifications of the CGHE, thermal physical properties of the ground, ground temperature gradient, mass flow rate, and operation time. The output is the outlet water temperature, which was then used for model validation. In Reference [32], the maximum relative error between the simulated and measured outlet temperatures was 4.9%. It can be considered that the simulation results agree with the measured data.

In this study, the outlet water temperature of the project is simulated by the developed numerical model, with the same input parameters as Reference [32]. The comparison results among the numerical simulation results of this study and those of Reference [32] and the measured data are shown in Figure 4. It can be seen from Figure 4 that the proposed numerical result in this study follows a trend similar to both the measured data and the simulated data from Reference [32]. The maximum relative temperature difference of the numerical simulation data for the referred model versus the developed model in this manuscript is less than 5%. Therefore, the numerical model is considered reliable for further analysis of the performance of the CGHE.

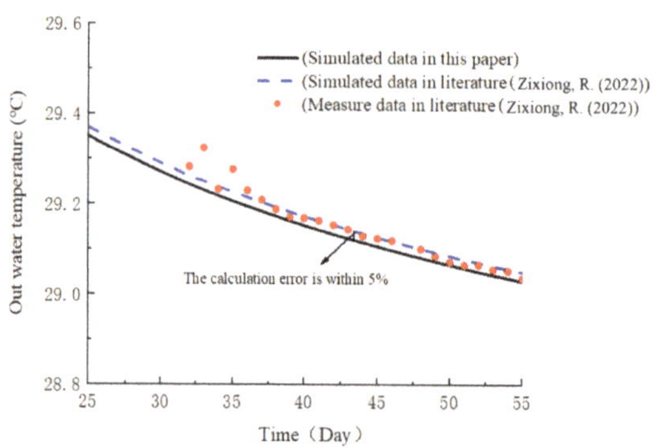

Figure 4. Comparison of simulated and measured outlet temperatures with the [32].

4. Thermal Performance Analysis of the Retrofitted CGHE

4.1. Parameter Settings of the Retrofitted CGHE

The heat transfer performance of the deep CGHE is affected by factors such as the subsurface thermal properties, geometric parameters of the abandoned well, and operational parameters. In this study, an abandoned oil well located in Northern China is investigated as a case study, to be retrofitted into a CGHE for space heating. The abandoned well has four sections with different borehole diameters along the borehole depth. According to the onsite measurement, the average thermal properties of each section are determined by sample measurement, using the hot-disk thermal conductivity tester, and the obtained property data are listed in Table 1. The accuracy of the instrument for the thermal properties is within ±10%. The basic geometric parameters of the well are also listed in Table 1.

Table 1. Basic parameters of the four different sections.

No.	Length of Borehole (m)	Borehole Diameter (m)	External Diameter of the Outer Pipe (m)	Inner Diameter of the Outer Pipe (m)	k of the Ground W/(m·K)	Specific Heat Capacity 10^6J/(Nm^3·K)
First section	0~455	0.44	0.34	0.32	2.8	1.10
Second section	455~2017	0.31	0.24	0.23	2.8	1.93
Third section	2017~2626	0.22	0.18	0.16	3.0	1.41
Forth section	2626~2829	0.15	0.14	0.12	3.5	2.24

According to the ground temperature gradient tested in the field and the thermal conductivity of each rock layer, the geothermal heat flux can be approximated to be 0.058 W/m^2, assuming a constant value. Other parameters are set as follows: the outer pipe is the steel pipe with a thermal conductivity of 43 W/(m·K), the circulating medium in the heat exchanger is water, the average temperature of the surface air is 15 °C, and the thermal conductivity of the backfill material is 1.5 W/(m·K). The inner and outer diameters of the inner pipe are set to be 110/90 mm.

4.2. Average Heat Extraction Rate in the Heating Season

The operating performance of the CGHE is related to the geometric and physical characteristics and, to a large extent, the working conditions. It is challenging to evaluate the heat transfer performance of CGHE comprehensively. Therefore, a quantitative index of the average heat extraction rate is defined to facilitate smooth communication between engineers and non-professionals. The average heat extraction rate of the CGHE refers to the mean heat extraction of a deep CGHE during the whole operation period, under specific operating conditions.

$$Q = \frac{\mathrm{Mc}\sum_{0}^{t}\Delta T}{1000t} \tag{27}$$

where $\sum_{0}^{t}\Delta T$ is the sum of the difference between the inlet and outlet water temperatures at all times (°C); and t is time, h.

In this study, the particular condition refers to the fact that the inlet temperature of the CGHE system shall not be lower than 5 °C during the heating period, and 5 °C, 10 °C, 15 °C, and 20 °C are, respectively, used for the simulation.

4.3. Orthogonal Test Scheme Design and Test Results

Design parameters, geological conditions, operating conditions, and other factors will significantly impact the heat transfer performance of the CGHE. Especially for abandoned oil wells, many different transformation schemes will influence the value of the average heat extraction rate. The major difference between the abandoned well and the conventional CGHE is the varied outer pipe diameter, along the borehole depth. The smaller diameter of the fourth section leads to increased flow resistance, so whether to consider the fourth section as the heat exchange section needs comprehensive consideration. In this study, four retrofitted configurations of CGHE are considered, as shown in Table 2. It is noticed that the constant d_o is selected to be the diameter of the fourth section. Therefore, this study focuses on the impact of four main factors on heat extraction, namely the configuration of the rebuilt abandoned oil well, the inlet temperature, the mass flow rate, and the internal pipe thermal conductivity.

Table 2. Schemes of orthogonal test.

Operating Condition	Modeling Parameter			
	A: Configuration of CGHE Retrofitted from Abandoned Oil Wells	B: Inlet Temperature (°C)	C: Flow Rate (m^3/h)	D: Thermal Conductivity of Inner Pipe k_{p2} (W/(M·K))
1	4-section with variant d_o of 2829 m	5	25	0.17
2	4-section with variant d_o of 2829 m	10	30	0.25
3	4-section with variant d_o of 2829 m	15	35	0.34
4	4-section with variant d_o of 2829 m	20	40	0.43
5	3-section with variant d_o of 2626 m	5	30	0.34
6	3-section with variant d_o of 2626 m	10	25	0.43
7	3-section with variant d_o of 2626 m	15	40	0.17
8	3-section with variant d_o of 2626 m	20	35	0.25
9	4-section with constant d_o of 2829 m	5	35	0.43
10	4-section with constant d_o of 2829 m	10	40	0.34
11	4-section with constant d_o of 2829 m	15	25	0.25
12	4-section with constant d_o of 2829 m	20	30	0.17
13	3-section with constant d_o of 2626 m	5	40	0.25
14	3-section with constant d_o of 2626 m	10	35	0.17
15	3-section with constant d_o of 2626 m	15	30	0.43
16	3-section with constant d_o of 2626 m	20	25	0.34

The orthogonal test is thus employed to analyze the influence parameters of the average heat extraction rate [33]. The orthogonal test method is widely used in scientific research because of its high efficiency and cost-effectiveness. It is necessary to clarify the significance of the influence of the factors.

The total influence factors of the orthogonal test are 4, and the level number of each factor is set to 4. The orthogonal test table, $L_{16}(4^4)$, is selected according to the factor level. The test scheme and test solution results corresponding to the level values of each factor are shown in Tables 2 and 3, and the average heat extraction is set as the target parameter.

Table 3. Results of orthogonal test.

Operating Condition	Combination of Factors	Average Heat Extraction (kW)
1	A1B1C1D1	422.18
2	A1B2C2D2	385.94
3	A1B3C2D2	344.84
4	A1B3C3D3	305.00
5	A2B1C2D3	383.09
6	A2B2C1D4	324.62
7	A2B3C4D1	319.11
8	A2B4C3D2	269.08
9	A3B1C3D4	420.76
10	A3B2C4D3	394.05
11	A3B3C1D2	326.41
12	A3B4C2D1	301.88
13	A4B1C4D2	393.67
14	A4B2C3D1	349.02
15	A4B3C2D4	292.01
16	A4B4C1D3	247.68

4.4. Analysis of Orthogonal Test Results

In the orthogonal test method, a certain factor may cause all the differences in the target parameters when only considering its influence. The factor which causes the most significant change in the average heat extraction rate is considered to be the main influencing factor. The calculation formula is as follows:

$$n_i = \frac{1}{s}\sum_{n=1}^{s} y_n$$

$$\mathrm{R} = \max(\mathrm{n_i}) - \min(\mathrm{n_i})$$

where n_i is the arithmetic average of orthogonal test results when the level i is taken for the column factors; s is the number of tests for these factors under level i; y_n is the index value of the k test; and R is the range value in this column.

The average heat extraction rate range analysis of the CGHE in abandoned well reconstruction is shown in Table 4.

Table 4. Analysis of average heat extraction rate of orthogonal test.

Analysis of Range	A	B	C	D
n_1	364.5	404.9	330.2	348.0
n_2	324.0	363.4	340.7	343.8
n_3	360.8	320.6	345.9	342.4
n_4	320.6	280.9	353.0	335.6
Ri	43.9	124	22.7	12.5

The range analysis shows that the configurations of CGHEs retrofitted from abandoned oil wells, inlet temperature, mass flow rate, and inner tube thermal conductivity are 43.9, 124, 22.7, and 12.5, respectively. Therefore, the significance of each factor in affecting the heat transfer performance of the CGHE is listed as follows: inlet water temperature > configuration of CGHE retrofitted from abandoned oil wells > flow rate > inner tube thermal conductivity.

As Figure 5 shows, the average heat extraction rate increases with higher flow rates. However, the average heat extraction rate decreases as the inlet temperature rises, which is attributed to the smaller temperature difference between the circulating fluid and the surrounding ground. Moreover, it also shows that the thermal insulation performance of the inner tube plays a vital role in preventing the thermal short circuit phenomenon between the inner tube and the outer tube and also improving the heat extraction performance of the system.

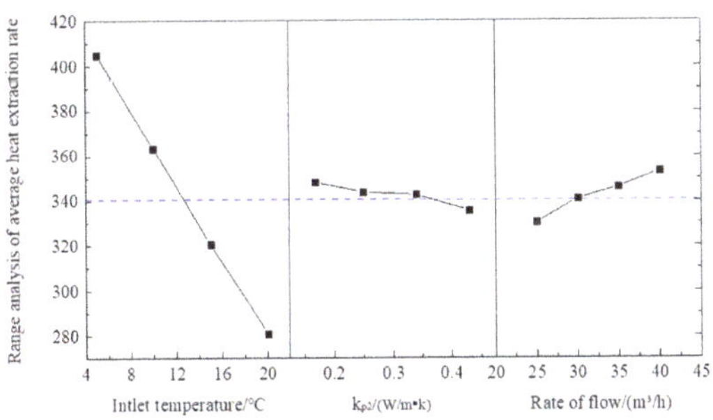

Figure 5. Analysis of average heat extraction of orthogonal test.

5. Case Study and Economic Analysis

In this section, the economic analysis is performed on the retrofitted CGHE from the abandoned oil well, aiming to disclose its practical feasibility and popularization potential of actual applications.

5.1. Actual Operation of the Ground Source Heat Pump (GSHP) System with Retrofitted CGHE

In this study, an abandoned oil well located in Jinan, China, is selected as a case to demonstrate the thermal performance of the GSHP system with a retrofitted CGHE. Based on the above discussion, the configuration of a three-section with a variant d_o of 2626 m is selected to be the optimal design scheme. The thermal conductivity of the inner pipe is set to be 0.17 W/(m·K). According to the real-time heating load of the concerned building in the project, as shown in Figure 6, the appropriate heat pump unit is selected, and its nominal performance under heating conditions is shown in Table 5.

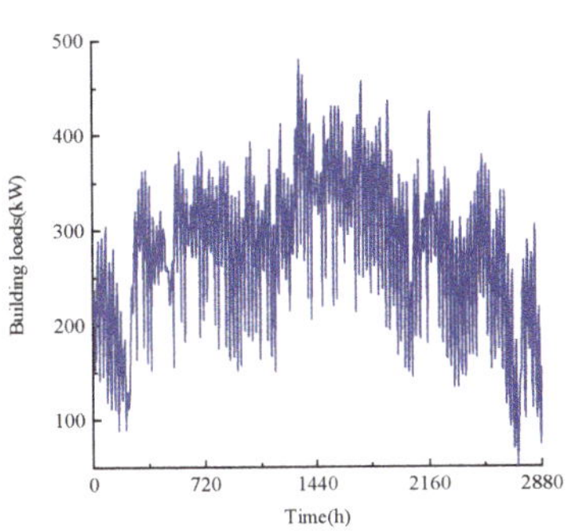

Figure 6. The space-heating load profile during the whole winter period.

Table 5. Technical performance parameter table of the heat pump unit.

Performance Parameter	Number
Nominal heating capacity	500 kW
Designed inlet and outlet water temperatures of the evaporator	15/7 °C
Design inlet and outlet water temperatures of the condenser	40/45 °C
Nominal power consumption	115 kW
Rated COP	4.48

The inlet water temperature variations in the CGHE under different water flow rates are simulated to facilitate the identification of the suitable circulation water flow. As shown in Figure 7, in certain conditions, with the circulating water flow rate of 25 m^3/h and 30 m^3/h, the inlet water temperature will drop below 5 °C, which is considered impractical for the actual application. However, when readjusting the circulating water flow to 35 or 40 m^3/h, the inlet water temperature consistently remains above 5 °C during the whole operation period, demonstrating its high reliability in practical applications.

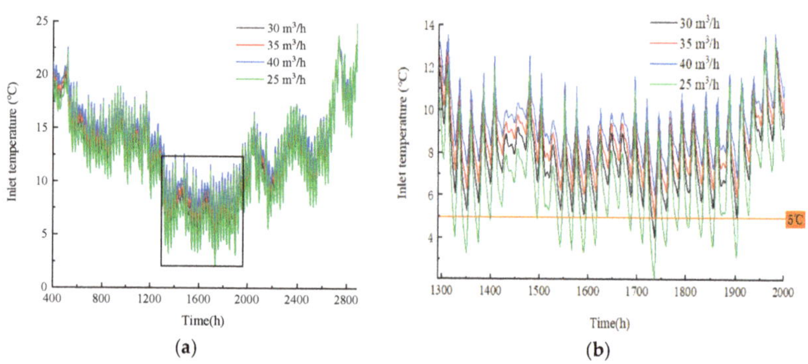

Figure 7. Inlet temperature variation in the CGHE under different flow rates. (**a**) Long-term temperature profiles (0–2880 h). (**b**) Short-term temperature profiles (1200–1900 h).

It is obvious that a higher fluid temperature can be obtained with a larger flow rate due to its better heat convection.

According to Equations (12)–(16), the convection heat exchange coefficients of the pipe sections can be calculated by the hydraulic diameter, Re, Pr, and Nu. Taking the flow rate of 25 m^3/h as an example, Table 6 can be obtained.

Table 6. Calculation parameters of the convection heat transfer coefficient (25 m^3/h).

Pipe Type	Hydraulic Diameter (m)	Re	Pr	Nu	h ($W/m^2\cdot K$)
Annular pipe section 1	0.32	16,287.79	9.02	112.02	315.78
Annular pipe section 2	0.23	20,760.25	9.02	136.02	705.69
Annular pipe section 3	0.16	25,844.40	9.02	162.07	2113.49
Inner pipe	0.09	83,431.80	9.02	413.88	2819.27

Similarly, the same calculation can be performed for other flow rates. As shown in Figure 8a, the heat transfer coefficient of the inner tube is significantly higher than that of the outer annular pipe. Furthermore, due to the decreasing diameter of the outer annular pipe at a greater depth, resulting in increased flow velocity, the heat transfer coefficient of the annular wall also exhibits an upward trend along the depth, reaching the highest value in the third (deepest) section. Table 7 is the mathematical relation of convection heat transfer coefficients and flow rates summarized according to Figure 8a.

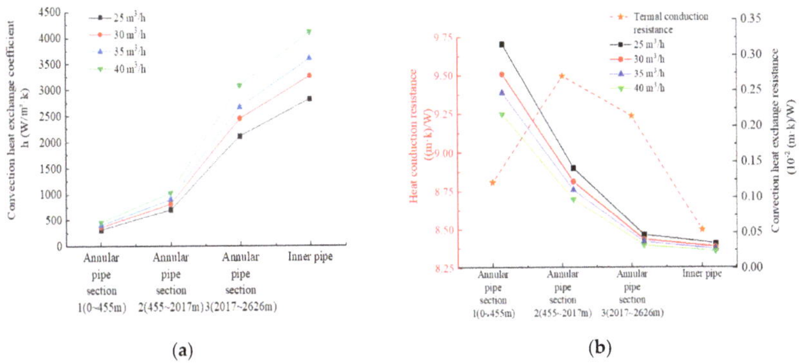

Figure 8. Heat transfer coefficients and resistances of the CGHE under different flow rates. (**a**) Convective heat transfer coefficient. (**b**) Heat transfer resistances per unit length of CGHE.

Table 7. Mathematical relation of convection heat transfer coefficients and flow rates.

Pipe Type	Fitted Curve	
Annular pipe section 1	y = 79.86 ± 15 + (9.43 ± 0.4) × x	y: convection heat transfer coefficient ($W/m^2 \cdot K$) x: flow rate (m^3/h)
Annular pipe section 2	y = 178.53 ± 34 + (21.07 ± 1.0) × x	
Annular pipe section 3	y = 548.87 ± 162 + (62.40 ± 4.9) × x	
Inner pipe	y = 712.90 ± 135 + (84.19 ± 4.0) × x	

It also can be seen in Figure 8b that the convective heat transfer resistances of the two pipe sides are much more insignificant compared to the heat conductivity resistance.

The surrounding ground temperatures after the operation of 20 days (480 h) and the end of the heating season (2880 h) are also simulated in the case of the flow rate of 35 m^3/h, as shown in Figure 9. It can be seen from the figures that, with the continuous operation of the CGHE, the ground temperature surrounding the CGHE notably decreases, highlighting the necessity of energy rejection during non-heating to maintain its heating capacity.

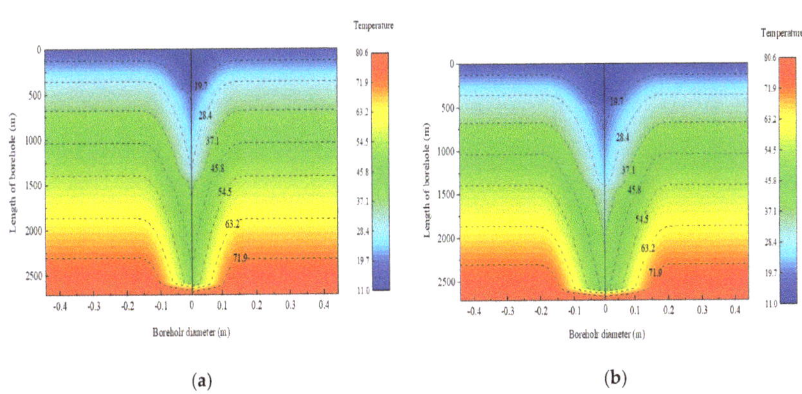

Figure 9. Ground temperature after different operation periods: (**a**) 480 h and (**b**) 2880 h.

5.2. Initial System Investment

(1) Drilling cost

The drilling cost of the deep CGHE in Shandong province, China, ranges from 500 to 600 CNY/m. This study estimates the drilling cost at 600 CNY/m. The cost of C_b at various drilling depths is shown in Table 8.

Table 8. Drilling cost of CGHE with different borehole depths.

Buried Depth	1000 m	1500 m	2000 m	2500 m	3000 m
Drilling cost (10^3CNY)	600	900	1200	1500	1800

Because the CGHE in this case is modified from the abandoned oil wells, the drilling cost savings can reach up to CNY 1800 thousand.

(2) Pipe material cost

According to the actual engineering material selection planning, the outer pipe material is a J55 special steel pipe with a cost of 400 CNY/m, and the inner pipe is a high-density polyethylene (HDPE) pipe with a cost of 40 CNY/m. This amounts to a total cost of 440 CNY/m for the deep casing-buried pipe. The cost of pipe material for various buried depths is calculated and summarized in Table 9.

Table 9. Cost of pipe material of CGHE with different borehole depths.

Buried Depth	1000 m	1500 m	2000 m	2500 m	3000 m
Pipe material cost (10^3CNY)	440	660	880	1100	1320

5.3. Total Operation Cost of the GSHP System with Retrofitted CGHE

This study uses the industrial electricity price in Jinan, China (116.98°, 36.67°), to calculate the operation cost of the water pump and heat pump.

(1) Water pump cost

The net power consumption of the water pump in this study is approximated by calculating the frictional pressure drop and local pressure drop of the CGHE.

The frictional pressure drop caused by the viscous force during the fluid flow is called the drag loss, which can be calculated by the following Equation (28) [29]

$$\Delta p_f = \lambda \times \frac{l \times \rho u^2}{d \times 2} \tag{28}$$

where λ is the friction factor; l is the tube length, m; and ρ is the density of the circulating fluid.

The equation for calculating the friction factor (λ) and Reynolds number (Re) along the path is as follows [29]:

$$\lambda = 0.11(\frac{K}{d} + \frac{68}{Re})^{0.25} \tag{29}$$

where K is the height of the equivalent coarse grain, m; and d is the equivalent diameter, m.

The weighted average method can calculate the equivalent roughness height of the annular area of the inner and outer tubes of the casing-buried tube heat exchanger. Therefore, the equivalent roughness height of the annular channel should be as follows [34].

$$K = \frac{K_1 \times d_{1i}{}^2 + K_2 \times d_{2o}{}^2}{d_{1i}{}^2 + d_{2o}{}^2} \tag{30}$$

where K_1 is the height of equivalent roughness on the inner wall of the outer tube, m; d_{1i} is the inner diameter of the outer tube, m; K_2 is the equivalent roughness height on the outer wall of the inner tube, m; and d_{2o} is the outer diameter of the inner tube, m.

According to the research results [31,32], there is a particular relationship between the local pressure drop and pressure drop along the pipeline. In this study, 2% of the pressure drop along the road is used for analysis.

The total pressure drop is the sum of the friction pressure drop and pressure drop,

$$\Delta p_z = \Delta p_f + \Delta p_s \tag{31}$$

where Δp_z is the total pressure drop of the heat exchanger, Pa; Δp_f is the friction pressure drop, Pa; and Δp_s is the local pressure drop.

Ignoring the resistance loss of the evaporator pipeline and the buried pipeline, the net power consumption of the circulating pump can be approximated to [29]

$$P = \Delta p_z \times q_f \tag{32}$$

where P is the net power consumption of the pump, W; and q_f is the mass flow rate of circulating water, kg/s.

Meanwhile, water pump efficiency is always a necessary concern in the regulation of the GSHP system with retrofitted CGHE. Thus, the water pump efficiency can be calculated by Equation (33) [35]:

$$\eta_{pump} = \frac{g q_f H_{pump}}{P_{pump}} \tag{33}$$

where g is the acceleration of gravity, m/s^2; H_{pump} is the pump lift, m; and P_{pump} is the rated power of the pump, kW, set to 18.5 kW.

Figure 10 shows the change in total pressure drops and pump efficiency in each section pipe under different flow rates. According to Figure 10, the mathematical relationship formula between the flow rate and total pressure drop can be obtained via linear fitting, which is shown in Table 10.

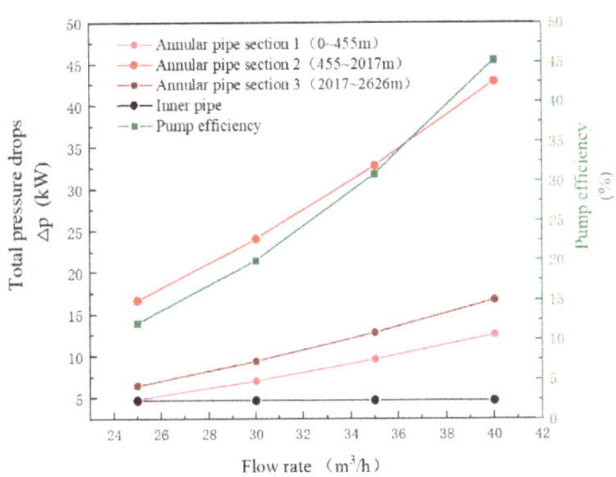

Figure 10. Total pressure drops of the CGHE under different flow rates.

Table 10. Mathematical relation of total pressure drops and flow rates.

Pipe Type	Fitted Curve	
Annular pipe section 1	$y = -8.07 \pm 0.82 + (0.51 \pm 0.02) \times x$	y: total pressure drop (m) x: flow rate (m^3/h)
Annular pipe section 2	$y = -27.69 \pm 2.82 + (1.75 \pm 0.09) \times x$	
Annular pipe section 3	$y = -10.80 \pm 1.10 + (0.68 \pm 03) \times x$	
Inner pipe	$y = 4.72$	

(2) Heat pump operation cost

The GSHP coefficient of performance (COP) represents the conversion efficiency between energy and heat, specifically the ratio of the heat energy produced by the heat pump to the electrical energy consumed. Figure 11a illustrates the fitted curve of the heat pump inlet water temperature and COP values. Figure 11b indicates the variations in heat pump COP with flow rates, demonstrating that the heat pump has a favorable heat extraction effect with larger flow rates. The GSHP exhibits the highest value of COP when the flow rate is 40 m^3/h.

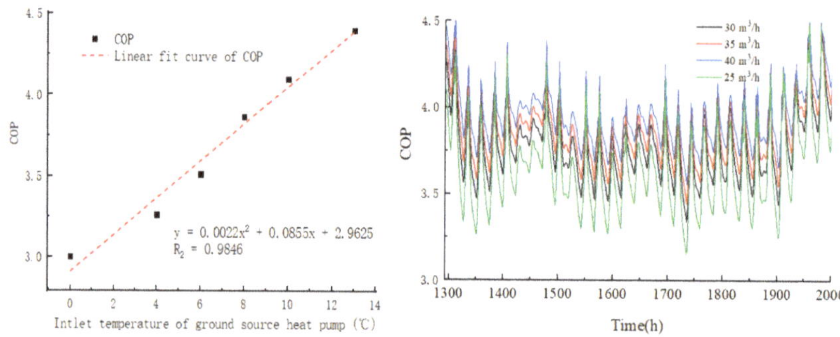

(a) (b)

Figure 11. Variation in heat pump COP with flow rates. (**a**) Linear fit curve of COP. (**b**) The COP variations in GSHP with different flow rates.

According to the latest implementation standard of the electricity price in Jinan in the peak, valley, and flat value, the average industrial electricity price is 0.6 CNY/(kW·h). The electricity cost required in one heating season (120 d) can be calculated based on the COP, as shown in Figure 12. As can be seen from the figure, the operation cost of CGHE retrofitted three-section abandoned oil wells. With an increase in the flow rate, the heat pump operation cost decreases rapidly.

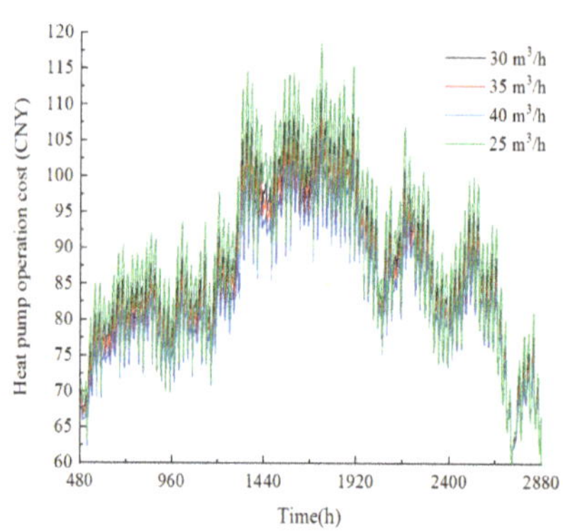

Figure 12. Variation in heat pump operation cost with flow rates.

In summary, Figure 13 enumerates that there are three kinds of total annual operation costs under four different flow rates, namely water pump operation cost, heat pump operation cost, and total operation cost. To guarantee the high operation efficiency of the whole system, it is advisable to maintain a flow rate of no less than 35 m^3/h. Meanwhile, the larger flow rate can improve the convection heat transfer coefficient, but it also increases the operation cost of the water pump.

From an economic perspective, 35 m^3/h is selected as the optimal flow rate, which ensures a relatively smaller distribution cost while maintaining a sufficiently high inlet water temperature of the CGHE to avoid system icing.

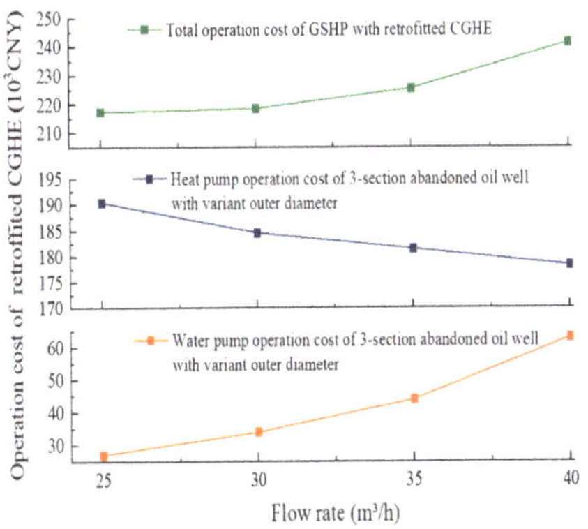

Figure 13. Operation costs of GSHP with retrofitted 3-section abandoned oil well.

6. Conclusions

This study demonstrates the practicality of repurposing abandoned wells into deep CGHEs, making use of the existing structures of both three-section and four-section abandoned oil wells. A numerical heat transfer model for CGHEs with varying outer pipe diameters is established, employing a finite difference method to handle complex calculations. The accuracy of the developed model is validated with experiment data, ensuring an accurate analysis of the heat transfer performance of the retrofitted heat exchangers from abandoned wells.

(1) A sensitive analysis was carried out by using the orthogonal experimental method to investigate the influence of four critical variables on the thermal performance of CGHE. The results show that the inlet water temperature plays the most important role in heat extraction rates, followed by types of CGHE retrofitted from abandoned wells, flow rate, and inner tube thermal conductivity.
(2) An economic analysis of the retrofit, considering the operation costs of the water pump and heat pump, was performed. Given the fact that the CGHE in this study was retrofitted from the abandoned oil wells, the drilling cost can be reduced by up to CNY 1800 thousand. The pump operation cost of CGHE retrofitted from three-section abandoned oil wells is significantly increase with the increase in the flow rate, while the heat pump operation cost decreases because the COP increases.
(3) This study integrates actual project considerations to determine the most appropriate circulation flow based on building load requirements. From an economic perspective, a flow rate of 35 m^3/h is selected as the optimal choice, which aligns with both the economic efficiency and operation stability of the middle and deep CGHEs.

In future research, the transformation project of the abandoned wells should be considered, and the interaction between the Wells should be deeply studied. Among them, the heat radius is the problem we must consider.

Author Contributions: Methodology, R.-J.L. and W.-S.Z.; writing—original draft, L.-R.J.; writing—review and editing, P.C., L.-R.J., M.-Z.Y. and X.-D.Z.; supervision, P.C. All authors have read and agreed to the published version of the manuscript.

Funding: This work was supported by the National Natural Science Foundation of China (52278115); the Natural Science Foundation of Shandong Province, China (ZR2020ME219); and the Leading Researcher Studio Fund of Jinan (Grant No.202333050).

Institutional Review Board Statement: The study did not require ethical approval.

Informed Consent Statement: The study did not involve humans.

Data Availability Statement: The study's data is unavailable due to privacy.

Conflicts of Interest: The authors declare no conflict of interest.

Abbreviations

AOGW	abandoned oil and gas wells
CGHE	coaxial geothermal heat exchanger
DBHE	deep borehole heat exchanger
EGS	enhanced geothermal systems
FDM	finite difference method
FVM	finite volume method
GSHP	ground source heat pump

Nomenclature

a	thermal diffusivity (m^2/s)
c, c_f, c_g	specific heat capacity of the circulating fluid (J/(kg·K))
C, C_1, C_2	heat capacity of the circulating fluid (J/(m·K))
$d, d_o, d_i, d_{1i}, d_{1o}, d_{2i}, d_{2o}$	pipe diameter (m)
Δp_f	frictional pressure drops (Pa)
Δp_s	local pressure drops (Pa)
H_j	bottom coordinate of the j-layer
k, k_g, k_{p1}, k_{p2}	thermal conductivity (W/(m·K))
q_g	geothermal heat flux (W/m^2)
h_1, h_2, ha	convective heat transfer coefficients ($W/(m^2 \cdot K)$)
K, K_1, K_2	height of the equivalent coarse grain (m)
l	pipe length (m)
M	circulating fluid flow rate (kg/s)
P	power consumption (W)
Q	total heat transfer rate of the CGHE (W)
r	radial coordinate (m)
r_b	borehole radius (m)
r_{bnd}	the radius of the radial boundary (m)
R_1, R_2	thermal resistance ((m·k)/W)
t_a	ambient air temperature (°C)
u	average section velocity (m/s)
v	kinematic viscosity (m^2/s)
ρ, ρ_f, ρ_g	density (kg/m^3)
λ	friction factor

References

1. Li, J.; Zhu, T.; Li, F.; Wang, D.; Bu, X.; Wang, L. Performance Characteristics of Geothermal Single Well for Building Heating *Energy Eng.* **2021**, *118*, 517–534. [CrossRef]
2. Rubio-Maya, C.; Ambríz Díaz, V.M.; Pastor Martínez, E.; Belman-Flores, J.M. Cascade Utilization of Low and Medium Enthalpy Geothermal Resources—A Review. *Renew. Sustain. Energy Rev.* **2015**, *52*, 689–716. [CrossRef]
3. Davis, A.P.; Michaelides, E.E. Geothermal Power Production from Abandoned Oil Wells. *Energy* **2009**, *34*, 866–872. [CrossRef]
4. Lund, J.W.; Toth, A.N. Direct Utilization of Geothermal Energy 2020 Worldwide Review. *Geothermics* **2021**, *90*, 101915. [CrossRef
5. Jia, L.; Lu, L.; Chen, J.; Han, J. A Novel Radiative Sky Cooling-Assisted Ground-Coupled Heat Exchanger System to Improve Thermal and Energy Efficiency for Buildings in Hot and Humid Regions. *Appl. Energy* **2022**, *322*, 119422. [CrossRef]
6. Jia, L.; Lu, L.; Chen, J. Exploring the Cooling Potential Maps of a Radiative Sky Cooling Radiator-Assisted Ground Source Hea Pump System in China. *Appl. Energy* **2023**, *349*, 121678. [CrossRef]
7. Chmielowska, A.; Tomaszewska, B.; Sowida, A. The Utilization of Abandoned Petroleum Wells in Geothermal Energy Sector Worldwide Trends and Experience. In Proceedings of the E3S Web of Conferences, Krynica, Poland, 12–14 June 2020; Volume 154 p. 05004. [CrossRef]
8. Morita, K.; Bollmeier, W.S.; Mizogami, H. Analysis of the Results from the Downhole Coaxial Heat Exchanger (DCHE) Experiment in Hawaii. *Trans. Geotherm. Resour. Counc.* **1992**, *16*, 17–23.
9. Morita, K.; Bollmeier, W.S.; Mizogami, H. An Experiment to Prove the Concept of the Downhole Coaxial Heat Exchanger (DCHE in Hawaii. *Geothermal Res. Council Trans.* 1992; 16, 9–16.
10. Kohl, T.; Brenni, R.; Eugster, W. System Performance of a Deep Borehole Heat Exchanger. *Geothermics* **2002**, *31*, 687–708. [CrossRef]

Kohl, T.; Salton, M.; Rybach, L. Data Analysis of the Deep Borehole Heat Exchanger Plant Weissbad (Switzerland). In Proceedings of the World Geothermal Congress, Kyushu-Tohoku, Japan, 28 May–10 June 2000.

Nian, Y.-L.; Cheng, W.-L. Insights into Geothermal Utilization of Abandoned Oil and Gas Wells. *Renew. Sustain. Energy Rev.* **2018**, *87*, 44–60. [CrossRef]

Cheng, W.-L.; Li, T.-T.; Nian, Y.-L.; Xie, K. Evaluation of Working Fluids for Geothermal Power Generation from Abandoned Oil Wells. *Appl. Energy* **2014**, *118*, 238–245. [CrossRef]

Cheng, W.-L.; Liu, J.; Nian, Y.-L.; Wang, C.-L. Enhancing Geothermal Power Generation from Abandoned Oil Wells with Thermal Reservoirs. *Energy* **2016**, *109*, 537–545. [CrossRef]

Cui, G.; Ren, S.; Zhang, L.; Ezekiel, J.; Enechukwu, C.; Wang, Y.; Zhang, R. Geothermal Exploitation from Hot Dry Rocks via Recycling Heat Transmission Fluid in a Horizontal Well. *Energy* **2017**, *128*, 366–377. [CrossRef]

Caulk, R.A.; Tomac, I. Reuse of Abandoned Oil and Gas Wells for Geothermal Energy Production. *Renew. Energy* **2017**, *112*, 388–397. [CrossRef]

Gharibi, S.; Mortezazadeh, E.; Hashemi Aghcheh Bodi, S.J.; Vatani, A. Feasibility Study of Geothermal Heat Extraction from Abandoned Oil Wells Using a U-Tube Heat Exchanger. *Energy* **2018**, *153*, 554–567. [CrossRef]

Kujawa, T.; Nowak, W.; Stachel, A.A. Utilization of Existing Deep Geological Wells for Acquisitions of Geothermal Energy. *Energy* **2006**, *31*, 650–664. [CrossRef]

Templeton, J.D.; Ghoreishi-Madiseh, S.A.; Hassani, F.; Al-Khawaja, M.J. Abandoned Petroleum Wells as Sustainable Sources of Geothermal Energy. *Energy* **2014**, *70*, 366–373. [CrossRef]

Fang, L.; Diao, N.; Shao, Z.; Zhu, K.; Fang, Z. A Computationally Efficient Numerical Model for Heat Transfer Simulation of Deep Borehole Heat Exchangers. *Energy Build.* **2018**, *167*, 79–88. [CrossRef]

Song, X.; Wang, G.; Shi, Y.; Li, R.; Xu, Z.; Zheng, R.; Wang, Y.; Li, J. Numerical Analysis of Heat Extraction Performance of a Deep Coaxial Borehole Heat Exchanger Geothermal System. *Energy* **2018**, *164*, 1298–1310. [CrossRef]

Saadi, M.S.; Gomri, R. Investigation of Dynamic Heat Transfer Process through Coaxial Heat Exchangers in the Ground. *Int. J. Hydrogen Energy* **2017**, *42*, 18014–18027. [CrossRef]

Yang, M.; Li, X.; Deng, J.; Meng, Y.; Li, G. Prediction of Wellbore and Formation Temperatures during Circulation and Shut-in Stages under Kick Conditions. *Energy* **2015**, *91*, 1018–1029. [CrossRef]

Bu, X.; Ma, W.; Li, H. Geothermal Energy Production Utilizing Abandoned Oil and Gas Wells. *Renew. Energy* **2012**, *41*, 80–85. [CrossRef]

Yekoladio, P.J.; Bello-Ochende, T.; Meyer, J.P. Design and Optimization of a Downhole Coaxial Heat Exchanger for an Enhanced Geothermal System (EGS). *Renew. Energy* **2013**, *55*, 128–137. [CrossRef]

Beier, R.A.; Acuña, J.; Mogensen, P.; Palm, B. Transient Heat Transfer in a Coaxial Borehole Heat Exchanger. *Geothermics* **2014**, *51*, 470–482. [CrossRef]

Nian, Y.-L.; Cheng, W.-L. Evaluation of Geothermal Heating from Abandoned Oil Wells. *Energy* **2018**, *142*, 592–607. [CrossRef]

Bertani, R. Geothermal Power Generation in the World 2010–2014 Update Report. *Geothermics* **2016**, *60*, 31–43. [CrossRef]

Winterton, R.H.S. Where did the Dittus and Boelter equation come from? *Int. J. Heat Mass Transfer* **1998**, *41*, 809–810. [CrossRef]

Gnielinski, V. On Heat Transfer in Tubes. *Int. J. Heat. Mass. Transf.* **2013**, *63*, 134–140. [CrossRef]

Jia, L.; Lu, L.; Cui, P.; Chen, J.; Pan, A. A Novel Study on Influence of Ground Surface Boundary Conditions on Thermal Performance of Vertical U-Shaped Ground Heat Exchanger. *Sustain. Cities Soc.* **2024**, *100*, 105022. [CrossRef]

Rao, Z.; Li, M.; Li, M.; Liu, P.; Zhao, L.; Lv, Y. Deep casing-buried pipe heat exchanger heat-economic analysis and application. *J. Cent. South Univ. (Nat. Sci. Ed.)* **2022**, *53*, 4749–4761.

Miansari, M.; Valipour, M.A.; Arasteh, H.; Toghraie, D. Energy and Exergy Analysis and Optimization of Helically Grooved Shell and Tube Heat Exchangers by Using Taguchi Experimental Design. *J. Therm. Anal. Calorim.* **2020**, *139*, 3151–3164. [CrossRef]

Zhang, H.; Yu, Y.; Hu, C. Buried pipe diameter selection of ground source heat pump. *HVAC Units* **2012**, *42*, 114–117.

Gan, X.; Pei, J.; Pavesi, G.; Yuan, S.; Wang, W. Application of Intelligent Methods in Energy Efficiency Enhancement of Pump System: A Review. *Energy Rep.* **2022**, *8*, 11592–11606. [CrossRef] sustainability MDPI

Review

A Review of Carbon Emission Reduction During the Operation Stage of Substations

Weixian Che [1], Yanfeng Wang [1], Wenwei Zhu [1], Lexin Hong [2], Can Fang [2], Huijun Wu [2,*] and Jia Liu [2]

[1] Power Grid Planning and Research Center, Guangdong Power Grid Corporation Limited, Guangzhou 511400, China
[2] School of Civil Engineering and Transportation, Guangzhou University, Guangzhou 510006, China
* Correspondence: wuhuijun@tsinghua.org.cn; Tel.: +86-20-3637-6477

Abstract: As an important part of the power industry, carbon reduction technology in substations plays a key role in supporting the implementation of national policy of carbon peak and reduction. In recent years, a series of carbon emission reduction measures have been developed in the operation phase of substations and promoted beneficial carbon emission reduction. This article summarizes the progress of related technologies and applications from the aspects of substation composition, carbon emission sources, carbon emission reduction technologies, and their effects. Firstly, the composition of carbon emissions from substations during operation is described. Secondly, various measures for carbon emission reduction are reviewed, including the selection of substation equipment types, the intelligent management of substation equipment, and the use of renewable energy. Finally, future technological directions for carbon reduction in substations are discussed, providing technical references and guidance for the sustainable construction and development of low-carbon substations.

Keywords: substation; carbon emission; carbon reduction measures; operation stage; sustainability

1. Introduction

1.1. Background—The Crucial Impact of the Power Industry and Substations on Global Carbon Emissions

With the continuous intensification of global warming, carbon emissions have attracted the attention of the international community [1–3]. In 2023, global energy-related carbon dioxide emissions grew by 1.1% (410 million tons) to 37.4 billion tons. In recent years, the international community has paid great attention to this issue. Therefore, some countries have taken action to reduce greenhouse gas emissions and encourage the rapid development of low-carbon economy and technology [4,5]. The evolution of the power sector has attracted special attention as it is one of the major contributors to global carbon emissions. At present, the power industry is developing towards reducing energy use and lowering carbon emissions. However, according to the International Energy Agency, the power generation and heating industry accounts for 41.2% of global carbon dioxide emissions. Moreover, the proportion in China is even higher, at 45% [6]. Substations are crucial to the power industry as they are key links in the process of receiving and distributing electricity. The operation of substations will inevitably lead to significant carbon emissions. The future development trend will be to build low-carbon substations.

1.2. Research Novelty and Contribution

At present, research on carbon emissions from substations mainly focuses on two aspects: the sources of carbon emissions from substations and the carbon reduction measures during the operation stage of substations. The research on the sources of carbon emissions from substations is based on the life cycle approach and can be divided into four stages: materialization, construction, operation, and demolition. On the other hand, most of the carbon reduction strategies during the operation stage of substations are aimed at the equipment and site of the substations. Among them, the research on carbon reduction technology for substation equipment mainly focuses on the selection of electrical equipment, such as transformers and sulfur hexafluoride equipment [7,8], and on using natural resources to meet the lighting and temperature requirements of substations [9,10]. In order to achieve the transformation of the energy structure, new energy sources mainly based on solar and wind energy are being combined with the construction of energy management systems to supervise and control equipment [11–13]. Meanwhile, research measures for substation construction are focused on establishing green spaces in the substation area to achieve carbon emission absorption for the operation of substation equipment [14,15].

Table 1 summarizes the previous articles on carbon emissions from substations, including the objects, important achievements, and innovations.

Table 1. Summary of previous review references related to carbon emissions in substations (√ indicates that the review is relevant to the topic, - indicates that the review is unrelated to the topic).

Reference	Carbon Emission Sources	Carbon Emission Reduction Measures	Main Conclusions and Innovations
Chen et al., 2023 [16]	√	-	Estimated carbon emissions of substations based on life cycle methods.
Liu et al., 2024 [17]	√	-	Analyzed the sources and amounts of carbon emissions during the construction phase based on actual cases of substations.
Zhao et al., 2023 [18]	-	√	Analyzed the actual carbon reduction effects of different carbon reduction measures based on climate conditions in different regions.
Singh et al., 2016 [19]	√	-	Using genetic algorithm to analyze the optimization configuration and carbon emissions of distributed power sources under different load models.
Chen et al., 2023 [20]	√	-	Calculated and analyzed the carbon emissions generated by each stage of the substation.
Wang et al., 2024 [21]	-	√	Analyzed and compared the carbon reduction effects of different cold and hot energy supply system schemes.
Tang et al., 2024 [22]	-	√	Analyzed the combination of substation construction and solar energy.
Wang et al., 2024 [23]	√	√	Analyzed green construction technology for substations based on multicamera vision.

However, after extensive investigation and analysis, several limitations were found in these references. Firstly, the above research mainly focused on the carbon emissions generated during the construction and operation phases of substations. Secondly, there is insufficient systematic review of the quantitative effects of various measures on carbon reduction.

1.3. Methodology

The overall framework of the review methodology is shown in Figure 1. The literature related to the topic was collected from the Web of Science, and the search methodology and results are shown in Figure 1a. This review adopts a keyword search method to retrieve research objectives, using terms such as substation, power transmission and distribution system, carbon emissions, low-carbon, and energy conservation, and retrieve the titles, abstracts, and conclusions of each piece of literature. It should be noted that this review only selects research in the engineering field related to sources on substation carbon emissions and measures for reducing substation carbon emissions and does not include potential technological research in other fields. In addition, this review only includes English articles, which may affect the comprehensive coverage of the topic.

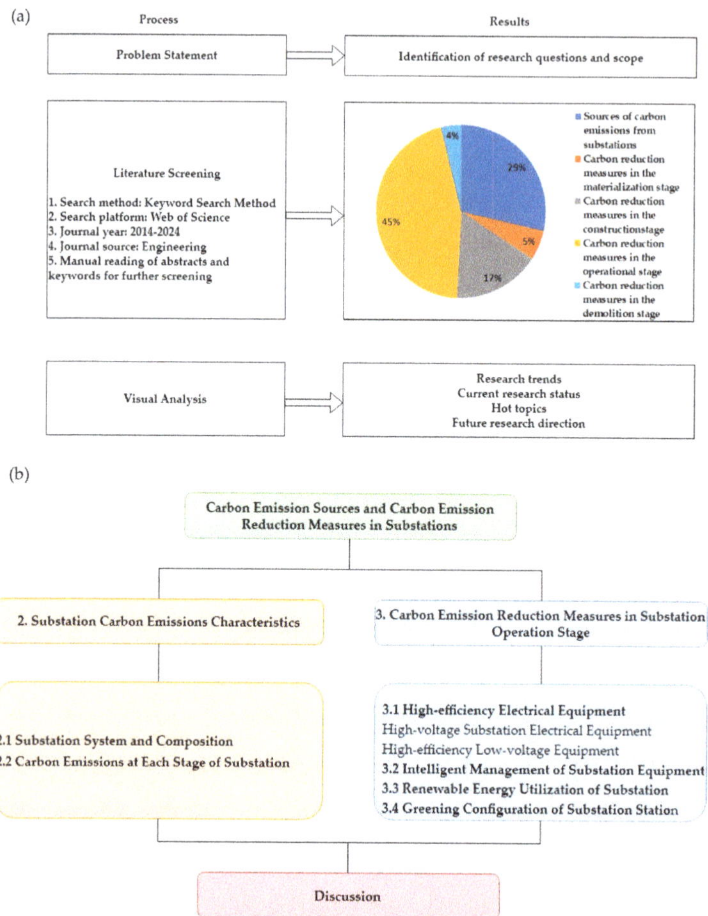

Figure 1. Overall framework of the review methodology. (**a**) Literature search from Web of Science. (**b**) Overall framework of the review.

The overall framework of the review is shown in Figure 1b. Firstly, the collected literature ise divided into the sources of carbon emissions from substations (Section 2) and carbon reduction measures during the operation phase of substations (Section 3). In Section 2, the overall composition of the substation is outlined; it is based on the life cycle approach, introducing the specific carbon emission sources of the four stages of substations—materialization, construction, operation, and demolition. In Section 3, the carbon reduction measures for the maximum carbon emission stage (operation stage) of substations and their impact on reducing substation carbon emissions are summarized and discussed. These measures mostly target the equipment and site of the substation. Section 4 analyzes progress and changes in the research, with a focus on the sources and measures of carbon emissions from substations. Finally, Section 5 summarizes the prospects, recommendations, and limitations for future research, providing future research directions for professionals and engineers and thereby accelerating the development of research on carbon reduction in substations.

2. Substation Carbon Emissions Characteristics

The second section mainly consists of two parts. The first part introduces the composition of the substation: (1) high-voltage equipment, (2) low-voltage equipment, and (3) substation buildings. The second part, based on the life cycle approach, introduces the specific carbon emission characteristics of substations during the materialization, construction, operation, and demolition stages.

2.1. Substation System and Composition

As shown in Figure 2, a typical substation composition is mainly composed of three parts: (1) high-voltage electrical equipment, including the main transformer system, cable and grounding system, power distribution device, reactive power compensation device, control and DC system, communication motor system, etc; (2) low-voltage equipment, including HVAC, fire protection, water supply and drainage, lighting, and other systems; And (3) the substation buildings, including the main control and communication building, distribution device building, duty room, integrated water pump room, and other buildings [24].

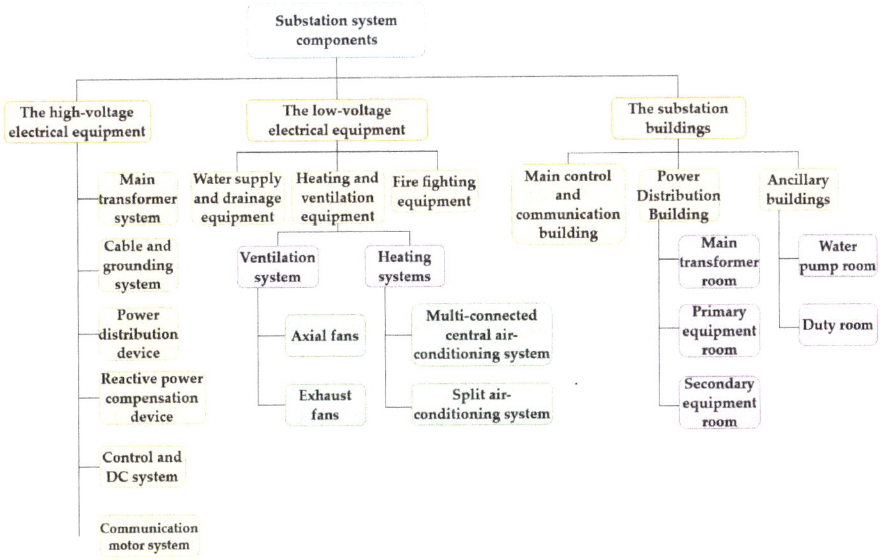

Figure 2. Substation system components.

2.2. Carbon Emissions Characteristics at Each Stage of Substation

Substation carbon emissions are usually classified into four stages based on the life cycle: materialization stage, construction stage, operation stage, and demolition stage [25–27]. Figure 3 illustrates the calculation process of carbon emissions generated by various carbon emission sources in each stage based on the life cycle approach.

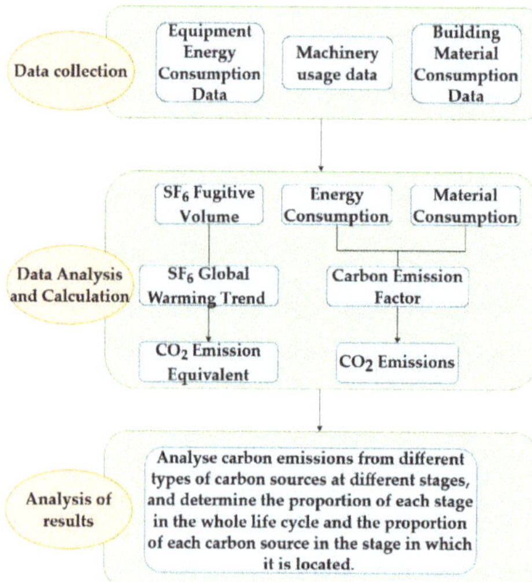

Figure 3. Calculation process of carbon emissions based on the life cycle approach.

Figure 4 shows the main sources of carbon emissions from substations and introduces the carbon emission composition of the four stages of the substation's entire life cycle, namely the materialization stage, construction stage, operation stage, and demolition stage.

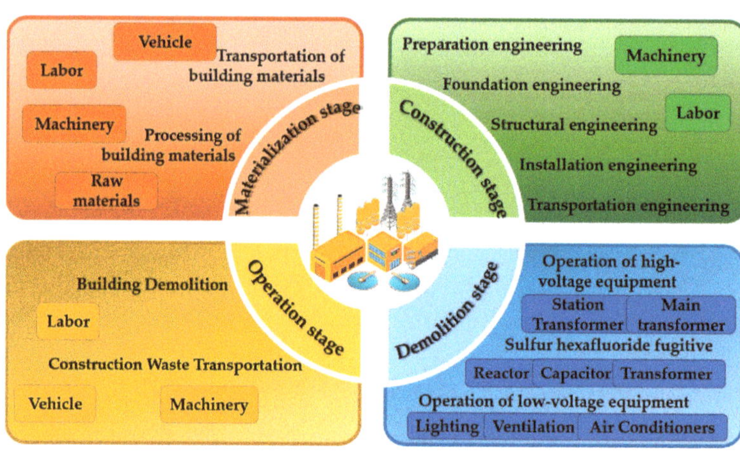

Figure 4. Sources of carbon emissions at each stage of a substation's life cycle.

In the first materialization stage of the substation, carbon emissions come from the production and transportation of building materials, accounting for about 20% of the entire life cycle carbon emissions [28]. Among them, the production of building materials and equipment accounts for about 86% of the total carbon emissions, while the transportation stage accounts for about 14% [29]. In a study on the carbon emissions of buildings using traditional and low-carbon materials at the building materials production stage, Myint et al. [30] pointed out in a study on carbon emissions during the production stage of building materials that bricks are the primary contributor (46%) in traditional structural buildings, followed by steel and cement (19% and 12%). On the other hand, in low-carbon structural buildings, steel and bricks account for 29% and 28% of the total carbon emissions. According to Luo et al. [31], the primary cause of carbon emissions during the transportation stage is the use of fossil fuels for equipment lifting, loading, unloading, and transport. These emissions can be calculated using the carbon emission factor technique.

In the second construction stage of the substation, the proportion of carbon emissions is relatively small, less than 10%, but the composition of its carbon emission sources is more complex. The construction stage can be divided into preparation engineering, foundation engineering, structural engineering, installation engineering, and on-site transportation engineering. The use of construction and transportation machinery is the primary source of carbon emissions during this stage [32,33]. Liu et al. [34] analyzed the substation's building stage and calculated its carbon emissions. The findings indicate that over 50% of the carbon emissions during the construction period are attributed to construction activities. Among these, the power distribution building's carbon emissions account for over 39% of the total, and the cable/grounding system accounts for 14%.

In the third operation stage of the substation, its carbon emissions account for the highest proportion throughout its entire life cycle. There are three sources of carbon emissions in this stage: operation of high-voltage equipment, operation of low-voltage equipment, and escape of sulfur hexafluoride.

(1) Carbon emissions generated by the operation of high-voltage equipment: Transformers are the most important equipment in high-voltage equipment, according to the research of Guo et al. [35]; after 40 years of operation, the carbon emissions from transformers can reach hundreds of thousands of tons. After researching transformer loss, Sima et al. [36] suggested that load loss and no-load loss make up the entirety of transformer power loss. No-load loss is caused by the magnetizing current required to energize the transformer core. Load loss is related to the load on the transformer. The proportion of both in the total losses of a transformer varies with the design and size of the transformer, the materials used, and its operating conditions. Generally, the share of no-load losses is relatively small [37,38].
(2) Carbon emissions generated by the operation of low-voltage equipment: Xue et al. [39] suggested that, during the operation stage of a substation, building energy consumption is mainly driven by the air conditioning system, ventilation system, lighting system, and heating system. Ding et al. [40] analyzed the substation power energy consumption using DeST software. The findings indicated that the main sources of power consumption were air conditioning (78.6%), ventilation (10.7%), electric heating (7.1%), and lighting (3.6%).
(3) Fugitive sulfur hexafluoride: According to a report by the National Energy Administration of China (NEA), by 2020, more than 96.5% of sulfur hexafluoride gas in China will be recovered from the power grid, with a total of 7323 tons of gas recovered. This represents a reduction in carbon dioxide emissions of 17,502,000 tons. Nonetheless, the issue of fugitive sulfur hexafluoride gas persists. Arvesen et al. [41,42] evaluated the transmission system and suggested that the main source of sulfur hexafluoride escape is high-voltage equipment such as circuit breakers.

In the fourth stage—dismantling the substation—its carbon emissions are estimated to be 10% of the carbon emissions produced during the construction stage [43]. According to research by Wang et al. [44], 54.62% of carbon emissions during the dismantling stage are primarily attributable to the on-site disposal of debris. The use of building machinery accounts for 31.74% of carbon emissions, followed closely by transportation machinery, which accounts for 9.55% of emissions. Therefore, the disposal of building waste should be the primary focus of investigations on carbon emission reduction strategies during the demolition stage.

The most significant stage in terms of carbon emissions is the substation operating stage, which is responsible for 70% of all carbon emissions. Therefore, the effectiveness of strategies to reduce carbon emissions during the operation stage of substations is analyzed and summarized in this review, as shown in Figure 5.

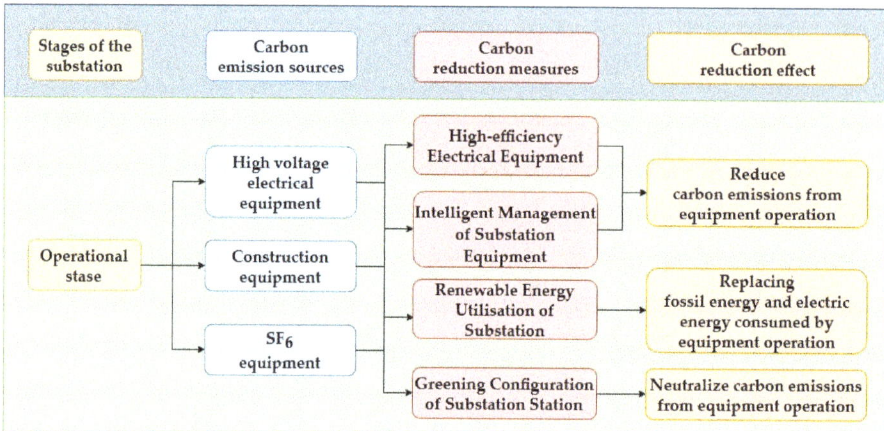

Figure 5. Relationship between carbon reduction measures and carbon emission sources at the operation stage of substations.

3. Carbon Emission Reduction Measures in Substation Operation Stage

The third section elaborates on the carbon reduction measures that can be taken by substations during the operation phase and focuses on four aspects: (1) use of clean high-pressure and low-pressure equipment; (2) intelligent management of substation equipment; (3) use of renewable energy for power generation; (4) arrangement of greenery in the vicinity of the substation to absorb the carbon emissions generated by its construction and operation.

3.1. *High-Efficiency Electrical Equipment*

Carbon emissions in the substation operation stage come from equipment operation. At present, transformers and sulfur hexafluoride equipment are the main sources of carbon emissions. Therefore, replacing cleaning equipment is the main way to reduce emissions. In addition, low-voltage electrical equipment, such as air conditioning, ventilation, and lighting, consume a large amount of energy, which can be reduced by utilizing natural resources.

3.1.1. High-Voltage Substation Electrical Equipment

(1) Replacement of transformer insulating oil

Oil-immersed transformers are one of the most widely used types and a major source of carbon emissions during substation operation. The insulating oil in these transformers helps with insulation, cooling, heat dissipation, and arc quenching. The most popular, cost-effective, and proven product is mineral oil, but it has a low rate of deterioration. Due to their fast rate of degradation and stable chemical characteristics, natural ester insulating oils and palm fatty acid esters are predicted to replace mineral oils.

According to measurements, the degradation rate of natural ester insulating oil exceeds 97%. The degradation rate of mineral insulating oil is less than 30% [45]. Zhou et al. [46] proposed a thin-film surface-modified insulating paper created using chemical vapor deposition technology to suppress the aging of natural ester oil. The experiment showed that the insulation paper significantly reduced the acid value of the insulation oil by 63%, reduced the dielectric loss by 60.50%, and increased the AC breakdown voltage by 20.35%. Yang et al. [47] studied the low-temperature performance improvement of natural ester insulated oil to enhance the low-temperature operational reliability of natural ester insulated oil-immersed transformers. By optimizing the stirring speed and temperature difference between the cooling medium and the natural ester during the crystallization distillation process, the pour point can be reduced to −20.3 °C. However, due to the decrease in the content of high melting point glycerides, the antioxidant activity of the natural ester decreases. However, by combining the CF process with poly (methyl methacrylate), the pour point can be reduced to −27.3 °C while ensuring good physical, chemical, and electrical properties. In addition, Gengadevi et al. [48] further investigated the electrical, mechanical, physical, and chemical parameters of various nonedible natural esters (such as jatropha oil and carrageenan oil) to determine their potential as insulation oils for power transformers. The results indicate that Karan's performance is superior to other fuels.

Transformer insulation oil derived from palm fatty acid esters is also an important substitute for mineral insulation oil [49]. An analysis of comparative studies conducted by Tokunaga et al. [50] revealed that the rate of decomposition of palm fatty acid ester insulating oil is approximately 70%. In the meantime, mineral oil and palm fatty acid ester were subjected to sealed SUS tank aging experiments. The findings indicate that the lifespan of products soaked in palm fatty acid ester mineral oil was 6.1 times longer than that of products soaked in mineral oil when aged at 110 °C. If a 50% retention rate of tensile strength is chosen as the end-of-life standard, the expected lifespan of high-density pressure plates impregnated with palm fatty acid ester mineral oil at 110 °C is 7.3 times that of high-density pressure plates impregnated with mineral oil. Kitt et al. [51] compared the physical and chemical properties of natural esters and palm oil as insulating liquids in distribution transformers. The experimental results show that, during the operation of transformers, the interfacial tension of natural ester insulating oil decreases by 18%–57%, and that of palm insulating oil decreases by 18%–46%, indicating that the interfacial tension of natural ester insulating oil is better. Therefore, natural ester insulating oils have better interfacial tension and lower degradation.

The main equipment of the substation—the transformer—contains insulating oils that need to be replaced and inspected on a regular basis. This comes with a risk of oil leaks. Due to their rapid biodegradation, natural ester insulating oils and palm fatty acid ester insulating oils are increasingly regarded as environmentally beneficial substitutes. Natural ester insulating oils are frequently regarded as the best option due to their superior degrading characteristics and insulation properties.

(2) Adoption of waste heat reuse technology

Furthermore, when transformers are operating, they emit a lot of heat. The waste heat recovery and reutilization technique uses a temperature difference power generator to transform waste heat into electrical energy. The heat is transferred through a heat exchanger from the transformer's oil to a medium like air or water. A high-pressure evaporator placed in the cooler circuit then absorbs the heat from the medium and transforms it into electrical energy [52]. Figure 6 shows schematic diagrams of three recovery systems. Among these configurations, there is a risk of oil pollution in configuration (a). Configuration (b) ensures the independent performance of the transformer cooling function in the event of a failure in the heat recovery system. However, the recycling efficiency is very low. Most of the heat in configuration (c) is captured by the oil-water heat exchanger, allowing for direct use of hot water without contamination and controlling the heat recovery process.

Figure 7 shows the research results of Lagoeiro et al. [53], which show that the theoretical waste heat potential of transformer substations in the UK ranges from 3.0 to 5.4 TWh a^{-1}, equivalent to 0.7% to 1.25% of annual heating demand. Compared with ground source heat pumps and gas turbines, this technology can reduce heating energy use and carbon emissions by about 80%. Furthermore, the operating cost of this system is 17% lower than that of air source heat pumps. The average COP of 4 indicates that heating equipment can effectively reduce carbon emissions. Stefan et al. [54] further investigates the practical application of the technique. First, a thermodynamic model of power transformers was developed to determine the amount of waste heat they generate. Next, the thermodynamic characteristics of the heat exchangers and heat pumps required to utilize the waste heat for district heating are analyzed. Finally, a geographic information system (GIS) analysis was performed to link the power transformers to a specific district heating network.

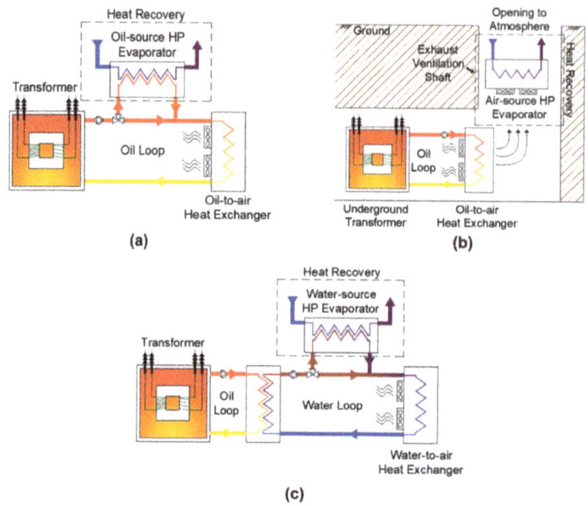

Figure 6. Potential heat recovery solutions for power transformers: (**a**) oil forced air (OFAF) cooling systems; (**b**) underground oil forced air forcing (OFAF) cooling system; (**c**) oil forced water forced (OFWF) cooling system [53].

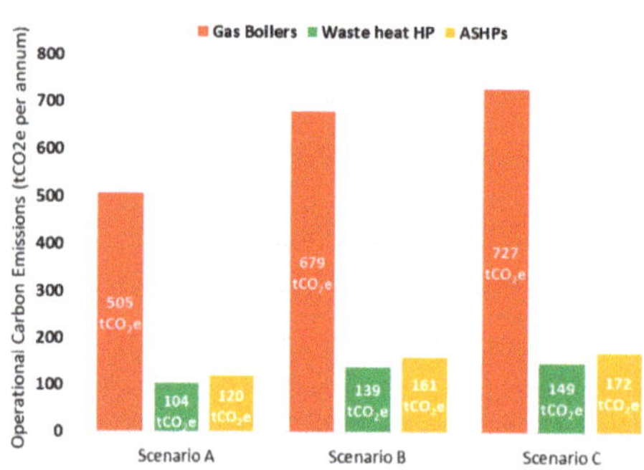

Figure 7. Comparison of carbon emissions from the operation of different heating schemes [53].

(3) Replacement of insulating gas for high-voltage equipment

The global warming potential of sulfur hexafluoride gas is 22,800 times greater than that of carbon dioxide. At present, the power industry uses 80% sulfur hexafluoride gas to fill important equipment such as transformers and circuit breakers. According to research conducted by Yang et al. [55], the new sulfur hexafluoride gas includes not only sulfur hexafluoride, but also many corrosive gases such as perfluoro alkanes, which shorten the lifespan of equipment. Naeem et al. [56] suggested using clean air, which is made up of nitrogen and oxygen, as an alternate insulating medium to address this issue. Cui et al. [57] further investigated the properties of alternative gases such as C_4F_8, CF_3I, $C_5F_{10}O$, $C_6F_{12}O$, C_4F_7N, and HFO, including their zero ozone depletion potential and insulating and arc extinguishing properties. Figure 8 shows the results of the study, where the performance parameters of HFO best meet the needs of high-voltage electrical equipment. When comprehensively evaluating the performance of sulfur hexafluoride and its alternative gases, Zhang et al. [58] pointed out that the immaturity of preparation technology is one of the disadvantages of alternative gases. Its strong stability and low environmental pollution are its advantages. In addition, a possible substitute is vacuum insulation. According to Hitoshi et al. [59] recommended using vacuum insulation as a substitute. Although vacuum insulation has superior insulation performance, electrode materials are very sensitive, and their breakdown strength can cause fouling effects.

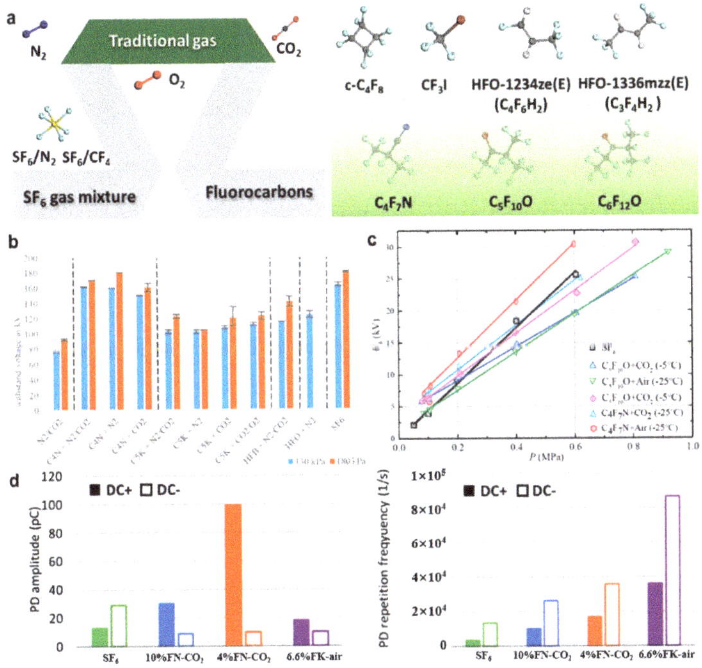

Figure 8. Physical property test results of SF_6 alternative gases [57]. (**a**) The three main categories of SF_6 substitute gases. (**b**) The insulation performance of various SF_6 substitute gases at 0.13 MPa, 0.18 MPa for MV-GIE. (**c**) The insulation performance of various SF6 substitute gases at 0.1–1.0 MPa for HV GIE. (**d**) The partial discharge (PD) amplitude (left) and the repetition frequency (right) of various SF6 substitute gases under direct current (DC) power.

3.1.2. High-Efficiency Low-Voltage Equipment

(1) Utilization of natural resources to regulate indoor temperature

The majority of air conditioning systems in substations are split and multi-connected air-conditioners, which are the main source of carbon emissions from low-voltage equipment in substation buildings. Due to their high energy consumption and carbon emissions, researchers are seeking more efficient cooling solutions.

Ground source heat pump systems use solar and geothermal energy from soil and groundwater to heat and cool buildings. The best places for this system are those with warm summers and cool winters, as well as balanced summer and winter cooling and heating loads. In areas with large differences in heat absorption and emission, solar ground source heat pump systems can be used to ensure system stability [60,61]. After analyzing the coupling between the photovoltaic system and the ground source heat pump system at the substation depicted in Figure 9, Bordignon et al. [62] proposed that the coupled system's energy efficiency ratio may reach 5.4.

In order to produce a cooling effect, sky radiation cooling technology passively releases long-wave radiation into space through atmospheric windows. Buildings have made extensive use of this technology in the form of cold panels, membranes, and radiant coatings [63]. Ferrara et al. [64] optimized the solar geothermal water storage system by optimizing the number of solar collectors, geothermal wells, thermal storage volume, and surface area of serpentine heat exchangers. The research results indicate that the system can reduce costs by 34%.

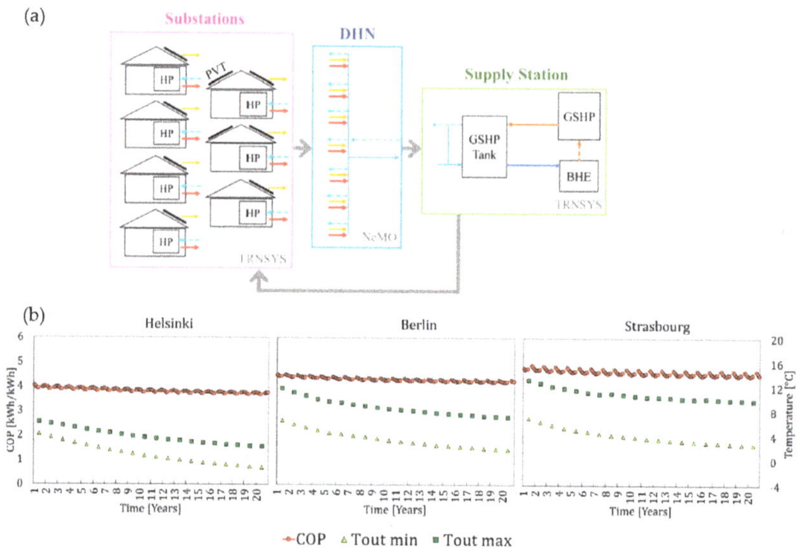

Figure 9. Operating schematic of a system coupled with a ground source heat pump system and a photovoltaic system and energy efficiency in different areas: (**a**) system schematic diagram [60]; (**b**) monthly COPs, annual minimum (Tout min) and maximum (Tout max) outlet fluid temperatures in Helsinki, Berlin, and Strasbourg [60].

Regional adaptability is a characteristic of carbon emissions generated by low-voltage equipment in substations. Using the five main climate zones of China as an example, by using sky radiation cooling technology, the internal temperature of substations can be effectively controlled in hot summers, warm winters, and mild regions. This helps to minimize energy consumption and carbon emissions that result from intense solar radiation. By utilizing sky radiation refrigeration technology and ground source heat pump air conditioning systems, hot summer and cold winter regions can benefit from ample groundwater resources and summertime solar radiation. This allows for energy-saving temperature control of substations. Therefore, in extremely cold and frigid places, due to the lower winter temperatures, large ground source heat pump systems can be used to provide stable heating for substations and ensure the normal operation of equipment under adverse weather conditions.

(2) Combination of natural and mechanical ventilation

In low-voltage electrical equipment in substations, ventilation systems are the second largest source of carbon emissions after air conditioning systems. At present, axial flow fans account for the majority of carbon emissions in the ventilation system of substations. The combination of natural ventilation and mechanical ventilation is a key strategy for reducing emissions [65]. Based on CFD simulation and modeling tools, Zhang et al. [66] discovered that staggered vents may cool the transformer by 1.3 °C, which lowers the need for air conditioning equipment.

In addition, a reasonable combination of natural ventilation and mechanical cooling systems can also effectively control indoor temperature [67]. Yang et al. [68] proposed a cooling and ventilation system combining an air-cooled/water-cooled chiller, night-time radiant panels and natural ventilation as shown in Figure 10. Compared with air-cooled chillers, water-cooled chillers, and natural ventilation, this system achieves

62.24% energy savings and an annual COP rate of over 2.7 times. Therefore, substations should modify the size and arrangement of vents to provide natural ventilation in order to satisfy the internal cooling needs. When natural ventilation cannot meet the temperature requirements, a combination of the refrigeration system, ventilation units, and window operation should be used to effectively manage the indoor temperature in high-temperature regions like the main transformer room and the high-voltage electrical room.

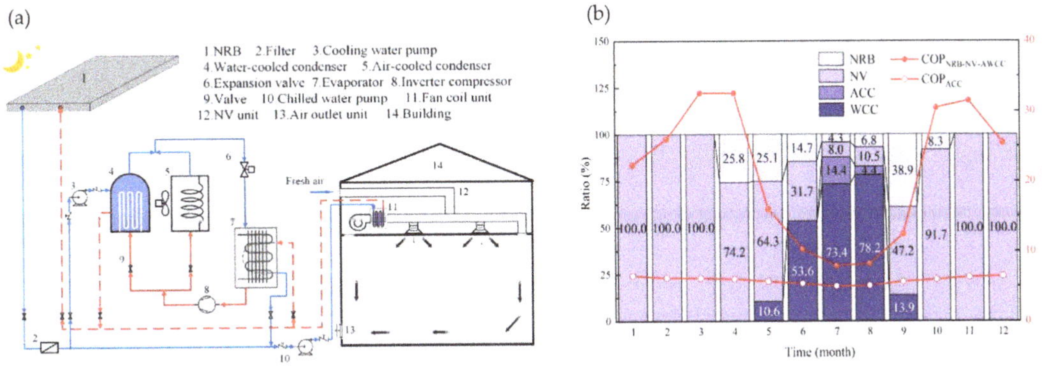

Figure 10. Schematic diagram of a refrigeration ventilation system and energy efficiency of different systems. (**a**) Principle of the novel combined refrigeration system [56]. (**b**) Annual cooling performance analysis of system [56].

(3) Combination of natural lighting and lighting systems

Substation support equipment's lighting equipment is a major source of carbon emissions. By optimizing building orientation to maximize natural illumination, lighting energy consumption and carbon emissions can be decreased [69,70]. The optimal orientation angle of a building usually depends on its geographical location and climatic conditions. For example, in the Northern Hemisphere, buildings oriented in a north-south direction are often advantageous for maximizing natural sunlight, as this orientation allows more sunlight to enter in winter and reduces direct sunlight in summer, thereby reducing cooling load. In the southern hemisphere, different strategies may be required. The east/west orientation may help with morning and afternoon lighting, thereby reducing carbon emissions from lighting equipment. In addition, the optimal angle may also be influenced by cultural customs, and urban planning.

However, outside variables like location and weather have an impact on how natural illumination works [71]. As a result, research on intelligent control systems has exploded. Two senseless dimming control techniques—one based on a building model and the other on solar panels—were examined by Lee et al. [72]. Figure 11 shows the results of the study, showing that the former is about 6% more energy efficient than the latter.

The structure should ideally be oriented southward to maximize natural lighting in order to ensure that the substation's interior space satisfies lighting standards and makes it easier to examine and maintain the equipment. Natural illumination may diminish as latitude rises, so it is best to combine artificial and natural lighting in an intelligent lighting system.

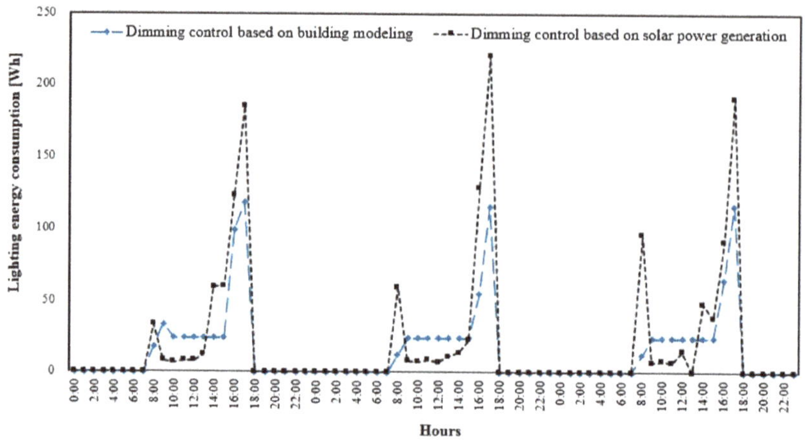

Figure 11. Comparison of lighting energy consumption between two different modes [72].

3.2. *Intelligent Management of Substation Equipment*

In order to prevent equipment failure, aging, and energy waste, the intelligent management of equipment comprises the energy management system and the operation and maintenance management system for electrical equipment. These systems analyze the energy consumption of the equipment and monitor its operation in real time.

By using modern monitoring technology to perform real-time equipment status detection on electrical equipment, it helps prevent both insufficient and excessive maintenance, improve system reliability, and save energy consumption. Ullah et al. [73] suggested using the HOMER Grid tool to analyze the annualized electricity production and consumption patterns of integrated loads in smart grids, in order to promote proactive energy management decisions. Meanwhile, monitoring the power parameters of substations and smart grids through IoT technology can effectively manage different types of loads. Hossain et al. [74] proposed a cloud-based deep analysis method to achieve real-time monitoring and deep data analysis, thereby conducting equipment life assessments and predictive maintenance. In addition, the system can also provide early warning of defects, accurately locate faults, and predict substation loads. Aftab [75] provides an overview of the performance and application of intelligent systems in substations and proposes that future intelligent systems in substations should develop towards achieving common language and interoperability between devices.

The environmental circumstances of substations influence the employment of low-voltage equipment, with the energy management system being the most frequently utilized. Figure 12 shows the schematic diagram of the system, which has powerful monitoring and regulation capabilities. Qian et al. [76] used tilt photogrammetry and remote sensing technology in conjunction with BIM software to develop an energy management system. The system used MATLAB and MySQL for data storage and intelligent screening, optimizing building energy use and carbon footprint monitoring. An intelligent energy management system incorporating data-driven prediction and MIP modules based on building networking was developed by Liang et al. [77]. This technology optimizes HVAC equipment scheduling through multi-objective optimization and precise prediction, thereby improving energy efficiency.

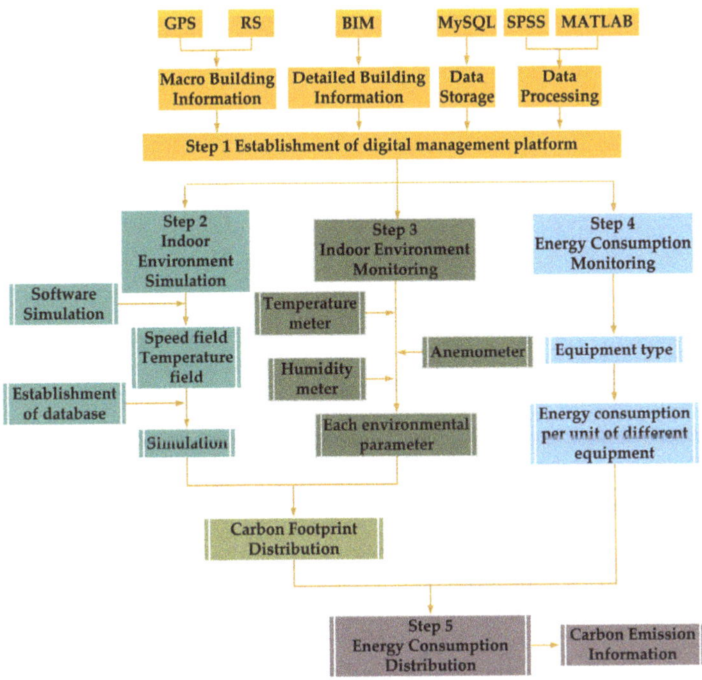

Figure 12. Flow chart of intelligent management system.

The energy consumption of the electrical equipment in substations can be tracked in real time by establishing an intelligent energy management system and an equipment operation and maintenance management system. In addition, transfer learning techniques can be used for energy consumption prediction to find potential opportunities for energy saving and emission reduction, optimizing energy use, and lower the substation's annual carbon emissions.

3.3. *Renewable Energy Utilization of Substation*

During the operation of substations, low-voltage equipment such as air conditioning, ventilation, and lighting will consume electricity, resulting in carbon emissions. To reduce these emissions, renewable energy sources such as solar, hydro, and wind can be utilized for energy generation and substitution.

The power generation of solar photovoltaic systems is one of the most widely used renewable energy sources. Due to the limited space and high-power consumption of installing photovoltaic modules in substations, independent photovoltaic systems are usually used to meet the power demand of the substation. However, photovoltaic systems may experience unstable power generation, such as insufficient or excessive power output, which requires appropriate energy storage systems to ensure stable operation of substations [78]. According to Liu et al. [79], energy storage devices can be broadly divided into three groups: mechanical, electrochemical, and electrical storage. Table 2 displays a comparison of each system.

Table 2. Comparison of Electric Energy Storage Systems.

EES Technology	Capacity	Efficiency (%)	Capital Cost ($/kWh)	Life Time (years)	Advantages	Disadvantages
Pump Hydro Energy Storage	130,000 TWh [80]	70–85 [80]	5–100	40–60	High energy; low cost; fast response; long life span	Large site; high environmental impact; long construction time
Flywheel Energy Storage	5–150 Wh/kg [81]	85–95 [81]	1000–5000 [82]	20+ [82]	High energy efficiency; fast response; low environmental impact	High initial cost; large site; long construction time
Compressed Air Energy Storage	3–400 MWh [83]	40–70	2–100	40+ [83]	Long life; low cost; low environmental impact	
Battery Energy Storage	100–260 Wh/kg [84]	90–95 [84]	1000–2000	5–30	Long life; high energy efficiency	High initial cost; low recyclability
Electric Vehicle	17 kWh-100 kWh [85]	70–90 [86]	300–600 [86]	3–15	Relatively cheap; recyclable	
Hydrogen Energy Storage	0–50 [87]	20–50	10–20	5–15	High energy density	Periodic inspection required; high initial cost; low energy density
Supercapacitor Energy Storage	0.3	90–95	2000	20+	Long life; high efficiency	

In order to accomplish the dual purpose of direct current system (DC) power and energy storage, Cavalcanti et al. [88] analyzed the electricity consumption during the operation stage of substations and suggested that photovoltaic energy storage and power station battery packs may work in concert. In order to achieve a clean energy alternative to power station electricity, the DC power from the PV system can be fed immediately into the DC system of the substation. The DC power can also be inverted into alternating current (AC) power and then supplied to AC loads such as lighting, ventilation, and air conditioning in the substation. Figure 13 shows the technical schematic of the system. Genetic algorithms were used by Gonçalves et al. [89] to optimize the photovoltaic battery system; the results are shown in Figure 14. As the power of the photovoltaic system grows, so does its contribution to the substation's energy consumption. However, due to the lack of solar radiation at night, the photovoltaic system is unable to generate electricity, at which time it needs to rely on the battery to provide the required power. Yahiaoui et al. [90] used Homer simulation and particle swarm optimization techniques to study photovoltaic battery systems and proposed that, when the photovoltaic system is combined with battery storage, the output of the photovoltaic panels can meet about 97% of the substation's electricity demand. Furthermore, researchers have thoroughly examined and proven the viability of this technology from the standpoints of photovoltaic panel power generation efficiency, battery life, and deterioration.

Buildings such as the substation's distribution unit building provide a platform for the installation of photovoltaic systems. The photovoltaic system in the substation can reduce energy consumption and carbon emissions by supplying power to the equipment. To further guarantee the stability of the power supply, the substation is outfitted with battery banks of the proper size. In addition to serving as a backup power supply in an emergency, these battery banks can also store energy when the capacity of the photovoltaic system exceeds.

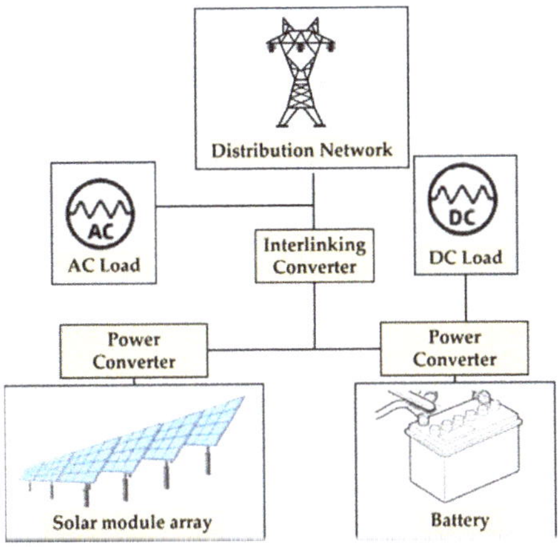

Figure 13. Working principle of PV battery system.

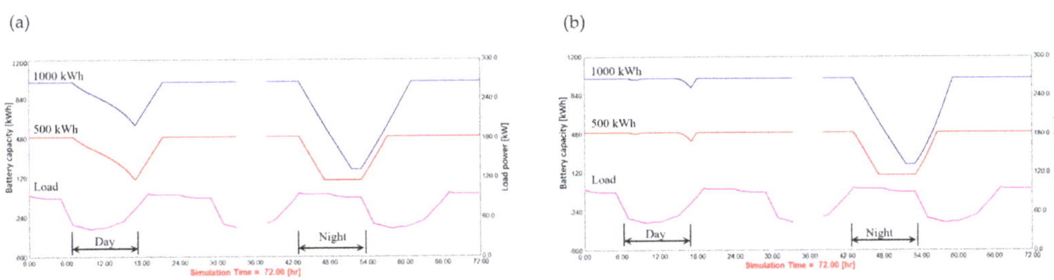

Figure 14. Comparison of the time–by–time power generation effect of different power PV systems and batteries. (**a**) The time response of 500 and 1000 kWh batteries (red and blue lines) and the load curve connected to a 16.2 kWp photovoltaic system (magenta line) [89]. (b) The time response of 500 and 1000 kWh batteries (red and blue lines) and the load curve connected to the 108 kWp photovoltaic system (magenta line) [89].

3.4. Greening Configuration of Substation Station

In order to absorb carbon dioxide, lower the outside air temperature, and cut down on carbon emissions from the air conditioning system, greenery can be grown in the substation's spare area [91]. Figure 15 shows the results of Shishiir et al.'s study on the impact of urban lawns and other ecosystems on building environment heat and carbon emissions [92]; it is proposed that each region should increase the diversity of greenery based on local climate conditions, public opinion, and other factors. Xi et al. [93] used a building complex as an example to study the impact of overall green layout on climate environment and carbon reduction. The results showed that street trees have a better cooling effect compared with green roofs and protective forests, and the maximum temperature can be reduced by 2 °C. Appropriate combination of street trees and shelterbelts can increase the cooling effect by at least 30%, thus improving the carbon emission reduction benefit. Wang [94] studied the effects of different biological communities and plant species on reducing carbon dioxide emissions. Among them, oleander, Japanese tar, and evergreen trees are plant species with high carbon sequestration efficiency.

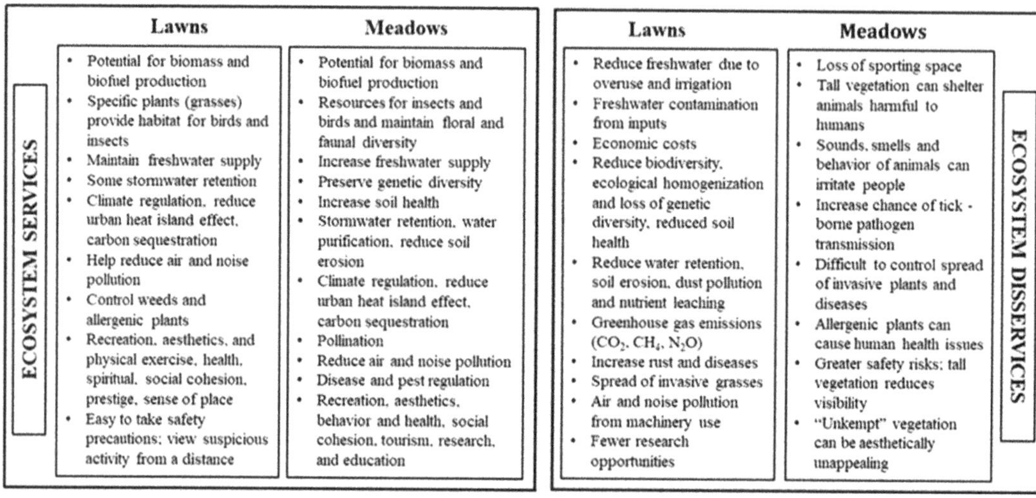

Figure 15. Advantages and disadvantages of several ecosystems of urban lawns and grasslands [92].

The presence of high-voltage equipment in the substation may pose safety risks, so tall trees should not be planted in the substation. Instead, regular turf should be planted on both sides of the roadways in the substation area to achieve "carbon neutrality".

4. Discussion

This section summarizes the advantages and disadvantages of existing research on carbon emissions in substations. Prospects and suggestions for the future development of carbon emissions in substations are also discussed.

(1) In the past decade, research within the engineering field on carbon emissions from substations has mainly focused on the sources of carbon emissions from substations and carbon reduction measures. Most studies focus on carbon reduction measures during the operation stage; future research needs to investigate the carbon reduction measures during the materialization and demolition stages.
(2) Based on the life cycle approach, there are currently many studies on the sources of carbon emissions across the different substation stages. However, these studies mostly focus on the carbon emissions generated by substation buildings. The carbon emissions generated by substation equipment during the materialization and dismantling stages are neglected. Future research can focus on the carbon emissions generated during the production, dismantling, or reuse of substation equipment.
(3) At present, the carbon reduction measures for the operation phase of substations are very comprehensive and can be divided into three types: reducing the energy consumption of substation buildings and equipment; increasing the application of renewable energy to replace traditional energy sources; and planting greenery to absorb the carbon dioxide generated by the operation of substation buildings and equipment. However, the carbon reduction effects of these measures have not been - actuallymeasured. In future work, on-site measurements can be used to calculate the carbon reduction emissions that the above measures can produce.

5. Conclusions

This review addresses the energy-efficient and low-emission construction and operation of substations. The sources of carbon emissions and their characteristics throughout the substation life cycle has been described. Various measures for carbon emission reduction were reviewed. The main conclusions include the following:

(1) Each stage of the substation's full life cycle's carbon emission sources and characteristics are introduced. The most significant stage (accounting for 70% of the total) is the substation operation stage, where electrical equipment, such as transformers, is the primary source of carbon emissions.
(2) Various technical measures to reduce carbon emissions from substations are summarized, including the selection of high-efficiency substation equipment, high-efficiency environmental control equipment, and the use of renewable energy. The impact of various carbon reduction measures is quantitatively evaluated for the sustainable development of substations.
(3) Concerning various carbon reduction measures in substations, it is necessary to thoroughly evaluate the regional adaptability of substation carbon reduction technologies for establishing technical guidelines for the construction and operation of low-carbon substations. Moreover, it is necessary to clarify the relationship between emission reductions and economic feasibility in the sustainable development of substations in the future.

Author Contributions: Project administration, original draft and review, W.C., Y.W., and W.Z.; conceptualization, original draft and visualization, L.H. and C.F.; supervision and funding acquisition, H.W. and J.L. All authors have read and agreed to the published version of the manuscript.

Funding: This research was supported by the Technology Consulting Project of China Southern Power Grid (grant No. 031000QQ00230023) and Tertiary Education Scientific research project of Guangzhou Municipal Education Bureau (grant No. 202235288).

Institutional Review Board Statement: This study did not require ethical approval.

Informed Consent Statement: Informed consent was obtained from all subjects involved in the study. Written informed consent has been obtained from the patient(s) to publish this article.

Data Availability Statement: No new data were created or analyzed in this study. Data sharing is not applicable to this article.

Conflicts of Interest: Author Weixian Che, Yanfeng Wang and Wenwei Zhu were employed by the company Guangdong Power Grid Corporation Limited. The remaining authors declare that the research was conducted in the absence of any commercial or financial relationships that could be construed as a potential conflict of interest.

References

1. Yi, Q.; Zhao, Y.; Huang, Y.; Wei, G.; Hao, Y.; Feng, J.; Mohamed, U.; Pourkashanian, M.; Nimmo, W.; Li, W. Life Cycle Energy-Economic-CO_2 Emissions Evaluation of Biomass/Coal, with and without CO_2 Capture and Storage, in a Pulverized Fuel Combustion Power Plant in the United Kingdom. *Appl. Energy* **2018**, *225*, 258–272. [CrossRef]
2. Rehfeldt, M.; Worrell, E.; Eichhammer, W.; Fleiter, T. A Review of the Emission Reduction Potential of Fuel Switch towards Biomass and Electricity in European Basic Materials Industry until 2030. *Renew. Sustain. Energy Rev.* **2020**, *120*, 109672. [CrossRef]
3. Gu, Y.; Chen, Q.; Wang, D.; Tang, Z. Strategy-Based Optimization and Life Cycle Environmental Impact Assessment of a Stable Photovoltaic Power-to-Hydrogen System. *Energy Convers. Manag.* **2024**, *315*, 118747. [CrossRef]
4. Wang, J.; An, Q.; Zhao, Y.; Pan, G.; Song, J.; Hu, Q.; Tan, C.-W. Role of Electrolytic Hydrogen in Smart City Decarbonization in China. *Appl. Energy* **2023**, *336*, 120699. [CrossRef]
5. Chang, X.; Wu, Z.; Wang, J.; Zhang, X.; Zhou, M.; Yu, T.; Wang, Y. The Coupling Effect of Carbon Emission Trading and Tradable Green Certificates under Electricity Marketization in China. *Renew. Sustain. Energy Rev.* **2023**, *187*, 113750. [CrossRef]
6. Wei, W.; Zhang, P.; Yao, M.; Xue, M.; Miao, J.; Liu, B.; Wang, F. Multi-Scope Electricity-Related Carbon Emissions Accounting: A Case Study of Shanghai. *J. Clean. Prod.* **2020**, *252*, 119789. [CrossRef]
7. Schmitz, W.I.; Feil, D.L.P.; Canha, L.N.; Abaide, A.R.; Marchesan, T.B.; Carraro, R. Operational Vulnerability Indicator for Prioritization and Replacement of Power Transformers in Substation. *Int. J. Electr. Power Energy Syst.* **2018**, *102*, 60–70. [CrossRef]
8. Humayun, M.; Sousa, B.J.O.; Safdarian, A.; Ali, M.; Degefa, M.Z.; Lehtonen, M.; Fotuhi-Firuzabad, M. Optimal Capacity Management of Substation Transformers Over Long-Run. *IEEE Trans. Power Syst.* **2016**, *31*, 632–641. [CrossRef]
9. Licklederer, T.; Zinsmeister, D.; Lukas, L.; Speer, F.; Hamacher, T.; Perić, V.S. Control of Bidirectional Prosumer Substations in Smart Thermal Grids: A Weighted Proportional-Integral Control Approach. *Appl. Energy* **2024**, *354*, 122239. [CrossRef]
10. Ramos, J.C.; Beiza, M.; Gastelurrutia, J.; Rivas, A.; Antón, R.; Larraona, G.S.; De Miguel, I. Numerical Modelling of the Natural Ventilation of Underground Transformer Substations. *Appl. Therm. Eng.* **2013**, *51*, 852–863. [CrossRef]

11. Efstratiadi, M.-A.; Tsakanikas, S.; Papadopoulos, P.; Salinas, D.R. A Novel Holistic Energy Management System Incorporating P Generation and Battery Storage for Commercial Customers. *IEEE Trans. Sustain. Energy* **2024**, *15*, 1475–1485. [CrossRef]
12. Teng, S.Y.; Cambier Van Nooten, C.; Van Doorn, J.M.; Ottenbros, A.; Huijbregts, M.A.J.; Jansen, J.J. Near Real-Time Predictions c Renewable Electricity Production at Substation Level via Domain Adaptation Zero-Shot Learning in Sequence. *Renew. Sustai Energy Rev.* **2023**, *186*, 113662. [CrossRef]
13. Toro-Cárdenas, M.; Moreira, I.; Morais, H.; Carvalho, P.M.S.; Ferreira, L.A.F.M. Net Load Disaggregation at Secondary Substatio Level. *Renew. Energy* **2023**, *207*, 765–771. [CrossRef]
14. Przybysz, A.; Popek, R.; Stankiewicz-Kosyl, M.; Zhu, C.Y.; Małecka-Przybysz, M.; Maulidyawati, T.; Mikowska, K.; Deluga, D Grižuk, K.; Sokalski-Wieczorek, J.; et al. Where Trees Cannot Grow—Particulate Matter Accumulation by Urban Meadows. *Sc Total Environ.* **2021**, *785*, 147310. [CrossRef] [PubMed]
15. Tomson, N. Removal of Particulate Air Pollutants by Australian Vegetation Potentially Used for Green Barriers. *Atmos. Pollu Res.* **2021**, *12*, 101070. [CrossRef]
16. Chen, X.; Ou, Y. Carbon emission accounting for power transmission and transformation equipment: An extended life cycl approach. *Energy Rep.* **2023**, *10*, 1369–1378. [CrossRef]
17. Liu, T.; Wu, Z.; Chen, C.; Chen, H.; Zhou, H. Carbon Emission Accounting during the Construction of Typical 500 kV Powe Transmissions and Substations Using the Carbon Emission Factor Approach. *Buildings* **2024**, *14*, 145. [CrossRef]
18. Zhao, S.; Zhu, Y.; Lou, P.; Hu, Y.; Xu, C.; Chen, Y. Optimization Model of Substation Building Envelope–Renewable Energ Utilization Based on Life-Cycle Minimum Carbon Emissions. *Buildings* **2023**, *13*, 1602. [CrossRef]
19. Singh, B.; Mukherjee, V.; Tiwari, P. Genetic algorithm for impact assessment of optimally placed distributed generations wit different load models from minimum total MVA intake viewpoint of main substation. *Renew. Sustain. Energy Rev.* **2016**, *57* 1611–1636. [CrossRef]
20. Chen, X.; Chen, X.; Chen, Z.; Wang, C.; Guo, W.; Lin, X. Analysis of the Substation Life Cycle Carbon Emission. In *Lecture Notes i Electrical Engineering, Proceedings of the 17th Annual Conference of China Electrotechnical Society. ACCES 2022, Beijing, China, 17–1 September 2022*; Li, J., Xie, K., Hu, J., Yang, Q., Eds.; Springer: Singapore, 2023; Volume 1013, pp. 282–290. [CrossRef]
21. Wang, Y.; Jin, X.; Zhang, J.; Zeng, C.; Gao, X.; Zhao, L.; Sha, S. Scheme Design and Energy-Saving Optimization of Cold and Hea Energy Supply System for Substation Main Control Building in Cold Area. *Appl. Sci.* **2024**, *14*, 1562. [CrossRef]
22. Tang, Z.; Zeng, Y.; Huang, H.; Zhang, W.; Mo, W. Structural Design and Simulation Analysis of New Photovoltaic Bracket fo Temporary Substation. *Integr. Ferroelectr.* **2023**, *233*, 67–80. [CrossRef]
23. Wang, C.; Yuan, H.; Bian, Q.; Zhao, C.; Zhou, B.; Chen, Z.; Basit, M.A. Multi-camera vision-based synchronous positioning anc mapping for green construction of electric substations. *Front. Energy Res.* **2024**, *12*, 1370873. [CrossRef]
24. Wei, W.; Wang, X.; Zhu, H.; Li, J.; Zhou, S.; Zou, Z.; Li, J.S. Carbon Emissions of Urban Power Grid in Jing-Jin-Ji Region Characteristics and Influential Factors. *J. Clean. Prod.* **2017**, *168*, 428–440. [CrossRef]
25. Roberts, M.; Allen, S.; Coley, D. Life Cycle Assessment in the Building Design Process—A Systematic Literature Review. *Build Environ.* **2020**, *185*, 107274. [CrossRef]
26. Wen, R.; Qi, S.; Jrade, A. Simulation and Assessment of Whole Life-Cycle Carbon Emission Flows from Different Residentia Structures. *Sustainability* **2016**, *8*, 807. [CrossRef]
27. Zhao, L.; Guo, C.; Chen, L.; Qiu, L.; Wu, W.; Wang, Q. Using BIM and LCA to Calculate the Life Cycle Carbon Emissions o Inpatient Building: A Case Study in China. *Sustainability* **2024**, *16*, 5341. [CrossRef]
28. Cheng, S.; Zhou, X.; Zhou, H. Study on Carbon Emission Measurement in Building Materialization Stage. *Sustainability* **2023** *15*, 5717. [CrossRef]
29. Dahiya, D.; Laishram, B. Life cycle energy analysis of buildings: A systematic review. *Build. Environ.* **2024**, *252*, 111160. [CrossRef]
30. Myint, N.N.; Shafique, M. Embodied Carbon Emissions of Buildings: Taking a Step towards Net Zero Buildings. *Case Stud. Constr Mater.* **2024**, *20*, e03024. [CrossRef]
31. Luo, W.; Zhang, Y.; Gao, Y.; Liu, Y.; Shi, C.; Wang, Y. Life Cycle Carbon Cost of Buildings under Carbon Trading and Carbon Tax System in China. *Sustain. Cities Soc.* **2021**, *66*, 102509. [CrossRef]
32. Zhang, Z.; Gao, Q.; Shao, S.; Zhang, Y.; Bao, Y.; Zhao, L. Carbon Emission Scenarios of China's Construction Industry Using a System Dynamics Methodology—Based on Life Cycle Thinking. *J. Clean. Prod.* **2024**, *435*, 140457. [CrossRef]
33. Li, X.; Wang, C.; Kassem, M.A.; Wu, S.-Y.; Wei, T.-B. Case Study on Carbon Footprint Life-Cycle Assessment for Construction Delivery Stage in China. *Sustainability* **2022**, *14*, 5180. [CrossRef]
34. Liu, X.; Zhang, J.; Hu, Y.; Liu, J.; Ding, S.; Zhao, G.; Zhang, Y.; Li, J.; Nie, Z. Carbon Emission Evaluation Method and Comparison Study of Transformer Substations Using Different Data Sources. *Buildings* **2023**, *13*, 1106. [CrossRef]
35. Guo, H.; Gao, Y.; Li, J. The Greenhouse Gas Emissions of Power Transformers Based on Life Cycle Analysis. *Energy Rep.* **2022**, *8* 413–419. [CrossRef]
36. Sima, L.; Miteva, N.; Dagan, K.J. A Novel Approach to Power Loss Calculation for Power Transformers Supplying Nonlinear Loads. *Electr. Power Syst. Res.* **2023**, *223*, 109582. [CrossRef]
37. Nnachi, G.U.; Akumu, A.O.; Richards, C.G.; Nicolae, D.V. Estimation of No-Load Losses in Distribution Transformer Design Finite Element Analysis Techniques in Transformer Design. In Proceedings of the 2018 IEEE PES/IAS PowerAfrica, Cape Town South Africa, 28–29 June 2018; pp. 1–532. [CrossRef]

3. Chen, X.; Yang, N.; Yang, Y. Comparison of Low-Frequency Non-Sinusoidal Measurement for Transformer No-Load Characteristics Based on Different Interpolation Methods. *Electr. Eng.* **2021**, *103*, 1807–1821. [CrossRef]
9. Xue, B.; Lu, F.; Guo, J.; Wang, Z.; Zhang, Z.; Lu, Y. Research on Energy Efficiency Evaluation Model of Substation Building Based on AHP and Fuzzy Comprehensive Theory. *Sustainability* **2023**, *15*, 14493. [CrossRef]
). Ding, B.; Liu, J.; Liu, J.; Hua, Y. Research and simulation study on energy use in indoor substation buildings. *Build. Energy Conserv.* **2021**, *49*, 85–90.
1. Arvesen, A.; Hauan, I.B.; Bolsøy, B.M.; Hertwich, E.G. Life Cycle Assessment of Transport of Electricity via Different Voltage Levels: A Case Study for Nord-Trøndelag County in Norway. *Appl. Energy* **2015**, *157*, 144–151. [CrossRef]
2. U.S. Environmental Protection Agency. Overview of SF_6 Emissions Sources and Reduction Options in Electric Power Systems. Available online: https://www.epa.gov/eps-partnership/overview-SF6-emissions-sources-and-reduction-options-electric-power-systems (accessed on 27 September 2024).
3. Hertwich, E.G.; Gibon, T.; Bouman, E.A.; Arvesen, A.; Suh, S.; Heath, G.A.; Bergesen, J.D.; Ramirez, A.; Vega, M.I.; Shi, L. Integrated Life-Cycle Assessment of Electricity-Supply Scenarios Confirms Global Environmental Benefit of Low-Carbon Technologies. *Proc. Natl. Acad. Sci. USA* **2015**, *112*, 6277–6282. [CrossRef]
4. Wang, J.; Wu, H.; Duan, H.; Zillante, G.; Zuo, J.; Yuan, H. Combining Life Cycle Assessment and Building Information Modelling to Account for Carbon Emission of Building Demolition Waste: A Case Study. *J. Clean. Prod.* **2018**, *172*, 3154–3166. [CrossRef]
5. Kiran, S.R.; Mariprasath, T.; Basha, C.H.; Murali, M.; Reddy, M.B. Thermal Degrade Analysis of Solid Insulating Materials Immersed in Natural Ester Oil and Mineral Oil by DGA. *Mater. Today Proc.* **2022**, *52*, 315–320. [CrossRef]
6. Zhou, J.; Wang, F.; Dan, L.; Qian, Y.; Shi, L.; Li, J.; Fofana, I.; Pompili, M. Advancing Thermal Stability in Natural Ester Oil-Paper Insulation Systems via Precision Nanostructuring with Parylene Films: Experimental and Molecular-Level Comprehensive Assessment. *Polym. Test.* **2024**, *140*, 108592. [CrossRef]
7. Yang, T.; Wang, F.; Yao, D.; Li, J.; Zheng, H.; Yao, W.; Lv, Z.; Huang, Z. Low-Temperature Property Improvement on Green and Low-Carbon Natural Ester Insulating Oil. *IEEE Trans. Dielectr. Electr. Insul.* **2022**, *29*, 1459–1464. [CrossRef]
8. Gengadevi, K.; Madavan, R. Aging analysis of non-edible natural ester oil–article insulation under various conditions. *Ind. Crops Prod.* **2023**, *25*, 117528. [CrossRef]
9. Ye, W.; Hao, J.; Zhang, J.; Li, H.; Zhang, J.; Liao, R. Lightning Impulse Breakdown Behavior of Natural Ester and Modified Natural Ester under Long Oil Gap: An Experiment and First-Principles Simulation Analysis. *Ind. Crops Prod.* **2024**, *210*, 118087. [CrossRef]
0. Tokunaga, J.; Koide, H.; Mogami, K.; Kawanishi, K.; Hikosaka, T. Lifetime Estimation of Cellulosic Insulation Material in Palm Fatty Acid Ester for Power Transformers. *IEEE Trans. Dielectr. Electr. Insul.* **2017**, *24*, 1010–1017. [CrossRef]
1. Kittikhuntharadol, Y.; Pattanadech, N.; Maneerot, S.; Jongvilaikasem, K.; Jariyanurat, K.; Jeenmuang, S.; Udomluksananon, P.; Chumpiboon, K.; Bunlaksananusorn, C. Physical and Chemical Properties' Comparison of Natural Ester and Palm Oil Used in a Distribution Transformer. *Energy Rep.* **2023**, *9*, 549–556. [CrossRef]
2. Michel, B.; Clausse, M. Design of Thermochemical Heat Transformer for Waste Heat Recovery: Methodology for Reactive Pairs Screening and Dynamic Aspect Consideration. *Energy* **2020**, *211*, 118042. [CrossRef]
3. Lagoeiro, H.; Davies, G.; Marques, C.; Maidment, G. Heat Recovery Opportunities from Electrical Substation Transformers. *Energy Rep.* **2023**, *10*, 2931–2943. [CrossRef]
4. Petrović, S.; Bühler, F.; Radoman, U.; McKenna, R. Power Transformers as Excess Heat Sources—A Case Study for Denmark. *Energy* **2022**, *239*, 122416. [CrossRef]
5. Yang, L.; Wang, S.; Chen, C.; Zhang, Q.; Sultana, R.; Han, Y. Monitoring and Leak Diagnostics of Sulfur Hexafluoride and Decomposition Gases from Power Equipment for the Reliability and Safety of Power Grid Operation. *Appl. Sci.* **2024**, *14*, 3844. [CrossRef]
6. Siddiqui, N.; Weeks, C.; Rogers, J. Advancements in Clean Air Insulation Technologies for Switchgear and Circuit Breakers. *IEEE Power Energy Mag.* **2022**, *20*, 132–138. [CrossRef]
7. Cui, Z.; Li, Y.; Xiao, S.; Tian, S.; Tang, J.; Hao, Y.; Zhang, X. Recent Progresses, Challenges and Proposals on SF_6 Emission Reduction Approaches. *Sci. Total Environ.* **2024**, *906*, 167347. [CrossRef] [PubMed]
8. Zhang, M.; Hou, H.; Wang, B. Theoretical Characterizations on the Eco-Friendly Gas Tetrafluoropropyne for Electrical Insulation to Replace Sulfur Hexafluoride. *J. Phys. Chem. A* **2023**, *127*, 7984–7996. [CrossRef]
9. Okubo, H.; Beroual, A. Recent Trend and Future Perspectives in Electrical Insulation Techniques in Relation to Sulfur Hexafluoride (SF_6) Substitutes for High Voltage Electric Power Equipment. *IEEE Electr. Insul. Mag.* **2011**, *27*, 34–42. [CrossRef]
0. Figueira, J.S.; García Gil, A.; Vieira, A.; Michopoulos, A.K.; Boon, D.P.; Loveridge, F.; Cecinato, F.; Götzl, G.; Epting, J.; Zosseder, K.; et al. Shallow Geothermal Energy Systems for District Heating and Cooling Networks: Review and Technological Progression through Case Studies. *Renew. Energy* **2024**, *236*, 121436. [CrossRef]
1. Walch, A.; Li, X.; Chambers, J.; Mohajeri, N.; Yilmaz, S.; Patel, M.; Scartezzini, J.-L. Shallow Geothermal Energy Potential for Heating and Cooling of Buildings with Regeneration under Climate Change Scenarios. *Energy* **2022**, *244*, 123086. [CrossRef]
2. Bordignon, S.; Quaggiotto, D.; Vivian, J.; Emmi, G.; De Carli, M.; Zarrella, A. A Solar-Assisted Low-Temperature District Heating and Cooling Network Coupled with a Ground-Source Heat Pump. *Energy Convers. Manag.* **2022**, *267*, 115838. [CrossRef]
3. Jeong, S.Y.; Tso, C.Y.; Zouagui, M.; Wong, Y.M.; Chao, C.Y.H. A Numerical Study of Daytime Passive Radiative Coolers for Space Cooling in Buildings. *Build. Simul.* **2018**, *11*, 1011–1028. [CrossRef]

64. Ferrara, M.; Fabrizio, E. Optimized design and integration of energy storage in Solar-Assisted Ground-Source Heat Pump systen *Build. Simul.* **2023**, *16*, 1933–1948. [CrossRef]
65. Chen, H.; Zhang, H.; Wu, S.; Liu, Y.; Zhong, H. Numerical Simulation and Optimisation Design for Ventilation and He Dissipation in High-Temperature and High-Load Indoor Substations. *Case Stud. Therm. Eng.* **2024**, *59*, 104502. [CrossRef]
66. Zhang, H.; Wang, L.; Yang, P.; Liu, Y.; Zhu, C.; Wang, L.; Zhong, H. Optimizing Air Inlet Designs for Enhanced Natural Ventilatic in Indoor Substations: A Numerical Modelling and CFD Simulation Study. *Case Stud. Therm. Eng.* **2024**, *59*, 104408. [CrossRef
67. Wang, L.; Greenberg, S. Window Operation and Impacts on Building Energy Consumption. *Energy Build.* **2015**, *92*, 313–32 [CrossRef]
68. Yang, H.; Hu, Z.; Ge, F.; Liu, X. Analysis of Cooling Performance and Energy Consumption of Nocturnal Radiation and Natur Ventilation Combined Mechanical Refrigeration System for Outdoor Prefabricated Substation. *Appl. Therm. Eng.* **2023**, *23* 120650. [CrossRef]
69. Habibi, S. The Effect of Building Orientation on Energy Efficiency. *Clean Technol. Environ. Policy* **2024**, *26*, 1315–1330. [CrossRe
70. Zheng, Z.; Xiao, J.; Yang, Y.; Xu, F.; Zhou, J.; Liu, H. Optimization of Exterior Wall Insulation in Office Buildings Based on Wa Orientation: Economic, Energy and Carbon Saving Potential in China. *Energy* **2024**, *290*, 130300. [CrossRef]
71. Choi, H.; Hong, S.; Choi, A.; Sung, M. Toward the Accuracy of Prediction for Energy Savings Potential and System Performanc Using the Daylight Responsive Dimming System. *Energy Build.* **2016**, *133*, 271–280. [CrossRef]
72. Lee, N.; Kim, J.; Jang, C.; Sung, Y.; Jeong, H. Comparison of Sensorless Dimming Control Based on Building Modeling and Sol Power Generation. *Energy* **2015**, *81*, 15–20. [CrossRef]
73. Ullah, Z.; Rehman, A.U.; Wang, S.; Hasanien, H.M.; Luo, P.; Elkadeem, M.R.; Abido, M.A. IoT-Based Monitoring and Control c Substations and Smart Grids with Renewables and Electric Vehicles Integration. *Energy* **2023**, *282*, 128924. [CrossRef]
74. Hossain, M.S.; Rahman, M.; Sarker, M.T.; Haque, M.E.; Jahid, A. A smart IoT based system for monitoring and controlling th sub-station equipment. *Internet Things* **2019**, *7*, 100085. [CrossRef]
75. Aftab, M.; Hussain, S.S.; Ali, I.; Ustun, T.S. IEC 61850 based substation automation system: A survey. *Int. J. Electr. Power Energ Syst.* **2020**, *120*, 106008. [CrossRef]
76. Qian, Y.; Leng, J.; Wang, H.; Liu, K. Evaluating Carbon Emissions from the Operation of Historic Dwellings in Cities Based on a Intelligent Management Platform. *Sustain. Cities Soc.* **2024**, *100*, 105025. [CrossRef]
77. Liang, X.; Chen, K.; Chen, S.; Zhu, X.; Jin, X.; Du, Z. IoT-Based Intelligent Energy Management System for Optimal Planning c HVAC Devices in Net-Zero Emissions PV-Battery Building Considering Demand Compliance. *Energy Convers. Manag.* **2023**, *29* 117369. [CrossRef]
78. Emrani, A.; Berrada, A. A Comprehensive Review on Techno-Economic Assessment of Hybrid Energy Storage Systems Integrate with Renewable Energy. *Energy Storage* **2024**, *84*, 111010. [CrossRef]
79. Liu, J.; Chen, X.; Cao, S.; Yang, H. Overview on Hybrid Solar Photovoltaic-Electrical Energy Storage Technologies for Powe Supply to Buildings. *Energy Convers. Manag.* **2019**, *187*, 103–121. [CrossRef]
80. Hunt, J.D.; Byers, E.; Wada, Y.; Parkinson, S.; Gernaat, D.E.H.J.; Langan, S.; van Vuuren, D.P.; Riahi, K. Global resource potenti of seasonal pumped hydropower storage for energy and water storage. *Nat Commun.* **2020**, *11*, 947. [CrossRef]
81. Xu, K.; Guo, Y.; Lei, G.; Zhu, J. A Review of Flywheel Energy Storage System Technologies. *Energies* **2023**, *16*, 6462. [CrossRef]
82. Amirante, R.; Cassone, E.; Distaso, E.; Tamburrano, P. Overview on Recent Developments in Energy Storage: Mechanica Electrochemical and Hydrogen Technologies. *Energy Convers. Manag.* **2017**, *132*, 372–387. [CrossRef]
83. Zhang, X.; Gao, Z.; Zhou, B.; Guo, H.; Xu, Y.; Ding, Y.; Chen, H. Advanced Compressed Air Energy Storage Systems: Fundamental and Applications. *Engineering* **2024**, *34*, 246–269. [CrossRef]
84. Sanjuan, L.F.R.; Lefebvre, G.; Peton, H.; Arkhangelski, J.; Tankari, M.A. Batteries Energy Storage Systems: Review of Mate rials, Technologies, Performances and Challenges. In Proceedings of the 2023 11th International Conference on Smart Gri (icSmartGrid), Paris, France, 4–7 June 2023; pp. 1–6. [CrossRef]
85. Hasan, M.K.; Mahmud, M.; Ahasan Habib, A.K.M.; Motakabber, S.M.A.; Islam, S. Review of Electric Vehicle Energy Storage an Management System: Standards, Issues, and Challenges. *J. Energy Storage* **2021**, *41*, 102940. [CrossRef]
86. Akbari, H.; Browne, M.C.; Ortega, A.; Huang, M.J.; Hewitt, N.; Norton, B.; McCormack, S.J. Efficient energy storage technologie for photovoltaic systems. *Solar Energy* **2019**, *192*, 144–168. [CrossRef]
87. Bovo, A.; Poli, N.; Trovò, A.; Marini, G.; Guarnieri, M. Hydrogen energy storage system in a Multi–Technology Microgrid:technica features and performance. *Int. J. Hydrogen Energy* **2023**, *48*, 12072–12088. [CrossRef]
88. Cavalcanti, M.D.M.; Costa, T.; Pereira, A.C.; Jatobá, E.B.; Filho, J.B.D.M.; Barreto, E.; Mohamed, M.A.; Ilinca, A.; Marinho, M.H.N Case Studies for Supplying the Alternating Current Auxiliary Systems of Substations with a Voltage Equal to or Higher tha 230 kV. *Energies* **2023**, *16*, 5396. [CrossRef]
89. Gonçalves, A.; Cavalcanti, G.O.; Feitosa, M.A.F.; Dias Filho, R.F.; Pereira, A.C.; Jatobá, E.B.; De Melo Filho, J.B.; Marinhc M.H.N.; Converti, A.; Gómez-Malagón, L.A. Optimal Sizing of a Photovoltaic/Battery Energy Storage System to Supply Electri Substation Auxiliary Systems under Contingency. *Energies* **2023**, *16*, 5165. [CrossRef]
90. Yahiaoui, A.; Benmansour, K.; Tadjine, M. Control, Analysis and Optimization of Hybrid PV-Diesel-Battery Systems for Isolate Rural City in Algeria. *Sol. Energy* **2016**, *137*, 1–10. [CrossRef]
91. Li, P.; Wang, Z.-H. Environmental Co-Benefits of Urban Greening for Mitigating Heat and Carbon Emissions. *J. Environ. Manag* **2021**, *293*, 112963. [CrossRef]

2. Paudel, S.; States, S.L. Urban Green Spaces and Sustainability: Exploring the Ecosystem Services and Disservices of Grassy Lawns versus Floral Meadows. *Urban For. Urban Green.* **2023**, *84*, 127932. [CrossRef]
3. Xi, C.; Ding, J.; Wang, J.; Feng, Z.; Cao, S.-J. Nature-Based Solution of Greenery Configuration Design by Comprehensive Benefit Evaluation of Microclimate Environment and Carbon Sequestration. *Energy Build.* **2022**, *270*, 112264. [CrossRef]
4. Wang, Y.; Chang, Q.; Li, X. Promoting Sustainable Carbon Sequestration of Plants in Urban Greenspace by Planting Design: A Case Study in Parks of Beijing. *Urban For. Urban Green.* **2021**, *64*, 127291. [CrossRef]

sustainability

MDPI

Article

Field Monitoring and Modeling of the Hygrothermal Performance of a Cross-Laminated Timber and Wood Fiber-Insulated Building Located in a Cold Climate †

Liam O'Brien [1], Ling Li [2,*], Benjamin Herzog [3], Jacob Snow [3] and Wilhelm A. Friess [4]

1 Graduate School, University of Maine, 5755 Nutting Hall, Orono, ME 04469, USA; liam.obrien@maine.edu
2 School of Forest Resources, University of Maine, 5755 Nutting Hall, Orono, ME 04469, USA
3 Advanced Structures and Composites Center, University of Maine, 35 Flagstaff Road, Orono, ME 04469, USA; benjamin.herzog@composites.maine.edu (B.H.); jacob.snow@composites.maine.edu (J.S.)
4 Department of Mechanical Engineering, University of Maine, 75 Long Road, Orono, ME 04469, USA; wilhelm.friess@maine.edu
* Correspondence: ling.li@maine.edu

† This article is a revised and expanded version of a paper entitled The Hygrothermal Performance and Durability of a One Story Cross-Laminated Timber and Wood Fiber Insulation School Building Located in Belfast, Maine, which was presented at the 14th World Conference on Timber Engineering 2025, Brisbane, Australia 22–26 June 2025.

Abstract: The increased complexity of buildings has led to rigorous performance demands from materials and building envelopes. As markets for low-carbon, renewable construction materials grow, cross-laminated timber and wood fiber insulation have emerged as promising alternatives to meet these rigorous demands. However, an investigation into the performance and interaction of materials within high-performance systems is necessary to determine the durability risks associated with increased complexity and the introduction of new materials. This is important in order to ensure that these materials can meet the required functions of the building while taking advantage of their environmental benefits. To do so, this case study investigated a building constructed of cross-laminated timber and wood fiber insulation in a cold climate (Zone 6A) (Belfast, ME, USA). During construction, the building was instrumented with temperature, relative humidity, and moisture content monitoring instrumentation through the envelope, i.e., wall and roof assemblies. The conditions within the envelope were monitored for a two-year period and used to calibrate a hygrothermal model, along with measured material properties. The calibrated model was used to conduct a 5-year simulation and mold risk assessment. Findings demonstrated that there was no moisture or mold risk throughout the monitoring period or simulation. This supports the integration of cross-laminated timber and wood fiber insulation in sustainable building practices, particularly in cold climates where moisture management is critical.

Keywords: hygrothermal; wood fiber insulation; cross-laminated timber; building monitoring; sustainable building materials

1. Introduction

The strive for total occupant comfort and increased energy efficiency has established demanding performance requirements of building systems. Building and construction design, materials, and techniques have evolved in an effort to meet these requirements. As these complex composite building envelopes are designed, material properties and interactions must be considered to ensure durable, efficient, and healthy buildings.

New building materials give designers more options to meet stakeholder objectives. Current innovations are centered around solutions that meet or exceed the performance of traditional materials, while being environmentally benign. Two such materials are cross-laminated timber (CLT) and wood fiber insulation (WFI). Both are well-suited for North American adoption, as there are ample wood baskets and a history of using wood and wood-based materials in construction.

CLT was developed in Austria during the latter half of the 20th century and was launched in European markets during the 1990s [1]. The focus on bio-based materials and the need for alternative wood-based products and systems was critical to the success of CLT in Europe and led to its introduction to North America in the early 2000s [2,3]. As of early 2025, over 500 CLT buildings have been constructed in North America, with more under development [4]. Throughout the acceptance process, research efforts have determined many benefits of CLT, including dimensional stability, high strength-to-weight ratio, thermal mass, and collective research has worked to provide information on how to implement these systems despite the lack of comprehensive code adoption in the U.S. [5].

WFI was developed in 1908 by the Heraklith company and was used as loose-fill insulation until advancements in the refinement and pressing processes led to greater manufacturing capacity nearly 50 years later [6,7]. North American manufacturing of WFI was not realized until 2023, and therefore, there has been limited investigation into WFI in North America. Previous barriers to acceptance in North American markets have been attributed to three factors: lack of manufacturing capacity, limited knowledge of material properties, and few proven systems within North American climates [8]. Efforts to address these challenges are underway, as North American manufacturing began in 2023, and material properties have recently begun to be compiled [9–12]. These efforts have been undertaken on a limited number of WFI materials, with limited consideration of performance under real-world conditions. Research has been conducted in European climates; however, climate, building, and construction methods are not directly comparable between the two continents.

North American regions experience a wide range of climatic conditions and typically employ light-frame residential construction, whereas European nations typically employ masonry structures, resulting in differing envelope composition and material interactions [13]. Additionally, code requirements must be navigated and can limit the use of WFI. For instance, WFI is restricted by combustibility requirements. Currently, WFI can be used within Type V construction in the U.S. without limitations; however, fire blocking and encapsulating the WFI are required in Type III or Type IV construction [14]. Further, previous assessment of performance has employed insulation materials of different types (loose-fill and board) and thicknesses, ranging from 20–300 mm, which can make conclusions difficult to relate to general use cases [15,16].

The utilization of insulation materials and their interaction with wooden structural materials can lead to increased building and envelope energy efficiency, thermal performance, and durability. However, the majority of available fossil-based insulation materials, for example, expanded polystyrene (EPS) and extruded polystyrene (XPS), on the market are vapor impermeable and can introduce moisture risks. This is of great importance when considering that failures relating to the building envelope, specifically moisture intrusion, are well documented as a leading cause of construction litigation [17]. Moisture-related issues are associated with reduced durability, material degradation, and mold formation, which have been linked to human health problems [18]. Using vapor-permeable insulation materials, such as WFI, in building envelopes would be a possible solution to effectively manage moisture migration and reduce associated risks.

WFI and CLT are hygroscopic in nature and therefore can absorb and desorb water vapor under real climactic conditions [19]. Therefore, assemblies constructed with entirely wood-based materials are considered vapor open so long as a vapor barrier is not included in the assembly. The ability of the assembly to react to environmental conditions in this manner can prevent the accumulation of moisture at any one location due to the inclusion of an impermeable layer. However, durability and thermal properties are affected by the presence of moisture in these materials, and it is necessary to understand the impact of this relationship on performance. It should be noted that consideration of the moisture risk should be undertaken regardless of the material type, as hygroscopic materials are of no greater risk to moisture intrusion than other material types. Although moisture intrusion is not necessarily more likely to occur in wood-based materials, they are sensitive to the presence of moisture and can support mold growth. Mold growth can occur under specific conditions: on wood surfaces with a MC ranging from 30 to 150 percent, a temperature between 0 and 40 °C, and where nutrients are accessible. It is important to ensure that these conditions are not maintained.

Previous work has established the best practices for moisture considerations and CLT-fossil-based insulation wall and roof systems. The U.S. edition of the CLT Handbook [20] has served as a reference for design and been used to inform building codes. This text does not include the consideration of bio-based insulation materials and only discusses the use of systems with XPS, EPS, polyisocyanurate, and mineral wool. However, useful information for designers is provided when considering the placement of insulation, differences in ventilation strategies, and which climates these strategies are effective in. In these examples, CLT is always insulated to the exterior, which protects it from temperature changes, and therefore, dimensional instability.

Additionally, the inclusion of a ventilation gap, often used as a drainage plane between siding and insulation layers, reduces the risk of moisture accumulation within the insulation. This reduces inward vapor drive, which can adversely affect the durability of the structure. Although WFI is not considered in the handbook, these principles can be applied to CLT and WFI systems to reduce risks of interstitial condensation. Traditionally, the efficacy of new materials and their system interaction has been determined based on empirical evidence. However, with the increased system demands, introduction of new materials, changing code requirements, and occurrence of catastrophic moisture-related failures, methods to supplement empirical assessment have been developed [21]. Hygrothermal simulation software programs, such as COMSOL, DELPHIN, and WUFI, can be used to assess the effects of boundary conditions and material selection on an envelope system. These programs have been widely used and validated using laboratory and field studies [21–25]. Utilization of these tools is especially useful during material development and design of the envelope system to understand moisture transfer, the effects of boundary conditions, and to determine potential risks. These software tools can be useful for designers and architects when there are few or no existing examples of a building system. These complex tools use a simple user interface that requires a limited number of standard material properties [21]. Despite validation and complex computational methods being incorporated into the simulation tools, there are still uncertainties stemming from model assumptions, material properties, and boundary conditions [24,26]. This can be further exacerbated when modeling hygroscopic materials, and careful consideration must be taken when using modeling tools to assess performance [27].

Material properties are often cited as one of the most important inputs to be accurately reported when using these modeling tools [27]. Material data sheets can be the first resource for obtaining information; however, these are not always available or provide sufficient information. Direct measurement is regarded as the preferred method to gather these input properties, but this can be costly and time-consuming [15]. Further, although measured properties are often the most representative, shortcomings of the measurement methodologies and uncertainty in measurement can still develop errors within the models [15,22]. Each material is subject to different transport mechanisms, which cannot all be accounted for in a simplified model. Measurement practices and prediction techniques have been found not to be representative of all materials, especially considering hygroscopic materials. For example, the measurement technique for the determination of the moisture storage function, which was developed using mineral-based materials, is not representative of hygroscopic materials, which can lead to an overprediction of the rate of water absorption in wood-based materials. This can be associated with water vapor diffusion being assumed negligible during water uptake measurement and is counted twice in the model [22]. Other limitations have been observed when using Fick's law to model the hygrothermal behavior of wood, where transient conditions and a retarded vapor sorption at high relative humidity (RH) can be seen [28–30].

Ideally, hygrothermal models should be compared to measured data to assess the quality of the model. This is necessary to assess what has the greatest influence on the model and can be used to demonstrate risks or shortcomings of the model and system. Material properties may be required to be adjusted to more accurately describe a material and envelope. Previous work has determined that material inputs may need to be adjusted to improve models, which is not necessarily indicative of the properties being incorrect but can be related to material interaction [22,24,26,28,31].

This study aimed to conduct a hygrothermal assessment of a building envelope comprised of CLT and WFI in a cold North American climate. Due to the fairly recent adoption of CLT and WFI in this region, there are only limited instances where these materials have been used, both on their own or in conjunction with one another. Development of accurate material property datasets can assist in the assessment of the moisture performance and ability to accurately represent the material within hygrothermal modeling tools, thereby helping to increase the adoption of these sustainable materials. As a part of this work, a 92 square meter building located in Belfast, Maine, U.S., was instrumented with 59 temperature, relative humidity (RH), and moisture content (MC) sensors throughout the wall and roof envelope. Monitoring data were collected over a 2-year period. Hygrothermal simulations were performed using on-site weather data and a one-dimensional hygrothermal simulation software tool (WUFI Pro 7.0). The inputs of WFI insulation material were measured by conducting laboratory experiments. The monitored and simulated data were compared and used to assess the mold growth index and durability of the building.

2. Materials and Methods

2.1. Case Study Building: Cornerspring Montessori School Annex

A one-story school building constructed in 2021 and located in Belfast, Maine, USA, was selected for this case study, as shown in Figure 1. This 92-square-meter building contained two classrooms, an entryway, and a bathroom. Panelized construction techniques were used to construct the building. Panels were manufactured using CLT and WFI. Both the CLT and WFI were of European origin, and samples of the WFI were used to determine material properties.

Figure 1. Rendering of the Cornerspring Montessori School Annex, courtesy of Opal Architecture.

The school was regularly occupied September through June on weekdays from 8:00 am to 4:00 pm by 5–10 persons. The building operated exclusively on electricity, using LED lighting, an electric hot water heater, and two mini-split heat pumps, accompanied by an energy recovery ventilator, to condition the building. The building employed large amounts of insulation, with insulation depths of 240 mm in the walls and 360 mm in the roof, which resulted in wall and roof nominal U-values of 0.12 $W/(m^2{\cdot}K)$ and 0.08 $W/(m^2{\cdot}K)$, respectively.

Wall and roof panels consisted of three- and five-ply CLT, respectively, and multiple layers of 120 mm thick WFI board. The WFI board had a nominal density of 140 kg/m^3, did not include wax, and was rated for internal use. Therefore, a water-resistive barrier (WRB) was applied to the exterior surface of WFI in wall assemblies. As part of the fabrication process, wall and roof panels were instrumented with temperature (°C), RH (%), and MC sensors (%) throughout the envelope, as shown in Figure 2. CLT was exposed to the interior and was finished using an oil-based finish.

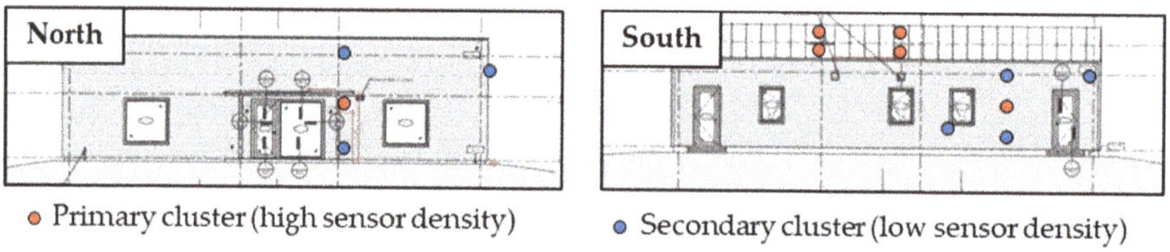

Figure 2. Location of sensor clusters throughout the building, distinguishing between primary and secondary locations. The front of the building faces north, while the rear of the building faces south. Instrumentation within the roof is shown in the south-facing diagram.

Sensors were installed at two densities, as shown in Figures 2 and 3. Primary cluster locations within the monitored walls included temperature and RH sensors at each WFI interface and MC sensors at the first and third lamella of the CLT, labeled P1 to P3 from interior to exterior. Secondary cluster locations include sensors at the interface between CLT and WFI and the WRB and WFI, labeled as P1 and P3, respectively. Similarly, the roof locations were instrumented with sensors in between each layer of WFI and in the first, third, and fifth lamella of the CLT, labeled R1 to R4 from interior to exterior. Measured data were collected over a 2-year period, beginning 1 January 2022.

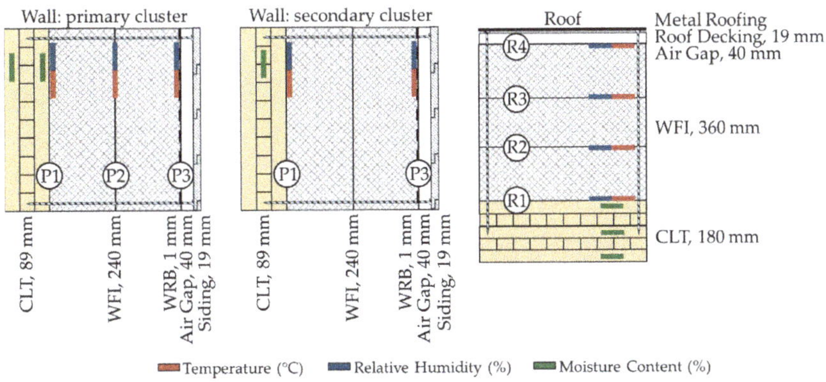

Figure 3. Diagram of wall and roof assemblies, with total dimensions of each material and sensor location indicated.

2.2. Monitoring Instruments

Building monitoring instrumentation recorded conditions within the building assembly at 5 min intervals, and weather data were recorded at 15 min intervals. A description of the technical specification of the monitoring equipment is presented in Table 1.

Table 1. List of technical specifications of monitoring equipment.

Instrument	Parameter(s)	Accuracy	Resolution	Range	Manufacturer	Model
T&RH	Temperature	±0.3 °C	0.1 °C	−40 to 120 °C	Omnisense	A-1 Humisense
	Relative Humidity	±2.0%	1%	0 to 100%		
MC	MC	-	0.1%	-		A-2 Moisture
Weather station	Temperature (Interior)	±0.3 °C	0.1 °C	0 to 60 °C	Davis	Vantage Vue, Wireless
	Relative Humidity (Interior)	±2%	1%	1 to 100%		
	Barometric Pressure	±10 Pa	10 Pa	54,000 to 110,000 Pa		
	Temperature (Exterior)	±0.1 °C	0.5 °C	−40 to 65 °C		
	Relative Humidity (Exterior)	±2%	1%	1 to 100%		
	Dewpoint	±1 °C	1 °C	−76 to 54 °C		
	Wind Speed	±1 m/s	0.5 m/s	0 to 89 m/s		
	Wind Direction	3°	1°	1 to 360°		
	Precipitation	±4%	0.2 mm	0 to 6553 mm, daily		

Temperature and RH sensors were probe-type sensors, and MC sensors were pin-type sensors. An example of these sensors and the methods of installation are presented in Figure 4. Temperature and RH sensors were situated between layers of WFI and/or WFI and CLT (P1, P2, R1, R2, and R3). At the exterior surface (P3 and R4), a small channel was routed in the WFI, which fit the sensor, and a piece of tape was placed over the bottom of the sensor, so as not to block the probe, which ensured that the sensor stayed in place during the panel manufacture and installation. MC screws were placed 3 cm apart, and sensors were calibrated based on lab measurements. Screw length was determined in order to reach the center of the lamella, which was being measured. The location and depth of the screw were determined based on the location and cluster density. Note that after initial installation, MC sensors at primary locations in the middle lamella, which were installed using nails as pins instead of screws, within the walls and roof, were found to be faulty.

Where accessible, these were replaced; otherwise, additional MC sensors were installed in the attic. However, as these were not installed at the same location and were not installed until 8 months into the monitoring period, they were not included in the analysis and only used to assess general trends. Roof assemblies did not have this issue, and MC was used for the calibration; however, one of the four roof sensor clusters was damaged during panel installation, and therefore, the data were not recorded at all sensors at that location.

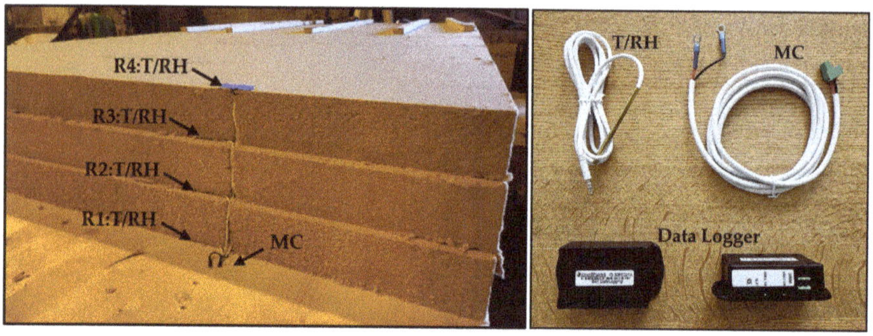

Figure 4. Example of sensor installation in a roof panel (**left**) and image of sensors used (**right**).

Hourly temperature and RH values were obtained by averaging 5 min readings. Intermittent outages were experienced throughout the 2-year period. Gaps in data were filled using the preceding data, e.g., if data were missing from 12 December 2022, at 2:00 pm, the previous hour would be used. Minimal outages were recorded over the 2 years: 5 h at the north wall, 56 h at the south wall, and 0 h at the roof location.

Belfast, Maine, USA (IECC Climate Zone 6A) is characterized by long, cold winters, with an average of 6230 annual heating degree days and 514 annual cooling degree days recorded from 1997–2024, using a base of 18.3 °C [32]. Cold temperatures were recorded throughout the monitoring period, with an average exterior temperature of 8.9 °C throughout the 2-year period.

Weather data, including both outdoor and indoor information, was recorded on a 15 min basis and averaged to obtain hourly measurements. Outages associated with weather data were more frequent than the measurements within the assemblies. Gaps in measured interior conditions were filled using the same approach as the assembly temperature and RH monitoring, by using the preceding time period. However, gaps in exterior conditions were filled using weather data from the closest available source, an airport approximately 39 km away (Knox County Regional Airport). Additionally, weather data were supplemented with solar radiation values from the National Renewable Energy Laboratory (NREL) National Solar Radiation Database (NSRDB) [33]. The weather data were used as the boundary conditions for the hygrothermal modeling.

2.3. Properties of Materials Used in Hygrothermal Simulations

Material properties were determined through a combination of manufacturer data sheets, laboratory evaluation, and materials that were available in the Fraunhofer IBP WUFI material database. The single condition thermal conductivity (21 °C, 55% RH), porosity, and water vapor permeability of WFI were empirically derived [10]. To support this work, further laboratory testing of WFI was conducted, including thermal conductivity (10 °C, oven-dry), moisture storage function, and water absorption coefficient. These were conducted using the following standards: ASTM C518, Standard [34] Test Method for Steady-State Thermal Transmission Properties by Means of the Heat Flow Apparatus; ASTM C1498 [35], Standard Test Method for Hygroscopic Sorption Isotherms of Building

Materials; and ASTM C1794 [36], Standard Test Methods for Determination of the Water Absorption Coefficient by Partial Immersion. Select material properties are presented in Table 2, and detailed material specifications are provided within the Supplementary Materials, Tables S1–S12.

Table 2. List of materials and selected properties.

Material	Density (kg/m^3)	Thermal Conductivity (W/mK)	Porosity (m^3/m^3)	Water Vapor Diffusion Resistance (-)	Specific Heat Capacity (J/kgK)
CLT [b]	426	0.1013	0.836	291	1237
WFI [c]	140	0.0346 [a]	0.905	2.4	2100 [e]
WRB [e]	210	2.3	0.001	85	2300
Siding [b]	426	0.1013	0.836	291	1237
ZIP System [d] (Roof Decking)	607	0.101	0.950	211	1880
ZIP System [d] (Roof Membrane)	130	2.3	0.001	2526	2300

[a] Laboratory measurement; [b] Glass et al., 2024 [22]; [c] Snow et al., 2024 [10]; [d] Fraunhofer IBP WUFI material database; [e] Manufacturer specification.

2.4. Hygrothermal Simulations and Data Analysis

Hygrothermal simulation was conducted for a two-year period using the one-dimensional WUFI Pro software (version 7.0). The WUFI software has been widely validated for assessment of building envelopes and materials [8,21,22] and is accessible and commonly used by both researchers and practitioners. The cross-section of the wall assembly was created in the software interface, with monitoring positions that matched the monitoring positions' locations in the case study building. Simulation outputs of temperature, RH, and MC at the selected positions were compared to measured data using the root mean square error (RMSE), Equation (1), and mean absolute error (MAE), Equation (2). The use of both metrics allowed for the determination of the magnitude of the errors in prediction, as RMSE is sensitive to large errors, and MAE is sensitive to small, continual errors [15,24,37]. These values were calculated for both the temperature and RH at each monitoring position, and MC where applicable. Scaled error (SE), Equation (3) (a-wall and b-roof), was also used to assess the goodness of fit across all measurements. This function was developed using the error associated with each measurement: temperature was measured to ±0.3 °C, and RH was measured to ±2%. This type of assessment has been conducted in previous work to assess the goodness of fit for hygrothermal models [32]. The RMSE was calculated at all positions and each measurement type, then divided by the representative error, and added, with equal weighting across all of the monitoring positions.

$$\mathrm{RMSE} = \sqrt{\frac{\sum_{i=1}^{N}(x_i - \hat{x}_i)^2}{N}} \quad (1)$$

$$\mathrm{MAE} = \frac{\sum_{i=1}^{N}|x_i - \hat{x}_i|}{N} \quad (2)$$

$$\begin{aligned}\text{Scaled Error}_{\text{wall}} &= \left(\frac{1}{6} * \frac{\mathrm{RMSE}_{T,P1}}{0.3}\right) + \left(\frac{1}{6} * \frac{\mathrm{RMSE}_{T,P2}}{0.3}\right) + \left(\frac{1}{6} * \frac{\mathrm{RMSE}_{T,P3}}{0.3}\right) + \left(\frac{1}{6} * \frac{\mathrm{RMSE}_{RH,P1}}{2}\right) + \left(\frac{1}{6} * \frac{\mathrm{RMSE}_{RH,P2}}{2}\right) \\ &+ \left(\frac{1}{6} * \frac{\mathrm{RMSE}_{RH,P3}}{2}\right)\end{aligned} \quad (3a)$$

$$
\begin{aligned}
\text{Scaled Error}_{\text{roof}} = & \left(\frac{1}{11} * \frac{\text{RMSE}_{\text{T,R1}}}{0.3}\right) + \left(\frac{1}{11} * \frac{\text{RMSE}_{\text{T,R2}}}{0.3}\right) + \left(\frac{1}{11} * \frac{\text{RMSE}_{\text{T,R3}}}{0.3}\right) + \left(\frac{1}{11} * \frac{\text{RMSE}_{\text{T,R4}}}{0.3}\right) + \left(\frac{1}{11} * \frac{\text{RMSE}_{\text{RH,R1}}}{2}\right) \\
& + \left(\frac{1}{11} * \frac{\text{RMSE}_{\text{RH,R2}}}{2}\right) + \left(\frac{1}{11} * \frac{\text{RMSE}_{\text{RH,R3}}}{2}\right) + \left(\frac{1}{11} * \frac{\text{RMSE}_{\text{RH,R4}}}{2}\right) + \left(\frac{1}{11} * \frac{\text{RMSE}_{\text{MC,Int}}}{1.5}\right) \\
& + \left(\frac{1}{11} * \frac{\text{RMSE}_{\text{MC,Mid}}}{1.5}\right) + \left(\frac{1}{11} * \frac{\text{RMSE}_{\text{MC,Ext}}}{1.5}\right)
\end{aligned}
\quad (3b)
$$

A second model was developed in order to perform a mold growth and durability assessment. This simulation was run for a 5-year period and utilized the calibrated material properties from the initial model. Weather conditions were used from typical meteorological year (TMY) data from Belfast, ME. The climate year was selected in accordance with ASHRAE 160, *Criteria for Moisture-Control Design Analysis in Buildings*.

2.5. Mold Growth Assessment

Mold growth assessment was performed using the WUFI Mold Index VTT software (Version 2.3), in accordance with ASHRAE 160, *Criteria for Moisture-Control Design Analysis in Buildings*. The model was developed from experimental observations, which developed the calculation of a mold index using temperature, RH, and material characteristics. The mold growth index is separated into three different levels: no or invisible mold growth (mold growth index less than 1), mold growth is visible only under a microscope (mold growth index between 1 and 2), and visible mold growth (mold growth index greater than 2). The mold growth assessment was conducted on both the measured and simulated datasets. The WFI was classified as a sensitive material and assigned a mold index decline coefficient of 0.1. The mold growth index was calculated at locations associated with wall and roof measurement positions, as well as the interior surface of the walls and roof.

3. Results

Measured and simulated results of the north wall monitoring position are presented below. A brief comparison of north and south monitored data is included below; however, all analyses of the south wall and roof monitoring and modeling are included within the Supplementary Materials, Figures S1–S9.

3.1. Monitored Data of Wall and Roof Assemblies

3.1.1. MC of CLT

MC was measured at locations throughout the building; data from the north and south walls, the north corner of the building, and the roof are presented in Figure 5. Recording from the middle lamella of the CLT is presented in the figure for the north and south walls. The north corner is representative of the center of the lamella, which is closest to the exterior of the building. As these sensors are pin-type, the measured data represent the maximum MC along the shank of the pin at a given monitoring location. The monitoring data presented for the roof include all three depths at the top left monitoring position of the roof (position closest to the peak of the roof, to be referred to as the north roof location, as shown in Figure 2). The North and South wall data were not included (dotted lines) for the months where sensor readings were unreliable due to the MC pin type. Limited differences in MC were evident between the monitoring locations and pin depths. In all cases, the weekly averaged MC fell in a narrow range, 7.5 to 10.2%. The highest recorded MC throughout the building was 13.4% in the south wall. MC of all panels was maintained below the 15–18% MC threshold, where fastener corrosion can occur [38,39]. Measured MC trends indicate annual seasonal changes, where the MC of the panels was lower during cooler months and higher during warmer months. Additionally, the MC trends indicate that the building envelope is drying towards an equilibrium, as the first year showed higher MC levels than the second year of monitoring.

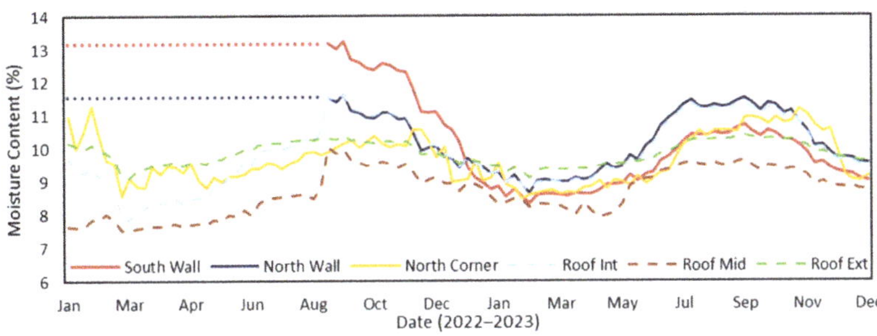

Figure 5. Weekly averaged MC of wall, roof, and corner. Dotted lines indicate time periods where data were not available.

3.1.2. Effects of Wall Orientation on Monitored Data of WFI

Measured temperature and RH of north and south wall sensors are presented in Figure 6. Distinct trends in monitoring data are evident. Conditions at the position closest to the exterior (P3) demonstrated greater seasonal variability than those at the positions closer to the interior (P2 and P1) for both temperature and RH. This is attributed to the thermal properties of the insulation and sensors being protected from extreme conditions as the depth of the insulation layer increases. Monitoring location P3 was protected from solar radiation and moisture; however, there was little to no thermal barrier. P2 and P1 were located under 120 and 240 mm of WFI, respectively, therefore reducing the temperature swings at interior positions. This difference in conditions under increasing insulation is represented through the difference between the maximum and minimum temperatures, which were 68, 37.7, and 23.6 °C at P3, P2, and P1, respectively.

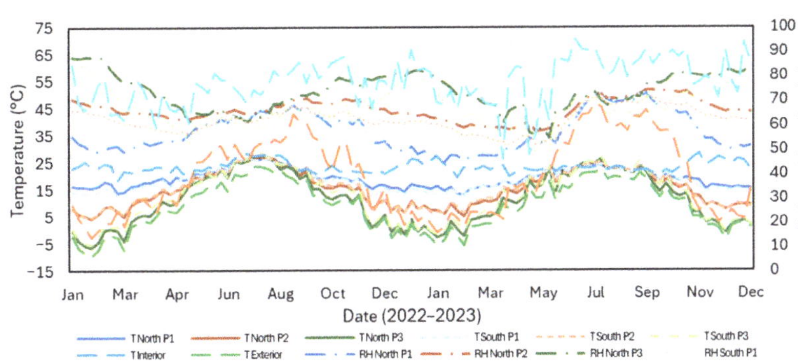

Figure 6. The north and south walls' weekly averaged temperature and relative humidity.

This trend did not hold true for RH, where fluctuation was greatest at P3 (45.9% RH), then P1 (33.2% RH), and smallest at P2 (22.1% RH). Measured RH recorded at P3 was closely related to temperature trends, which follow exterior climatic conditions of low RH during warm months and increased RH during cold months. Seasonal variability of RH at P3 was associated with the relative ability of air to hold moisture at a given temperature, and the increase in RH during cold periods is indicative of air being able to hold less moisture than during warm periods. This phenomenon can explain why RH stabilizes across all positions during warm periods, where the temperature at all positions is similar. Additionally, greater variability in RH at interior positions was observed and associated with changes in the interior conditions. The RH at P3 is more closely related to the interior conditions of the building than the exterior conditions. CLT acts as a vapor retarder, which slows the movement of moisture throughout the assembly; however, it does not stop movement. This allows for inward vapor drive. As there is no humidity control in the building, this means that air is heated, and low humidity conditions draw moisture into the building through the envelope.

Wall measurements indicate limited variability associated with orientation. The south-oriented wall experienced higher temperatures than the north wall at all positions except for P1. RH measurements demonstrated similar trends, where the south wall had lower relative humidities than the corresponding north wall monitoring positions. This was expected, as the south wall was exposed to greater levels of solar radiation than the north wall. However, this was not true for P1, where recordings in the south wall were lower for temperature and higher for RH than those of the north wall. This was associated with the greater effect of interior conditions at P1 than exterior conditions. The greater variation in RH than that of temperatures can be attributed to seasonal variability and the effects of temperature on RH. When interior and exterior temperatures were similar, there were fewer differences in the RH distribution. This occurred during the summer months, when temperatures at the interior and exterior of the building were similar. During cold months, when the building was heated, the differences in relative humidity were greater. This can be attributed to the temperature dependency of RH, when, as temperatures decrease, so does the ability of air to hold moisture. Additionally, there was no RH control within the building, and during the winter months, the building was heated, further exacerbating the differences in RH. Temperature and RH of the exterior were closely related to P3, which experienced greater seasonal variation than P2 and P1. Instances where rainfall occurred led to an increase in RH, leading to periods where exterior RH is upwards of 20% RH higher than at P3. Trends in interior and exterior temperature are more closely related to monitoring positions within the wall assembly.

3.1.3. Effects of Wall Height on Monitored Data

Monitored temperature and RH at P3 in three vertical positions (top, middle, and bottom) of the north wall are presented in Figure 7. Temperature and RH stratification were minimal across vertical positions. Temperature was consistent across the top and middle positions (average 0.4 °C difference), with a minor decrease at the bottom position (average 1.6 °C difference). Similarly, RH was consistent at the top and middle positions (average 1.5% RH difference), with an increase at the bottom position (average 8.6% RH difference). This can be attributed to ground proximity and contact. Soil conditions are cooler and are constant moisture sinks, which is reflected in the slight difference between the bottom monitoring position. Additionally, P3 sensors were installed under the building's WRB, which was directly underneath a 40 mm air gap, which acted as a rainscreen. The 40 mm air cavity was protected from insects using a mesh screen and was open to air and moisture exchange at the bottom, and an approximately 13 mm gap was left at the top of the wall between the cladding and fascia boards. This allowed air and water vapor to enter and exit the cavity and can explain the differences in the measured data.

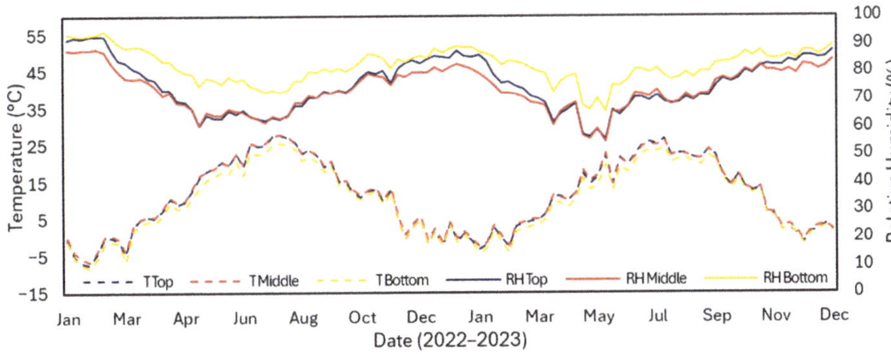

Figure 7. The north wall's weekly averaged temperature and relative humidity at three vertical monitoring clusters.

3.1.4. Effects of Roof Slope on Monitored Data of Roof Assembly

The monitored temperature and RH at two roof locations (Figure 2, left) and all depths (R1 through R4) are presented in Figure 8. Trends similar to the wall monitoring data are evident, where the vertical location had a limited impact on the measured temperature and RH. Temperature was consistent across the monitoring positions, with the average difference at R1 being 1 °C, R2 being 0.74 °C, R3 being 0.21 °C, and R4 being 0.35 °C. This trend with low average differences was also evident in the monitored RH, with average differences being 4.3, 4.4, 2.9, and 0.1% RH at monitoring positions R1 through R4, respectively. Interior and exterior temperature and RH trends are similar to the wall monitoring data. Interior and exterior temperatures are closely related to the conditions within the roof assembly than RH.

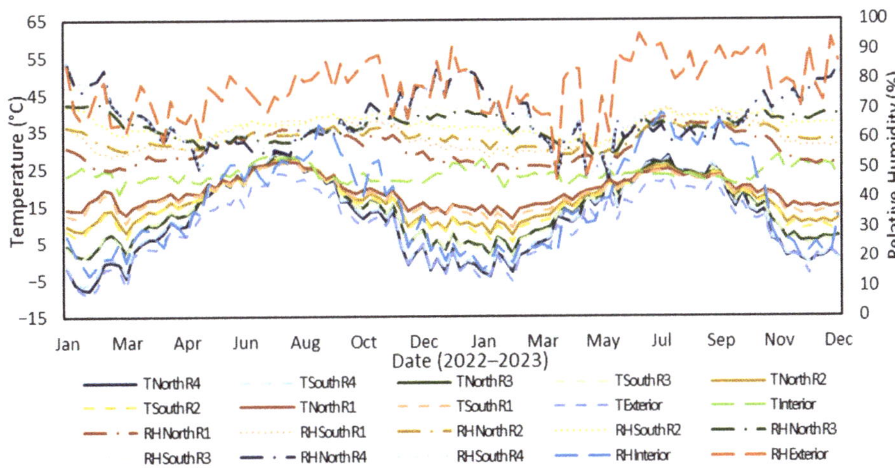

Figure 8. The roof's weekly averaged temperature and relative humidity at two locations on the slope of the roof.

3.2. *Hygrothermal Model Development and Calibration*

The north wall's hygrothermal behavior was simulated using WUFI Pro software (Version 7.0). The results of MC in CLT and temperature and RH in WFI are reported separately. An initial model was first developed, and calibration was conducted to improve the model accuracy in predicting the temperature and RH in WFI.

3.2.1. Hygrothermal Model Development—MC in CLT

Measured MC at wall locations was not captured throughout the entire two-year period and was therefore not included in the SE calculation or calibration considerations. However, a comparison of the available data is presented in Figure 9. The simulation period began on 1 January 2021, and the measurement of MC was first available in August 2022. The RMSE and MAE values presented for MC use the shortened time period (25 August 2022–31 December 2023). The initial and calibrated models refer to different versions of the simulation. The initial model was simulated utilizing the measured material properties, and the calibrated model was simulated utilizing the results of the sensitivity analysis to adjust material properties and interior conditions. The initial model was able to represent trends in MC well, and the RMSE was 2.07% MC. Considering the low error rate and inability to accurately adjust initial parameters of the model in order to improve the error rate, this was excluded from the sensitivity analysis. MC was not considered in the SE, and therefore, limited improvement of the model is evident when considering MC.

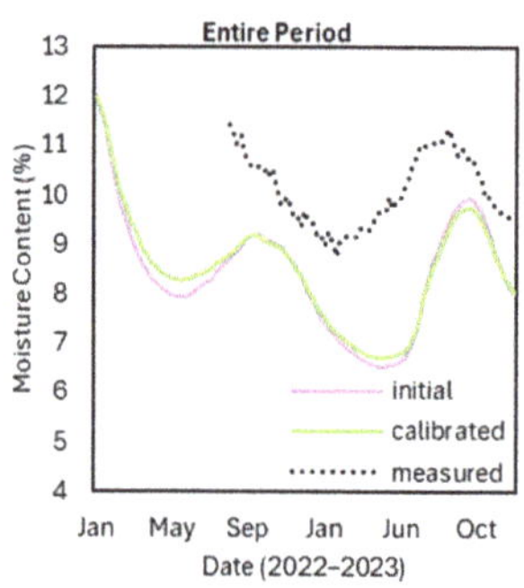

Figure 9. The north wall's P3 measured and simulated MC results.

3.2.2. Hygrothermal Model Development—Temperature and RH in WFI

Initial model results of the temperature and RH at P1, P2, and P3 are presented in Figures 9–12. Overall, good agreement was evident at all positions, with the simulated results following general patterns in measured data at all positions. Temperature predicted aligned more closely with measured data at all positions when compared to RH results. Temperature and RH at P3 were predicted with greater agreement than at P1. This can be attributed to large variability in exterior conditions and limited separation between P3 and exterior conditions, where small errors are evident, and there are a few instances of larger errors. This is followed by the greater separation between interior conditions and P1, which is located at the interface between CLT and WFI, and an overprediction in the temperature and underprediction of RH. As previously noted, increased temperatures cause reduced RH values, which is represented in the model, where greater disagreement between measured and simulated RH was evident when temperature was at greater disagreement.

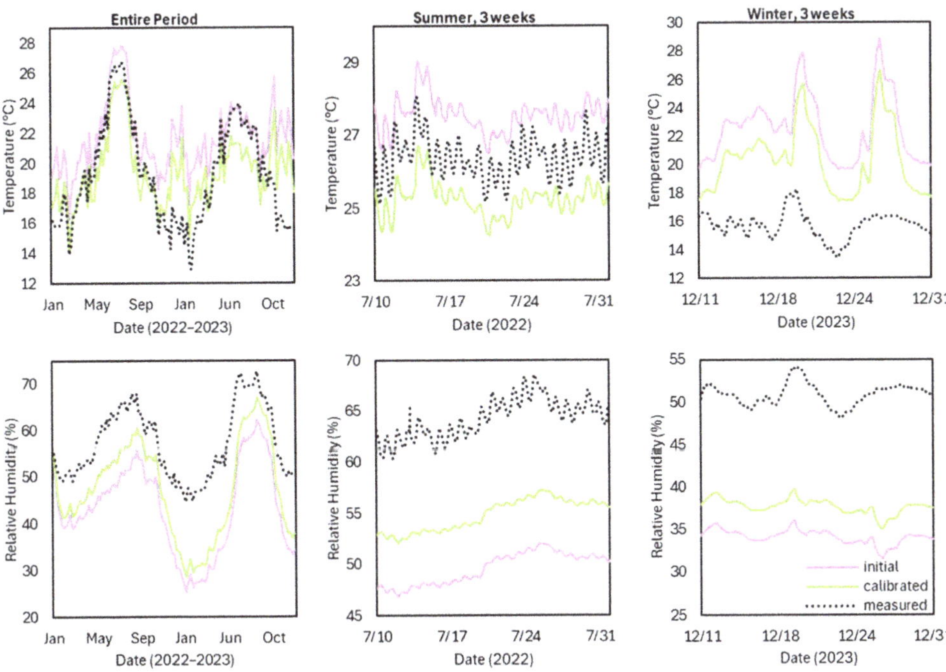

Figure 10. North wall P1's measured and simulated temperature and relative humidity results.

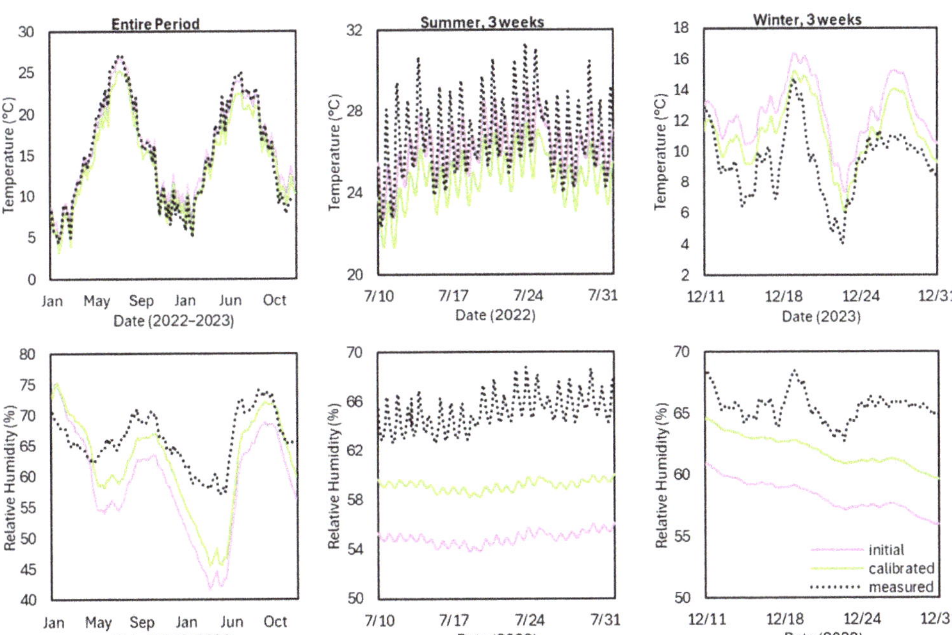

Figure 11. P2 north wall's measured and simulated temperature and relative humidity results.

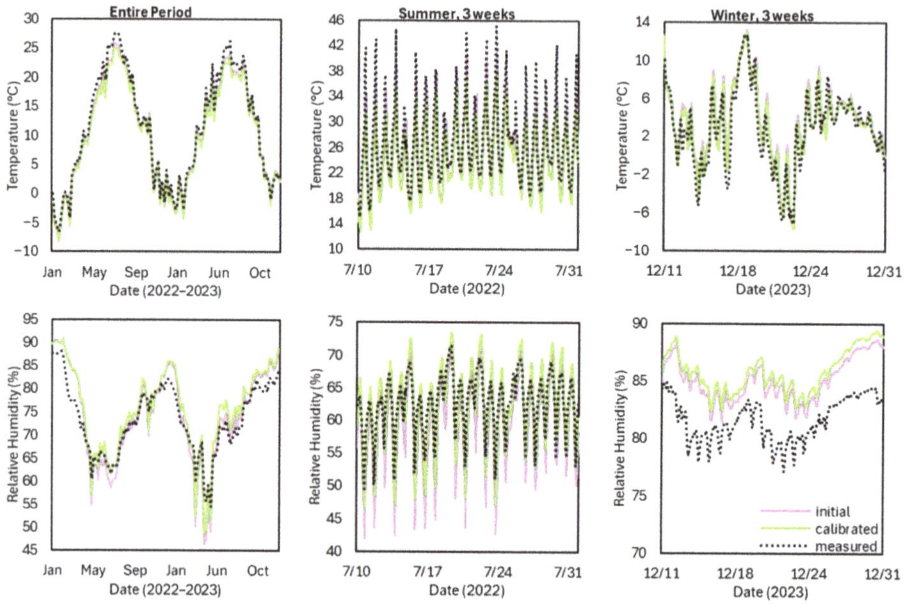

Figure 12. North wall P3's measured and simulated temperature and relative humidity results.

Observations were substantiated quantitatively through the use of RMSE, MAE, and SE, as shown in Table 3. Calculated RMSE of temperature and RH were greatest at P1, followed by P3, and then P2. This indicated that the greatest error was evident at P1, and P3 was influenced by the constant variability of exterior temperature, which led to errors. Summary of the errors between measured and simulation results align with the qualitative observations presented above. RMSE and MAE are similar at all locations; however, a comparison of these metrics at P3 demonstrated the key differences in what is represented. P3 RH RMSE was 4.50% RH, whereas MAE was 3.51% RH. This is indicative of errors associated with a few instances of greater variability, and not continual small errors. However, the discrepancy between calculated RMSE and MAE was small; therefore, only RMSE was selected to calibrate the model and included in the SE.

Table 3. North wall's summary of error between measured (M) and simulated (S) results.

Location	Metric	Data Type	Initial Model Ave	Initial Model RMSE	Initial Model MAE	Calibrated Model Ave.	Calibrated Model RMSE	Calibrated Model MAE
P1	Temperature (°C)	M S	19.2 21.7	3.38	2.56	19.2 19.5	2.32	1.84
	Relative Humidity (%)	M S	57.2 43.4	14.5	13.9	57.2 47.5	10.6	9.7
P2	Temperature (°C)	M S	15.3 15.8	1.68	1.31	15.3 14.4	1.90	1.51
	Relative Humidity (%)	M S	66.0 59.1	8.92	7.94	66.0 62.4	6.08	5.09
P3	Temperature (°C)	M S	11.4 10.4	3.18	2.41	11.4 14.4	3.23	2.39
	Relative Humidity (%)	M S	73.4 74.0	4.50	3.51	73.4 75.6	4.50	3.74
Scaled Error				6.91			5.90	

3.2.3. Model Calibration

Additionally, alternative methods were utilized to determine the source of error for calibrating the model, including adjustment of boundary conditions and sensitivity analysis (see detailed discussion in Section 4.1). A series of discussions with the building manager revealed that the two classrooms within the building were not occupied at the same rate. Interior conditions were measured using the temperature and RH sensor of the weather station, which was only representative of a single location in the room where it was placed. Further, the ceilings are sloped, with the ceiling height ranging from 3 to 4 m, which can lead to vertical temperature stratification. Therefore, the measured interior conditions at a single location were insufficient in representing the actual conditions near the wall surfaces at three monitoring positions with varying heights, resulting in a potential source of error. A series of simulations was run using the measured P1 and P3 as boundary conditions and only included the materials that are in between these monitoring positions (see Section 4.1). In order to address this temperature discrepancy effect, temperature shifts were conducted for the interior conditions, based on the P1 average temperatures. The differences between measured and simulated temperatures at P1 led to a difference and then a decrease of 2.5 °C. The absolute humidity was maintained the same during these shifts, which preserved the moisture profile of the assembly. The temperature shift is responsible for 51% of the error reduction at P1, or 15.4% of the 31.6% difference. The adjustment of interior conditions was then identified as a main source of error, and sensitivity analysis was also performed on the model to determine the effects of selected properties.

Another simulation was conducted using the transformed interior conditions in the calibrated model. The temperature and RH results of the calibrated model are presented in Figures 10–12. Additionally, different time periods are displayed to provide detailed information about the improvements made to the model. The left column is full of two years of data; the middle and right columns are three-week periods during the summer and winter. This format was adopted for all positions, P1, P2, and P3. Temperature and RH tracks well at all locations; however, minor discrepancies are still evident. During the winter, the temperature is consistently overpredicted. However, during the summer, it is underpredicted. The simulation was able to accurately capture trends at all three positions.

In Table 4, greater agreement was evident when considering the RMSE, MAE, and SE. Error at P1 decreased greatly when compared to initial results, with a RMSE of 2.32 °C. RMSE increased at P3 when compared to initial results; however, the magnitude of improvement at P1 led to an overall reduction in SE. Errors associated with P2 remained similar to those of the initial results. Overall, changes to the model were better able to predict the conditions within the wall assembly, and the model was considered sufficiently calibrated for the mold growth assessment. In addition, the acceptance criteria of hygrothermal simulation for field monitoring are discussed in Section 4.2.

Table 4. Summary of error between measured (M) and simulated (S) results using the calibrated model.

Assembly	Location	Metric	Averaged Results		Calibrated Model		
			M	S	RMSE	MAE	SE
North Wall	P1	T (°C)	19.2	19.5	2.32	1.84	5.90
		RH (%)	57.2	47.5	10.6	9.7	
	P2	T (°C)	15.3	14.4	1.90	1.51	
		RH (%)	66.0	62.4	6.08	5.09	
	P3	T (°C)	11.4	14.4	3.23	2.39	
		RH (%)	73.4	75.6	4.50	3.74	
	MC	MC (%)	10.1	8.20	2.07	4.26	

Table 4. *Cont.*

Assembly	Location	Metric	Averaged Results M	Averaged Results S	Calibrated Model RMSE	Calibrated Model MAE	SE
South Wall	P1	T (°C) RH (%)	18.5 58.0	18.9 48.2	3.09 10.7	2.54 9.7	8.28
	P2	T (°C) RH (%)	15.8 62.0	15.3 58.5	2.78 5.44	2.14 4.39	
	P3	T (°C) RH (%)	12.7 66.6	15.3 68.6	5.87 4.99	3.82 4.05	
	MC	MC (%)	11.2	8.39	2.99	4.88	
Roof Panel	R1	T (°C) RH (%)	18.6 56.7	18.7 57.8	3.72 4.6	3.70 3.9	5.93
	R2	T (°C) RH (%)	16.5 60.2	16.5 65.1	3.76 4.39	3.66 3.79	
	R3	T (°C) RH (%)	14.1 63.8	14.5 71.8	3.86 4.19	3.61 3.58	
	R4	T (°C) RH (%)	11.4 66.3	12.4 75.1	5.13 2.87	3.46 2.08	
	Int Mid Ext	MC (%) MC (%) MC (%)	9.9 8.7 9.9	7.21 8.90 10.2	1.30 0.99 1.13	1.10 0.81 0.92	

3.3. Mold Growth Assessment Using Calibrated Model

Mold growth assessment was conducted using two data sets: the monitoring data and a 5-year simulation. Simulation was conducted using the calibrated materials and a TMY weather file for Belfast, ME. The assessment included all monitoring locations (P1, P2, and P3) as well as the conditions at the interior surface of the CLT, which was in contact with the occupant. As stated in ASHRAE 160, different criteria are specified for surfaces that the occupant has direct contact with and those within the wall envelope. Passing criteria for interior surfaces requires the mold growth index to remain at zero throughout the entire period. Monitoring positions within the wall envelope have a greater tolerance but must not cross a mold growth index of 3 at any point during the time period. Results for the measured and simulated mold growth assessment are presented in Table 5.

Table 5. Mold growth assessment of the north wall, south wall, and roof panel.

Assembly	Position	Measured (2 Years) Maximum Mold Growth Index	Measured (2 Years) Pass/Fail	Simulation (5 Years) Maximum Mold Growth Index	Simulation (5 Years) Pass/Fail
North Wall	Interior Surface	-	-	0.00	Pass
	P1	0.00	Pass	0.00	Pass
	P2	0.00	Pass	0.00	Pass
	P3	0.06	Pass	0.56	Pass
South wall	Interior Surface	-	-	0.00	Pass
	P1	0.00	Pass	0.00	Pass
	P2	0.00	Pass	0.00	Pass
	P3	0.00	Pass	0.05	Pass
Roof	Interior Surface	-	-	0.00	Pass
	R1	0.00	Pass	0.00	Pass
	R2	0.00	Pass	0.00	Pass
	R3	0.00	Pass	0.00	Pass
	R4	0.03	Pass	0.92	Pass

Mold growth assessment thresholds were not crossed at any time during the measurement or simulation periods. A mold growth index above zero was recorded at P3 in both instances; however, neither instance caused failure due to not crossing the thresholds, and the short duration, where the index was above 0. The mold growth index above zero at P3 was associated with rapid changes in temperature and RH, where conditions for mold growth were met for short periods of time. Although the mold growth index at P3 was greater than 0 for both measured and simulated data, this correlates with mold growth occurring at levels that are not visible without a microscope. Increases in the mold growth index occurred cyclically during cold months, with a decreasing magnitude, where the greatest potential for mold growth occurred during the first winter and decreased as the assembly reached equilibrium. Additionally, the predicted mold growth index was greater when assessing the simulation data than the measured data. This can be attributed to the overestimation of RH at P3 in the simulation results, as well as the harsh weather conditions associated with the weather file used in the mold growth assessment.

4. Discussion

Monitored data were collected over a two-year period and demonstrated that there were minor differences between monitoring locations in the walls and roof. This led to the development of a hygrothermal model to assess the performance of the system. The RMSE between the modeled and measured data were reduced to 5.87 °C and 10.7% RH at the greatest across all models. Further, the SE in all models decreased. The largest improvement was evident in the roof model, where SE was reduced from 9.35 to 5.93. At this threshold, we considered the model to be calibrated. The developed model was considered effective in predicting hygrothermal performance and assessing mold growth using historical weather data. However, the authors did not find any existing standardized criteria for assessing model effectiveness. Further discussion is given to the sensitivity analysis of model inputs carried out by the authors and the criteria for model accuracy derived from the literature, aiming to provide stronger evidence of the model's effectiveness and clarify the criteria applicable for field studies.

4.1. Sensitivity Analysis of Model Inputs

The hygrothermal model inputs can be categorized as (1) material properties of the components in wall and roof assemblies, e.g., permeability of WFI and WRB, and (2) boundary conditions, i.e., interior temperature and exterior solar absorption. Sensitivity analyses can reveal how changes in input variables affect a model's output accuracy. In this study, the sensitivity analysis was performed to determine the effects of material selection (where other similar materials were available within the Fraunhofer IBP WUFI database) and weather data, including solar absorption and permeability, as shown in Figure 13. This yielded minimal changes in simulation results; however, changes to the exterior solar absorption and permeability of the exterior and interior surfaces were accepted and included in the final model. In most cases, the changes to material properties led to greater error within the model; however, two changes were accepted: an increase in the air change rate of the gap behind the siding and a reduction in the solar absorptivity of the siding. The model was of greatest sensitivity to the air change rate within the air cavity, which affected the direction of moisture movement. Results of the sensitivity analysis are included within the Supplementary Materials.

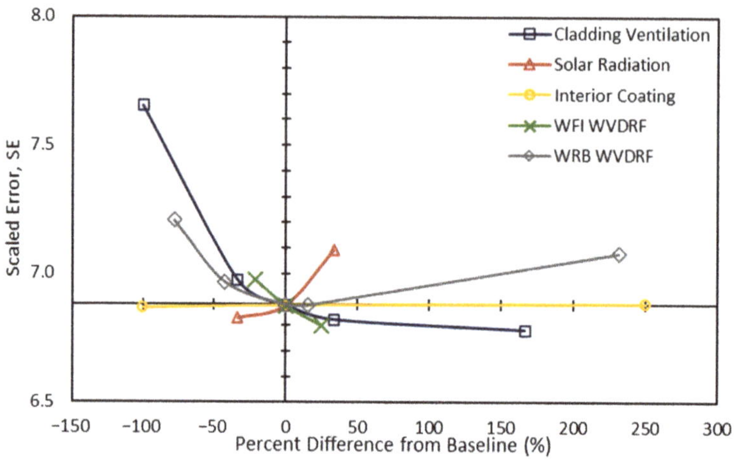

Figure 13. Sensitivity analysis results of the north wall, with the x-axis set to the scaled error of the initial model (SE North, 6.91).

4.1.1. Effects of Material Properties on Simulation of T, RH, and MC

The hygrothermal simulation tools use basic material properties to predict heat, air, and moisture transfer. Although these methods have been validated, consideration of the material properties and understanding of the shortcomings of these tools needs to be considered. First, the models are based on the assumptions in Fick's law to model vapor transport and Darcy's law to model liquid transport under a vapor pressure difference or MC gradient, as well as coefficients that describe the material's permeability. This can create discrepancies when considering porous materials, especially at high RH and transient conditions [28–30]. Researchers have found that sorption is reduced at higher RH, which can lead to overestimation of moisture in wood when not modeled with a dynamic sorption curve. Further, this can impact the influence of predicted RH, as variability in the changes of vapor pressure can also affect the behaviors. The limitations within the assumptions of the modeling tools can explain errors associated with RH during periods with high RH and cold temperatures.

The moisture storage function is one of the most influential material properties to classify. In wood and wood-based materials, the relationships between the equilibrium MC and RH are often referred to as the sorption and desorption isotherms. Wood materials exhibit what is called the hysteresis effect, where the desorption isotherm is greater than that of the adsorption isotherm [40]. Current hygrothermal modeling tools do not consider the effects of hysteresis [41]. In an effort to capture this phenomenon, an averaged isotherm was used in the simulation. Furthermore, the effect of the hysteresis loop was considered in the sensitivity analysis; however, differences between the averaged, adsorption, and desorption isotherms in WFI were found to be negligible on the model results. This aligns with previous works assessing the hygrothermal behavior of wood fiber insulation, where the differences between EMC during adsorption and desorption ranged from 0.85% to 1% MC at 10% and 97% RH, respectively [12]. When implemented, differences of approximately 1% MC have a limited impact on the model.

Although there were shortcomings associated with the modeling tools and ability to represent the building envelope accurately, there were key trends that can help to aid design and consideration when using WFI. Previous works have demonstrated that vapor open assemblies are influenced by interior conditions more than they are influenced by exterior conditions. This is evident through the requirement to shift interior conditions to increase the accuracy of the model. Freimanis et al. found that systems that were vapor open, with a low vapor diffusion resistance, align their RH behavior with temperature profiles. Vapor-tight systems exhibit reduced RH amplitudes as vapor diffusion resistance decreases [23]. This same study found that when both indoor RH and outdoor temperature oscillate, vapor-open systems with low vapor diffusion resistance resemble indoor humidity profiles more closely. Additionally, investigation of cellulose fiber insulation demonstrated that the interior conditions could impact the conditions within the wall in vapor open assemblies, as there is the potential to dry out both the interior and exterior [42]. Connections between the interior conditions of the building and P1 in the wall are evident in the monitoring data. P1 temperature and RH were influenced by the interior conditions more than the positions to the exterior (P2 and P3). This was also evident during the modeling process, where shifts in the interior conditions translated to the greatest source of error across the monitoring positions.

4.1.2. Effects of Boundary Conditions on Simulation of T, RH, and MC

Interior boundary conditions were found to have the greatest impact on improving the accuracy of the model; however, adjustment of solar radiation was necessary in all cases. A solar absorptivity of 0.6, which is representative of grey paint, was selected for the siding material. Additionally, the black metal roof was assumed to have an absorptivity of 0.8. In all cases, reducing the solar absorptivity increased the accuracy of the model. This demonstrates that the exterior boundary conditions can impact the results of the model and have minor implications on the performance of the building. The small differences can be related to the thermal properties of the envelope, which are not sensitive to small changes in climatic conditions. Many of the effects of the boundary conditions are associated with the local climate, where cold temperature extremes occur during the winter and warm temperatures are recorded in the summer. This led to similar trends throughout the two-year period. Temperatures within the assembly are similar at all positions during the summer months, as the temperature difference between the interior and exterior conditions was small. Conversely, large temperature differences between interior and exterior conditions led to a large temperature gradient throughout the wall assembly. This is also mirrored within the RH data.

The high thermal resistance of the envelope can also lead to differences in the hygric performance in response to exterior conditions. The model was able to predict general trends related to MC in the assemblies; however, the assumption of instantaneous equilibrium can explain the discrepancy between monitored and measured results. Wood and wood-based materials can take longer time horizons to reach equilibrium, often weeks or months. This is relevant for the exterior MC monitoring positions, which are in direct contact with the WFI. This position saw the largest discrepancy between measured and simulated results. The exterior MC monitoring position was not subjected to large variations in climate, which led to minimal changes in MC.

4.2. *Acceptance Criteria of Hygrothermal Simulation*

Acceptance criteria for determining goodness of fit between measured and simulation data were assessed both qualitatively and quantitatively. Previous research has used acceptance criteria that varied greatly, with both qualitative and quantitative comparisons made, which are summarized in Table 6. None of the researchers explicitly stated their acceptance criteria; however, a comparison of the results was described by most studies. In instances where qualitative statements of good agreement were the only indicator of the ability of the model to predict conditions, no quantitative metrics were provided, exclusive of graphic evidence [43–46]. In the available literature, RMSE and/or MAE metrics were used the most often and were included in 15 of the papers [15,24,32,37,47–55]. The measured values corresponded with temperature, RH, or MC and depended on the scope of the paper. Wide ranges in acceptance criteria were found in RMSE from 0.11–9.27 °C [3,15,24,32,37,47–50,52,53], 0.97–16.7% RH [15,24,32,37,47–50,52,53], and 0.1–4.7% MC [32,51,53–55]. This demonstrates a wide range of criteria to establish what is considered acceptable. Further, it can be seen that when quantitative measurement was used to determine goodness of fit, there was a range, which often depended on the conditions used for measurement and modeling. Under steady state or laboratory conditions, differences between measured and modeled results were lower than those taken under real-world conditions. Laboratory or steady state conditions were seen in [32,47,48,50,51,54,55], where RMSE ranged from 0.11–1.53 °C, 0.97–8.68% RH, and 0.83–1.6% MC. Studies which were collected under real-world conditions were seen in papers [15,24,37,49,53,56], where RMSE ranged from 0.2–9.27 °C, 1–16.7% RH, and 0.1–0.5% MC. Limited variation was seen in MC in building settings due to the limited use in studies, where only one building-level study included the MC in its analysis [53]. Additionally, the use of differences (percent, maximum, and average) was used in [25,55–59]; however, the basis of measurement was not often discussed. Some papers used the maximum difference in percentage, and others used the average; however, no cohesive measurement was described. Thus, there were no formal acceptance criteria for this study; the MAE and RMSE were used to determine the goodness of fit, and sensitivity analysis was conducted to reduce any error associated with the model.

Table 6. Summary of model and measurement error from literature [15,24,32,37,47–59].

Measurement	Error Type		All	Laboratory	Building
Temperature (°C)	MAE	Min.	0.2	0.2	0.24
		Max.	2.01	0.3	2.01
	RMSE	Min.	0.11	0.11	0.29
		Max.	9.27	1.53	9.27
	Other [a]	Min.	1.2	1.2	2
		Max.	5	5	3
Relative Humidity (%)	MAE	Min.	0.89	3.38	0.89
		Max.	16.5	8	16.5
	RMSE	Min.	0.97	0.97	1
		Max.	16.7	9.3	16.7
	Other [a]	Min.	5	5	5
		Max.	30	30	15
Moisture Content (%)	MAE	Min.	-	-	-
		Max.	-	-	-
	RMSE	Min.	0.1	1	0.1
		Max.	4.7	4.7	0.5
	Other	Min.	-	-	-
		Max.	-	-	-

[a] Includes difference, percent difference, and percent error.

A review of literature demonstrated that the type of experiment, laboratory, or building, led to differences in the error within the model results. Laboratory experiments utilize well-defined and controlled boundary conditions, and RMSE below 1.53 °C, 9.3% RH, and 4.7% MC, whereas building scale experiments recorded RMSE below 2.01 °C, 16.7% RH, and 0.5% MC. In all cases, excluding MC, the error that was considered allowable was less in laboratory experiments when compared to building-scale experiments.

The parameters of the model were altered in an effort to evaluate the discrepancy in results and determine the ability of the model to predict conditions within WFI. To do so, P1 and P3 monitoring data were used as boundary conditions, with P2 simulated data being compared to monitored data. The model was conducted over the same time period; however, the CLT, air gap, and siding layers were not included. Additionally, wind-driven rain and solar radiation were not considered on the exterior surface. Results demonstrated good agreement with measured data, indicating errors were not associated with the material properties of WFI. The results of the model are included within the Supplementary Materials, as shown in Figures S10–S19. It should be noted that this assessment did not account for potential sources of error associated with material interaction within the wall envelope.

These models demonstrated better agreement between measured and monitored data. The RMSE at 7 of the 11 positions for temperature and 11 of 11 positions for RH was below the laboratory thresholds listed in the literature. Considering this, it was determined that the model was able to accurately predict the conditions within the assembly and represent the materials.

4.3. *Effects of Simulation of RH Discrepancy on Mold Growth Assessment*

Although the model accurately predicts the temperature and RH profile of the wall assembly, discrepancies between measured and modeled results were evident. These discrepancies, especially in RH, carry into the mold growth assessment, leading to an overprediction in the likelihood of mold growth occurring. In this case, overprediction of RH increases the likelihood of mold growth in the assessment, which would lead to a conservative prediction.

There are periods in the simulations in which the RH was both over- and underpredicted. The RH at P1 was generally underpredicted; however, the measured RH remained below 55% RH for the entire period. This low RH indicates that there was no potential for mold growth throughout the measurement period, and there is little concern with underprediction. The model overpredicted the RH at P3. This trend was beneficial for the mold growth assessment, as the WFI and WRB interface was the greatest likelihood for mold growth. This means that the risk associated with mold growth occurring at this position was overpredicted, leading to a conservative mold growth index prediction.

5. Conclusions

A hygrothermal assessment of building envelopes and materials can identify risks associated with use. A one-story building constructed of CLT and WFI was instrumented with temperature, RH, and MC sensors and monitored over a two-year period. A hygrothermal model was calibrated using measured material properties and monitoring data. Several conclusions can be drawn from this assessment:

1. The hygrothermal model was able to sufficiently represent the hygroscopic materials;
2. Under conditions within the study, mold growth assessment demonstrated no risk of degradation or attack;
3. Vapor open wall assemblies are greatly influenced by the interior conditions of the building;
4. Ventilation within an air cavity behind siding can mitigate risks of mold growth at the exterior surface of the insulation layer;
5. WFI and CLT show no moisture or mold risk in the cold climate of Belfast, ME;
6. A material property dataset of WFI was developed and is included in the Supplementary Materials to act as an aid for future design considerations and research efforts.

Overall, the simulated and monitored data showed that in a cold North American climate, natural materials demonstrate good hygrothermal performance. Future work will include supplementing monitoring equipment within the building to reduce gaps in measured interior conditions. Monitoring of the building will continue to assess performance and long-term durability. The wall assembly considered within this study is representative of a small section of building typologies in North American climates, and further research should be conducted to assess the performance of additional types of WFI, as well as their integration with other structural elements.

Supplementary Materials: The following supporting information can be downloaded at: https://www.mdpi.com/article/10.3390/su17177879/s1. This document includes the full set of material properties used in the hygrothermal model, monitoring results of the roof system, hygrothermal model results from the south wall and roof, and mold growth assessment of the south wall and roof.

Author Contributions: L.O.: Writing—original draft, data acquisition, processing, and analysis; writing—review and editing, investigation, and conceptualization. L.L.: Writing—review and editing, supervision, project administration, and funding acquisition. B.H.: Writing—review and editing, project administration, and funding acquisition. J.S.: Writing—review and editing. W.A.F.: Writing—review and editing and supervision. All authors have read and agreed to the published version of the manuscript.

Funding: This material is based upon work supported by the Department of Agriculture, Agricultural Research Service, under agreement No. 58–0204-1–180. Any opinions, findings, conclusions, or recommendations expressed in this publication are those of the authors and do not necessarily reflect the view of the Department of Agriculture. This material is based upon work supported by the U.S. Department of Energy's Office of Energy Efficiency and Renewable Energy (EERE) under the Building Technology Office (BTO), Award Number DE-EE0010137. This report was prepared as an account of work sponsored by an agency of the United States Government. Neither the United States Government nor any agency thereof, nor any of their employees, makes any warranty, express or implied, or assumes any legal liability or responsibility for the accuracy, completeness, or usefulness of any information, apparatus, product, or process disclosed, or represents that its use would not infringe privately owned rights. Any reference herein to any specific commercial product, process, or service by trade name, trademark, manufacturer, or otherwise does not necessarily constitute or imply its endorsement, recommendation, or favoring by the United States Government or any agency thereof. The views and opinions of the authors expressed herein do not necessarily state or reflect those of the United States Government or any agency thereof.

Institutional Review Board Statement: Not applicable.

Informed Consent Statement: Not applicable.

Data Availability Statement: Data used in this study are available within Supplementary Documents; further information may be available upon request.

Acknowledgments: The authors would like to acknowledge OPAL and NotchSB for enabling and helping with the instrumentation of the building. We would also like to acknowledge the Cornerspring Montessori School for allowing data collection to occur, as well as providing continued access for maintenance and data collection. Additionally, we would like to acknowledge the help and expertise provided by Samuel Glass and Natalia Farkas at the USDA Forest Products Laboratory, Building and Fire Sciences Group during the hygrothermal modeling process. This article is a revised and expanded version of paper [60], which was presented at the 14th World Conference on Timber Engineering 2025, Brisbane, Australia, 22–26 June 2025.

Conflicts of Interest: The authors declare no conflicts of interest.

Abbreviations

The following abbreviations are used in this manuscript:

C	Celsius
CLT	Cross-laminated timber
EPS	Expanded polystyrene
M	Measured
MAE	Mean absolute error
MC	Moisture content
ME	Maine
NREL	National Renewable Energy Laboratory
NSRDB	National Solar Radiation Database
RH	Relative humidity
RMSE	Root-mean square error
S	Simulated
SE	Scaled error
TMY	Typical meteorological year
WFI	Wood fiber insulation
WRB	Water-resistive barrier
WVDRF	Water vapor diffusion resistance factor
XPS	Extruded polystyrene

References

1. Krzosek, S.; Kłosińska, T. CLT—Material for the measure of the future. *Ann. Wars. Univ. Life Sci.* **2021**, *114*, 76–85. [CrossRef]
2. Mohammad, M.; Gagon, S.; Douglas, B.K.; Podesto, L. Introduction to cross laminated timber. *Wood Des. Focus.* **2012**, *22*, 3–12.
3. Pei, S.; Rammer, D.; Popovski, M.; Williamson, T.; Line, P.; van de Lindt, J.W. An overview of CLT research and implementation in North America. In Proceedings of the World Conference on Timber Engineering, Vienna, Austria, 22–25 August 2016.
4. WoodWorks. Available online: https://www.woodworks.org/resources/mapping-mass-timber/ (accessed on 25 April 2025).
5. Kurzinski, S.; Crovella, P.; Kremer, P. Overview of cross-laminated timber (CLT) and timber structures standards across the world. *Mass Timber Constr. J.* **2022**, *5*, 1–13.
6. Bozsaky, D. Nature based thermal insulation materials from renewable resources-a state-of-the-art review. *Slovak J. Civ. Eng.* **2019**, *27*, 52–59. [CrossRef]
7. Winkler, A. *Farostlemezek (Fiberboards)*; Mezőgazdasági Szaktudás Kiadó: Villány, Hungary, 1999.
8. O'Brien, L.; Li, L.; Friess, W.; Snow, J.; Herzog, B.; O'Neill, S. Review of bio-based wood fiber insulation for building envelopes: Characteristics and performance assessment. *Energy Build.* **2025**, *328*, 115114. [CrossRef]
9. Veitmans, K.; Grinfelds, U. Wood fiber insulation material. In Proceedings of the Research for Rural Development, Jelgava, Latvia, 18–20 May 2016.
10. Snow, J.; Herzog, B.; O'Brien, L.; Li, L. Characterization of wood fiber insulation for the development of wood fiber-insulated panels (WIPs) for use in building envelope. *BioResources* **2024**, *19*, 6142–6159. [CrossRef]
11. Asdrubali, F.; Evangelisti, L.; Guattari, C.; Roncone, M.; Milone, D. Experimental analysis of the thermal performance of wood fiber insulating panels. *Sustainability* **2023**, *15*, 1963. [CrossRef]
12. Asli, M.; Sassine, E.; Brachelet, F.; Antczak, E. Hygrothermal behavior of wood fiber insulation, numerical and experimental approach. *Heat Mass Transfer.* **2021**, *57*, 1069–1085. [CrossRef]
13. Callegher, C.Z.; Grazieschi, G.; Wilczynski, E.; Oberegger, U.F.; Pezzutto, S. Assessment of building materials in the European residential building stock: An analysis at EU27 level. *Sustainability* **2023**, *15*, 8840. [CrossRef]
14. International Code Council. Types of Construction. In *2024 International Building Code*; International Code Council: Country Club Hills, IL, USA, 2023.
15. Bastien, D.; Winther-Gaasvig, M.; Andersson, J.Z.; Xiao, Z.; Ge, H. Hygrothermal performance of natural building materials: Simulations and field monitoring of a case study home made of wood fiber insulation and clay. *J. Build. Phys.* **2023**, *47*, 249–283. [CrossRef]
16. Rahim, M.; Djedjig, R.; Wu, D.; Bennacer, R.; EL Ganaoui, M. Experimental investigation of hygrothermal behavior of wooden-frame house under real climate conditions. *Energy Built Environ.* **2023**, *4*, 122–129. [CrossRef]
17. VanDemark, L.; Clevenger, C.; Click, M. Building envelope issues within construction-defect litigation. *J. Leg. Aff. Disput. Resolut. Eng. Constr.* **2021**, *13*, 03721003. [CrossRef]

18. Zhang, X.; Sahlberg, B.; Wieslander, G.; Janson, C.; Gislason, T.; Norback, D. Dampness and moulds in workplace building Associations with incidence and remission of sick building syndrome (SBS) and biomarkers of inflammation in 10 year follow-u study. *Sci. Total Environ.* **2012**, *430*, 75–81. [CrossRef] [PubMed]
19. Rafidiarison, H.; Remond, R.; Mougel, E. Dataset for validating 1-D heat and mass transfer models within building walls wit hygroscopic materials. *Build. Environ.* **2015**, *89*, 356–368. [CrossRef]
20. Karacabeyli, E.; Douglas, B. *CLT Handbook: Cross-Laminated Timber*, U.S. ed.; FPInnovations and Binational Softwood Lumbe Council: Point-Claire, QC, Canada, 2013.
21. Karagiozis, A.; Künzel, H.; Holm, A. WUFI-ORNL/IBP-a North American hygrothermal model. In Proceedings of the Perfo mance of Exterior Envelopes of Whole Buildings VIII, Clearwater Beach, FL, USA, 2–7 December 2001.
22. Glass, S.V.; Boardman, C.R.; Farkas, N.; Lazarcik, E.Q.D.; Zelinka, S.L. Improved engineering model for water absorption i softwoods. *Sci. Technol. Built Environ.* **2024**, *30*, 694–708. [CrossRef]
23. Freimanis, R.; Vanaga, R.; Balodis, V.; Zundans, Z.; Blumberga, A. Hygrothermal assessment of insulation systems for interna insulation of solid masonry walls under various conditions. *Buildings* **2023**, *13*, 2511. [CrossRef]
24. Pianco, S.; Larcher, M.; Marincioni, V.; Troi, A.; Baglivo, C.; Congedo, P.M. Identifying key parameters through a sensitivit analysis for realistic hygrothermal simulations at wall level supported by monitored data. *Build. Environ.* **2023**, *229*, 10996 [CrossRef]
25. Simo-Tagne, M.; Remond, R.; Kharchi, R.; Bennammoun, L.; Ndukwu, M.C.; Rogaume, Y. Modeling, numerical simulation an validation of the hygrothermal transfer through a wooden building wall in Nancy, France. *Therm. Sci. Eng. Prog.* **2021**, *22*, 10080 [CrossRef]
26. Pianco, S.; Larcher, M.; Avellanosa, D.H.; Baglivo, C.; Troi, A.; Congedo, P.M. Hygrothermal simulation challenges: Assessin boundary condition choices in retrofitting historic European buildings. *Energy Build.* **2023**, *297*, 113464. [CrossRef]
27. Berher, J.; Busser, T.; Reddy, S.; Dulikravich, G.S. Evaluation of the reliability of a heat and mass transfer model in hygroscopi material. *Int. J. Heat Mass Transfer.* **2019**, *142*, 118258. [CrossRef]
28. Håkansson, H. Retarded Sorption in Wood: Experimental Study, Analyses and Modeling. Ph.D. Thesis, Lund University, Lund Sweden, 25 May 1998.
29. Peuhkuri, R. Moisture Dynamics in Building Envelopes. Ph.D. Thesis, Technical University of Denmark, Kongens Lyngby Denmark, October 2003.
30. Wang, J.; Mukhopadhyaya, P.; Morris, P.I. Sorption and capillary condensation in wood and the MC of red pine. *J. Build. Phys* **2012**, *37*, 327–347. [CrossRef]
31. Boardman, C.R.; Glass, S.V. Improving the accuracy of a hygrothermal model for wood-frame walls: A cold-climate stud *Buildings* **2020**, *10*, 236. [CrossRef]
32. U.S. Energy Information Administration (EIA). *U.S. Regional Energy Data: Short-Term Energy Outlook Data Browser*; U.S. Energy Information Administration (EIA): Washington, DC, USA, 2025.
33. Sengupta, M.; Xie, Y.; Lopez, A.; Habte, A.; Maclaurin, G.; Shelby, J. The national solar radiation data base (NSRDB). *Renew Sustain. Energy Rev.* **2018**, *89*, 51–60. [CrossRef]
34. *ASTM C518-21*; Standard Test Method for Steady-State Thermal Transmission Properties by Means of the Heat Flow Apparatus ASTM International: West Conshohocken, PA, USA, 2021.
35. *ASTM C1498-04a*; Standard Test Method for Hygroscopic Sorption Isotherms of Building Materials. ASTM International: West Conshohocken, PA, USA, 2023.
36. *ASTM C1794-19*; Standard Test Methods for Determination of the Water Absorption Coefficient by Partial Immersion. ASTM International: West Conshohocken, PA, USA, 2019.
37. Costa-Carrapiço, I.; Croxford, B.; Raslan, R.; González, J.N. Hygrothermal calibration and validation of vernacular dwellings: A genetic algorithm-based optimisation methodology. *J. Build. Eng.* **2022**, *55*, 104717. [CrossRef]
38. Dennis, J.K.; Zou, C.; Short, N.R. Corrosion behavior of zinc and zinc alloy coated steel preservative treated timber. *Trans. Inst Met. Finish.* **1995**, *75*, 96–101. [CrossRef]
39. Short, N.R.; Dennis, J.K. Corrosion resistance of zinc-alloy coated steel in construction industry environments. *Trans. Inst. Met Finish.* **1997**, *75*, 47–52. [CrossRef]
40. Glass, S.V.; Zelinka, S.L. Moisture relations and physical properties of wood. In *Wood Handbook-Wood as an Engineering Material* United States Forest Products Laboratory: Madison, WI, USA, 2021; pp. 95–116.
41. Fredriksson, M.; Thybring, E.E. On sorption hysteresis in wood: Separating hysteresis in cell wall water and capillary water in the full moisture range. *PLoS ONE* **2019**, *14*, e0225111. [CrossRef] [PubMed]
42. Ojanen, T.; Laaksonen, J. Hygrothermal performance benefits of cellulose fibre thermal insulation structures. In Proceedings of the 41st IAHS World Congress: Sustainability and Innovation for the Future, Albuferia, Portugal, 13–16 September 2016.
43. Alev, Ü.; Uus, A.; Teder, M.; Miljan, M.; Kalamees, T. Air leakage and hygrothermal performance of an internally insulated log house. In Proceedings of the 10th Nordic Symposium on Building Physics, Lund, Sweden, 15–19 June 2014.

4. Andersen, H.V.; Morelli, M.; Hansen, T.K.; Peuhkuri, R. In-situ characterization of the relative humidity at the surface of building materials. *J. Phys. Conf. Ser.* **2023**, *2654*, 012043. [CrossRef]
5. Birjukovs, M.; Apine, I.; Jakovics, A. Establishing material hygrothermal characteristics via long-term monitoring and best-fit numerical models. In Proceedings of the 12th Nordic Symposium on Building Physics, Tallinn, Estonia, 6–9 September 2020.
6. Latz, S.; Scholzen, F.; Thewes, A.; Maas, S. Comparison of high-performance and conventional internal insulation materials based on hygrothermal analysis using in situ measurements and simulation. *Bauphysik* **2022**, *44*, 9–20. [CrossRef]
7. Boudreaux, P.R.; Salonvaara, M.H.; Florian, A.; Desjarlais, A.O. *Comparing Wall Performance Predicted from Hygrothermal Simulations to Wall Performance Measured from Environmental Chamber Experiments*; Technical Report No. ORNL/TM-2019/1369; Oak Ridge National Laboratory (ORNL): Oak Ridge, TN, USA, 2019. [CrossRef]
8. Dong, W.; Chen, Y.; Bao, Y.; Fang, A. A validation of dynamic hygrothermal model with coupled heat and moisture transfer in porous building materials and envelopes. *J. Build. Eng.* **2020**, *32*, 101484. [CrossRef]
9. Frasca, F.; Verticchio, E.; Cornaro, C.; Siani, A.M. Performance assessment of hygrothermal modelling for diagnostics and conservation in an Italian historical church. *Build. Environ.* **2021**, *193*, 107672. [CrossRef]
0. Goto, Y.; Wakili, K.G.; Ostermeyer, Y.; Kalagasidis, A.S.; Wallbaum, H. Hygrothermal performance of a vapor-open envelope for subtropical climate, field test and model validation. *Build. Environ.* **2016**, *110*, 55–64. [CrossRef]
1. Kordziel, S.; Glass, S.V.; Boardman, C.R.; Munson, R.A.; Zelinka, S.; Pei, S.; Tabares-Velasco, P.C. Hygrothermal characterization and modeling of cross-laminated timber in the building envelope. *Build. Environ.* **2020**, *177*, 106866. [CrossRef]
2. Panico, S.; Larcher, M.; Troi, A.; Codreanu, I.; Baglivo, C.; Congedo, P.M. Hygrothermal analysis of a wall isolated from the inside: The potential of dynamic hygrothermal simulation. *IOP Conf. Ser. Earth Environ. Sci.* **2021**, *863*, 012053. [CrossRef]
3. Rahiminejad, M.; Räber, V.; Wakili, K.G.; Geyer, C.; Zöllig, S.; Renfer, C. Wooden basements; hygrothermal performance analysis using in-field measurements and numerical simulations. *Build. Environ.* **2024**, *256*, 111475. [CrossRef]
4. Stöckl, B.; Zirkelbach, D.; Künzel, H.M. Hygrothermal simulation of green roofs–new models and practical application. In Proceedings of the 10th Nordic Symposium on Building Physics, Lund, Sweden, 15–19 June 2014.
5. Wang, L.; Ge, H.; Wang, J. Model validation and 2-D hygrothermal simulations of wetting and drying behavior of cross-laminated timber. *J. Build. Phys.* **2023**, *46*, 737–761. [CrossRef]
6. Kramer, M.A.; Finch, G.; Hubbs, B.; Teetaert, J. Monitoring of historic structures for whole-building improvements. In Proceedings of the Thermal Performance of Exterior Envelopes of Whole Buildings XI, Clearwater Beach, FL, USA, 5–9 December 2010.
7. Moujalled, B.; Ouméziane, Y.A.; Moissette, S.; Bart, M.; Lanos, C.; Samri, D. Experimental and numerical evaluation of the hygrothermal performance of a hemp lime concrete building: A long term case study. *Build. Environ.* **2018**, *136*, 11–27. [CrossRef]
8. Mundt-Petersen, S.O.; Harderup, L. Predicting hygrothermal performance in cold roofs using a 1D transient heat and moisture calculation tool. *Build. Environ.* **2015**, *90*, 215–231. [CrossRef]
9. Qin, M.; Belarbi, R.; Aït-Mokhtar, A.; Nilsson, L. Coupled heat and moisture transfer in multi-layer building materials. *Constr. Build. Mater.* **2009**, *23*, 967–975. [CrossRef]
0. O'Brien, L.; Li, L.; Herzog, B.; Snow, J. The hygrothermal performance and durability of a one-story cross-laminated timber and wood fiber insulation school building located in Belfast, Maine. In Proceedings of the 14th World Conference on Timber Engineering 2025, Brisbane, Australia, 22–26 June 2025. [CrossRef]

sustainability

MDPI

Review

Towards Passive Building Thermal Regulation: A State-of-the-Art Review on Recent Progress of PCM-Integrated Building Envelopes

Kai Jiao [1], Lin Lu [1,*], Liang Zhao [2] and Gang Wang [2]

1 Department of Building Environment and Energy Engineering, The Hong Kong Polytechnic University, Hung Hom, Kowloon, Hong Kong; 19046688r@connect.polyu.hk
2 Sinopec (Dalian) Research Institute of Petroleum and Petrochemicals Co., Ltd., Dalian 116045, China
* Correspondence: vivien.lu@polyu.edu.hk

Abstract: The building envelope serves as a barrier against climatic conditions and as insulation to prevent energy waste within buildings. As global energy shortages become more pressing, the requirements for building envelopes are becoming increasingly stringent. Among the available technologies, phase change materials (PCMs) stand out for their high latent thermal energy storage and temperature stabilization capabilities. This paper reviews the recent advancements in PCM technology for building envelopes, starting with an overview of organic, inorganic, and eutectic PCMs, along with their respective advantages and disadvantages. The paper explores various incorporation methods such as shape stabilization, macroencapsulation, micro/nanoencapsulation, and solid–solid transition techniques. The integration of PCMs enhances thermal inertia, reduces thermal fluctuations, and delays heat peaks, presenting several multifunctional benefits. However, challenges such as fire hazards, potential toxicity, pollution, reduced mechanical performance, and higher initial costs persist. In light of these challenges, criteria for PCM integration in building applications are introduced. Additionally, the paper reviews recent hybrid technologies that combine PCMs with other novel technologies for building envelopes, including radiant temperature regulation systems, thermochromic windows, passive radiative cooling coatings, and others. It is shown that these PCM-integrated hybrid technologies significantly improve energy savings and indoor comfort. PCMs offer substantial potential for modern green building strategies and have further applications in other building contexts. Finally, the paper provides future prospects for studies in this field, aiming towards a green and energy-saving future.

Keywords: phase change material; building envelope; indoor comfort; zero carbon building

1. Introduction

The evolution of building envelopes has closely paralleled the progression of human civilization, historically functioning as protective barriers against harsh climatic conditions. The energy crisis of 1973 acted as a catalyst, significantly amplifying the focus on research related to building envelopes, which are responsible for approximately 36% of energy consumption in human activities [1]. The concept of the "building envelope" encompasses a diverse range of components, including opaque envelopes, structural materials, insulation, and glazing [2,3]. Numerous studies have investigated various materials suitable for building envelopes across different climatic conditions, extending from traditional materials like wood and earth [4–6] to cutting-edge hybrid polymers that boast enhanced properties such as superior thermal insulation, radiative sky cooling, fire resistance, and radiant cooling capabilities [7–10].

Within this spectrum, phase change materials (PCMs) have been recognized as particularly promising for their passive thermal regulation properties when integrated into building envelopes, owing to their latent thermal energy storage (LTES) capabilities. Unlike conventional sensible heat storage methods, PCMs can maintain a relatively stable temperature range while they absorb and release heat during their phase transitions, typically from solid to liquid, as illustrated in Figure 1. This distinctive capability enables PCMs to act as buffers to mitigate temperature fluctuations or as thermal energy reservoirs that can store surplus heat or cooling energy for subsequent use. Such dynamic processes significantly enhance a building's thermal inertia, thereby reducing the reliance on heating and cooling systems and consequently lowering the building's carbon footprint. Recent literature has extensively discussed these aspects [11–15]. For instance, Rathore et al. [16] provided a comprehensive summary concerning the integration techniques, performance evaluations, and underlying principles of PCMs in building materials. Li et al. [17] explored the thermal and optical performance of PCMs specifically for glazing applications, acknowledging the substantial potential for energy savings but also noting challenges such as diminished heat gain in colder regions, scattering effects caused by the porous structures of solid-state PCMs, and the ongoing need for refined models to simulate dynamic processes. Al-Yasiri and Szabó [18] highlighted a research gap in PCM applications for extreme climates and emphasized the necessity for extended durability testing. They also noted that while the low thermal conductivity of PCMs provides excellent insulation, it simultaneously impedes the efficiency of energy charge–discharge cycles, impacting the overall energy storage capacity.

Despite these challenges, the domain of PCMs in building envelopes is rapidly expanding, with innovative technologies emerging at an unprecedented rate. For example, the transformation of PCMs from solid–liquid to solid–solid states through chemical modifications is being investigated as a viable approach to stabilize their shape and prevent leakage [19]. Moreover, the integration of PCMs with other functional materials such as radiative cooling agents, photothermal, and photocatalytic materials is enhancing the multifunctional capabilities of PCM composites beyond simple thermal energy storage [20–23]. In light of these advancements, it is essential to continually update and disseminate the latest research findings to further promote the application of PCMs in building envelopes, steering toward a zero-carbon future.

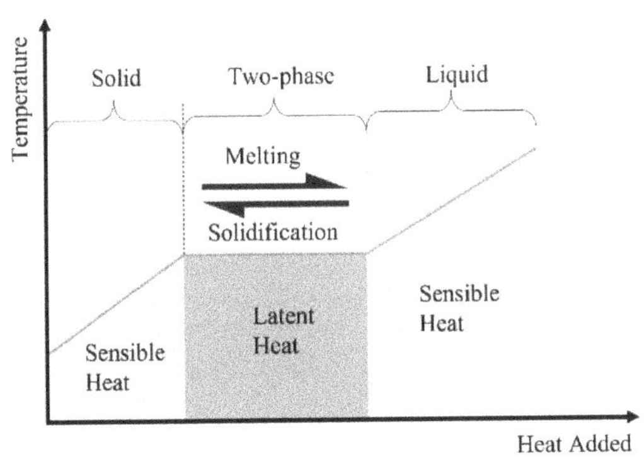

Figure 1. Temperature–enthalpy diagram of a PCM.

This paper begins by introducing different types of PCMs, emphasizing their distinct properties and functionalities. Special attention is given to recent advances in eutectic alloys with room temperature melting points, which are often overlooked in many PCM review articles. Examples of these PCMs are discussed, particularly those with appropriate phase change temperatures for building applications. We also explore PCM incorporation methods and anti-leakage techniques, including solid–solid PCMs (SSPCMs) that enhance the durability and reliability of PCMs in building contexts. Unlike existing review works on traditional building envelopes, this paper focuses on the innovative integration of PCMs with other energy-saving technologies such as radiant temperature control and radiative cooling. It addresses both successes and challenges, offering insights into future research directions. The review aims to present a comprehensive view of the state of the art in PCM technology, underscoring its potential to significantly impact energy conservation strategies in multifunctional building envelopes.

2. Features and Classification of PCMs

PCMs used in building envelopes must satisfy a range of critical criteria, including an appropriate phase-transition range, high latent heat capacity, efficient heat transfer capabilities, stable phase equilibrium, high density, minimal volume change during phase changes, low vapor pressure to circumvent containment issues, resistance to supercooling, a rapid crystallization rate, long-term chemical stability, compatibility with construction materials, non-toxicity, safety from fire hazards, and economic viability. The energy storage capacity of a PCM is determined by a combination of its sensible and latent thermal energy storage capacities. However, despite their potential, PCMs inherently exhibit some limitations such as tendencies toward supercooling, phase separation, leakage, and low thermal conductivity. These challenges necessitate careful selection and engineering to optimize the performance of PCMs in practical applications.

Assuming the heat capacity C_p is constant, the sensible heat Q_s of an ideal PCM with mass m is given by

$$Q_s = mC_p(T_2 - T_1) \quad (1)$$

where T_1 and T_2 are the start and end temperatures during the heat storage process. In contrast, when a phase change is completed, the total stored thermal energy by the PCM is calculated by

$$Q = mC_p(T_2 - T_1) + H \quad (2)$$

where H signifies the latent heat of the PCM. It is important to note that most commercialized PCMs do not exhibit a fixed melting point as one might assume. Instead, due to material impurities, they typically display a melting range as indicated in Figure 1 [24–26]. Figure 2 illustrates different types of differential scanning calorimetry (DSC) results, which reveal various melting patterns. For instance, paraffin wax is a mixture of n-alkanes, and the presence of different types of alkanes and variations in the lengths of molecular chains not only results in a non-fixed melting temperature but also contributes to solid–solid phase changes due to alterations in crystalline structures [27,28].

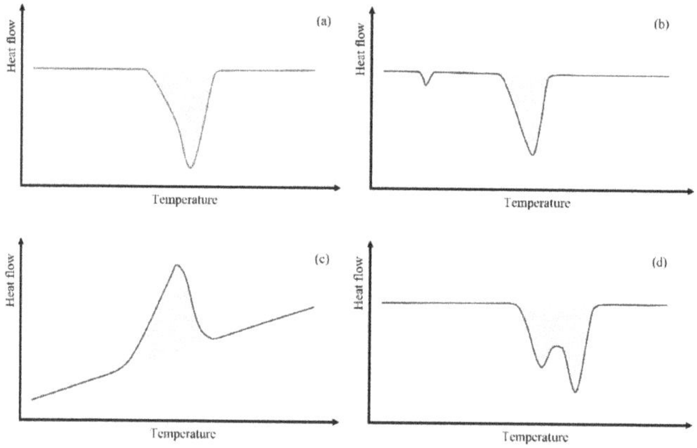

Figure 2. DSC patterns of different PCMs with (**a**) single-phase change process; (**b**) more than one phase change process; (**c**) temperature-dependent heat capacity; (**d**) merged phase change processes.

PCMs can be categorized based on their chemical composition into three types, organic, inorganic, and eutectic, each demonstrating unique properties that make them suitable for integration into building envelopes. Each type possesses advantages over others in different aspects as shown in Figure 3. Organic PCMs are particularly valued for their congruent melting behavior, minimal supercooling, excellent compatibility with construction materials, and stability, making them a preferred choice for many applications. In contrast, inorganic PCMs are chosen for their high latent heat capacity and higher thermal conductivity, although they pose challenges such as phase separation and potential corrosion, which can limit their usability. Eutectic PCMs, which are combinations of different types of PCMs, are engineered to combine the advantageous properties of the composed materials. This hybrid approach allows for the customization of melting points and latent heat capacities to meet specific requirements, thus broadening their applicability in diverse scenarios [15,29].

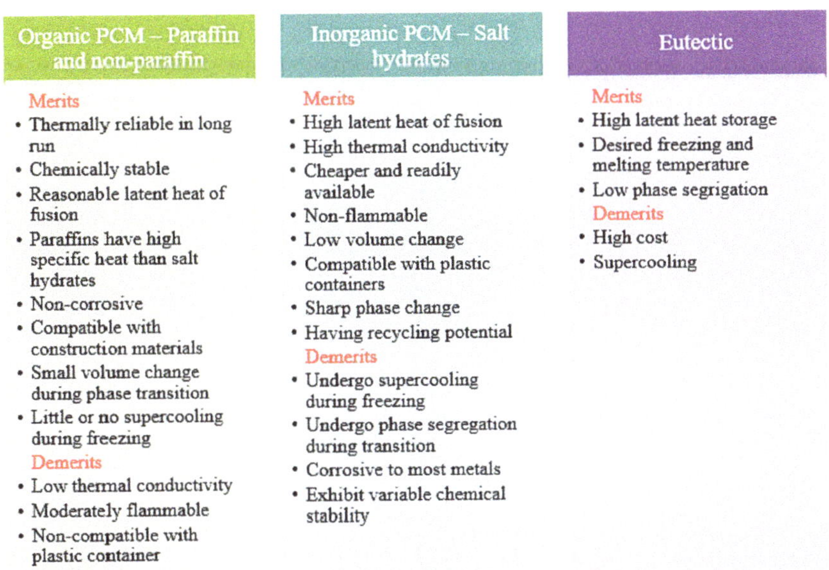

Figure 3. Merits and demerits of different types of PCMs. Figures reprinted from [30].

2.1. Organic PCMs

Organic PCMs are generally divided into two main categories: paraffins and non-paraffins. Paraffins, which are mixtures of n-alkanes, exhibit a range of melting points that increase with the length of the molecule's chain, making them highly versatile for use in building envelopes [31]. They are characterized by multiple allotropic forms, which vary in physical characteristics and crystal structures [32]. These allotropes can transition between states in response to temperature changes, with the conversion being entirely reversible. The advantages of using paraffins for heat storage are manifold: they offer a high latent heat of fusion, minimal supercooling, low vapor pressure in their liquid state, chemical inertness, long-term stability, self-nucleating properties, and an absence of phase segregation [33,34]. Additionally, as byproducts of the oil industry, paraffins are readily available at reasonable costs, enhancing their economic viability [35,36].

Non-paraffins encompass a diverse group that includes bio-based PCMs (BPCMs) such as fatty acids and alcohols, and polymer PCMs represented by polyethylene glycol (PEG). BPCMs are typically derived from bioproducts like plant and animal fats [37,38]. They are characterized by their recyclability, degradability, and lower melting points, making them more eco-friendly and suitable for low-temperature applications [39,40]. Furthermore, BPCMs exhibit high surface tension in their liquid state, which helps confine the material within the host matrix, thereby addressing leakage concerns [41]. PEG, on the other hand, offers benefits including ease of synthesis, non-toxicity, biocompatibility, low cost, flexibility, compatibility with various additives, and the potential for environmentally friendly fabrication processes [42].

Despite their numerous advantages, organic PCMs commonly suffer from low thermal conductivity, necessitating the use of conductivity-enhancing techniques for effective application. Additionally, the potential fire hazards associated with organic PCMs should not be underestimated. Recent studies, such as those by Weng et al. [43], have explored the integration of fire retardants to mitigate the flammability risks associated with these materials. The ongoing development and refinement of organic PCMs for use in building envelopes continue to promise further optimizations in energy efficiency and sustainability within the construction sector. Table 1 provides the thermophysical properties (melting point/range T_m, average thermal conductivity k, and latent heat H) of some representative organic PCMs applicable to building envelopes.

Table 1. Thermophysical properties of some organic PCM candidates for building envelopes.

PCM	T_m (°C)	H (kJ/kg)	k (W/(m·K))
n-Octadecane [44]	24.90	204.91	0.192
Tetradecane [45]	5.8	227	-
Pentadecane [45]	9.9	206	-
Hexadecane [45]	18.1	236	-
n-Eicosane [46]	36.58	248.5	0.25
Paraffin [47]	21.7–22.7	161.6	0.21
Paraffin [25]	27–29	245	-
Paraffin [48]	27–33	165	0.13
Paraffin [49]	44	174.12	-
Paraffin [50]	70	173.03	0.197
Lauric acid [51]	43.93	178.11	-
Myristic acid [51]	54.28	191.27	-
Palmitic acid [51]	62.73	206.16	-
Stearic acid [51]	69.62	217.62	-
PEG [52]	21.3–31.5	187.03	-
PEG [53]	18.51	118.22	-
Lauryl alcohol [54]	24	217	0.221
Oleic acid [55]	5–7	140	-
Butyl stearate [55]	21	88	-
Isopropyl palmitate [55]	11	117	-

2.2. Inorganic PCMs

Inorganic PCMs include salt hydrates, molten salts, and metallic PCMs (encompassing metals and eutectic alloys). These inorganic PCMs generally exhibit higher energy storage capacities and superior thermal conductivities compared to their organic counterparts [56]. Notably, the melting points of some inorganic PCMs, particularly molten salts, often exceed 200 °C, which makes them less suitable for building envelope applications due to the high operational temperatures required. Table 2 provides the thermophysical properties of some inorganic PCM candidates. Decades ago, metallic PCMs were often excluded from building applications due to their excessively high melting temperatures [57]. However, advancements in materials science have facilitated the discovery of low-temperature metals and alloys, sparking renewed interest in these materials for their significantly higher thermal conductivities compared to other PCM types.

Salt hydrates, composed of ionic compounds incorporating water molecules within their crystal lattice, offer high energy storage density at a relatively low cost [58]. They are often preferred over organic PCMs due to their non-flammable and odorless properties, which enhance safety and comfort in inhabited spaces. Nevertheless, salt hydrates face several challenges that limit their broader application. One major issue is incongruent melting, where the solid phase, typically a lower hydrate of the same salt, separates and settles at the bottom of the container, leading to chemical instability and irreversible loss of thermal storage capability [32]. Additionally, supercooling in salt hydrates can delay heat release, which is a disadvantage in many building applications despite its potential benefits for long-term energy storage [59,60]. To address these issues, the addition of nucleating and thickening agents is a common practice [61–63]. The corrosive nature of salt hydrates also poses significant challenges as they can degrade metals, especially when encapsulated in metal containers. Although the thermal conductivity of salt hydrates is higher than that of organic PCMs—for example, the thermal conductivity of solid $CaCl_2 \cdot 6H_2O$ is 1.08 W/(m·K), which is four times higher than that of OP42E paraffin [64]—the overall thermal energy storage efficiency remains low, necessitating further heat transfer enhancement techniques.

Metallic PCMs for building envelopes, including low melting point metals and metal eutectics, are increasingly considered for efficient thermal energy storage due to their excellent thermal conductivity. This intrinsic property eliminates the need for additional heat transfer enhancement, making metallic PCMs highly efficient. They also tend to have higher overall energy densities and offer stability advantages compared to salt hydrates [65]. However, metallic PCMs face challenges such as containment issues and high corrosivity in their liquid state, which can compromise safety and reduce system longevity [66]. Despite these challenges, ongoing developments in metallic PCMs hold promise for advancing high-temperature energy storage solutions, which are crucial for the next generation of energy-efficient buildings.

Table 2. Thermophysical properties of some inorganic PCM candidates for building envelopes.

PCM	Classification	T_m (°C)	H (kJ/kg)	k (W/(m·K))
$Na_2SO_4 \cdot 10H_2O$ [67]	Salt hydrate	32.24	139.9	0.34
$CaCl_2 \cdot 6H_2O$ [68]	Salt hydrate	29.6	212	0.82
$Na_2HPO_4 \cdot 12H_2O$ [69]	Salt hydrate	35.1	209.5	1.25
$CH_3COONa \cdot 3H_2O$ [70]	Salt hydrate	58.3	286.3	-
$Na_2CO_3 \cdot 10H_2O$ [71]	Salt hydrate	33	247	0.6
$FeCl_3 \cdot 6H_2O$ [71]	Salt hydrate	37	223	-
$Mg(NO_3)_2 \cdot 6H_2O$ [72]	Salt hydrate	91.5	146	0.4
$Zn(NO_3)_2 \cdot 6H_2O$ [73]	Salt hydrate	34.6	140	-
Gallium [74]	Metal	29.78	80.16	32
Bi-Sn-In [75]	Alloy	60.9	29.18	16.4
Bi-21In-18Pb-12Sn [76]	Alloy	56.83	28.98	32.2~10.6
Ga-13.5Sn [76]	Alloy	20	78.29	30
Bi–Cd–Sn–Pb [77]	Alloy	70	32.9	18
$In_{51}Bi_{32.5}Sn_{16.5}$ [78]	Alloy	61.53	28.98	-
In-Sn-Bi [79]	Alloy	54.9	25	16.85
50Bi/25Pb/13Sn/12Cd [80]	Alloy	73	30.6	23

2.3. *Eutectic PCMs*

Eutectic PCMs are homogeneous mixtures composed of two or more different PCMs, which can be combinations of organic–organic, inorganic–inorganic, or organic–inorganic materials. The primary objective in synthesizing eutectic PCMs is to moderate their melting temperatures, typically achieving a melting point that is lower than that of each individual component, thereby aligning with desired operational temperatures [81]. The melting point of eutectic PCMs can be determined using the Schroeder equation by analyzing the phase diagram of the binary eutectic system [82]. As illustrated in Figure 4a, the phase diagram features two solubility lines, labeled "a" and "a′", which converge at an equilibrium state known as the eutectic point, denoted as $X_{1,e}$. This point marks the highest degree of melting point depression achievable in the system, allowing for the precise quantification of the melting points of eutectic PCMs by understanding the thermal behavior of the constituent PCMs.

However, one issue with eutectic PCMs is the presence of impurities, which can introduce thermal instability, often leading to latent heat loss after multiple cycles of charging and discharging [83]. Despite this challenge, recent advancements in eutectic PCM technology have significantly minimized these issues [84–86]. Another concern with eutectic PCMs, particularly those containing an inorganic component, is the problem of supercooling as shown in Figure 4b [87]. Ideally, a eutectic PCM should not suffer from phase separation if the composition is maintained at the equilibrium point indicated on the phase diagram. However, if one of the components undergoes supercooling, it disrupts this phase equilibrium, leading to phase separation and a corresponding loss in latent heat [88]. Table 3 displays the thermophysical properties of some recently developed eutectic PCMs with phase change temperatures suitable for building envelope applications.

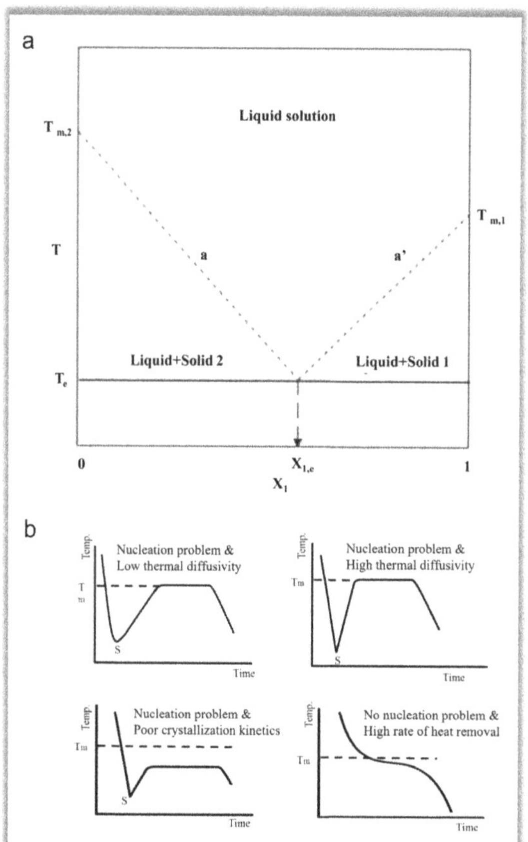

Figure 4. (**a**) Phase diagram of a simple binary eutectic system, reprinted from [89]. (**b**) Different thermal behaviors of PCMs suffering from supercooling, reprinted from [90].

Table 3. Thermophysical properties of some recently developed eutectic PCMs.

Classification	Composition (in wt%)	T_m(°C)	H (kJ/kg)	k (W/(m·K))
Inorganic–organic [91]	$CH_3COONa{\cdot}3H_2O$, CH_4N_2O, 6:4	31.4	205.3	-
Binary organic [92]	Lauryl alcohol, stearyl alcohol, 9:1	22.93	205.79	0.18
Binary organic [93]	Lauric acid, stearyl alcohol, 7:3	39.87	186.94	-
Binary inorganic [94]	$CaCl_2{\cdot}6H_2O$, $Na_2HPO_4{\cdot}12H_2O$, 78.3:21.7	22.6	143.6	-
Ternary inorganic [94]	$Na_2SO_4{\cdot}10H_2O$, $Na_2CO_3{\cdot}10H_2O$, $Na_2HPO_4{\cdot}12H_2O$, 34.8:46.2:19	20.2	207.65	-
Binary organic [95]	Palmitic acid, stearic acid, 61:39	54.8	223.6	0.8602
Binary inorganic [96]	$Na_2SO_4{\cdot}10H_2O$, $Na_2HPO_4{\cdot}12H_2O$, 62:38	27.1	202.4	0.464
Ternary organic [97]	Adipic acid, sebacic acid, stearic acid, -	61.7	193.3	-
Binary organic [98]	Paraffin, stearic acid, 45:35	56.2	192.5	0.368
Binary organic [98]	Palmitic acid, stearic acid, 32:68	55.8	213.3	0.323
Binary organic [98]	Palmitic acid, myristic acid, 14:86	51.3	225.8	0.285
Binary organic [99]	Decanoic acid, PEG, 1:1	22.9	173.9	0.2025

3. Incorporation Methods

As PCMs undergo state transitions between solid and liquid during phase change processes, several methods of incorporation are required to prevent the leakage of liquid PCMs in applications. Generally, there are four types of incorporation methods: the first involves using highly porous materials to contain the liquid PCM, which is commonly referred to as shape stabilization or form stabilization [100,101]. The second method involves encapsulating PCMs within a macroscale container [102,103], while the third method focuses on constructing core–shell structures to create micro/nanoencapsulated PCMs [104]. Finally, the fourth method converts solid–liquid PCMs into SSPCMs through chemical linking [105]. Each of these methods aims to enhance the application feasibility of PCMs by addressing the challenges associated with the liquid phase of the material.

3.1. Shape-Stabilized PCMs

PCMs can be incorporated into porous materials such as graphene [106], a matrix of carbon nanotubes [107], and expanded graphite [54]. These materials provide space for the volume change of PCMs, and the liquid PCMs are held within the porous structure due to the combined effects of capillary force, surface tension, and van der Waals forces [46,108]. Not only do these porous materials serve as supporting matrices for PCMs, but they also modulate the thermal conductivities of the PCMs to either enhance heat storage efficiency or improve thermal insulation capabilities [109,110]. The fabrication process for these composites is generally straightforward [111], involving the mixing of the substrate foam material and PCM at a temperature above the PCM melting point, followed by cooling and shaping in molds; vacuum impregnation may be employed to achieve a more intimate combination, allowing the PCM to thoroughly impregnate the porous structure and form a shape-stabilized composite.

Yang et al. [112] prepared a shape-stabilizing PCM by incorporating lauryl alcohol/stearic acid eutectic PCM and Al_2O_3 nanoparticles as thermal conductivity enhancers into ceramsite, creating a ceramsite-based shape-stabilized PCM as shown in Figure 5a. It was found that this shape-stabilized PCM effectively prevents leakage with a melting point of 22.5 °C, and the enthalpy of the PCM composite is 133.4 kJ/kg. By integrating the PCM composite into concrete blocks, satisfying mechanical performance was achieved, and temperatures on both the hot side and cold side of the PCM-integrated building envelope were reduced as shown in Figure 5b,c. Kumar et al. [113] experimentally studied the effectiveness of incorporating a lauric acid/zeolite/graphite PCM composite into gypsum plaster, observing maximum temperature reductions of 13.86% and 7.78% in peak indoor surface temperatures at the roof and south wall, respectively. Atinafu et al. [114] studied the performance of four different carbon-based materials—biochar, activated carbon, carbon nanotubes, and expanded graphite—on n-heptadecane. It was found that expanded graphite has the highest PCM loading ratio of 94.5%, a result attributed to the intermolecular interactions between the PCM and the carbon materials, including hydrogen bonding, pore characteristics, and surface functionality.

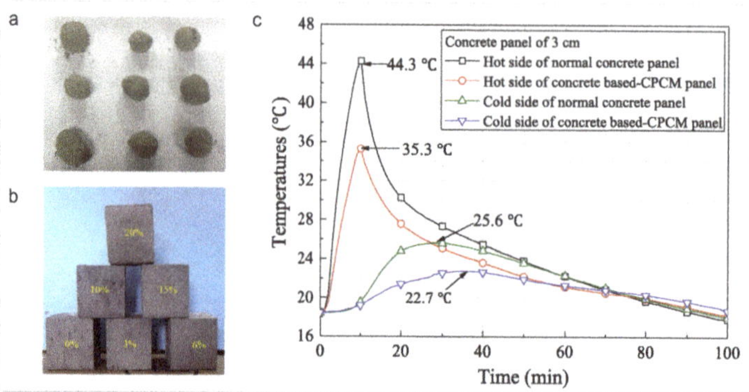

Figure 5. (**a**) The ceramsite-based shape-stabilized PCM. (**b**) Concrete incorporated with different mass fractions of shape-stabilized PCM. (**c**) Temperature variation on the hot and cold sides of a 10 cm × 10 cm × 3 cm thick concrete panel integrated with 15 wt% PCM compared to a normal concrete panel. Figures reprinted from [112].

3.2. *Macroencapsulated PCMs*

PCMs can be encapsulated within macroscale containers, allowing these macroencapsulated PCMs to be further integrated with building materials such as bricks and concrete as illustrated in Figure 6. Compared to other types of PCM incorporation methods, macroencapsulation offers superior mechanical strength and a higher PCM loading ratio. Moreover, because the encapsulation isolates the PCM from the environment, it also enhances the stability of the PCM itself. This method not only ensures structural integrity but also optimizes the thermal performance of the building materials by incorporating the thermal management capabilities of PCMs [115,116].

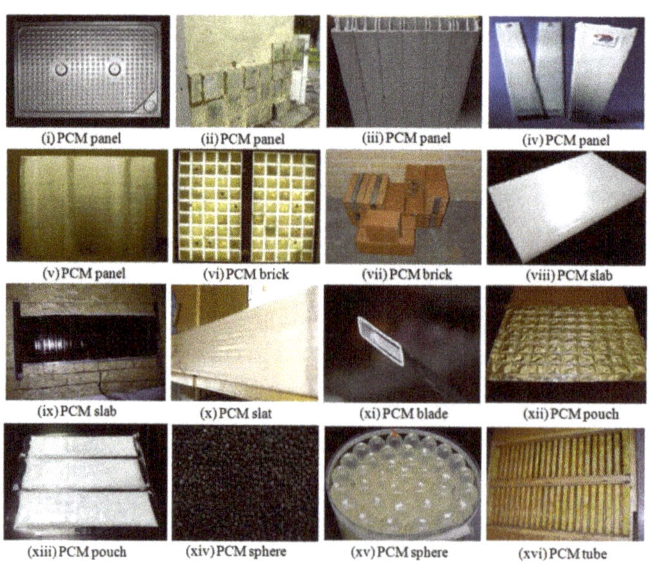

Figure 6. PCM macroencapsulation methods for building envelopes. Figure reprinted from [117].

Abass and Muthulingam [118] developed a reinforced concrete roof integrated with macroencapsulated PCMs to enhance passive cooling in buildings. Their findings indicate that the PCM-integrated roof slab significantly improves thermal performance by reducing interior temperatures by up to 7.2 °C during sunny hours, decreasing heat transfer into the building by up to 60.6%, and reducing the thermal load by up to 54%. This demonstrates the effectiveness of macroencapsulation in mitigating heat ingress, particularly during peak solar radiation periods. Rida and Hoffmann [119] investigated the impact of the exposed surface area of a macroencapsulated PCM (coconut oil) panel when applied to ceilings. They discovered that increasing the exposed surface area by 50% reduced the time needed for the PCM to melt completely by 20%, highlighting the importance of surface area in the thermal performance of PCM applications. Hu et al. [120] developed a casing pipe wall system with macroencapsulated PCM, characterized by enhanced anti-leakage properties, mechanical strength, and cost effectiveness. This innovative design involves filling the inner and outer pipes with PCMs of different melting points, allowing the wall to provide both passive heating and cooling. This dual functionality is particularly beneficial in regions of China that experience both hot summers and cold winters. Under optimal operating conditions, the wall surface temperatures were observed to be approximately 20.5 °C in summer and 26.8 °C in winter, demonstrating substantial improvements in maintaining comfortable indoor temperatures throughout the year.

3.3. *Micro/Nanoencapsulated PCMs*

Compared to macroencapsulation, PCMs can also be encapsulated at the microscale (diameter above 1 mm) or nanoscale (diameter less than 1 mm) by constructing a core–shell structure. The smaller size of micro/nanoencapsulated PCMs facilitates their integration into various building materials. These micro/nanoencapsulated PCMs offer a high surface area for heat transfer, which is advantageous for thermal management. However, their thermal conductivity is not particularly remarkable due to limitations in enhancing heat transfer at such small scales [104].

The encapsulation also provides insulation from the surrounding environment, enabling high thermal stability. Al-Absi et al. [121] compared the thermal performance of microencapsulated PCM-integrated cement render (CR) and foamed concrete (FC) on building walls with conventional CR and FC as shown in Figure 7a. The study demonstrated that PCM panels can significantly reduce temperature fluctuations over three heating and cooling cycles. The integration of PCMs results in a lower temperature profile for PCM-integrated walls during the heating period and higher profiles during the cooling period, attributed to the phase change processes of PCMs. During the heating period, the use of PCM panels can lead to a maximum temperature reduction of 4.8 °C for the external surface and 7.35 °C for the internal surface. Additionally, the peaks of internal surface temperature can be reduced by up to 3.95 °C. The micro/nanoencapsulation of PCMs also provides a protective layer that enhances the stability of PCMs. Maleki et al. [122] developed a plaster wallboard integrated with nanoencapsulated PCM as shown in Figure 7b. It was observed that the thermal and mechanical performance of the nanoencapsulated PCM remained stable even after 500 thermal cycles. Cabeza et al. [123] studied the long-term performance of a concrete wall integrated with microencapsulated PCMs. Their findings revealed that after 10 years of operation, the PCM-integrated wall continued to effectively reduce temperature fluctuations, with no significant degradation of the PCMs noted. Although there was a slight reduction in the mechanical performance of the wall, it still functioned comparably to its original performance, underscoring the potential of microencapsulated PCMs in sustainable building practices that require long-term thermal management solutions.

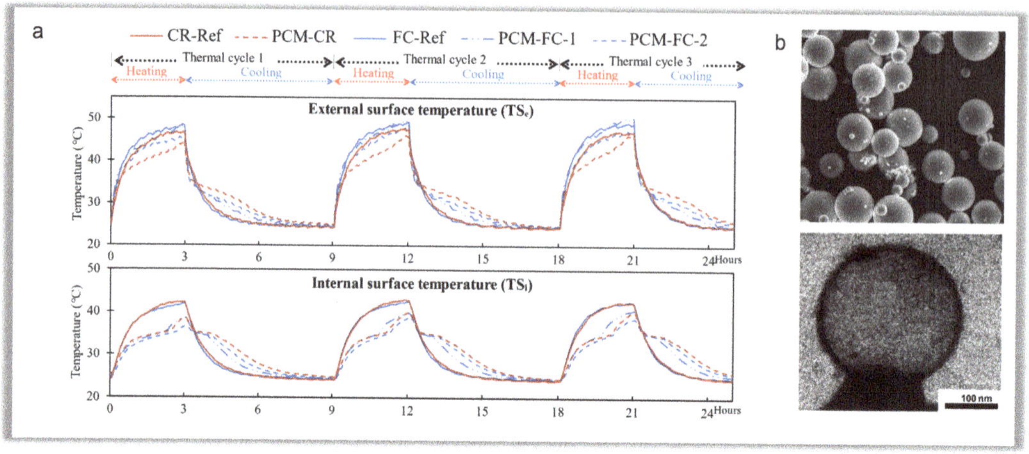

Figure 7. (**a**) Exterior surface temperature and interior surface temperature of a building wall integrated with/without PCM panels in three heating and cooling cycles. Figure reprinted from [121]. (**b**) SEM image (**top**) and TEM image (**bottom**) of the nanoencapsulated PCM. Figure reprinted from [122].

3.4. SSPCMs

SSPCMs remain in the solid state during their phase change processes. There are two types of SSPCMs. The first type consists of natural SSPCMs, which undergo a phase change by altering their crystalline structure, such as certain salt hydrates and paraffin [64,124–126]. The second type comprises chemically modified SSPCMs derived from conventional solid–liquid PCMs to promote a solid–solid phase transition. Unlike some PCMs that transition between crystalline structures, such as paraffin [127], chemically modified SSPCMs typically transition from a crystalline structure to an amorphous phase [19]. This transformation is primarily observed in polymer-based PCMs with hydroxyl-terminated structures.

PEG is a commonly studied base material for SSPCMs. A distinctive feature of PEG is its repeating ethyl–ether segments, which facilitate grafting or crosslinking with other polymers, thereby enhancing its structural integrity and functional adaptability [128]. Despite these benefits, the chemical stabilization process can reduce the latent heat capacity and alter the melting point [129]. Additionally, the absence of natural convection in SSPCMs necessitates the use of heat transfer enhancement techniques such as fins and metal foams [130–133]. Nevertheless, SSPCMs substantially mitigate leakage issues, presenting a significant advantage, particularly in applications where the containment of the PCM is critical.

Gao et al. [134] explored the application of PEG-based SSPCMs in building windows. They developed a novel, translucent PCM window that maintained consistent optical properties. Implementing a 3 mm thick layer of this PCM enabled energy savings of up to 9.4% for heating, 6.7% for ventilation, and 3.2% for air conditioning across the entire building. This demonstrates the potential of SSPCMs in reducing energy consumption in building applications. Harlé [135] investigated the use of SSPCM integrated with plaster to create a composite material for building applications. The study found that while the addition of PCM reduced the mechanical performance of the plaster, a concentration of 20 wt% PCM achieved an optimal balance between mechanical strength and thermal performance. This composite material exhibited satisfactory thermal regulation capabilities and stability, indicating its viability for enhancing building energy efficiency while maintaining structural integrity.

4. Criteria for PCM Integration in Building Applications

For the incorporation of PCMs in buildings, various issues need to be considered during the design process. For instance, the cost of the building increases when using such materials to replace conventional insulation materials. Moreover, integrating PCMs into building materials in any form can reduce their mechanical strength. This section discusses the criteria for PCM incorporation in buildings, providing a thorough evaluation of the main considerations for PCM selection and design.

4.1. Thermophysical Properties

The thermophysical properties that need to be considered for PCM selection in building applications mainly include latent heat, thermal conductivity, and melting point (or range). For each of these properties, there is no single criterion in the design principles that definitively states whether higher values are better. For example, some studies suggest that an ideal PCM for buildings should have high latent heat (or high energy storage density for metallic PCMs). While this can be true for applications requiring high energy storage capacity, it is not universally applicable. Chen et al. [136] showed that for PCM-integrated building envelopes in summer, using a PCM with higher latent heat can save more energy under high cooling energy consumption. However, for low cooling energy use, a higher latent heat PCM makes the stored heat more difficult to dissipate, leading to more significant energy waste.

Another factor to consider is thermal conductivity. In most cases, the application of PCMs in buildings requires some heat transfer enhancement techniques to ensure that the PCM can quickly respond to temperature changes, thereby increasing its energy storage efficiency [137,138]. However, the low thermal conductivity of PCMs (such as organic PCMs and salt hydrates) can be beneficial as a thermal insulation material [139].

The melting point of PCMs should be selected according to their applications. Generally, for heat supply applications, the melting point of PCM should lie between 29 and 60 °C; for building envelope applications, it should be 22–28 °C; and for refrigeration applications, it should be below 21 °C [140]. Overall, the selection of PCMs in terms of their thermophysical properties, amount of use, and incorporation position should be optimized based on the local climate conditions and the design of the buildings [141–143].

4.2. Chemical and Thermal Stability

Buildings are long-term investments that stand for decades, requiring PCMs with good stability. This means that PCMs should maintain their thermal performance after a large amount of cycles of charging and discharging. As a result, the PCM should retain its chemical composition and not react by itself during heating and cooling. Additionally, PCMs should not react with encapsulating materials and other surrounding building materials. Such issues are usually observed in inorganic salt hydrates and metallic PCMs. Studies have reported that inorganic PCM-integrated concrete can lead to corrosion problems with steel rebars [144,145]. This issue is typically solved by building a barrier between the corrosive PCM and the materials prone to corrosion, either through encapsulation or coating [146]. Furthermore, PCMs should exhibit minimal supercooling effects, which are usually observed in micro/nanoencapsulated PCMs and inorganic PCMs as discussed in Section 2. The degree of supercooling can be mitigated by increasing the roughness of the encapsulation or adding additives as nucleation sites for solidification [147].

To date, many PCMs with great stability and reliability have been prepared in recent studies. Emeema et al. [148] prepared paraffin/carbon quantum dot (CQD) composites. Thermogravimetric analysis (TGA) results indicated that adding CQD to paraffin raised the onset degradation temperature of the composites to as high as 297 °C from 269 °C, implying a longer lifespan of the composite. In a year-long test with 300 charging and discharging cycles, it was shown that the latent heat of the paraffin/QOD composite decreased slightly by 0.6%, demonstrating great potential for long-term operations.

4.3. Safety Issues

Despite their non-corrosive, low-cost, chemically stable, and non-supercooling advantages, fire safety is a crucial concern for organic PCMs in building envelopes, which must be considered in the design of buildings. Mclaggan et al. [149] showed that different applications of PCMs in buildings exhibit different levels of fire risks. When integrating PCMs in wall linings, fire risks can be quantified and mitigated by designers through proper fire risk assessment. However, using PCMs as insulation material at very high loadings results in extremely high heat release rates, even if the burning duration is short. Present methods to reduce the flammability of organic PCMs include the incorporation of flame retardants, chemical transformations, and surface coatings [150].

Toxicity of PCMs is another safety concern. Many studies indicate that low toxicity to humans can be found in many salt hydrates, with different salt hydrates exhibiting different toxicity levels [151–153]. Although this does not cause considerable issues if they are carefully encapsulated, the disposal of these salt hydrate wastes can cause significant environmental problems. For organic PCMs, although they are non-toxic, they still release toxic gases if they catch fire [43,154], highlighting the need for anti-flammability techniques in applications.

4.4. Economic Considerations

The decision to integrate PCMs in buildings is influenced by the economic benefits relative to the cost. The cost consists of the initial investment and annual cost for the PCM-integrated system. The initial investment cost is determined by the pricing of the PCMs and the PCM incorporation technique [155]. PCM pricing varies significantly across different regions, irrespective of the PCM category. Moreover, macroencapsulation is considered the most cost-friendly PCM incorporation technique, while micro/nanoencapsulations are the most expensive [156,157]. In calculating the payback time, the benefits of using PCM can be directly evaluated based on the saved energy, as the energy and price have a clear relation. Another consideration is the degradation in PCM performance, which can be evaluated by exergy [158].

4.5. Recyclability and Environmental Impact

The recycling of PCMs in building material waste presents both challenges and opportunities from environmental and economic perspectives. The recyclability of PCMs is primarily determined by their incorporation methods. For instance, it is almost impossible to recycle PCMs in concrete waste containing microencapsulated PCMs. Conversely, it is much easier to collect macroencapsulated PCMs. Specifically, great recyclability has been observed in many SSPCMs through simple shredding, heating, and remoulding processes. Yang et al. [159] prepared a PEG-based SSPCM exhibiting excellent recyclability, with almost no difference between the original and the recycled PCM. Similar achievements have been reported by Yang et al. [160], Bai et al. [161], and Yang et al. [162].

The environmental impact of a PCM-integrated building envelope should be evaluated from two aspects: first, how much greenhouse gas emission is reduced by the integrated PCM, and second, the pollution and greenhouse emissions caused from manufacturing to installation of the PCM-integrated building envelope [163]. A well-accepted method for analyzing the environmental impact of PCMs in buildings is to conduct a life cycle assessment (LCA) or life cycle cost analysis (LCCA). By conducting an LCA in Puigverd de Lleida (Spain), Gracia et al. [164] showed that the benefits of PCMs increase in locations where the weather conditions are consistent throughout the year. It is better to use salt hydrates than paraffins to reduce the manufacturing/disposal impact. Zhang et al. [165] conducted an LCCA of residential building walls enhanced with PCMs and showed that in Xi'an, China, for PCM Brick exterior walls, the thicker the phase change layer, the better the economic and environmental benefits. The payback period in the Xi'an area is between 10.13 and 13.41 years, the carbon payback period is between 3.3 and 4.3 years, and the carbon reduction is between 244.69 and 82.60 kg CO_2/m^2.

5. Novel Applications of PCMs for Building Envelopes

PCMs and their composites can be integrated into various building materials and components. Traditional applications involve PCM-integrated building materials such as bricks [13,166], panels [167,168], slabs [118,169], and more, which are then incorporated into roofs, floors, and walls to serve as thermal dampers. These materials help reduce and delay thermal fluctuations, thereby acting as a passive thermal regulation technique to achieve a comfortable indoor temperature [170,171] based on their energy charge and discharge cycles.

There are also hybrid applications of PCMs with other building envelopes as shown in Table 4. A notable example is the integration of PCMs with building-integrated photovoltaics (BIPVs). In this application, PCMs function as a heat sink with large heat capacity to cool down the PV panels and simultaneously reduce the heat load to the building introduced by the PV panels [172–174]. Moreover, the heat stored in the PCM can be utilized by auxiliary devices like heat pumps and thermoelectric generators, which further improves the energy efficiency of the building [175,176]. Another well-known integration is the combination of PCMs with Trombe walls [177]. The marriage of these two passive temperature regulation techniques can provide more stable temperature stabilization performance, thanks to the improved energy density, stabilized temperature variations, and high thermal insulation performance introduced by the PCM [178–180].

Moreover, new energy-saving techniques in buildings, such as radiant cooling, radiative sky cooling, and thermochromic coatings have emerged in recent years. This section introduces the novel integration techniques of PCMs with these hybrid technologies.

Table 4. Some hybrid PCM applications in building envelopes with different incorporation methods.

PCM	Incorporation	Integrated Application
Poly (Octadecyl Methacrylate) (PSMA) [181]	SSPCM	Thermo-optical coating for windows
Paraffin [182]	Macroencapsulated PCM	BIPV facade
Binary eutectic PCM [183]	Macroencapsulated PCM	Trombe wall
$CaCl_2 \cdot 6H_2O$ [184]	Shape-stablized PCM	Radiant cooling panel system
n-dodecane [185]	Microencapsulated PCM	Radiative cooling coating
PEG [186]	Shape-stablized PCM	Photothermal functional concrete
n-octadecane [187]	Shape-stablized PCM	Electromagnetic interference shielding wall
Binary eutectic PCM [188]	Microencapsulated PCM	Icephobic coating
OM32 (salt hydrate) [189]	Macroencapsulated PCM	Solar photocatalytic ventilation wall

5.1. PCM-Integrated Radiant Cooling/Heating Systems

Radiant cooling (RC) and radiant heating (RH) systems are increasingly recognized for their ability to provide high thermal comfort due to the significant role of radiative heat transfer between humans and indoor building envelopes [190]. Moreover, RC systems are instrumental in advancing low-carbon buildings, with reports indicating energy savings of up to 41% compared to traditional air conditioners and 34% compared to Variable Air Volume (VAV) systems [191–193]. The integration of PCMs into these systems enhances their energy storage capability, providing more stable surface temperatures and improved energy efficiency [194].

Bogatu et al. [195] analyzed the performance of a macroencapsulated PCM panel embedded with water pipes, designed for use as an RC ceiling (as shown in Figure 8a). The system, actively cooled by the discharge of water, successfully maintained the indoor thermal environment. Findings indicated that the average specific cooling power of the system was 11.3 W/m^2, surpassing that of commercial microencapsulated PCM gypsum panels but falling below the performance of traditional RC panels. González and Prieto [196] conducted a numerical study on the thermal performance of an RH floor integrated with PCM bands. They discovered that incorporating PCM into the concrete core somewhat reduced the heat transfer efficiency. However, during the discharge phase when the heating was turned off, the PCM RH floor maintained a stable heat flux ranging from

31.4 W/m^2 to 44.6 W/m^2 for more than 24 h, aligning closely with the time-average value. Liu et al. [197] studied a PCM-integrated RH Trombe wall (as illustrated in Figure 8b). Their research revealed that this integration lowered the peak inner surface temperature of the south wall by 3 °C compared to a traditional Trombe wall and also reduced indoor heat gain by 55.2%, indicating superior energy efficiency. Cesari et al. [198] explored the integration of two types of PCMs with different melting points (17 °C and 27 °C) in a radiant floor system, designed to offer both cooling and heating capabilities (as depicted in Figure 8c). During summer, the system could be turned off, allowing the PCM to absorb excess heat and shift energy use to off-peak hours. In winter, the system maintained an indoor temperature of approximately 20 °C for 9 h without heating, achieving a 13% energy saving compared to an air handling unit (AHU) alone.

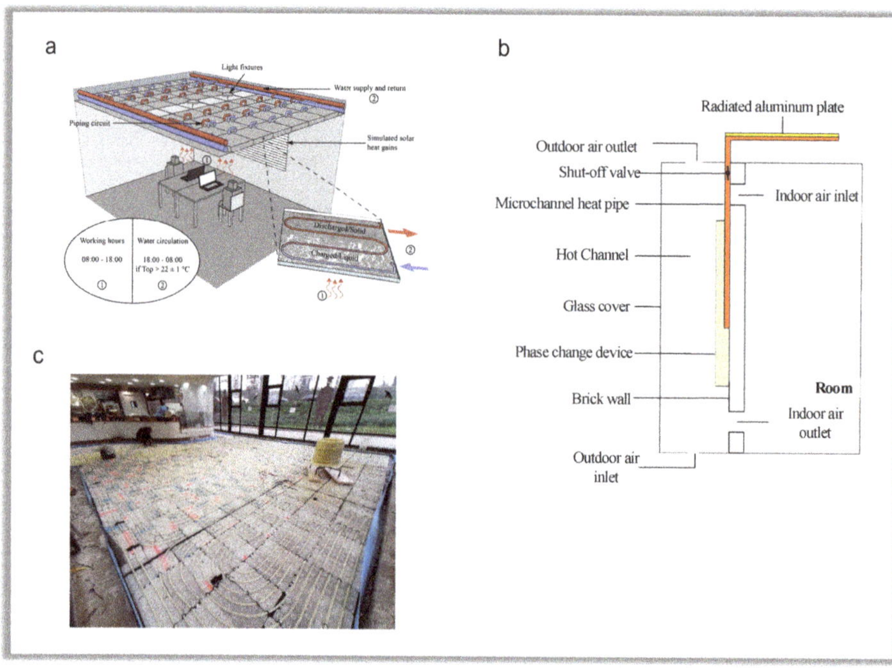

Figure 8. (**a**) Schematic of an RC ceiling with PCM. Figure reprinted from [195]. (**b**) Schematic of a PCM-integrated RH Trombe wall. Figure reprinted from [197]. (**c**) Radiant floor integrated with macroencapsulated PCM. Figure reprinted from [198].

5.2. PCM Assisted Passive Radiative Cooling

Passive radiative cooling (PRC) has emerged as a promising technology for reducing cooling loads and mitigating the urban heat island effect. Advances in material science over the past decade have significantly contributed to the development of daytime PRC materials characterized by their emissivity in the mid-infrared region, specifically within the atmospheric window (mainly between 8–13 μm), which enhances their cooling capabilities. The integration of PCMs with PRC materials not only improves thermal stability but also enhances the ability to save and shift cooling capacity [199].

Atiganyanun et al. [200] explored a method for incorporating microencapsulated PCMs into PRC paint (specifically $BaSO_4$) to boost its cooling performance. Their findings indicated that the addition of microencapsulated PCM had little impact on the emissivity of the paint. Additionally, the PCM was able to absorb excess heat from the interior, thereby maintaining the paint's cooling effectiveness. Yang et al. [199] developed a bi-layer phase change material-enhanced radiative cooler (PCMRC) as shown in Figure 9. The PCMRC demonstrated adaptive temperature regulation capabilities: during the day, the PCM layer absorbs excess heat from the RC layer, which is above the melting temperature of the PCM. At night, the PCM releases this stored heat to the surroundings. This cycle effectively provided an average ambient temperature reduction of 6.3 °C during the day and an average temperature increase of 2.1 °C at night, showcasing its dual cooling and warming functions. Su et al. [201] introduced a dynamic radiative cooler incorporating a top layer of paraffin. In this innovative design, the paraffin not only serves as a PCM for energy storage but also acts as a self-switchable cover. This transition occurs as the paraffin changes from solid to liquid; the spectral transmittance of solid paraffin remains below 5%, whereas it exceeds 90% when the paraffin is in its liquid state. This property allows for an adaptive response to varying thermal conditions, enhancing the overall functionality of the radiative cooler.

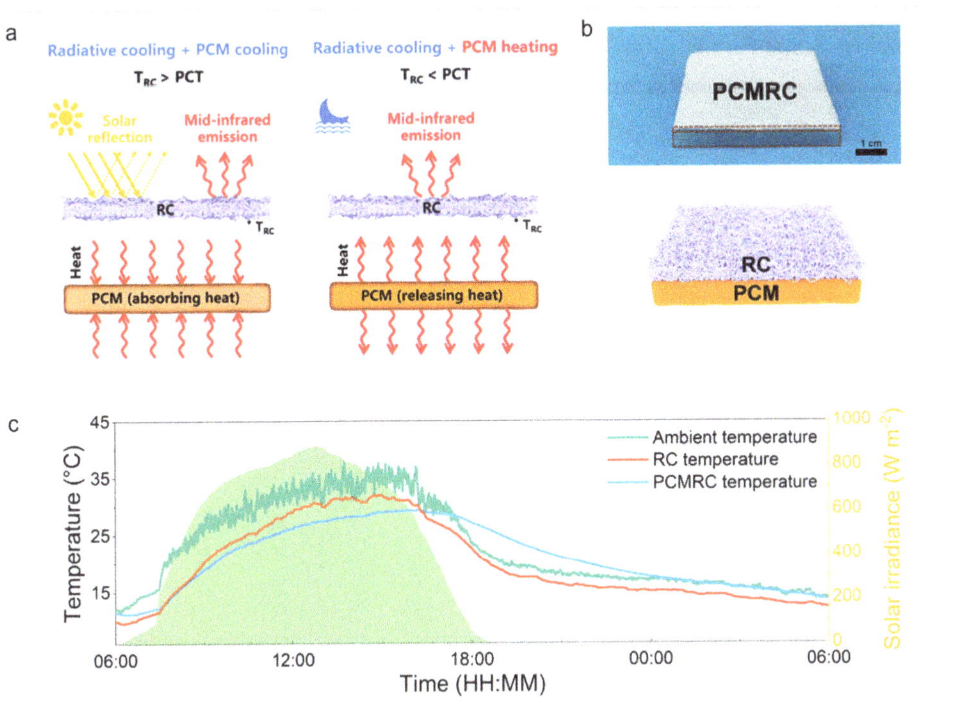

Figure 9. A bi-layer phase change material-enhanced radiative cooler (PCMRC). (**a**) Working principle of the PCMRC. (**b**) The structure of the PCMR. (**c**) Cooling performance comparison using RC and the PCMRC. The data were measured from 4 to 5 April 2022, in Shenzhen, China. Figures reprinted from [199].

5.3. PCM-Integrated Thermochromic Coatings

Thermochromic coatings are a novel energy-saving solution that regulate solar irradiation penetration to minimize radiative heat exchange between buildings and the environment [202]. These coatings respond dynamically to changes in temperature, adjusting their properties to either block or allow solar heat, thus helping to maintain indoor thermal comfort more efficiently. Hydrogels, perovskite, and vanadium dioxide (VO_2) are predominant materials for thermochromic coatings [203]. VO_2, as a type of PCM, is particularly noted for its ability to shift from a monoclinic to a rutile phase at a transition temperature of 341 K, changing its transparency in the NIR region and thereby enabling smart thermal regulation [204]. However, the high phase change temperature of VO_2 limits its practical applications in cooler climates or during milder weather conditions. To address this, researchers have explored doping strategies to modify the transition temperature. For example, Shen et al. [26] successfully reduced the phase change temperature of VO_2 by 21.6 °C through tungsten doping, making it more suitable for a broader range of climatic conditions.

Further advancements in thermochromic technologies include the development of multifunctional coatings. Pi et al. [205] prepared a smart window coating consisting of core–shell VO_2 nanoparticles that are both thermochromic and hydrophobic, providing antifouling properties. This superhydrophobic surface is resistant to a broad range of liquids and exhibits robust mechanical strength, making it ideal for long-term applications. Geng et al. [206] developed a self-templated method to produce ultrahigh transparent VO_2 nanoparticles. By controlling the grain size, they achieved a luminous transmittance of up to 82.9% for a single-sided VO_2 coating, enhancing both the aesthetic and functional properties of the material. The integration of thermochromic coatings with other types of PCMs can further enhance thermal performance. Jin et al. [207] numerically analyzed the thermal performance of window structures with various glazings. They found that triple glazing windows integrated with PCM, VO_2 coating, and low-E coating were the most energy efficient, reducing heat gain by up to 32% on sunny days and 40% on cloudy days during summer compared to standard double glazing. Ji and Li [208] explored a combined thermochromic–PCM system aimed at improving thermal performance in extreme environmental conditions as shown in Figure 10. Their simulations indicated that this system could reduce annual energy demand by 42.9 kWh/m^2, achieving an energy saving rate of up to 39%. Moreover, Imghoure et al. [209] assessed the thermal performance of a smart wall with five different configurations, incorporating W-doped VO_2 and two layers of PCMs with distinct melting temperatures as shown in Figure 10. This configuration demonstrated optimal performance, achieving comfortable indoor temperatures in both summer and winter seasons.

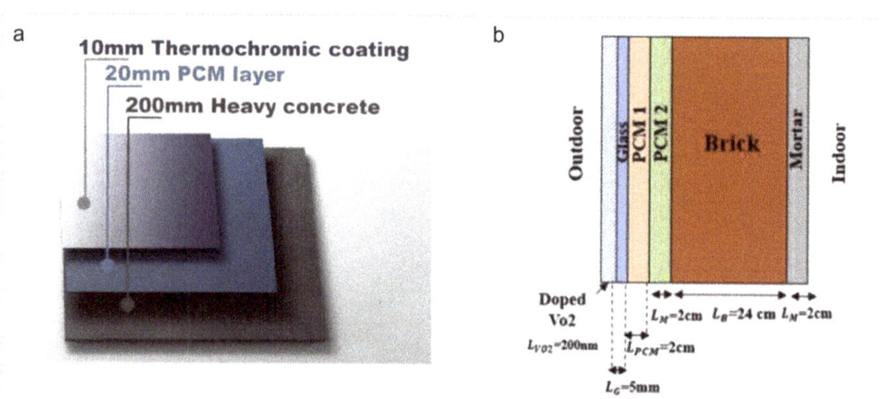

Figure 10. (**a**) A combined thermochromic–PCM system for extreme weather conditions. Figure reprinted from [208]. (**b**) Configuration of a PCM thermochromic smart wall towards indoor comfort for both summer and winter seasons. Figures reprinted from [209].

5.4. Other Applications

PCMs offer significant benefits in thermal regulation and energy storage, making them ideal for integration with various building envelope systems. Čurpek et al. [182] investigated the dynamic thermal response of a PCM-integrated BIPV system as illustrated in Figure 11a. Their study revealed that by incorporating PCMs, the BIPV/PCM system could reduce the peak operating temperature of photovoltaic panels by approximately 3 °C. This temperature reduction is crucial, as it enhances the electrical efficiency of the PV panels, which typically decrease in performance as temperature increases. Additionally, the integration of PCMs also resulted in reduced temperature fluctuations on the wall side, contributing positively to maintaining a more consistent and comfortable indoor environment. Izadpanah et al. [210] assessed the thermal efficiency of a double skin facade (DSF) integrated with PCMs and a green roof. Their findings indicated that this combination significantly reduced the cooling load in three different cities in Iran as depicted in Figure 11b. The synergy between the DSF, which provides an additional layer of thermal insulation and solar shading, and the PCM, which modulates the heat gains and losses, coupled with the natural cooling effect of a green roof, creates an effective barrier against heat penetration, thereby enhancing the overall energy efficiency of the building. Németh et al. [211] advanced the functionality of PCMs by developing an antimicrobial microencapsulated PCM, which involves encapsulating coconut oil in a calcium alginate shell embedded with silver nanoparticles. This novel PCM demonstrated significant antimicrobial properties, effectively eliminating both bacteria and fungi by the antimicrobial effect introduced by the Ag nanoparticles as shown in Figure 11c. The experimental results indicated that the highest loading of Ag nanoparticles (1.3%wt related to the total capsule) dispersed on the shell was particularly effective. This development holds promise for applications in office and home buildings, where it can provide both thermal regulation and antimicrobial functions, ensuring a healthier indoor environment. These applications indicate that PCMs not only contribute to thermal management but also broaden the scope of their utility in creating more adaptive and health-conscious building environments.

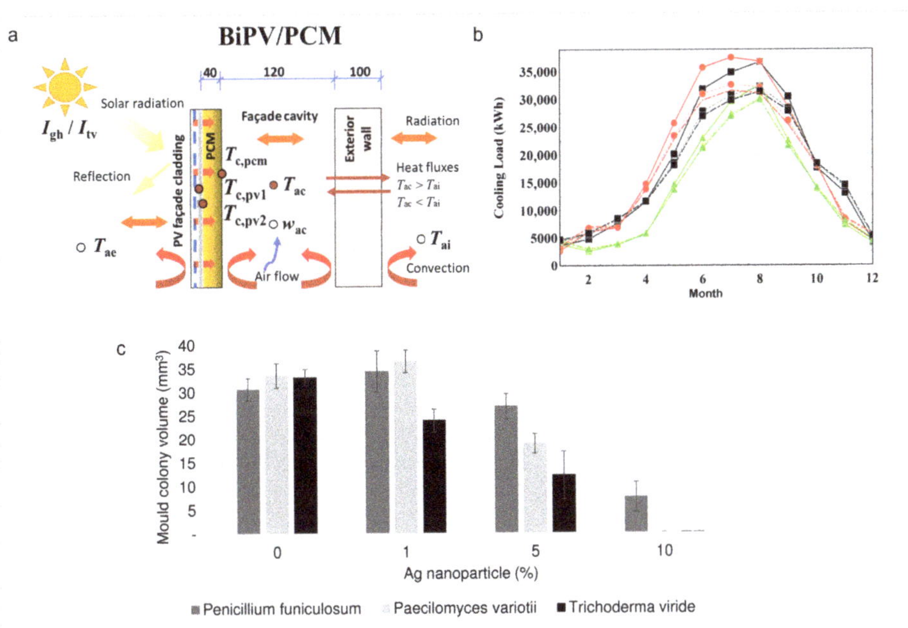

Figure 11. (**a**) Schematic of a BIPV/PCM facade. Figure reprinted from [182]. (**b**) Cooling load of a building with different envelope configurations in three Iranian cities, Amol (green curves), Tehran (black curves), and Yazd (red curves). The bold solid line, dashed line, and bold dashed line represent the cooling load by the DSF, DSF with PCM, and DSF with PCM and a green roof, respectively. Figure reprinted from [210]. (**c**) Antifungal effect with varying Ag nanoparticle concentration. Figure reprinted from [211].

6. Conclusions and Future Prospects

This paper provides a comprehensive review of the state-of-the-art PCM technology for building envelope applications. Various types of PCMs, including organic, inorganic, and eutectic candidates, are surveyed, and their advantages and disadvantages are analyzed. Different PCM incorporation methods such as shape stabilization, macroencapsulation, micro/nanoencapsulation, and solid–solid transition techniques are also discussed. These methods are crucial in addressing challenges like leakage and phase separation, thereby enhancing the practical viability of PCMs for construction uses. Furthermore, this paper conducts a review of the criteria for integrating PCMs into buildings and explores the novel applications of PCMs. The integration of PCMs into various building envelopes demonstrates their potential to meet specific environmental and functional requirements, making them a versatile and essential component of modern green building strategies. The main findings of this review are summarized below:

- In building envelopes, PCMs can increase the thermal inertia of a building, reduce thermal fluctuations, and delay heat peaks. Although some materials, such as SSPCMs and eutectic alloys, show great potential, they currently have limited studies in building envelopes due to their underdevelopment.
- The novel applications of PCMs in buildings leverage their temperature-stabilizing and energy storage capabilities, providing multifunctional benefits and significant potential for use in various building applications.
- Despite their advantages, integrating PCMs can introduce several issues, including fire hazards, potential toxicity, pollution, reduced mechanical performance, and higher initial costs. These issues should be considered during the building design process and are often overlooked in novel applications discussed in this paper.
- It is shown that integrating PCMs in buildings can reduce the energy consumption of HVAC systems. However, in some regions, traditional insulation materials may suffice, and the payback time for PCM integration may exceed the building's lifespan. Therefore, conducting a LCA or LCCA is necessary for PCM-integrated buildings.
- There are numerous choices for PCMs in building envelope design, each with distinct advantages in different aspects. Selection should be optimized based on integration circumstances, building design, sustainability considerations, and local climate.

These are some promising areas for future study to further advance PCM applications in building envelopes, summarized below:

- Currently, there are few studies on the applications of eutectic alloy PCMs in building envelopes. Although these PCMs are more costly than organic PCMs and salt hydrates, their high thermal conductivity and volumetric energy density warrant further investigation.
- Research on SSPCMs has primarily focused on material preparation rather than applications. Techniques for enhancing heat transfer in SSPCMs require further development.
- The long-term performance of PCMs currently relies on thermal stability and reliability analysis, yet there is a lack of experimental data over extended periods.
- Further investigations are needed to integrate PCMs with other functional materials, such as photothermal materials for building envelopes in cold regions, noise-canceling materials, or electromagnetic wave insulation materials for buildings with specialized uses.
- Studies should be conducted on the large-scale impacts of PCM-integrated envelopes on the urban heat island effect.

Author Contributions: Conceptualization, L.L.; methodology, L.Z. and G.W.; formal analysis, L.Z.; investigation, K.J.; resources, L.L.; data curation, K.J. and L.L.; writing—original draft preparation, K.J.; writing—review and editing, L.L.; supervision, L.L.; project administration, G.W.; funding acquisition, L.L. and L.Z. All authors have read and agreed to the published version of the manuscript.

Funding: This study is partially financially supported by a collaborative scheme between the Hong Kong Polytechnic University and the China Petrochemical Technology Company Limited, which is sponsored by the China Petrochemical Technology Company Limited (Project No.P0050065). This paper is also partially supported by a grant from the Research Grants Council of the Hong Kong Special Administrative Region, China (Project No. PolyU/RGC Project No 15219323).

Institutional Review Board Statement: Not applicable.

Informed Consent Statement: Not applicable.

Data Availability Statement: Not applicable.

Conflicts of Interest: Liang Zhao and Gang Wang are employed by the company Sinopec (Dalian) Research Institute of Petroleum and Petrochemicals Co. Ltd. The remaining authors declare that the research was conducted in the absence of any commercial or financial relationships that could be construed as a potential conflict of interest.

eferences

Jia, L.; Lu, L.; Chen, J. Exploring the cooling potential maps of a radiative sky cooling radiator-assisted ground source heat pump system in China. *Appl. Energy* **2023**, *349*, 121678. [CrossRef]

Kumar, D.; Alam, M.; Memon, R.A.; Bhayo, B.A. A critical review for formulation and conceptualization of an ideal building envelope and novel sustainability framework for building applications. *Clean. Eng. Technol.* **2022**, *11*, 100555. [CrossRef]

Cristino, T.M.; Neto, A.F.; Wurtz, F.; Delinchant, B. The evolution of knowledge and trends within the building energy efficiency field of knowledge. *Energies* **2022**, *15*, 691. [CrossRef]

. Pihelo, P.; Kalamees, T. The effect of thermal transmittance of building envelope and material selection of wind barrier on moisture safety of timber frame exterior wall. *J. Build. Eng.* **2016**, *6*, 29–38. [CrossRef]

Stazi, F.; Tomassoni, E.; Bonfigli, C.; Di Perna, C. Energy, comfort and environmental assessment of different building envelope techniques in a Mediterranean climate with a hot dry summer. *Appl. Energy* **2014**, *134*, 176–196. [CrossRef]

Sitzia, F.; Lisci, C.; Mirao, J. Accelerate ageing on building stone materials by simulating daily, seasonal thermo-hygrometric conditions and solar radiation of Csa Mediterranean climate. *Constr. Build. Mater.* **2021**, *266*, 121009. [CrossRef]

. Kapoor, G.; Singhal, M. Impact of innovative thermal insulation materials in the building envelope on energy efficiency of residential buildings. *Mater. Today Proc.* **2024**, *in press* . [CrossRef]

. Ma, Z.; Liu, X.; Xu, X.; Liu, L.; Yu, B.; Maluk, C.; Huang, G.; Wang, H.; Song, P. Bioinspired, highly adhesive, nanostructured polymeric coatings for superhydrophobic fire-extinguishing thermal insulation foam. *ACS Nano* **2021**, *15*, 11667–11680. [CrossRef] [PubMed]

. Karimi, F.; Soltani, P.; Zarrebini, M.; Hassanpour, A. Acoustic and thermal performance of polypropylene nonwoven fabrics for insulation in buildings. *J. Build. Eng.* **2022**, *50*, 104125. [CrossRef]

0. Guo, Y.; Ruan, K.; Shi, X.; Yang, X.; Gu, J. Factors affecting thermal conductivities of the polymers and polymer composites: A review. *Compos. Sci. Technol.* **2020**, *193*, 108134. [CrossRef]

1. Li, C.; Wen, X.; Cai, W.; Yu, H.; Liu, D. Phase change material for passive cooling in building envelopes: A comprehensive review. *J. Build. Eng.* **2023**, *65*, 105763. [CrossRef]

2. Wang, P.; Liu, Z.; Zhang, X.; Hu, M.; Zhang, L.; Fan, J. Adaptive dynamic building envelope integrated with phase change material to enhance the heat storage and release efficiency: A state-of-the-art review. *Energy Build.* **2023**, *286*, 112928. [CrossRef]

3. Gao, Y.; Meng, X. A comprehensive review of integrating phase change materials in building bricks: Methods, performance and applications. *J. Energy Storage* **2023**, *62*, 106913. [CrossRef]

4. Rathore, P.K.S.; Shukla, S.K.; Gupta, N.K. Potential of microencapsulated PCM for energy savings in buildings: A critical review. *Sustain. Cities Soc.* **2020**, *53*, 101884. [CrossRef]

5. Wang, X.; Li, W.; Luo, Z.; Wang, K.; Shah, S.P. A critical review on phase change materials (PCM) for sustainable and energy efficient building: Design, characteristic, performance and application. *Energy Build.* **2022**, *260*, 111923. [CrossRef]

6. Rathore, P.K.S.; Gupta, N.K.; Yadav, D.; Shukla, S.K.; Kaul, S. Thermal performance of the building envelope integrated with phase change material for thermal energy storage: An updated review. *Sustain. Cities Soc.* **2022**, *79*, 103690. [CrossRef]

7. Li, D.; Wu, Y.; Wang, B.; Liu, C.; Arıcı, M. Optical and thermal performance of glazing units containing PCM in buildings: A review. *Constr. Build. Mater.* **2020**, *233*, 117327. [CrossRef]

8. Al-Yasiri, Q.; Szabó, M. Incorporation of phase change materials into building envelope for thermal comfort and energy saving: A comprehensive analysis. *J. Build. Eng.* **2021**, *36*, 102122. [CrossRef]

9. Raj, C.R.; Suresh, S.; Bhavsar, R.; Singh, V.K. Recent developments in thermo-physical property enhancement and applications of solid solid phase change materials: A review. *J. Therm. Anal. Calorim.* **2020**, *139*, 3023–3049. [CrossRef]

0. Mousavi, S.; Rismanchi, B.; Brey, S.; Aye, L. PCM embedded radiant chilled ceiling: A state-of-the-art review. *Renew. Sustain. Energy Rev.* **2021**, *151*, 111601. [CrossRef]

1. Li, D.; Ma, Y.; Zhang, S.; Yang, R.; Zhang, C.; Liu, C. Photothermal and energy performance of an innovative roof based on silica aerogel-PCM glazing systems. *Energy Convers. Manag.* **2022**, *262*, 115567. [CrossRef]

22. Zhao, A.; Xiao, X.; Hu, Z.T.; Zhu, W.; Yang, J.; Yang, E.H. Multifunctional F-doped TiO_2 PCM microcapsules for visible-light-drive photocatalysis and latent heat storage. *Appl. Energy* **2024**, *359*, 122674. [CrossRef]
23. Wu, Y.; Zhao, H.; Sun, H.; Duan, M.; Lin, B.; Wu, S. A review of the application of radiative sky cooling in buildings: Challeng and optimization. *Energy Convers. Manag.* **2022**, *265*, 115768. [CrossRef]
24. Ushak, S.; Gutierrez, A.; Galleguillos, H.; Fernandez, A.G.; Cabeza, L.F.; Grágeda, M. Thermophysical characterization of by-product from the non-metallic industry as inorganic PCM. *Sol. Energy Mater. Sol. Cells* **2015**, *132*, 385–391. [CrossRef]
25. Sun, X.; Lee, K.O.; Medina, M.A.; Chu, Y.; Li, C. Melting temperature and enthalpy variations of phase change materials (PCM A differential scanning calorimetry (DSC) analysis. *Phase Transitions* **2018**, *91*, 667–680. [CrossRef]
26. Shen, Z.; Kwon, S.; Lee, H.L.; Toivakka, M.; Oh, K. Enhanced thermal energy storage performance of salt hydrate phase chan material: Effect of cellulose nanofibril and graphene nanoplatelet. *Sol. Energy Mater. Sol. Cells* **2021**, *225*, 111028. [CrossRef]
27. Louanate, A.; El Otmani, R.; Kandoussi, K.; Boutaous, M. Non-isothermal crystallization kinetics of paraffin wax as a pha changing energy storage material. *Phys. Scr.* **2020**, *95*, 105003. [CrossRef]
28. Nazari Sam, M.; Caggiano, A.; Mankel, C.; Koenders, E. A comparative study on the thermal energy storage performance bio-based and paraffin-based PCMs using DSC procedures. *Materials* **2020**, *13*, 1705. [CrossRef]
29. Aftab, W.; Usman, A.; Shi, J.; Yuan, K.; Qin, M.; Zou, R. Phase change material-integrated latent heat storage systems f sustainable energy solutions. *Energy Environ. Sci.* **2021**, *14*, 4268–4291. [CrossRef]
30. Tripathi, B.M.; Shukla, S.K. A comprehensive review of the thermal performance in energy efficient building envelope incorp rated with phase change materials. *J. Energy Storage* **2024**, *79*, 110128. [CrossRef]
31. Atinafu, D.G.; Yun, B.Y.; Kwon, E.E.; Chang, S.J.; Kim, S. Unveiling the effect of molecular chain length on the thermal energ storage capacity and transition temperature of alkane-based phase change composites. *Chem. Eng. J.* **2023**, *462*, 142303. [CrossRe
32. Abhat, A. Low temperature latent heat thermal energy storage: Heat storage materials. *Sol. Energy* **1983**, *30*, 313–332. [CrossRe
33. Chang, Z.; Wang, K.; Wu, X.; Lei, G.; Wang, Q.; Liu, H.; Wang, Y.; Zhang, Q. Review on the preparation and performance paraffin-based phase change microcapsules for heat storage. *J. Energy Storage* **2022**, *46*, 103840. [CrossRef]
34. Vakhshouri, A.R. Paraffin as phase change material. In *Paraffin Overview*; InTech Open: London, UK, 2020; pp. 1–23.
35. Jia, F.R.; Jing, W.T.; Liu, G.X.; Yue, Q.; Wang, H.M.; Shi, L. Paraffin-based crude oil refining process unit-level energy consumptio and CO2 emissions in China. *J. Clean. Prod.* **2020**, *255*, 120347. [CrossRef]
36. Adlan, N.A.; Sabri, S.; Masomian, M.; Ali, M.S.M.; Rahman, R.N.Z.R.A. Microbial biodegradation of paraffin wax in Malaysia crude oil mediated by degradative enzymes. *Front. Microbiol.* **2020**, *11*, 565608. [CrossRef] [PubMed]
37. Nazari, M.; Jebrane, M.; Terziev, N. Bio-based phase change materials incorporated in lignocellulose matrix for energy storage i buildings—A review. *Energies* **2020**, *13*, 3065. [CrossRef]
38. Mishra, R.K.; Verma, K.; Mishra, V.; Chaudhary, B. A review on carbon-based phase change materials for thermal energy storag *J. Energy Storage* **2022**, *50*, 104166. [CrossRef]
39. Yu, S.; Jeong, S.G.; Chung, O.; Kim, S. Bio-based PCM/carbon nanomaterials composites with enhanced thermal conductivit *Sol. Energy Mater. Sol. Cells* **2014**, *120*, 549–554. [CrossRef]
40. Mehrizi, A.A.; Karimi-Maleh, H.; Naddafi, M.; Karimi, F. Application of bio-based phase change materials for effective he management. *J. Energy Storage* **2023**, *61*, 106859. [CrossRef]
41. Sarı, A.; Karaipekli, A. Preparation, thermal properties and thermal reliability of palmitic acid/expanded graphite composite a form-stable PCM for thermal energy storage. *Sol. Energy Mater. Sol. Cells* **2009**, *93*, 571–576. [CrossRef]
42. Kumar, A.; Jain, H.; Tripathi, B.P. Synthesis and nanoencapsulation of poly (ethylene glycol)-distearates phase change materia for latent heat storage and release. *ACS Appl. Energy Mater.* **2020**, *3*, 5965–5976. [CrossRef]
43. Weng, J.; Huang, Q.; Li, X.; Zhang, G.; Ouyang, D.; Chen, M.; Yuen, A.C.Y.; Li, A.; Lee, E.W.M.; Yang, W.; et al. Safety issue o PCM-based battery thermal management: Material thermal stability and system hazard mitigation. *Energy Storage Mater.* **202** *53*, 580–612. [CrossRef]
44. Raju, K.; Kumar, G.N.; Ram, V.V.; Parameshwaran, R. Preparation, thermal and structural properties of n-octadecane/melamin formaldehyde nanocapsules embedded cement mortar for energy storage application in buildings. *Mater. Today Proc.* **202** *56*, 1424–1431. [CrossRef]
45. Bo, H.; Gustafsson, E.M.; Setterwall, F. Tetradecane and hexadecane binary mixtures as phase change materials (PCMs) for coc storage in district cooling systems. *Energy* **1999**, *24*, 1015–1028. [CrossRef]
46. Li, C.; Zhang, B.; Liu, Q. N-eicosane/expanded graphite as composite phase change materials for electro-driven thermal energ storage. *J. Energy Storage* **2020**, *29*, 101339. [CrossRef]
47. Liu, Y.; Wang, M.; Cui, H.; Yang, L.; Liu, J. Micro-/macro-level optimization of phase change material panel in building envelop *Energy* **2020**, *195*, 116932. [CrossRef]
48. Hasan, M.I.; Basher, H.O.; Shdhan, A.O. Experimental investigation of phase change materials for insulation of residentia buildings. *Sustain. Cities Soc.* **2018**, *36*, 42–58. [CrossRef]
49. de Albuquerque Landi, F.F.; Fabiani, C.; Pisello, A.L. Palm oil for seasonal thermal energy storage applications in buildings: Th potential of multiple melting ranges in blends of bio-based fatty acids. *J. Energy Storage* **2020**, *29*, 101431. [CrossRef]
50. Aslfattahi, N.; Saidur, R.; Arifutzzaman, A.; Sadri, R.; Bimbo, N.; Sabri, M.F.M.; Maughan, P.A.; Bouscarrat, L.; Dawson, R.J Said, S.M.; et al. Experimental investigation of energy storage properties and thermal conductivity of a novel organic phas change material/MXene as A new class of nanocomposites. *J. Energy Storage* **2020**, *27*, 101115. [CrossRef]

51. Fan, Z.; Zhao, Y.; Liu, X.; Shi, Y.; Jiang, D. Thermal properties and reliabilities of lauric acid-based binary eutectic fatty acid as a phase change material for building energy conservation. *ACS Omega* **2022**, *7*, 16097–16108. [CrossRef]
52. Sun, J.; Zhao, J.; Zhang, W.; Xu, J.; Wang, B.; Wang, X.; Zhou, J.; Guo, H.; Liu, Y. Composites with a novel core–shell structural expanded perlite/polyethylene glycol composite PCM as novel green energy storage composites for building energy conservation. *Appl. Energy* **2023**, *330*, 120363. [CrossRef]
53. Mokhtari, S.; Madhkhan, M. The performance effect of PEG-silica fume as shape-stabilized phase change materials on mechanical and thermal properties of lightweight concrete panels. *Case Stud. Constr. Mater.* **2022**, *17*, e01298. [CrossRef]
54. Wu, M.; Liu, C.; Rao, Z. Experimental study on lauryl alcohol/expanded graphite composite phase change materials for thermal regulation in building. *Constr. Build. Mater.* **2022**, *335*, 127400. [CrossRef]
55. Dinesh, R.; Hussain, S.I.; Roseline, A.A.; Kalaiselvam, S. Experimental investigation on heat transfer behavior of the novel ternary eutectic PCM embedded with MWCNT for thermal energy storage systems. *J. Therm. Anal. Calorim.* **2021**, *145*, 2935–2949. [CrossRef]
56. Mohamed, S.A.; Al-Sulaiman, F.A.; Ibrahim, N.I.; Zahir, M.H.; Al-Ahmed, A.; Saidur, R.; Yılbaş, B.; Sahin, A. A review on current status and challenges of inorganic phase change materials for thermal energy storage systems. *Renew. Sustain. Energy Rev.* **2017**, *70*, 1072–1089. [CrossRef]
57. Soares, N.; Costa, J.J.; Gaspar, A.R.; Santos, P. Review of passive PCM latent heat thermal energy storage systems towards buildings' energy efficiency. *Energy Build.* **2013**, *59*, 82–103. [CrossRef]
58. Önder, E.; Sarier, N. Thermal regulation finishes for textiles. In *Functional Finishes for Textiles. Improving Comfort, Performance and Protection*; Woodhead Publishing: Cambridge, UK, 2015; p. 17.
59. Sandnes, B. *Exergy Efficient Production, Storage and Distribution of Solar Energy*; U.S. Department of Energy Office of Scientific and Technical Information: Oak Ridge, TN, USA, 2003.
60. Sandnes, B.; Rekstad, J. Supercooling salt hydrates: Stored enthalpy as a function of temperature. *Sol. Energy* **2006**, *80*, 616–625. [CrossRef]
61. Beaupere, N.; Soupremanien, U.; Zalewski, L. Nucleation triggering methods in supercooled phase change materials (PCM), a review. *Thermochim. Acta* **2018**, *670*, 184–201. [CrossRef]
62. He, Z.; Ma, H.; Lu, S. Preparation and thermal performance study of a novel hydrated salt composite PCM for space heating. *J. Energy Storage* **2024**, *90*, 111906. [CrossRef]
63. Cabeza, L.F.; Castell, A.; Barreneche, C.d.; De Gracia, A.; Fernández, A. Materials used as PCM in thermal energy storage in buildings: A review. *Renew. Sustain. Energy Rev.* **2011**, *15*, 1675–1695. [CrossRef]
64. Jiao, K.; Lu, L.; Wen, T.; Wang, Q. A modified mixture theory for one-dimensional melting of pure PCM and PCM/metal foam composite: Numerical analysis and experiment validation. *Int. J. Heat Mass Transf.* **2022**, *186*, 122461. [CrossRef]
65. Yagi, J.; Akiyama, T. Storage of thermal energy for effective use of waste heat from industries. *J. Mater. Process. Technol.* **1995**, *48*, 793–804. [CrossRef]
66. Copus, M.; Fraser, B.; Reece, R.; Hands, S.; Cuskelly, D.; Sugo, H.; Reed, S.; Bradley, J.; Post, A.; Kisi, E. On-sun testing of Miscibility Gap Alloy thermal storage. *Sol. Energy* **2019**, *177*, 657–664. [CrossRef]
67. Liu, L.; Li, J.; Deng, Y.; Yang, Z.; Huang, K.; Zhao, S. Optimal design of multi-layer structure composite containing inorganic hydrated salt phase change materials and cement: Lab-scale tests for buildings. *Constr. Build. Mater.* **2021**, *275*, 122125. [CrossRef]
68. Rezvanpour, M.; Borooghani, D.; Torabi, F.; Pazoki, M. Using $CaCl_2 \cdot 6H_2O$ as a phase change material for thermo-regulation and enhancing photovoltaic panels' conversion efficiency: Experimental study and TRNSYS validation. *Renew. Energy* **2020**, *146*, 1907–1921. [CrossRef]
69. Xie, N.; Gao, X.; Zhong, Y.; Ye, R.; Chen, S.; Ding, L.; Zhong, T. Enhanced thermal performance of $Na_2HPO_4 \cdot 12H_2O$ composite phase change material supported by sepiolite fiber for floor radiant heating system. *J. Build. Eng.* **2022**, *56*, 104747. [CrossRef]
70. Xiao, Q.; Fan, J.; Fang, Y.; Li, L.; Xu, T.; Yuan, W. The shape-stabilized light-to-thermal conversion phase change material based on $CH_3COONa \cdot 3H_2O$ as thermal energy storage media. *Appl. Therm. Eng.* **2018**, *136*, 701–707. [CrossRef]
71. Man, X.; Lu, H.; Xu, Q.; Wang, C.; Ling, Z. Review on the thermal property enhancement of inorganic salt hydrate phase change materials. *J. Energy Storage* **2023**, *72*, 108699. [CrossRef]
72. Gupta, N.; Kumar, A.; Dhasmana, H.; Kumar, V.; Kumar, A.; Shukla, P.; Verma, A.; Nutan, G.V.; Dhawan, S.; Jain, V. Enhanced thermophysical properties of Metal oxide nanoparticles embedded magnesium nitrate hexahydrate based nanocomposite for thermal energy storage applications. *J. Energy Storage* **2020**, *32*, 101773. [CrossRef]
73. Dixit, P.; Reddy, V.J.; Dasari, A.; Chattopadhyay, S. Preparation of perlite based-zinc nitrate hexahydrate composite for electric radiant floor heating in model building and numerical analysis. *J. Energy Storage* **2022**, *52*, 104804. [CrossRef]
74. Bouhal, T.; Meghari, Z.; El Rhafiki, T.; Kousksou, T.; Jamil, A.; Ghoulam, E.B. Parametric CFD analysis and impact of PCM intrinsic parameters on melting process inside enclosure integrating fins: Solar building applications. *J. Build. Eng.* **2018**, *20*, 634–646. [CrossRef]
75. Chen, K.; Ding, J.; Wang, W.; Lu, J. Shape-stable Bi-Sn-In alloy/Ag/copper foam composite phase change material for thermal storage and management. *Chem. Eng. J.* **2023**, *454*, 140087. [CrossRef]
76. Pan, A.; Wang, Y.; Cao, Y.; Zhang, H.; Zhao, J.; Hao, Z.; Wu, W. Research on thermal properties of phase change thermal control materials Bi-21In-18Pb-12Sn and Ga-13.5 Sn with low melting point. *J. Mater. Res. Technol.* **2020**, *9*, 3897–3906. [CrossRef]

77. Ding, B.; Zhu, X.; Wang, H.; He, X.Y.; Tan, Y. Experimental study on phase change heat transfer characteristics of alloys. *Int. J. Heat Mass Transf.* **2017**, *105*, 261–269. [CrossRef]
78. Deng, C.; Zhang, X.; Peng, J.; Yang, X.; Shang, B.; Luo, X. In51Bi32. 5Sn16. 5@ SiO2 microcapsules-based composite phase change materials with high thermal conductivity and heat storage density for electronics thermal management. *J. Energy Storage* **2024**, *86*, 111432. [CrossRef]
79. Ding, B.; Feng, W.C.; Fang, J.; Li, S.Z.; Gong, L. How natural convection affect cooling performance of PCM heat sink. *Int. J. Heat Mass Transf.* **2022**, *184*, 122272. [CrossRef]
80. Huang, P.; Wei, G.; Cui, L.; Xu, C.; Du, X. Numerical investigation of a dual-PCM heat sink using low melting point alloy and paraffin. *Appl. Therm. Eng.* **2021**, *189*, 116702. [CrossRef]
81. Singh, P.; Sharma, R.; Ansu, A.; Goyal, R.; Sarı, A.; Tyagi, V. A comprehensive review on development of eutectic organic phase change materials and their composites for low and medium range thermal energy storage applications. *Sol. Energy Mater. Sol. Cells* **2021**, *223*, 110955. [CrossRef]
82. Sun, M.; Liu, T.; Sha, H.; Li, M.; Liu, T.; Wang, X.; Chen, G.; Wang, J.; Jiang, D. A review on thermal energy storage with eutectic phase change materials: Fundamentals and applications. *J. Energy Storage* **2023**, *68*, 107713. [CrossRef]
83. Qiu, W.; Bai, Y.; Fang, Y.; Wang, S. Preparation and thermal properties of novel inorganic-organic eutectic composite material with high latent heat and thermal conductivity based on aluminum sulfate salt. *J. Energy Storage* **2022**, *55*, 105364. [CrossRef]
84. Jacob, J.; Pandey, A.; Abd Rahim, N.; Selvaraj, J.; Paul, J.; Samykano, M.; Saidur, R. Quantifying thermophysical properties, characterization, and thermal cycle testing of nano-enhanced organic eutectic phase change materials for thermal energy storage applications. *Sol. Energy Mater. Sol. Cells* **2022**, *248*, 112008. [CrossRef]
85. Chen, C.; Wang, X.; Ma, F.; Wang, Y.; Jiu, S.; Chen, Y. Preparation and characterization of modified activated carbon-based shape stabilized eutectic phase change materials for gypsum composites application. *Constr. Build. Mater.* **2023**, *369*, 130551. [CrossRef]
86. Sang, G.; Zeng, H.; Guo, Z.; Cui, H.; Zhang, Y.; Cui, X.; Zhang, L.; Han, W. Studies of eutectic hydrated salt/polymer hydrogel composite as form-stable phase change material for building thermal energy storage. *J. Build. Eng.* **2022**, *59*, 105010. [CrossRef]
87. Zahir, M.H.; Mohamed, S.A.; Saidur, R.; Al-Sulaiman, F.A. Supercooling of phase-change materials and the techniques used to mitigate the phenomenon. *Appl. Energy* **2019**, *240*, 793–817. [CrossRef]
88. Gunasekara, S.N.; Martin, V.; Chiu, J.N. Phase equilibrium in the design of phase change materials for thermal energy storage: State-of-the-art. *Renew. Sustain. Energy Rev.* **2017**, *73*, 558–581. [CrossRef]
89. Chakraborty, S.; Chormale, J.H.; Bansal, A.K. Deep eutectic systems: An overview of fundamental aspects, current understanding and drug delivery applications. *Int. J. Pharm.* **2021**, *610*, 121203. [CrossRef]
90. Garg, H.P.; Mullick, S.C.; Bhargava, V.K. *Solar Thermal Energy Storage*; Springer Science & Business Media: Berlin/Heidelberg, Germany, 2012.
91. Li, X.; Fu, Z.; Qiao, Y.; Zhang, Z.; Zhou, Y.; Hai, C.; Shen, Y.; Sun, Y.; Zeng, J.; Ren, X. Preparation, characterization, and modification of sodium acetate trihydrate-urea binary eutectic mixtures as phase change material. *Energy Fuels* **2020**, *34*, 6439–6447. [CrossRef]
92. Philip, N.; Veerakumar, C.; Sreekumar, A. Lauryl alcohol and stearyl alcohol eutectic for cold thermal energy storage in buildings: Preparation, thermophysical studies and performance analysis. *J. Energy Storage* **2020**, *31*, 101600. [CrossRef]
93. Liu, J.; Jiang, D.; Fei, H.; Xu, Y.; Zeng, Z.; Ye, W. Preparation and energy storage properties of a lauric acid/octadecanol eutectic mixture. *ACS Omega* **2021**, *6*, 23542–23550. [CrossRef]
94. Kalidasan, B.; Pandey, A.; Saidur, R.; Tyagi, S.; Mishra, Y.K. Experimental evaluation of binary and ternary eutectic phase change material for sustainable thermal energy storage. *J. Energy Storage* **2023**, *68*, 107707.
95. Zhang, D.; Sani, B.M.; Xu, P.; Liu, K.; Gu, F. Preparation and characterization of binary eutectic phase change material laden with thermal conductivity enhancer for cooling steel slag asphalt pavement. *Constr. Build. Mater.* **2023**, *388*, 131688. [CrossRef]
96. Kalidasan, B.; Pandey, A.; Saidur, R.; Kothari, R.; Sharma, K.; Tyagi, V. Eco-friendly coconut shell biochar based nano-inclusion for sustainable energy storage of binary eutectic salt hydrate phase change materials. *Sol. Energy Mater. Sol. Cells* **2023**, *262*, 112534. [CrossRef]
97. Dai, J.; Ma, F.; Sangiorgi, C.; Tarsi, G.; Fu, Z.; Tataranni, P.; Li, C.; Hou, Y. Assessment of high-enthalpy composite eutectic phase change materials efficiency in asphalt binders for cooling pavements. *J. Clean. Prod.* **2024**, *442*, 140999. [CrossRef]
98. Chang, S.; Zhang, L.; Li, X.; Liu, B.; Meng, Y.; Hu, H. Experimental study of novel paraffin-fatty acid eutectic mixtures for thermal management of electronic devices. *J. Energy Storage* **2024**, *84*, 110846. [CrossRef]
99. Yu, K.; Jia, M.; Liu, Y.; Yang, Y. Binary decanoic acid/polyethylene glycol as a novel phase change material for thermal energy storage: Eutectic behaviors and energy conservation evaluation. *J. Energy Storage* **2023**, *68*, 107663. [CrossRef]
100. Gandhi, M.; Kumar, A.; Elangovan, R.; Meena, C.S.; Kulkarni, K.S.; Kumar, A.; Bhanot, G.; Kapoor, N.R. A review on shape-stabilized phase change materials for latent energy storage in buildings. *Sustainability* **2020**, *12*, 9481. [CrossRef]
101. Wu, M.; Wu, S.; Cai, Y.; Wang, R.; Li, T. Form-stable phase change composites: Preparation, performance, and applications for thermal energy conversion, storage and management. *Energy Storage Mater.* **2021**, *42*, 380–417. [CrossRef]
102. Laasri, I.A.; Es-sakali, N.; Charai, M.; Mghazli, M.O.; Outzourhit, A. Recent progress, limitations, and future directions of macro-encapsulated phase change materials for building applications. *Renew. Sustain. Energy Rev.* **2024**, *199*, 114481. [CrossRef]
103. Lachheb, M.; Younsi, Z.; Youssef, N.; Bouadila, S. Enhancing building energy efficiency and thermal performance with PCM-Integrated brick walls: A comprehensive review. *Build. Environ.* **2024**, *256*, 111476. [CrossRef]

4. Ghasemi, K.; Tasnim, S.; Mahmud, S. PCM, nano/microencapsulation and slurries: A review of fundamentals, categories, fabrication, numerical models and applications. *Sustain. Energy Technol. Assessments* **2022**, *52*, 102084. [CrossRef]
5. Usman, A.; Xiong, F.; Aftab, W.; Qin, M.; Zou, R. Emerging solid-to-solid phase-change materials for thermal-energy harvesting, storage, and utilization. *Adv. Mater.* **2022**, *34*, 2202457. [CrossRef] [PubMed]
6. Abu-Hamdeh, N.H.; Khoshaim, A.; Alzahrani, M.A.; Hatamleh, R.I. Study of the flat plate solar collector's efficiency for sustainable and renewable energy management in a building by a phase change material: Containing paraffin-wax/Graphene and Paraffin-wax/graphene oxide carbon-based fluids. *J. Build. Eng.* **2022**, *57*, 104804. [CrossRef]
7. Sarrafha, H.; Kasaeian, A.; Jahangir, M.H.; Taylor, R.A. Transient thermal response of multi-walled carbon nanotube phase change materials in building walls. *Energy* **2021**, *224*, 120120. [CrossRef]
8. Meng, Y.; Zhao, Y.; Zhang, Y.; Tang, B. Induced dipole force driven PEG/PPEGMA form-stable phase change energy storage materials with high latent heat. *Chem. Eng. J.* **2020**, *390*, 124618. [CrossRef]
9. Zhang, M.; Xiao, Q.; Chen, C.; Li, L.; Yuan, W. Developing a heat-insulating composite phase change material with light-to-thermal conversion performance from graphene oxide/silica hybrid aerogel. *Appl. Therm. Eng.* **2020**, *174*, 115303. [CrossRef]
10. Rathore, P.K.S.; kumar Shukla, S. Improvement in thermal properties of PCM/Expanded vermiculite/expanded graphite shape stabilized composite PCM for building energy applications. *Renew. Energy* **2021**, *176*, 295–304. [CrossRef]
11. Mishra, A.K.; Lahiri, B.; Philip, J. Carbon black nano particle loaded lauric acid-based form-stable phase change material with enhanced thermal conductivity and photo-thermal conversion for thermal energy storage. *Energy* **2020**, *191*, 116572. [CrossRef]
12. Yang, Y.; Wu, W.; Fu, S.; Zhang, H. Study of a novel ceramsite-based shape-stabilized composite phase change material (PCM) for energy conservation in buildings. *Constr. Build. Mater.* **2020**, *246*, 118479. [CrossRef]
13. Kumar, N.; Rathore, P.K.S.; Sharma, R.; Gupta, N.K. Integration of lauric acid/zeolite/graphite as shape stabilized composite phase change material in gypsum for enhanced thermal energy storage in buildings. *Appl. Therm. Eng.* **2023**, *224*, 120088. [CrossRef]
14. Atinafu, D.G.; Yun, B.Y.; Wi, S.; Kang, Y.; Kim, S. A comparative analysis of biochar, activated carbon, expanded graphite, and multi-walled carbon nanotubes with respect to PCM loading and energy-storage capacities. *Environ. Res.* **2021**, *195*, 110853. [CrossRef]
15. Alam, T.E.; Dhau, J.S.; Goswami, D.Y.; Stefanakos, E. Macroencapsulation and characterization of phase change materials for latent heat thermal energy storage systems. *Appl. Energy* **2015**, *154*, 92–101. [CrossRef]
16. Rathore, P.K.S.; Shukla, S.K. An experimental evaluation of thermal behavior of the building envelope using macroencapsulated PCM for energy savings. *Renew. Energy* **2020**, *149*, 1300–1313. [CrossRef]
17. Liu, Z.; Yu, Z.J.; Yang, T.; Qin, D.; Li, S.; Zhang, G.; Haghighat, F.; Joybari, M.M. A review on macro-encapsulated phase change material for building envelope applications. *Build. Environ.* **2018**, *144*, 281–294. [CrossRef]
18. Abass, P.J.; Muthulingam, S. Thermal energy storage performance of biaxial voided RCC roof slab integrated with macroencapsulated PCM for passive cooling of buildings. *J. Energy Storage* **2024**, *88*, 111478. [CrossRef]
19. Rida, M.; Hoffmann, S. The influence of macro-encapsulated PCM panel's geometry on heat transfer in a ceiling application. *Adv. Build. Energy Res.* **2022**, *16*, 445–465. [CrossRef]
20. Hu, Z.; Li, W.; Yang, C.; Huang, H.; Guo, Y.; Ge, F.; Zhang, Y. Thermal performance of an active casing pipe macro-encapsulated PCM wall for space cooling and heating of residential building in hot summer and cold winter region in China. *Constr. Build. Mater.* **2024**, *422*, 135831. [CrossRef]
21. Al-Absi, Z.A.; Hafizal, M.I.M.; Ismail, M. Experimental study on the thermal performance of PCM-based panels developed for exterior finishes of building walls. *J. Build. Eng.* **2022**, *52*, 104379. [CrossRef]
22. Maleki, B.; Khadang, A.; Maddah, H.; Alizadeh, M.; Kazemian, A.; Ali, H.M. Development and thermal performance of nanoencapsulated PCM/plaster wallboard for thermal energy storage in buildings. *J. Build. Eng.* **2020**, *32*, 101727. [CrossRef]
23. Cabeza, L.F.; Navarro, L.; Pisello, A.L.; Olivieri, L.; Bartolomé, C.; Sánchez, J.; Álvarez, S.; Tenorio, J.A. Behaviour of a concrete wall containing micro-encapsulated PCM after a decade of its construction. *Sol. Energy* **2020**, *200*, 108–113. [CrossRef]
24. Palacios, A.; Navarro, M.E.; Barreneche, C.; Ding, Y. Hybridization of Salt Hydrates with Solid–Solid Phase Change Materials: A Novel Pathway to Sorption Thermochemical Materials Manufacturing. *Adv. Sustain. Syst.* **2023**, *7*, 2200184. [CrossRef]
25. Gough, R.; Primm, K.; Rivera-Valentín, E.; Martínez, G.; Tolbert, M. Solid-solid hydration and dehydration of Mars-relevant chlorine salts: Implications for Gale Crater and RSL locations. *Icarus* **2019**, *321*, 1–13. [CrossRef]
26. Broadhurst, M.G. An analysis of the solid phase behavior of the normal paraffins. *J. Res. Natl. Bur. Stand. Sect. A Phys. Chem.* **1962**, *66*, 241. [CrossRef]
27. Paneliya, S.; Khanna, S.; Singh, A.P.; Patel, Y.K.; Vanpariya, A.; Makani, N.H.; Banerjee, R.; Mukhopadhyay, I. Core shell paraffin/silica nanocomposite: A promising phase change material for thermal energy storage. *Renew. Energy* **2021**, *167*, 591–599. [CrossRef]
28. Shi, J.; Aftab, W.; Liang, Z.; Yuan, K.; Maqbool, M.; Jiang, H.; Xiong, F.; Qin, M.; Gao, S.; Zou, R. Tuning the flexibility and thermal storage capacity of solid–solid phase change materials towards wearable applications. *J. Mater. Chem. A* **2020**, *8*, 20133–20140. [CrossRef]
29. Wang, R.; Xiao, Y.; Lei, J. A solid-solid phase change material based on dynamic ion cross-linking with reprocessability at room temperature. *Chem. Eng. J.* **2020**, *390*, 124586. [CrossRef]

130. Hariti, Y.; Hader, A.; Faraji, H.; Boughaleb, Y. Scaling Law of Permeability and Porosity for Fluid Transport Phenomena in Poro PCM Media. *J. Appl. Comput. Mech.* **2021**, *7*, 84–92. [CrossRef]
131. Midhun, V.; Maroliya, M.; Saha, S.K. Numerical investigation and optimisation of solid–solid phase change material composi based plate-fin heat sink for thermal management of electronic package. *Appl. Therm. Eng.* **2024**, *248*, 123183. [CrossRef]
132. Maroliya, M.; Midhun, V.; Saha, S.K. Comparative Study of Solid-Solid and Solid-Liquid Phase Change Materials for a Pin-Finn Heat Sink in Avionics Thermal Management. In Proceedings of the 27th National and 5th International ISHMT-ASTFE Heat ar Mass Transfer Conference, Bihar, India, 14–17 December 2023; Begel House Inc.: Danbury, CT, USA, 2024.
133. Zhi, M.; Yue, S.; Zheng, L.; Su, B.; Fu, J.; Sun, Q. Recent developments in solid-solid phase change materials for thermal ener storage applications. *J. Energy Storage* **2024**, *89*, 111570. [CrossRef]
134. Gao, Y.; Zheng, Q.; Jonsson, J.C.; Lubner, S.; Curcija, C.; Fernandes, L.; Kaur, S.; Kohler, C. Parametric study of solid-sol translucent phase change materials in building windows. *Appl. Energy* **2021**, *301*, 117467. [CrossRef]
135. Harlé, T.; Hebert, R.L.; Nguyen, G.T.; Ledésert, B.A. A composite of cross-linked polyurethane as solid–solid phase chan material and plaster for building application. *Energy Build.* **2022**, *262*, 111945. [CrossRef]
136. Chen, X.n.; Xu, B.; Xie, X.; Pei, G. Evaluating and optimizing the energy saving benefits of latent heat in phase change materia with new indices. *Appl. Therm. Eng.* **2023**, *228*, 120479. [CrossRef]
137. Liu, L.; Su, D.; Tang, Y.; Fang, G. Thermal conductivity enhancement of phase change materials for thermal energy storag A review. *Renew. Sustain. Energy Rev.* **2016**, *62*, 305–317. [CrossRef]
138. Wu, M.; Li, T.; He, Q.; Du, R.; Wang, R. Thermally conductive and form-stable phase change composite for building therm management. *Energy* **2022**, *239*, 121938. [CrossRef]
139. Zakaria, N.M.; Omar, M.A.; Mukhtar, A. Numerical Study on the Thermal Insulation of Smart Windows Embedded with Lo Thermal Conductivity Materials to Improve the Energy Efficiency of Buildings. *CFD Lett.* **2023**, *15*, 41–52. [CrossRef]
140. Pasupathy, A.; Velraj, R.; Seeniraj, R. Phase change material-based building architecture for thermal management in residenti and commercial establishments. *Renew. Sustain. Energy Rev.* **2008**, *12*, 39–64. [CrossRef]
141. Zahir, M.H.; Irshad, K.; Shafiullah, M.; Ibrahim, N.I.; Islam, A.K.; Mohaisen, K.O.; Sulaiman, F.A.A. Challenges of the applicatio of PCMs to achieve zero energy buildings under hot weather conditions: A review. *J. Energy Storage* **2023**, *64*, 107156. [CrossRe
142. Nurlybekova, G.; Memon, S.A.; Adilkhanova, I. Quantitative evaluation of the thermal and energy performance of the PC integrated building in the subtropical climate zone for current and future climate scenario. *Energy* **2021**, *219*, 119587. [CrossRe
143. Kitagawa, H.; Asawa, T.; Del Rio, M.A.; Kubota, T.; Trihamdani, A.R. Thermal energy simulation of PCM-based radiant floo cooling systems for naturally ventilated buildings in a hot and humid climate. *Build. Environ.* **2023**, *238*, 110351. [CrossRef]
144. Kim, Y.R.; Khil, B.S.; Jang, S.J.; Choi, W.C.; Yun, H.D. Effect of barium-based phase change material (PCM) to control the heat hydration on the mechanical properties of mass concrete. *Thermochim. Acta* **2015**, *613*, 100–107. [CrossRef]
145. Porisini, F.C. Salt hydrates used for latent heat storage: Corrosion of metals and reliability of thermal performance. *Sol. Energ* **1988**, *41*, 193–197. [CrossRef]
146. Liu, M.; Zhang, X.; Ji, J.; Yan, H. Review of research progress on corrosion and anti-corrosion of phase change materials in therm energy storage systems. *J. Energy Storage* **2023**, *63*, 107005. [CrossRef]
147. Shamseddine, I.; Pennec, F.; Biwole, P.; Fardoun, F. Supercooling of phase change materials: A review. *Renew. Sustain. Energy Re* **2022**, *158*, 112172. [CrossRef]
148. Emeema, J.; Murali, G.; Reddi, B.V.; Mangesh, V. Investigations on paraffin wax/CQD composite phase change material-Improve latent heat and thermal stability. *J. Energy Storage* **2024**, *85*, 111056. [CrossRef]
149. McLaggan, M.S.; Hadden, R.M.; Gillie, M. Flammability assessment of phase change material wall lining and insulation materia with different weight fractions. *Energy Build.* **2017**, *153*, 439–447. [CrossRef]
150. Png, Z.M.; Soo, X.Y.D.; Chua, M.H.; Ong, P.J.; Suwardi, A.; Tan, C.K.I.; Xu, J.; Zhu, Q. Strategies to reduce the flammability c organic phase change Materials: A review. *Sol. Energy* **2022**, *231*, 115–128. [CrossRef]
151. Wong-Pinto, L.S.; Milian, Y.; Ushak, S. Progress on use of nanoparticles in salt hydrates as phase change materials. *Renew. Sustai Energy Rev.* **2020**, *122*, 109727. [CrossRef]
152. Du, K.; Calautit, J.; Wang, Z.; Wu, Y.; Liu, H. A review of the applications of phase change materials in cooling, heating an power generation in different temperature ranges. *Appl. Energy* **2018**, *220*, 242–273. [CrossRef]
153. Purohit, B.; Sistla, V. Inorganic salt hydrate for thermal energy storage application: A review. *Energy Storage* **2021**, *3*, e21 [CrossRef]
154. Xu, Z.; Chen, W.; Wu, T.; Wang, C.; Liang, Z. Thermal management system study of flame retardant solid–solid phase chang material battery. *Surfaces Interfaces* **2023**, *36*, 102558. [CrossRef]
155. Cascone, Y.; Capozzoli, A.; Perino, M. Optimisation analysis of PCM-enhanced opaque building envelope components for th energy retrofitting of office buildings in Mediterranean climates. *Appl. Energy* **2018**, *211*, 929–953. [CrossRef]
156. Cárdenas-Ramírez, C.; Jaramillo, F.; Gómez, M. Systematic review of encapsulation and shape-stabilization of phase chang materials. *J. Energy Storage* **2020**, *30*, 101495. [CrossRef]
157. Raj, A.; Srinivas, M.; Jayaraj, S. A cost-effective method to improve the performance of solar air heaters using discrete macr encapsulated PCM capsules for drying applications. *Appl. Therm. Eng.* **2019**, *146*, 910–920. [CrossRef]
158. Rezaei, M.; Anisur, M.; Mahfuz, M.; Kibria, M.; Saidur, R.; Metselaar, I. Performance and cost analysis of phase change material with different melting temperatures in heating systems. *Energy* **2013**, *53*, 173–178. [CrossRef]

159. Yang, S.; Du, X.; Deng, S.; Qiu, J.; Du, Z.; Cheng, X.; Wang, H. Recyclable and self-healing polyurethane composites based on Diels-Alder reaction for efficient solar-to-thermal energy storage. *Chem. Eng. J.* **2020**, *398*, 125654. [CrossRef]
160. Yang, Y.; Cai, X.; Kong, W. A novel intrinsic photothermal and flexible solid–solid phase change materials with super mechanical toughness and multi-recyclability. *Appl. Energy* **2023**, *332*, 120564. [CrossRef]
161. Bai, S.; Zhang, K.; Zhang, Q.; Zhu, Y.; Wang, W.; Zhang, J.; Li, X.; Zhang, X.; Wang, R. Intrinsic Flame Retardancy and Flexible Solid–Solid Phase Change Materials with Self-Healing and Recyclability. *ACS Appl. Mater. Interfaces* **2023**, *15*, 48613–48622. [CrossRef] [PubMed]
162. Yang, B.; Zhang, X.; Ji, J.; Hua, W.; Jiang, M. Multifunctional phase change film with high recyclability, adjustable thermal responsiveness, and intrinsic self-healing ability for thermal energy storage. *J. Energy Chem.* **2024**, *97*, 216–227. [CrossRef]
163. Anika, U.A.; Kibria, M.G.; Kanka, S.D.; Mohtasim, M.S.; Paul, U.K.; Das, B.K. Exergy, exergo-economic, environmental and sustainability analysis of pyramid solar still integrated hybrid nano-PCM, black sand, and sponge. *Sol. Energy* **2024**, *274*, 112559. [CrossRef]
164. De Gracia, A.; Rincón, L.; Castell, A.; Jiménez, M.; Boer, D.; Medrano, M.; Cabeza, L.F. Life Cycle Assessment of the inclusion of phase change materials (PCM) in experimental buildings. *Energy Build.* **2010**, *42*, 1517–1523. [CrossRef]
165. Zhang, L.; Zhang, Q.; Jin, L.; Cui, X.; Yang, X. Energy, economic and environmental (3E) analysis of residential building walls enhanced with phase change materials. *J. Build. Eng.* **2024**, *84*, 108503. [CrossRef]
166. Gupta, M.K.; Rathore, P.K.S.; Kumar, R.; Gupta, N.K. Experimental analysis of clay bricks incorporated with phase change material for enhanced thermal energy storage in buildings. *J. Energy Storage* **2023**, *64*, 107248. [CrossRef]
167. Yang, S.; Zhang, Y.; Zhao, Y.; Torres, J.F.; Wang, X. PCM-based ceiling panels for passive cooling in buildings: A CFD modelling. *Energy Build.* **2023**, *285*, 112898. [CrossRef]
168. Kumar, D.; Alam, M.; Doshi, A.J. Investigating the influence of thermal conductivity and thermal storage of lightweight concrete panels on the energy and thermal comfort in residential buildings. *Buildings* **2023**, *13*, 720. [CrossRef]
169. Rahi, M.F.; Arıcı, M.; Abdulateef, A.M.; Niyas, H. Numerical investigation and optimization of macro-encapsulated phase change material capsules in building roof slab. *J. Energy Storage* **2023**, *68*, 107806. [CrossRef]
170. Babaharra, O.; Choukairy, K.; Faraji, H.; Hamdaoui, S. Improved heating floor thermal performance by adding PCM microcapsules enhanced by single and hybrid nanoparticles. *Heat Transf.* **2023**, *52*, 3817–3838. [CrossRef]
171. Zhang, Q.; Liu, Y.; Zhang, Q.; Wang, G.; Lü, X. Research on the indoor temperature regulation characteristics of a nearly zero energy building with the nonlinear heat capacity building components: A simple model. *J. Build. Eng.* **2023**, *76*, 106850. [CrossRef]
172. El-Nagar, D.H.; Emam, M.; El-Betar, A.; Nada, S.A. Performance improvement of building-integrated photovoltaic panels using a composite phase change material-carbon foam heat sink: An experimental study. *J. Build. Eng.* **2024**, *91*, 109623. [CrossRef]
173. Gao, L.; Zhang, X.; Hua, W. Recent progress in photovoltaic thermal phase change material technology: A review. *J. Energy Storage* **2023**, *65*, 107317. [CrossRef]
174. Maghrabie, H.M.; Mohamed, A.; Fahmy, A.M.; Samee, A.A.A. Performance enhancement of PV panels using phase change material (PCM): An experimental implementation. *Case Stud. Therm. Eng.* **2023**, *42*, 102741. [CrossRef]
175. Kang, Y.K.; Joung, J.; Kim, M.; Jeong, J.W. Energy impact of heat pipe-assisted microencapsulated phase change material heat sink for photovoltaic and thermoelectric generator hybrid panel. *Renew. Energy* **2023**, *207*, 298–308. [CrossRef]
176. Gallero, F.J.G.; Siles, G.G.; Maestre, I.R.; Blázquez, J.L.F.; Bottarelli, M. Performance analysis of a novel façade-based building integrated photovoltaic-thermal system with phase change material for domestic hot water. *Appl. Therm. Eng.* **2024**, *240*, 122321. [CrossRef]
177. Elsaid, A.M.; Hashem, F.A.; Mohamed, H.A.; Ahmed, M.S. The energy savings achieved by various Trombe solar wall enhancement techniques for heating and cooling applications: A detailed review. *Sol. Energy Mater. Sol. Cells* **2023**, *254*, 112228. [CrossRef]
178. Li, J.; Zhang, Y.; Zhu, Z.; Zhu, J.; Luo, J.; Peng, F.; Sun, X. Thermal comfort in a building with Trombe wall integrated with phase change materials in hot summer and cold winter region without air conditioning. *Energy Built Environ.* **2024**, *5*, 58–69. [CrossRef]
179. Shen, J.; Lassue, S.; Zalewski, L.; Huang, D. Numerical study on thermal behavior of classical or composite Trombe solar walls. *Energy Build.* **2007**, *39*, 962–974. [CrossRef]
180. Chan, H.Y.; Riffat, S.B.; Zhu, J. Review of passive solar heating and cooling technologies. *Renew. Sustain. Energy Rev.* **2010**, *14*, 781–789. [CrossRef]
181. Xiao, Z.; Mishra, P.; Nejad, A.M.; Tao, M.; Granados-Focil, S.; Van Dessel, S. Thermal optimization of a novel thermo-optically responsive SS-PCM coatings for building enclosures. *Energy Build.* **2021**, *247*, 111129. [CrossRef]
182. Čurpek, J.; Čekon, M.; Šikula, O.; Slávik, R. Thermodynamic responses of adaptive mechanisms in BiPV façade systems coupled with latent thermal energy storage. *Energy Build.* **2023**, *279*, 112665. [CrossRef]
183. Duan, S.; Wang, L.; Zhao, Z.; Zhang, C. Experimental study on thermal performance of an integrated PCM Trombe wall. *Renew. Energy* **2021**, *163*, 1932–1941. [CrossRef]
184. Zhang, J.; Li, H.; Du, Y.; Xu, T.; Zou, T.; Wei, S.; Zhang, D. Development of $CaCl_2\cdot 6H_2O$-mannitol/SiO_2 shape-stabilized composite phase change material for the radiant cooling panel system. *Energy Build.* **2024**, *313*, 114224. [CrossRef]
185. Long, H.; Lei, S.; Wang, F.; Yang, S.; Ju, H.; Ou, J. Superhydrophobic daytime radiative cooling coating incorporated with phase change microcapsules for building thermal regulation. *J. Mater. Sci.* **2024**, *59*, 6459–6475. [CrossRef]

186. Zheng, C.; Huang, Z.; Wang, D.; Ju, Z.; Zhang, Z.; Zhang, H.; Zhu, Y.; Jiang, Z. Synthesis and properties of biomass derive carbon/PEG composite as photothermal conversion effective phase change material for functional concrete. *Cem. Concr. Compo* **2024**, *149*, 105495. [CrossRef]
187. Yue, H.; Ou, Y.; Wang, J.; Wang, H.; Du, Z.; Du, X.; Cheng, X. Ti_3C_2Tx MXene/delignified wood supported flame-retardar phase-change composites with superior solar-thermal conversion efficiency and highly electromagnetic interference shielding fc efficient thermal management. *Energy* **2024**, *286*, 129441. [CrossRef]
188. Shamshiri, M.; Jafari, R.; Momen, G. An intelligent icephobic coating based on encapsulated phase change materials (PCM *Colloids Surfaces A Physicochem. Eng. Asp.* **2022**, *655*, 130157. [CrossRef]
189. He, J.W.; Huang, X.Y.; Shu, Z.Y.; Huang, R.N.; Cai, Y.; Lv, Y.; Zhao, F.Y. Preliminary study of a dual channel solar photocatalyti ventilation wall system for year-round air purification and building energy savings. *J. Build. Eng.* **2024**, *82*, 108261. [CrossRef]
190. Hu, R.; Niu, J. A review of the application of radiant cooling & heating systems in Mainland China. *Energy Build.* **2012**, *52*, 11–1
191. Sastry, G. VAV vs. radiant: Side-by-side comparison. *ASHRAE J.* **2014**, *56*, 16.
192. Taweekun, J. An experimental and simulated study on thermal comfort. *Int. J. Eng. Technol.* **2013**, *5*, 177.
193. Zhang, F.; Guo, H.A.; Liu, Z.; Zhang, G. A critical review of the research about radiant cooling systems in China. *Energy Buil* **2021**, *235*, 110756. [CrossRef]
194. Rebelo, F.; Figueiredo, A.; Vicente, R.; Almeida, R.; Ferreira, V. Innovation and progress in the incorporation of phase chang materials into radiant floor systems. *J. Energy Storage* **2023**, *74*, 109495. [CrossRef]
195. Bogatu, D.I.; Kazanci, O.B.; Olesen, B.W. An experimental study of the active cooling performance of a novel radiant ceilin panel containing phase change material (PCM). *Energy Build.* **2021**, *243*, 110981. [CrossRef]
196. González, B.; Prieto, M. Radiant heating floors with PCM bands for thermal energy storage: A numerical analysis. *Int. J. Thern Sci.* **2021**, *162*, 106803. [CrossRef]
197. Liu, H.; Tan, Q.; Shi, Y.; Yu, B.; Zhang, M. Enhancing Indoor Thermal Comfort and Energy Efficiency: A Comparative Study o RC-PCM Trombe Wall Performance. *Renew. Energy* **2024**, *227*, 120542. [CrossRef]
198. Cesari, S.; Baccega, E.; Emmi, G.; Bottarelli, M. Enhancement of a radiant floor with a checkerboard pattern of two PCMs fo heating and cooling: Results of a real-scale monitoring campaign. *Appl. Therm. Eng.* **2024**, *246*, 122887. [CrossRef]
199. Yang, M.; Zhong, H.; Li, T.; Wu, B.; Wang, Z.; Sun, D. Phase change material enhanced radiative cooler for temperature-adaptiv thermal regulation. *ACS Nano* **2023**, *17*, 1693–1700. [CrossRef] [PubMed]
200. Atiganyanun, S.; Hasuchon, C.; Prithan, K.; Wongwan, K.; Kumnorkaew, P. Inclusion of microencapsulated phase chang materials in waterborne paints to enhance radiative cooling performance. *Mater. Today Proc.* **2023**. [CrossRef]
201. Su, W.; Kang, R.; Cai, P.; Hu, M.; Kokogiannakis, G.; Darkwa, J.; Chen, J.; Xu, S.; Wang, L. Development of spectrally self switchable cover with phase change material for dynamic radiative cooling. *Sol. Energy Mater. Sol. Cells* **2023**, *251*, 11212 [CrossRef]
202. Aburas, M.; Ebendorff-Heidepriem, H.; Lei, L.; Li, M.; Zhao, J.; Williamson, T.; Wu, Y.; Soebarto, V. Smart windows–Transmittanc tuned thermochromic coatings for dynamic control of building performance. *Energy Build.* **2021**, *235*, 110717. [CrossRef]
203. Zhang, Z.; Zhang, L.; Zhou, Y.; Cui, Y.; Chen, Z.; Liu, Y.; Li, J.; Long, Y.; Gao, Y. Thermochromic energy efficient windows Fundamentals, recent advances, and perspectives. *Chem. Rev.* **2023**, *123*, 7025–7080. [CrossRef] [PubMed]
204. Alsaadi, M.; Lerttraikul, K.; Chatraphorn, S.; Kittiwatanakul, S. Strategies for Enhanced Optical Performance and Durability o VO_2-based Thermochromic Windows. *Adv. Eng. Mater.* **2024**, *26*, 2302224. [CrossRef]
205. Pi, J.; Li, C.B.; Sun, R.Y.; Li, L.Y.; Wang, F.; Song, F.; Wu, J.M.; Wang, X.L.; Wang, Y.Z. Superhydrophobic and thermochromi VO2-Based composite coatings for energy-saving smart windows. *Compos. Commun.* **2022**, *32*, 101167. [CrossRef]
206. Geng, C.; Dou, S.; Zhao, J.; Ren, F.; Gu, J.; Wei, H.; Guan, H.; Liang, S.; Li, L.; Li, Y.; et al. Self-templated method to fabricat VO2 nanoparticles with ultrahigh luminous transmittance for energy-efficient thermochromic windows. *Appl. Surf. Sci.* **2022** *592*, 153267. [CrossRef]
207. Jin, Q.; Long, X.; Liang, R. Numerical analysis on the thermal performance of PCM-integrated thermochromic glazing systems *Energy Build.* **2022**, *257*, 111734. [CrossRef]
208. Ji, R.; Li, X. Numerical analysis on the energy performance of the PCMs-integrated thermochromic coating building envelopes *Build. Environ.* **2023**, *233*, 110113. [CrossRef]
209. Imghoure, O.; Belouaggadia, N.; Ezzine, M.; Lbibb, R.; Younsi, Z. Evaluation of phase change material and thermochromic layers in a "smart wall" in different climates for improving thermal comfort in a building. *J. Build. Eng.* **2022**, *56*, 104755. [CrossRef]
210. Izadpanah, S.; Fazelpour, F.; Eftekhari Yazdi, M. Comparative study of simultaneous use of a double or a triple skin facade with phase change materials, green roof, and photovoltaics in residential buildings of Iran. *Environ. Prog. Sustain. Energy* **2023** *42*, e13935. [CrossRef]
211. Németh, B.; Németh, Á.S.; Ujhidy, A.; Tóth, J.; Trif, L.; Jankovics, H.; Kriszt, B.; Dobolyi, C.; May, Z.; Gyenis, J.; et al. Antimicrobial functionalization of Ca alginate-coconut oil latent heat storing microcapsules by Ag nanoparticles. *Int. J. Energy Res.* **2020** *44*, 11998–12014. [CrossRef]

sustainability MDPI

eview

Advancing Sustainable Development: Broad Applications of Passive Radiative Cooling

in Liang, Shengxi Bai, Kaixin Lin, Chui Ting Kwok, Siru Chen, Yihao Zhu and Chi Yan Tso *

School of Energy and Environment, City University of Hong Kong, Kowloon Tong, Hong Kong, China; martin.i2cool@my.cityu.edu.hk (Y.Z.)
* Correspondence: chiytso@cityu.edu.hk

Abstract: With the increasing demand for energy worldwide, researchers from different fields have been striving to improve the sustainability and proper utilization of energy resources. Passive radiative cooling, as a natural energy transport method, can achieve cooling without additional external energy input. This review provides a comprehensive examination of passive radiative cooling, including its fundamental theories and latest development. A particular emphasis is placed on the diverse range of fields where passive radiative cooling has been applied, notably including but not limited to construction and architecture. The current state of applications, potential challenges that may arise with wider adaption and promising research directions for each field are thoroughly discussed. This review emphasizes the extensive potential and practical viability of passive radiative cooling in diverse applications and identifies pressing challenges and future research directions aimed at scaling up real-world implementation.

Keywords: radiative cooling; energy sustainable development; building energy efficiency; smart textile; metamaterials; micro/nano structures

1. Introduction

As global energy demand continues to rise, researchers across diverse disciplines are relentlessly exploring avenues to enhance the sustainability of energy production and its efficient usage. Active cooling technologies, mainly based on vapor compression cycles, currently dominate the cooling effect generation scenarios across various industries. Vapor compression refrigeration systems are mature, stable, and capable of high cooling output [1]. However, these widely used vapor compression systems, such as heating, ventilation, and air conditioning (HVAC) units, consume significant amounts of energy. Although novel refrigerants with low Global Warming Potential (GWP) and Ozone Depletion Potential (ODP), such as R134a [2], R152a [3], R744 [4], R1234yf [3,5], and R290 [5], have been introduced, they only represent a compromise under the ultimate goal of zero energy consumption. Thus, it is critical to develop alternative energy-saving and environmentally friendly cooling solutions considering long-term sustainability. Passive radiative cooling, a naturally occurring energy transfer process, offers a promising solution by providing cooling effect without additional external energy input [6]. Fundamentally, in the passive radiative cooling process, a sky-facing surface reflects solar irradiance with a wavelength range of 0.2–2.5 μm and dissipates its own heat in the form of thermal radiation to the cold outer space through the atmospheric window lying within 8–13 μm of the electromagnetic spectrum. Due to the large temperature difference between the sky-facing surface and the outer space (3 K), the outer space can act as a heat sink with nearly unlimited capacity, providing a cost-free and sustainable cooling solution [7]. In this study, we aim to systematically review the principle of this innovative cooling technology and its potential applications for a more sustainable energy future.

Systematic studies of passive radiative cooling can be dated back to the 1960s, with focuses limited on nighttime passive cooling in the 1970s and 1980s [7,8]. For nocturnal cooling, the performance mainly depends on the thermal radiation efficiency of the radiative cooler, which is determined by the emissivity in the mid-infrared (MIR) range. However, higher MIR emissivity comes with a higher absorptivity, which increases radiative absorption from the surroundings and hence reduces the cooling effectiveness. Thus, a selective radiative surface is required for optimal cooling efficiency in order to achieve daytime cooling. Ideally, a selective radiative cooling surface should achieve high emissivity only at the highly transparent atmospheric window for maximizing heat dissipation towards the outer space while close other atmospheric channels for minimizing radiative heat received. In addition, high reflection and low absorption for solar light is needed for daytime passive radiative cooling. For the first time in California USA, Raman used a passive radiative cooler with a multilayer photonic structure to achieve daytime cooling, which shows a cooling power of 40.1 W/m^2 and a temperature drop of ~5 °C [9]. Since then, various materials have been proposed to achieve daytime radiative cooling with sub-ambient temperatures and great potential for practical applications.

One common material design concept is to enhance MIR emissivity and solar reflectance for high cooling efficiency [10]. Wide absorption bands of metal–O, nonmetal–O, and nonmetal–nonmetal bonds of inorganic dielectric materials can be used to achieve strong IR emission. Among these, SiO_2 [11,12], Si_xN_yO [12–14], and other silicon-related materials have been widely used in radiative cooling applications. Apart from silicon-based coatings, other inorganic coatings have also been explored for their potential in radiative cooling applications. For instance, coatings developed by Berdahl using magnesium oxide (MgO) and/or lithium fluoride (LiF) as radiators have demonstrated a remarkable cooling power of 85 W/m^2 under a clear night sky [15]. Polymer-based materials have also been used for passive radiative cooling, such as polymethyl methacrylate (PMMA), polyvinylidene fluoride (PVDF), and Polydimethylsiloxane (PDMS) [16–20]. With the rapid development of micro/nanomaterials and advanced manufacturing technologies, high solar reflectance of over 97% and high selective MIR emissivity of up to 99% have been achieved [21,22]. A diverse range of passive radiative cooling material designs have been investigated, which show exciting application prospects. In addition to the aforementioned organic or inorganic film structure radiative coolers, pigment paint represents another viable option for nocturnal radiative cooling. Many oxides elements such as titanium (TiO_2) [23,24], aluminum (Al_2O_3) [25,26], barium ($BaSO_4$) [27,28], silicon (SiO_2, SiO_xN_y) [29], and zinc (ZnS, ZnO, ZnSe) [25,30–33] exhibit high emissivity in the atmospheric window wavelength range and high reflectivity in the visible range. Part of the advanced radiative cooling technologies is summarized in Table 1.

Table 1. Summary of advanced radiative cooling technology.

Literature	Key Materials	Location	Cooling Power	Temperature Drop
Raman et al. [9]	HfO_2, SiO_2	Stanford, CA, USA	40.1 W/m^2	4.9 °C
Atiganyanun et al. [11]	SiO_2	Albuquerque, NM, USA	-	4.7 °C
Chae et al. [12]	Al_2O_3, Si_3N_4, and SiO_2	Seoul, South Korea	>60 W/m^2	8.2 °C
Chen et al. [14]	Si_3N_4, Si, Al	Stanford, CA, USA	-	37.4 °C (in winter)
Gao et al. [16]	PMAA	-	-	7.5 °C
Aili et al. [17]	PVDF, Ag	Boulder, CO, USA	-	6 °C (nighttime) 9 °C (daytime)
Zhai et al. [20]	PMDS, ZrO_2	Wuhan, Hubei, China	-	16.1 °C
Jeong T et al. [23]	TiO_2, SiO_2	Hong Kong, China	136.3 W/m^2	7.2 °C
Chae al. [26]	Al_2O_3, SiO_2	Seoul, South Korea	100 W/m^2	7.9 °C
Li et al. [28]	$BaSO_4$, Si	West Lafayette, IN, USA	117 W/m^2	>4.5 °C
Huang et al. [31]	ZnO, SiO_2	Nanjing, Jiangsu, China	-	5.3 °C (nighttime) 4.1 °C (daytime)
Lv et al. [32]	$ZnTiO_3$	Riverside, CA, USA	-	14.9 °C

The field of passive radiative cooling is witnessing a notable expansion in research that portends its broader application across various sectors, complementing existing industrial applications with potential for integration into everyday life. While several review papers have thoroughly examined the different facets of passive radiative cooling, including working principles, material science, and structural innovation, there remains a critical need for a comprehensive review that encapsulates the recent innovative advancements in real-world applications of this technology. When transitioning radiative cooling technology from laboratory to practical applications, certain technical and economic challenges must be addressed. For instance, developing a technology that can dynamically adjust the cooling power to facilitate intelligent and flexible thermal management suitable for various climates and conditions presents a challenge. In addition, for outdoor applications, radiative coolers must be designed with durability and self-cleaning capabilities to endure adverse environmental conditions while maintaining efficiency. In order to meet aesthetic requirements (for textiles and architecture) and the specific optical needs of integrated systems (such as photovoltaics), the spectral optical performance of radiative coolers must be individually tailored to balance functionality with cooling efficiency. there is a need to optimize and develop simple and scalable manufacturing technologies to reduce production costs, thus addressing the economic challenges of marketization.

This review aims to deepen our understanding of radiative cooling and to underscore the feasibility of this naturally occurring energy-saving technology across a variety of practical applications. By identifying the unique challenges and requirements within each application, we also outline several directions for future research that support the transition to more energy-efficient and sustainable cooling solutions. In the forthcoming sections, we will start with the fundamental principle of passive radiative cooling, followed by a comprehensive overview of the real-world applications of radiative cooling under various circumstances including thermal regulation for buildings, human beings, PV panels, greenhouse and food storage as well as dew water collection. In each application scenario, we will highlight the specific characteristics that radiative cooling technology needs to be possessed to function effectively within that particular environment. Conclusively, we will identify the critical issues that should be addressed for large-scale practical applications of passive radiative cooling technology and further propose potential directions for future research.

2. Methodological Approaches

This review adopts an integrative methodological approach to collate, scrutinize, and synthesize relevant scientific literature on passive radiative cooling, its underlying principles, advancements, and applications across various domains. The methodology encompasses comprehensive literature searches, methodical selection, and critical examination of scholarly works pertaining to passive radiative cooling, with a particular focus on its contribution to sustainable energy practices and its integration into diverse fields such as building, personal thermal management, and other renewable energy systems.

The research methodology involves systematic searches across multidisciplinary and peer-reviewed databases, employing advanced search algorithms paired with Boolean logic to refine the scope and relevance of the retrieved literature (shown in Figure 1). The search encompasses scientific articles, technical reports, books, and authenticated online resources, with the criteria for inclusion being relevance to passive radiative cooling, contribution to sustainability, and applicability across various industries. Selection is limited to documents published in English, which align with the search algorithms and meet the established inclusion and exclusion criteria.

A robust evaluation process further distills the literature to ensure quality and relevance. This process follows the Preferred Reporting Items for Systematic Reviews and Meta-Analyses (PRISMA) guidelines, as shown in Figure 1, providing a transparent and replicable framework for literature appraisal. The PRISMA flow diagram facilitates the illustration of the literature screening process, from identification through to final inclusion [34].

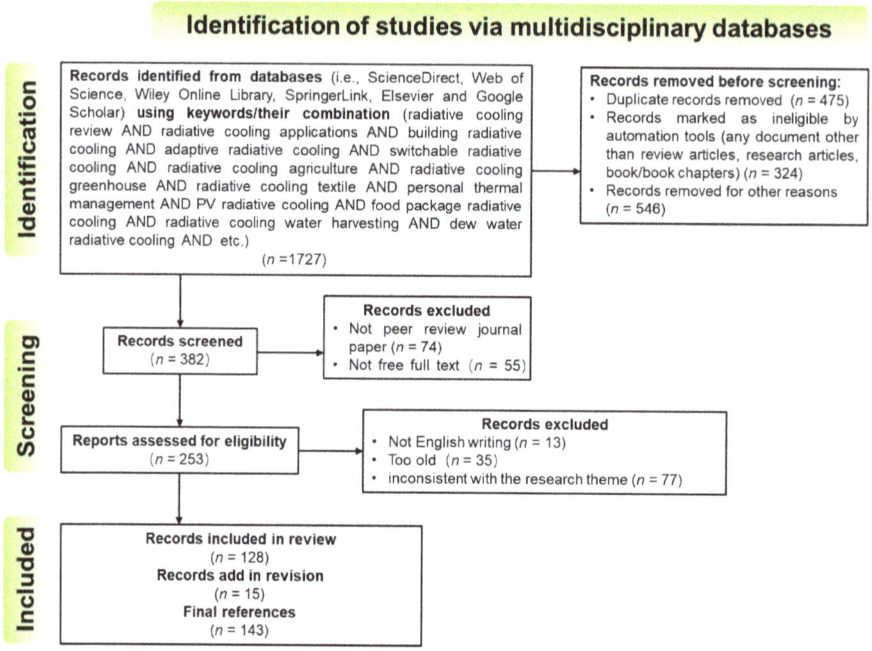

Figure 1. Literature selection depicted in a PRISMA 2020 flow diagram.

3. Fundamentals of Passive Radiative Cooling

In the field of thermodynamics, it is understood that any object with a temperature above absolute zero (0 K) has the capability to spontaneously absorb and emit electromagnetic waves [35]. This interaction between objects of different temperatures results in a form of heat exchange among objects at varying temperatures through the process of absorption and emission of these electromagnetic waves. On Earth, the most critical example of radiative heat exchange takes place between the Earth's surface and outer space, which is a process moderated by the transparency of the Earth's atmosphere in the wavelength range of 8–13 μm [36], a phenomenon known as the atmospheric long-wave infrared (LWIR) transmission window and playing an integral role in maintaining the Earth's temperature within a range that supports life.

The net cooling power of a typical passive radiative cooler in daytime is determined by outgoing thermal radiation from the cooler, solar radiation absorption and atmospheric thermal radiation absorption by the cooler, and non-radiative heat transfer (shown in Figure 2a), and can be expressed as follows:

$$q_{\text{net}} = q_{\text{cool}} - q_{\text{atm}} - q_{\text{solar}} - q_{\text{non-rad}} \tag{1}$$

where q_{net} is the net cooling power of the cooler; q_{cool} is the radiative power of the cooler which can be calculated as: $q_{\text{cool}} = \int d\Omega \cos(\theta) \int d\lambda I_{bb}(T_c, \lambda)\varepsilon(\lambda)$, where $\int d\Omega = 2\pi \int_0^{\pi/2} d\theta \sin\theta$ is the angular integral over a hemisphere, $I_{bb} = \frac{2hc^2}{\lambda^5} \frac{1}{\exp\left(\frac{hc}{\lambda k_B T_c}\right) - 1}$ is the spectral blackbody emission at a cooler temperature of T_c, h is Planck's constant, k_B is the Boltzmann constant, c is the speed of an electromagnetic wave in a vacuum, and $\varepsilon(\lambda)$ is the spectral emissivity of the cooler, which is equal to its absorptivity based on Kirchhoff's radiation law. The emissivity of the cooler is considered consistent across the hemisphere of its surface. q_{solar} is the absorbed power from solar radiation, which can be calculated as $\int_0^{\infty} d\lambda I_{solar}(\lambda)\varepsilon(\lambda)$, where $\varepsilon(\lambda)$ can be obtained as $1 - R(\lambda)$. q_{atm} is the absorbed power from the atmospheric radiation, which can be calculated as $\int d\Omega\cos(\theta)\int d\lambda I_{bb}(T_{atm},\lambda)\varepsilon(\lambda)\varepsilon_{atm}(\lambda)$, where T_{atm} and $\varepsilon_{atm}(\lambda)$ are the atmospheric temperature and spectral emissivity, respectively. $\varepsilon_{atm}(\lambda)$ can be obtained as $1 - \tau^{1/cos\theta}$, where τ is the atmospheric transmission at a zenith angle of zero. $q_{\text{non-rad}}$ is the power of all non-radiative heat transfer processes, i.e., convection and conduction, occurring between the radiative cooler and its surrounding environment. It can be determined as $h(T_{atm} - T_c)$, where h is the overall conduction and convection coefficient for heat exchange between the cooler and ambient air.

To maximize q_{net}, it is advisable to amplify q_{cool} as much as possible and concurrently reduce $q_{\text{non-rad}}$. With regards to q_{cool}, the utilization of spectrally selective surfaces is crucial. Specifically, a daytime radiative cooler should be designed to exhibit low solar absorption while simultaneously maintaining high MIR emissivity (shown in Figure 2b). The high MIR emissivity is typically achieved through the use of materials with specific phonon resonance or plasmonic nanostructures, which strongly emit in the MIR wavelength range [37]. The low solar absorption is accomplished by using materials with a high index of refraction or patterning the surface at the nanoscale to create destructive interference for solar wavelengths. This characteristic facilitates the achievement of radiative cooling effect during the daytime. In terms of minimizing $q_{\text{non-rad}}$, it is paramount to effectively insulate the radiative cooler from its surrounding environment, which can minimize the heat transfer from convection and conduction.

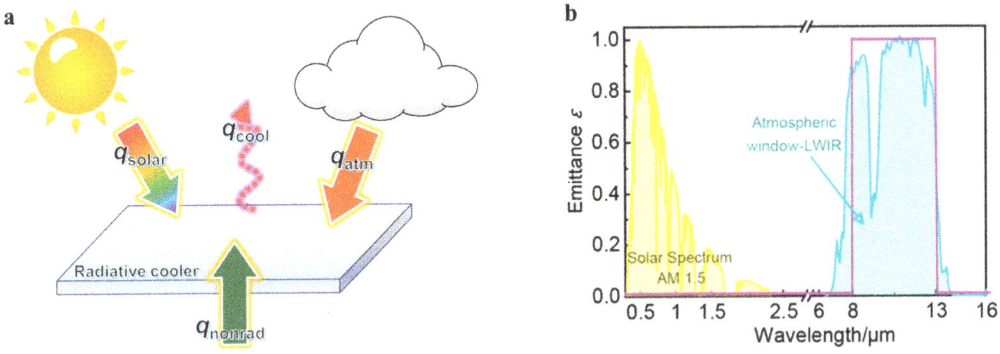

Figure 2. Fundamentals of passive radiative cooler. (**a**) Schematic diagram of the heat transfer processes of a typical passive radiative cooler. (**b**) The ideal solar reflectivity and MIR emissivity of daytime passive radiative cooling.

4. Promising Applications of Passive Radiative Cooling

As people delve deeper into the realm of daytime radiative cooling, it becomes apparent that this technology holds significant potential for transformative applications across various sectors. In this section, we will explore several application scenarios in detail, highlighting the recent advances in daytime radiative cooling materials, devices, and systems that are not only paving the way for sustainable energy usage but are also redefining the landscape of temperature regulation.

4.1. Building Cooling

In the realm of energy consumption pertinent to modern societies, cooling demand constitutes a substantial 40% of the overall energy utilization [38]. Within this significant proportion, it is interesting to note that conventional HVAC systems in architectural structures occupy a significant 65% of the energy usage [39]. Recent study illuminating the global status of HVAC systems suggests that by 2030, the annual electricity expenditure on HVAC systems could exceed hundreds of billions of dollars [40,41]. This underscores the critical role HVAC systems play in the larger context of energy consumption and the potential energy savings that could be realized in this sector [42]. Against the backdrop of significant energy expenditure by traditional HVAC systems, passive radiative cooling emerges as a promising alternative in the field of building cooling. This technology, requiring no additional energy input, has the potential to achieve substantial cooling effects. The utilization of radiative cooling materials in the construction of building rooftops, often referred to as 'cooling roofs', is a well-established applications of radiative cooling technologies within the field of architecture. In periods of high ambient temperatures, the implementation of cooling roofs can considerably diminish the cooling load of a building. These specially designed roofs can reflect a large fraction of the incoming solar radiation, while concurrently emitting thermal radiation towards the outer space. This dual mechanism facilitates an efficient cooling process, making a substantial contribution to overall energy saving within buildings. Indeed, the film-type [9,12,15–17,23,28,33] and paint-type [11,24–28,31] materials introduced in the introduction section can all serve as potential candidates for building cooling. However, beyond the basic requirements of optical properties that enable radiative cooling, additional performance criteria should be evaluated for materials intended for building roofs, including durability and scalability, aesthetic functions (color), and cooling power self-adaptive. These factors play crucial roles in determining the practical applicability and long-term performance of these materials in real-world architectural contexts.

Many meticulously designed radiative coolers necessitate intricate manufacturing processes and are vulnerable to a variety of factors that can challenge their durability during outdoor applications. Instances such as material structural degradation due to prolonged and intense ultraviolet (UV) exposure, surface contamination or erosion caused by wind-borne sand and dust, and surface damage from accidental impacts or continuous friction can adversely affect their cooling efficiency. Researchers are actively engaged in efforts to mitigate these challenges. Titanium dioxide (TiO_2) has high refractive index and excellent hydrophobic and UV-blocking properties [43], leading to radiative coolers using TiO_2 nanoparticles exhibit high solar reflectance and dirt resistance. Song et al. [44] demonstrated a strategy for realizing a durable radiative cooling coating through the evaporation-driven assembly of TiO_2 nanoparticles. The UV resistance of the designed anti-aging cooling paint (AACP) is significantly enhanced by the formation of a porous layer using rutile TiO_2 nanoparticles, without the incorporation of polymeric binders. When subjected to an accelerated aging test equivalent to one year of natural sunlight exposure, it was observed that the solar reflectance of the coating only diminished by a slight 0.5% (from 92.5% to 92.0%), while LWIR emissivity and contact angles were almost unchanged (shown in Figure 3a), maintaining reliable performance compared to the original state. In a similar vein, taking advantage of the intrinsic properties of TiO_2, Li M et al. [45] developed a triple-layer structure radiative cooler based on polyethersulfone (PES) with TiO_2 and Al_2O_3, using a phase change method (shown in Figure 3b). The top layer, consisting of high-refractive-index ($n \approx 1.8$) Al_2O_3 nanoparticles treated with fluorides for enhanced hydrophobicity (with a contact angle of 168°), reflects most of the UV light and is waterproof, thereby protecting the underlying structure. This design exhibits efficient and durable daytime radiative cooling, even in harsh climates. Acknowledging the critical importance of enhancing the mechanical strength of materials used in passive radiative cooling, Li T et al. [46] designed a wood-based passive radiative cooling material with enhanced tensile strength through delignification and re-pressing. This material features cellulose nanofibers that scatter solar irradiance and emit strongly in the MIR range, yielding an average daytime cooling power of 54 W/m^2. Significantly, the absence of lignin from the wood matrix results in an enhanced tensile strength, measured at 404.3 MPa—representing an 8.7-fold increase over conventional wood (shown in Figure 3c). This substantial reinforcement of mechanical properties confers resistance to erosive forces such as wind-sand abrasion and incidental impacts, ensuring persistent and stable cooling performance. Additionally, to address challenges in cooling performance degradation caused by outdoor contamination, Wang et al. [47] prepared a polymer composite coating of polystyrene, polydimethylsiloxane, and poly(ethyl cyanoacrylate) (PS/PDMS/PECA) using a simple two-step spray method. The ultra-hydrophobic self-cleaning characteristics of this coating can maintain the optical performance of the coating surface. During anti-fouling tests, it was observed that when mud was applied to the material's surface, it slid off instead of adhering to the surface due to the surface's ultra-hydrophobicity (shown in Figure 3d), thus not compromising the material's cooling efficiency. Meng et al. [48] reported a robust superhydrophobic self-cleaning porous coating characterized by the presence of nano-globules. This coating is synthesized through the thermal polymerization of methyl methacrylate (MMA) and ethylene glycol dimethacrylate (EDMA). The interconnected nano-globules structure contributes to increased surface roughness, endowing the surface with significant superhydrophobic properties, evidenced by a high contact angle of 165° (shown in Figure 3e). Moreover, the superhydrophobicity of the coating remained stable after rigorous testing, including tape peeling, abrasion with sandpaper, immersion in strong acids, and UV aging experiments. These findings suggest that the coating is highly resistant to dirt contamination, mechanical wear, chemical corrosion, and prolonged exposure to direct sunlight.

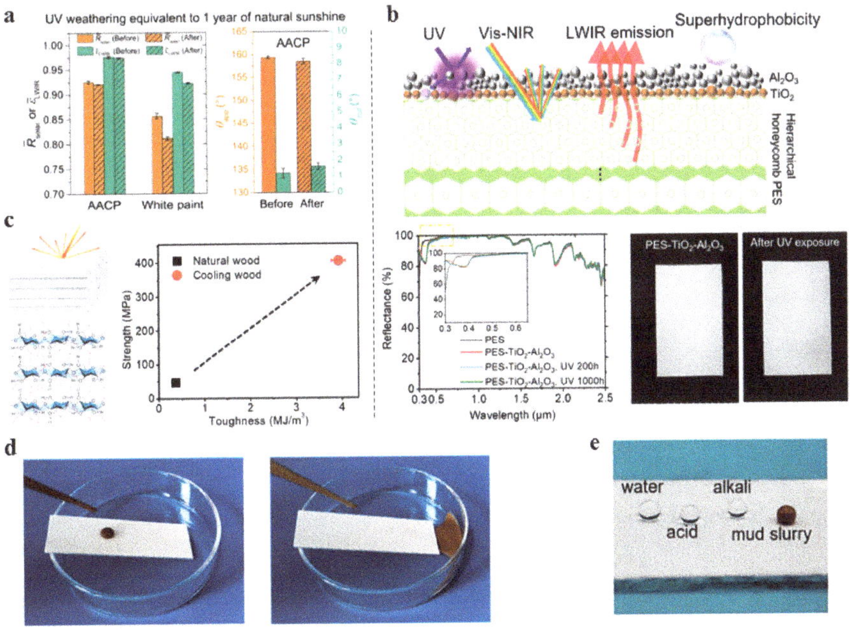

Figure 3. High durability radiative cooler design. (**a**) The almost unchanged optical and wetting properties of AACP coating after 1000 h of UV exposure. Ref. [44] is licensed under CC BY-NC 4.0. (**b**) Cross-sectional schematic of the trilayer PES-TiO_2-Al_2O_3 cooler and its UV resistance capability. Figures by [45] is licensed under CC BY-NC 4.0, with the permission of John Wiley and Sons. (**c**) Schematic diagram of the cooling wood structure and its strength comparison with ordinary wood. Ref. [46] is licensed under CC BY-NC 4.0. (**d**) Photograph of a mud slurry droplet on a PS/PDMS/PECA coating. Reprinted (adapted) with permission from [47]. Copyright 2024 American Chemical Society. (**e**) Photograph of the spherical droplets of water, acid, alkali, and mud slurry on the coating. Reprinted from [48], with permission from Elsevier [OR APPLICABLE SOCIETY COPYRIGHT OWNER].

Reducing light pollution and glare is another challenge of radiative cooling for building applications. The development of radiative cooling materials with rich colors can not only solve the issue of glare but also improve visual aesthetics, enhancing the overall acceptability and appeal of colored radiative cooling technologies for building applications.

The dye-rendering strategy is the simplest and most direct method to address color issue. Chen et al. [49] combined commercial dyed coating layers with porous P(VdF-HFP) or TiO_2/polymer composite films (shown in Figure 4a). The top layer selectively absorbs specific visible wavelengths to exhibit a particular color, while the bottom layer maximally reflects near-short-wavelength infrared (NSWIR) light to reduce solar heating. This approach accomplished a temperature drop of 15.6 °C during the hot summer. Tao et al. [50] engineered a colored radiative cooling emitter composed of a multilayer structure consisting of a high reflective substrate, a selective emission coating, and a sprayed pigment layer (shown in Figure 4b). A silver-aluminum alloy sheet serves as the high-reflectance substrate, with quinaldine red, phthalocyanine green, and phthalocyanine blue employed to achieve the primary colors of red, yellow, and blue, respectively. The results indicate that the emitter is capable of reflecting over 90% of near-infrared (NIR) light and demonstrates an infrared (IR) emission of over 85% within the atmospheric window. Compared to conventional colored polyurethane (PU) panels, the emitter achieves a temperature reduction of 5–10 °C during the daytime and 3–5 °C at night. In the pursuit of alternatives to conventional commercial dyes, phosphor dyes have attracted increasing attention from researchers. Xu et al. [51] employed luminescent diode phosphor materials such as $Y_3Al_5O_{12}$:Ce, $Y_3(Al,Ga)_5O_{12}$:Ce, $SrSiO_4$:Eu, and $CaAlSiN_3$:Eu as pigments to impart yellow, yellow-green, green, and red hues, respectively (shown in Figure 4c). These phosphors were applied to a dual-layer structured radiative cooler composed of TiO_2 and $BaSO_4$ powders embedded in a polyacrylic resin. Owing to the high selectivity of the phosphor dyes, which possess impurity bands within a large bandgap to circumvent extensive solar absorption and facilitate narrow-band visible light absorption, the colored radiative cooler achieved an optimization of color display while minimizing visible light absorption. It was demonstrated that the solar reflectance and the mid-to-far-IR emissivity of the cooler both exceed 90%. Under peak solar irradiation of 1000 W/m^2 at noon, the device achieved a sub-ambient temperature of 2.7 °C.

However, the color-displaying dye material absorbs not only specific wavelengths of visible light but also intercepts a fraction of IR waves, which hinders the efficiency radiative cooling [52]. Besides dye-rendering strategy, another strategy is the structural color strategy [53], which achieves color decoration of the cooler through structural design. By meticulously arranging nano-scale or micro-scale structures, coloration is achieved through the interference of scattered light without absorbing extra energy in the IR wavelengths [54], thus enhancing the cooling performance. Through numerical simulation, Yalçın et al. [55] demonstrated that the spectral position of surface plasmon resonances can be tuned by adjusting the sizes of plasma core-shell nanoparticles (with a silica core and a silver shell) and ordinary silver nanoparticles embedded within a silica and polydimethylsiloxane matrix, thereby resulting in the manifestation of various colors. This coloration strategy offers the potential for obtaining radiative cooling structures with colored coatings without affecting the solar reflectance. Building upon this concept of tuning nanoparticle sizes within radiative cooler structures to experimentally achieve different colorations, Yu et al. [56] engineered a flexible, colored radiative cooler whose coloration is derived from structural color induced by interference backward reflection (shown in Figure 4d). The chromaticity can be finely tuned by modulating the diameter of the polystyrene (PS) microspheres responsible for the interference, thus rendering a broad color palette. Owing to its unique structural design, the cooler exhibits high solar reflectance (>90%) and high MIR emissivity (96%). Zhu et al. [57] proposed an efficient structurally colored radiative cooling film with a photonic nanostructure, which is fabricated from natural cellulose nanocrystals (CNC). In this work, while achieving intense and fade-resistant coloration, they further diminished the sunlight absorption to approximately 3%, while achieving an emissivity of over 90% within the atmospheric window. Additionally, this film can be produced using a roll-to-roll (R2R) deposition process, thus suggesting its commercial viability.

All of the above implementations are static passive radiative coolers, which maintain a continuous cooling effect regardless of the surrounding environmental conditions. However, the continues cooling during cold seasons or nighttime can exacerbate heating costs and personal thermal discomfort. In response to this, adaptive radiative cooling technology has been proposed. In accordance with the stimulation mechanism that switches between cooling and non-cooling (heating) modes, the previously reported switchable radiative cooling systems can be categorized into two groups: active systems and passive adaptive systems (shown in Table 2).

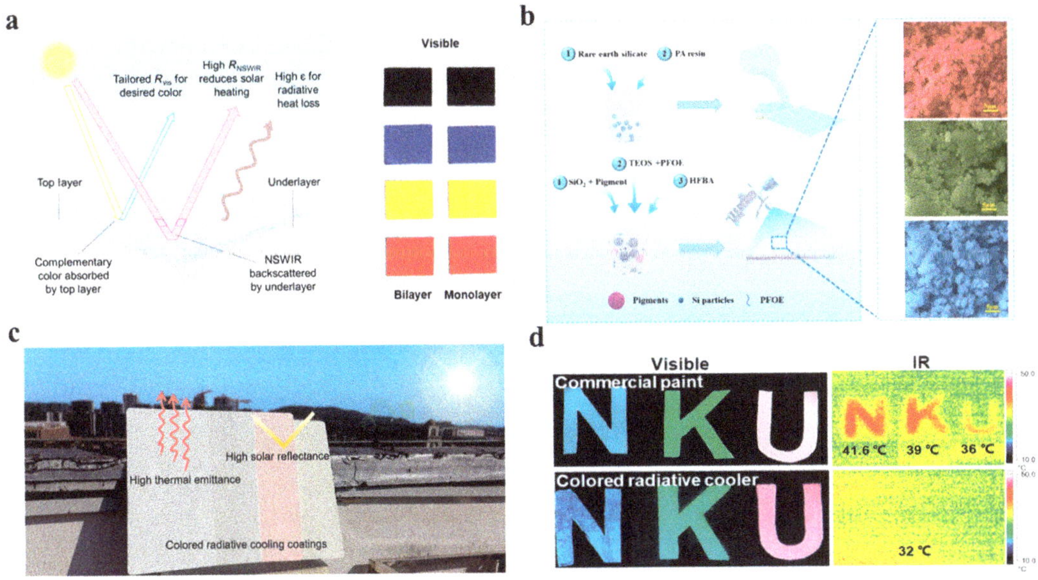

Figure 4. Colored radiative coolers. (**a**) Bilayer colored radiative coolers s for enhanced NSWIR reflectance. From [49]. Reprinted with permission from AAAS. (**b**) The schematic of colorful radiative cooling emitters. Reprinted from [50], with permission from Elsevier [OR APPLICABLE SOCIETY COPYRIGHT OWNER]. (**c**) Colored radiative cooling coatings using phosphor dyes. Reprinted from [50,51], with permission from Elsevier [OR APPLICABLE SOCIETY COPYRIGHT OWNER]. (**d**) Photonic-Structure Colored Radiative Coolers. Reprinted (adapted) with permission from [56]. Copyright 2024 American Chemical Society.

Table 2. Summary of adaptive radiative cooling technology.

Literature	Stimulation Mechanism	Structure/Materials	Control Performance
Yang et al. [58]	Active	Janus-structured bilayer Aerogel, MXene-CNF	α_{solar} changes from ~97.5% to ~12%
Li et al. [59]	Active	Rollable structure powered by electricity Polyimide, PDMS, CuO, Ag	Hot state: 71.6 W/m^2 cooling; Cold state: 643.4 W/m^2 heating
Zhao et al. [18]	Active	Porous bilayer film PDMS, Silicone	Heating mode: α_{solar} = 95%; Cooling mode: R_{solar} = 93%
Fan et al. [60]	Passive	Planar photonic multilayer system VO_2	Cooling "off" state: ~0 °C temperature drop; Cooling "on" state: ~9 °C temperature drop
Tang et al. [61]	Passive	Mechanically flexible coating structure VO_2 doped with tungsten $W_xV_{1-x}O_2$	ε changes from 0.20 (when T_{amb} < 15 °C) to 0.90 (when T_{amb} > 30 °C)
Wang J et al. [62]	Passive	Bilayer coating bottom layer: P(VdF-HFP) top layer: thermochromic microcapsule	R_{solar} changes from 91.25% to 72.71%
Wang T et al. [63]	Passive	Reversible thermochromic chameleon microcapsules	Modulation capacity of $\Delta T_{cooling\text{-}heating}$ = 9.5 °C
Xue et. al. [64]	Passive	Sandwich structure PNIPAm, PVDF	Modulation capacity of ΔR_{vis} = 70.0% and ΔT_{vis} = 86.3%
Dastidar et al. [65]	Passive	Janus structure film Cellulose, carbon nanotube	Dry state: R_{solar} = 88%; wet state: α_{solar} = 60%

Active control systems, which combine external stimuli with responsive materials or structures, require extra energy input for the dynamic control of thermal performance. Yang et al. [58] proposed a Janus-structured bilayer aerogel composed of a photothermal MXene-cellulose nanofibers (CNF) layer and a radiative cooling CNF layer (shown in Figure 5a). By simply manually flipping the two sides, cooling power control can be achieved. At cold state, the MXene-CNF side was exposed to sunlight, harvesting solar energy with a high solar absorption (≈97.5%). At a hot state, the CNF side was exposed to sunlight, reflecting solar energy with low absorption (≈12%) and porous structure, thereby creating a comfortable cooling effect. Besides simply toggling cooling and heating materials, mode switching can be triggered under electrical and mechanical stimuli. Li et al. [59] designed a rollable structure powered by electricity that utilizes a pair of electronically controlled rotary actuators enabling the flipping of both sides of a thin film (shown in Figure 5b), thereby providing a seamless switch between solar heating and radiative cooling modes. In addition, Zhao et al. [18] achieved a spectrum switch through the induction of mechanical force. They induced cavitation in a dynamic porous silicon bilayer film by applying external compression and stretching (shown in Figure 5c). This enables a transition between a transparent solid state, which absorbs 95% of sunlight for heating, and a highly porous state, which reflects about 93% of solar radiation and allows for approximately 94% of LWIR thermal emission, facilitating solar reflection and radiative cooling.

Passive control method, which operates without additional energy and uses materials' innate reaction to environmental changes, offers a promising solution for more efficient and self-regulating radiative cooling. Designing adaptive radiative coolers with phase change materials is an effective approach. Taking advantage of the unique temperature-dependent emissivity adaption of VO_2, Fan et al. [60] proposed a concept based on a planar photonic multilayer system that incorporates VO_2 and a spectrally selective filter (shown in Figure 5d). However, intrinsic VO_2 exhibits a high phase transition temperature of ~68 °C, far exceeding the threshold of human thermal comfort. To achieve practically applicable adaptive cooling performance, Tang et al. [61] found that doping VO_2 with tungsten significantly reduced its transition temperature. Thus, they fabricated a temperature-adaptive radiative coating based on $W_xV_{1-x}O_2$, which switches thermal emittance from 0.20 (when the ambient temperature is lower than 15 °C) to 0.90 (when the ambient temperature is above 30 °C). Other temperature-responsive materials such as germanium antimony telluride (GST) [66–68], and perovskites ($LMnO_3$) [69,70], undergo a phase transition from metallic to insulating states near room temperature. Their ability to modulate emissivity in the atmospheric window makes them ideal candidates for dynamically tuning daytime radiative cooling materials. In addition to altering the emissivity within the infrared band (including atmospheric window), since the visible light spectrum also makes a significant contribution to solar heating, modulating emissivity in this range is an effective strategy as well. Thermochromic materials that exhibit temperature-sensitive reflectivity within the visible light spectrum due to thermally reversible transformation of the molecular structure of the pigments are gaining significant scholarly attention. These materials undergo a decolorization process at elevated temperatures, thereby increasing solar reflectivity and minimizing solar absorption. Conversely, at lower temperatures, these materials display color, enhancing solar absorption to diminish the cooling power [71]. Wang J et al. [62] developed a temperature-adaptive radiative cooling coating consisting of a P(VdF-HFP) bottom layer and a thermochromic microcapsule top layer (shown in Figure 5e). Via color transitions at high and low temperatures (transition temperature @ 33 °C), this material demonstrates a capacity for ~20% modulation (91.25%–72.71%) in solar reflectivity. Further, Wang T et al. [63] incorporated a small quantity of reversible thermochromic chameleon microcapsules into environmentally friendly and low-cost hollow glass beads as a radiative cooling base (shown in Figure 5f). This facilitated the possibility for color diversity and achieved a temperature modulation capacity of $\Delta T_{\text{cooling-heating}}$ = 9.5 °C. Organic thermochromic hydrogels have also been investigated and applied in self-adaptive radiative cooling technologies. At low temperatures, the refractive index of the hydrogel is similar to that of water, resulting in high solar light transmission. However, when temperature increases, the hydrogen bonds within the hydrogel break, leading to phase separation and polymer aggregation. This causes severe internal scattering and results in a low solar light transmission rate [72]. Xue et. al. [64] designed a self-adaptive film with a sandwich structure, based on a thermochromic poly(N-isopropylacrylamide) (PNIPAm) hydrogel and PVDF film with high IR emissivity (Figure 5g). This film exhibits a relatively low transition temperature (31 °C) and achieves a substantial modulation in visible light reflectance/transmittance (ΔR_{vis} = 70.0% and ΔT_{vis} = 86.3%). In addition to temperature-driven, Dastidar et al. [65] reported a humidity-induced cellulose-carbon nanotube film of the Janus structure, which reflects most sunlight (R = 88%) in a dry state and switches to absorbing sunlight (α = 60%) for rapid evaporation in a wet state.

4.2. Personal Thermal Management

While the implementation of radiative cooling in building envelopes has enhanced indoor comfort within architectural structures, it is equally imperative to address personal thermal management in expansive outdoor environments. The application of radiative cooling technologies to textiles emerges as a potential and promising solution, facilitating the creation of wearable and adaptive thermal management apparel [73–75]. The human body temperature is around 37 °C, which can emit MIR thermal radiation in a wavelength range of 7–14 μm with a peak at 9.5 μm, and emissivity of 0.98 [76]. Considering this characteristic, radiative cooling textiles can be conceptualized following two design perspectives: high MIR transparency and high MIR emissivity.

The fundamental premise behind the design guideline of high MIR transparent textiles is to allow, as much as possible, the dissipation of human body heat radiation—which is already situated within the atmospheric window—into outer space. Cai et al. [77] developed a spectrally selective nanocomposite textile by embedding zinc oxide nanoparticles into nanoporous polyethylene (ZnO-PE). This textile reflects over 90% of solar irradiance while maintaining approximately 80% high transmittance in the 7–14 μm range, where human body heat radiation is concentrated (shown in Figure 6a). Subsequently, they achieved coloring of the IR-transparent PE textile by utilizing unique inorganic pigment nanoparticles, enriching its visual appeal while maintaining negligible IR absorption, non-toxicity, and low cost [78]. To go a step further, Hsu et al. [79] proposed an IR-transparent textile composed of nanoporous polyethylene (nanoPE). This material features interconnected pores with diameters ranging from 50 to 1000 nm, comparable to the wavelengths of visible light (400–700 nm) but considerably smaller than those of IR light. Consequently, it exhibits an IR transmittance of over 90% and a visible light opacity exceeding 99%. Experimental validation, conducted with a device engineered to mimic skin heat output, indicated that when covered with the nanoPE fabric, the resulting temperature was 2.7 °C lower than that covered with cotton fabric. Due to the inherent high MIR transmissivity of PE, it is a straightforward and effective approach to integrate it with other conventional textiles. Liu et al. [80] developed an ultra-high molecular weight polyethylene (UHMWPE)/polyester composite fabric using the thermally induced phase separation method. This fabric demonstrates a weighted average human body MIR transmissivity of 82.8% and an opacity to visible light >98%, while also maintaining an adequate moisture wicking rate, breathability, and mechanical strength. Considering the large-scale applications, Peng et al. [81] reported the large-scale extrusion of uniform and continuous nanoPE microfibers for industrial mass production. They claimed that the proposed nanoPE fabric achieves a great radiative cooling power compared to cotton fabric with the same thickness, reducing the human skin temperature by 2.3 °C and leading to over 20% energy saving on indoor energy consumption for cooling demand.

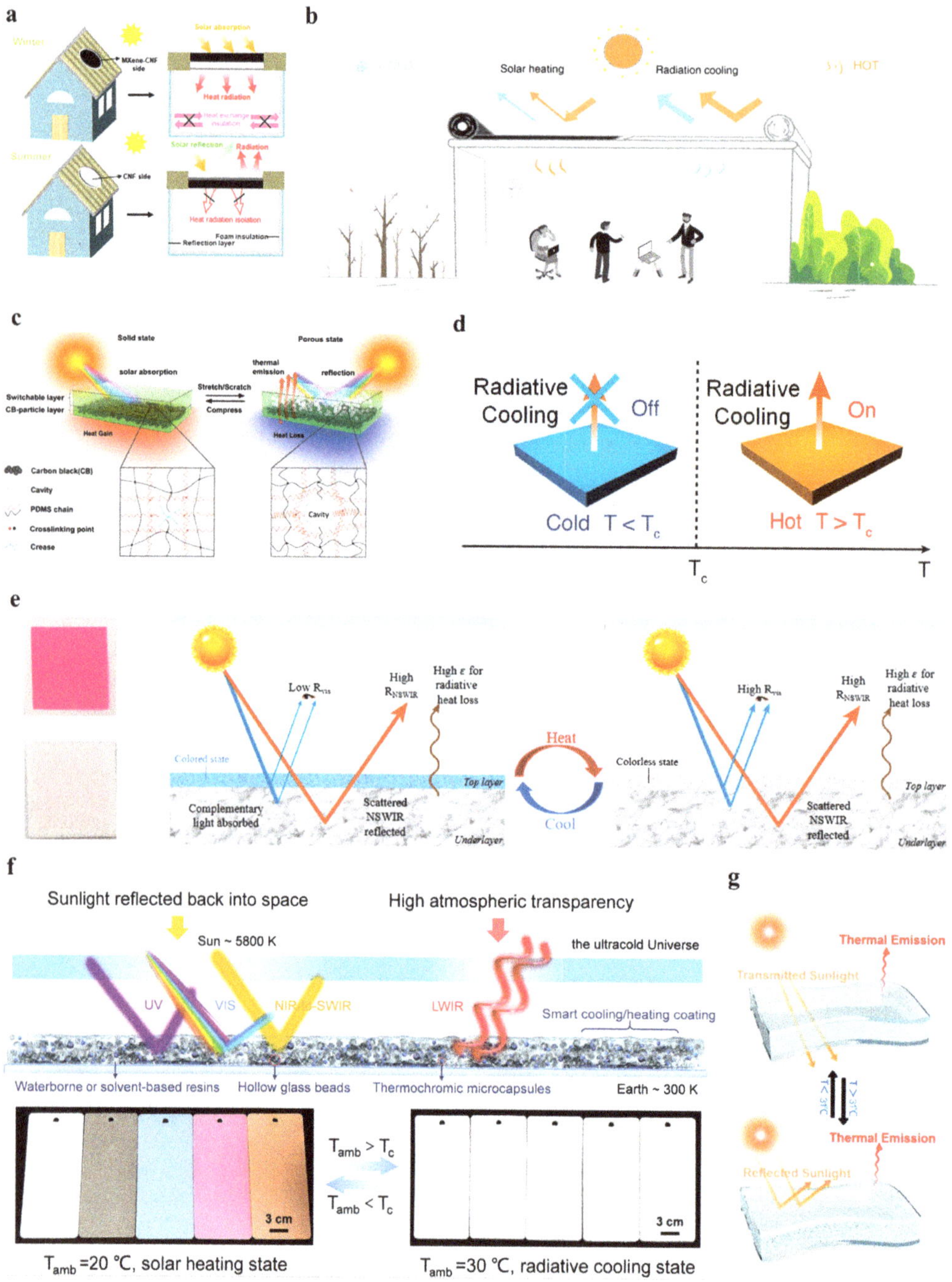

Figure 5. Adaptive radiative cooling design. (**a**) Schematic of an engineering structural Janus MXene-nanofibrils aerogels. Figure by [58] is licensed under CC BY-NC 4.0, with the permission of John Wiley and Sons. (**b**) Schematic of the dual-mode device at heating (left) mode and cooling (proper) mode. Ref. [59] is licensed under CC BY-NC 4.0. (**c**) Schematic of the bilayer structure consisting of a switchable silicone top layer and carbon black particle-embedded bottom layer. Figure by [18] is licensed under CC

BY-NC 4.0, with the permission of John Wiley and Sons. (**d**) Concept of a temperature-dependent adaptive radiative cooler containing VO_2. Ref. [60] is licensed under CC BY-NC 4.0. (**e**) Schematic of a low-cost and scalable temperature adaptive radiation-cooling coating Reprinted from [50,62], with permission from Elsevier [OR APPLICABLE SOCIETY COPYRIGHT OWNER]. (**f**) Schematic of a TiO_2-free coating with self-adaptive switching between cooling. Reprinted from [50,63], with permission from Elsevier [OR APPLICABLE SOCIETY COPYRIGHT OWNER]. (**g**) Schematic representation of the PVDF@PNIPAm film. Used with permission of Royal Society of Chemistry (Great Britain), from [64]; permission conveyed through Copyright Clearance Center, Inc.

Apart from MIR transparent textiles, the design of textiles with elevated surface IR emissivity presents a viable strategy for realizing personal thermal management. Such designs typically function by facilitating effective heat conduction between the textile and skin, thereby enabling the textile to serve as a radiator, effectively radiating heat towards the external sky. Song et al. [82] fabricated a novel porous PVDF fiber and the woven fabric via an in-situ microfiber method. This fabric demonstrates high MIR emissivity (94.5%) and opacity to solar radiation (90.3% in the NIR range and 94.2% in visible light range), thereby avoiding a simulated human body overheating effect of 17.7 °C under direct solar irradiance. Building on this accomplishment, the team subsequently developed a radiative cooling textile composed of a three-layer structure to further enhance the cooling effect and streamline the manufacturing process [83]. This structure is made of PVDF as the base layer, with nylon (PA) and PE stacked on top, fabricated using electrospinning. The textile presents substantial emissivity within the atmospheric window and over 90% reflectance to visible light, resulting in a human body cooling effect of 6.5 °C. In terms of large-scale applications, Zeng et al. [73] produced large-scale woven metafabrics with high MIR emissivity (94.5%) and high solar reflectivity (92.4%) due to the hierarchical-morphology design of the randomly dispersed scatterers in the metafabric (shown in Figure 6b). Their experiments demonstrated that a human body covered by the designed metafabric could be ~4.8 °C cooler than that covered by commercial cotton fabrics. They claimed that the metafabrics also possess preferred mechanical strength, waterproofness, and breathability through a scalable industrial textile manufacturing process.

To further meet the thermal comfort requirements under more practical conditions, the combination of passive radiative cooling and evaporative cooling is proposed. Zhang et al. [84] designed and demonstrated a nanofiber membrane-based moisture-wicking passive cooling hierarchical metafabric, which couples selective radiative cooling and wick-evaporation cooling, for effective temperature and moisture regulation (shown in Figure 6c). Due to the optical properties of the material and hierarchical morphology design, the proposed hierarchical metafabric possessed high sunlight reflectivity (99.16% in the 300–760 nm range and 88.60% in the 760–2500 nm range), high MIR emissivity (78.13%), and preferred moisture permeability. Their experiments showed that covering simulated skin with the hierarchical metafabric prevented overheating by 16.6 °C compared to conventional textiles, where the management of the humidity contributed around ~8.2 °C. Alberghini et al. [85] engineered PE fibers, yarns and fabrics for efficient water wicking and fast-drying performance. Their high-performance PE fabrics are made from fibers melt spun and woven using standard equipment in the worldwide textile industry and do not need any chemical coatings. They claim that the evaporative cooling function of the engineered PE fabrics can not only reduce the tumble-drying cycle but also provide an additional passive cooling mechanism for both indoor and outdoor thermal comfort. Hu et al. [86] developed a bilayer nanoPE membrane with anisotropic wettability, achieving superior radiative cooling ability (~2.6 °C lower compared to cotton) without perspiration. Simultaneously, the proposed membrane has efficient sweat drainage and good evaporation cooling characteristics (~1.0 °C lower compared to cotton) in perspiration to minimize sticky and hot sensations (shown in Figure 6d).

It is noteworthy that apart from thermal comfort, it is essential to meet the criteria of wear comfort for the practical application of radiative cooling textiles. Zhu et al. [74]

proposed radiative cooling textiles by modifying silk with high wearability, which was accomplished using a coupling agent (tetrabutyl titanate) to bind high refractive index inorganic oxide nanoparticles (Al_2O_3) with the silk. The nanoprocessed silk not only exhibits high cooling performance (with a temperature reduction of 3.5 °C), but also retains the high wearability and durability characteristics intrinsic to natural silk. In addition, from the perspective of long-term thermal regulation performance, superhydrophobic and self-cleaning properties of radiative cooling textiles are significant to enhance their durability and practicality, as the thermal regulation function of radiative cooling textiles may be compromised due to dampness and accumulation of surface stains. Zhong et al. [87] modified the cotton fabric to be superhydrophobic (contact angle ~152°) by introducing a PDMS top layer (shown in Figure 6e), allowing the fabric to maintain a cooling effect of 4.2 °C even after 20 washes. On the other hand, Cui et al. [88] employed a combination of template and etching techniques to create self-cleaning radiative cooling textiles using a porous coating of poly(vinylidene fluoride-co-hexafluoropropylene) (P(VDF-HFP)). The modified-coating fabric exhibits a hydrophobic contact angle of 124.9°, preventing contamination from aqueous pigments and maintaining a high solar reflectance of 0.93 and an IR emissivity of 0.98.

Additionally, similar to applications in buildings, the cooling demand from the human body is not always needed but dependent on the ambient season and environment. Therefore, adaptive textiles that can switch between cooling and heating mode as per the external conditions, also referred to as Janus textiles, are of significant interest. Active or passive switching methods have been proposed. In active switching methods, the most common one is wearing the textile inside out. Based on previous research on nanoPE [79], Hsu et al. [89] demonstrated a dual-mode textile for heating and cooling using a bilayer thermal emitter embedded within the IR-transparent nanoPE. This dual-mode textile enables an easy transition between a heating mode, which can increase skin temperature by 3.4 °C, and a cooling mode, which can reduce skin temperature by 3.1 °C, thereby extending the thermal comfort zone by 6.5 °C through a simple inversion of the fabric from inside to outside. Abebe et al. [90] designed Janus yarns for a dual-mode double-sided thermoregulating fabric using asymmetric yarn composition with dual-emissivity properties. The strong emissivity difference between the two surfaces of the designed fabric is ~0.72, due to the metallic and dielectric fibers within the yarn. Thus, a wide set-point temperature window of 13.1 °C can be achieved, with the wearer maintaining thermal comfort between 11.3 °C and 24.4 °C. On the other hand, passive switching methods include dynamic switching the MIR emittance or transmittance, dynamic switching the thermal conductance, etc. Zhang et al. [91] constructed a humidity-responsive, IR-adaptive textile composed of polymer fibers coated with carbon nanotubes (shown in Figure 6f). Humidity-driven expansion and contraction of the textile yarn modulates inter-fiber meta-element proximity, triggering resonant electromagnetic coupling to elevate emissivity for improved thermal exchange. In arid conditions, the yarn's inverse reaction diminishes heat loss. Experimental results showed that when the humidity level increased from 10% to 75%, the spacing between the yarns decreased by 100 μm, thereby increasing the fabric's IR transmittance by 35.4%.

In practical applications, low cost and environmentally friendless are two important requirements considering feasibility and sustainability. First, using low-cost raw materials and cost-effective fabrication or manufacturing process can lead to low-cost textiles. Catrysse et al. [92] proposed large-scale radiative cooling textiles using industry-standard particle-free non-porous micro-structured fibers which are adaptable to conventional textile materials and production methods, leading to advantages of low cost and large-scale volume production. Second, biodegradable materials are usually adopted taking environmentally friendless and long-term sustainability into consideration. Xu et al. [93] demonstrated an eco-friendly radiative cooling fiber membrane derived from biomass-derived silk fibroin and polylactic acid, which can be degraded in soil within one month. In addition, a simple and cost-effective electrospinning-based fabrication method was used for large-scale production [93]. Recently, Chen et al. [94] proposed spectrally selective fabrics using chitosan (CS) fibers and SiO_2 microparticles for all-day personal thermal management with simple and scalable wet-spinning technique. The low cost and biodegradability of CS result in a scalable and inexpensive wearable textile [94]. In terms of transition from research stage to practical applications, we would like to emphasize the inclusion of technology transfer data as a requirement for the comprehensive development and application of passive radiative cooling technologies for adaptive thermal management apparel. Based on Crauit et al. [95], technology transfer can be conducted in several ways, such as scientific papers, educational and governmental initiatives, and commercialization.

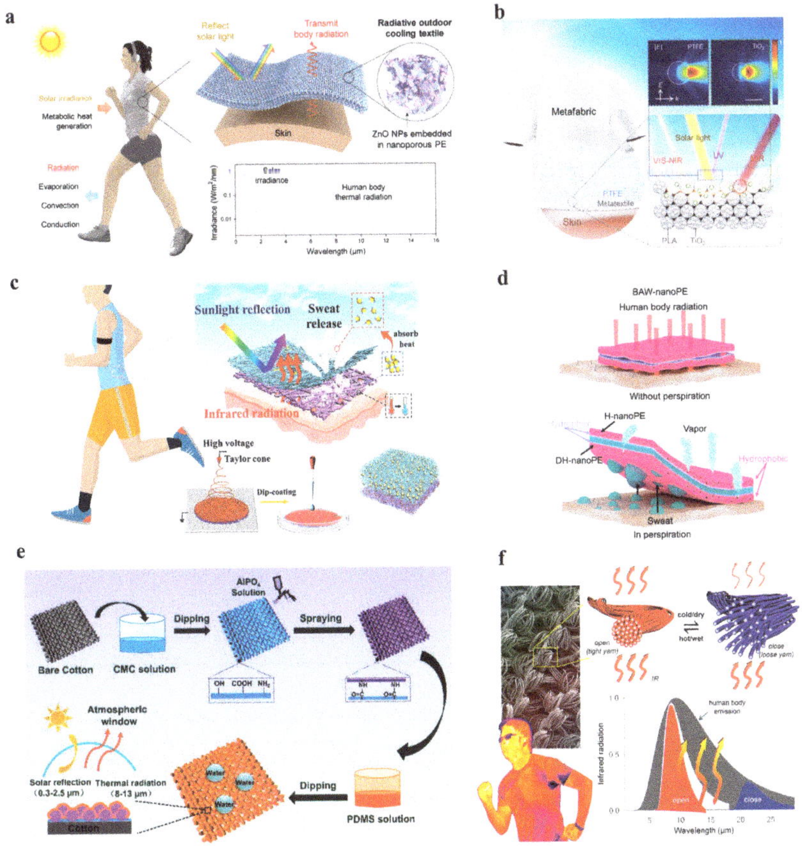

Figure 6. Passive radiative cooling textiles. (**a**) Schematic of the ZnO nanoparticle–embedded nanoPE radiative outdoor cooling textile. Figure by [77] is licensed under CC BY-NC 4.0, with the permission of John Wiley and Sons. (**b**) Schematic of the structure and simulated properties of the daytime radiative cooling metafabric. From [73]. Reprinted with permission from AAAS. (**c**) Schematic of the structure design and preparation process of the hierarchical metafabric. Reprinted (adapted) with permission from [84]. Copyright 2024 American Chemical Society. (**d**) Schematic of the moisture/thermal management of the bilayer nanoPE. Reprinted (adapted) with permission from [86]. Copyright 2024 American Chemical Society. (**e**) Schematic of the fabrication process of self-cleaning and spectrally selective cotton fabric with daytime radiative cooling. Reprinted from [87], with permission from Elsevier [OR APPLICABLE SOCIETY COPYRIGHT OWNER]. (**f**) Schematic of design principles of an IR-adaptive textile. From [91]. Reprinted with permission from AAAS.

4.3. Other Applications

4.3.1. Photovoltaics Cooling

Photovoltaics (PV) is a novel renewable energy method that can convert incident radiation into electrical energy, and is considered as one of the most promising solutions to the current energy and environmental challenges we are facing [96]. However, despite its ability to absorb most of the solar energy, a significant portion of the absorbed sunlight cannot be converted into electrical energy, but instead generates heat. This heat generation can reduce the efficiency of PV systems and significantly shorten their lifespan [97–99]. Nowadays, silicon solar cells are predominantly used in PV systems. For crystalline silicon solar cells, every temperature rise of 1 K leads to a relative efficiency decline of about 0.45% [100]. Hence, it is imperative to engineer advanced cooling methodologies tailored for PV systems. Among the potential strategies, passive radiative cooling emerges as a feasible approach. Radiative coolers integrated with PV systems are typically designed with two key optical properties in order to balance cooling efficiency and energy generation performance. Firstly, they should maintain high absorption in the photovoltaic conversion band (0.3–1.1 μm) [101,102] to maximize the transmission of sunlight for electricity generation. Additionally, they should exhibit high emissivity in the mid-infrared range (4–25 μm) to maximize heat dissipation, thereby reducing the temperature of the solar cells. Although the silicon-based cells already exhibit substantial emissivity, roughly 85% within the atmospheric window [103], further research to boost MIR emissivity is essential. Such studies would be crucial in achieving superior radiative cooling effects.

Through microstructural modifications to the silica on top of PV systems, the emissivity spectrum can be altered. Zhu et al. [104] optimized a pyramidal structure through numerical simulation, which significantly increased the emissivity of the outer surface of the PV across the entire IR spectrum. This enhancement passively lowered the operating temperature of the PV system by 18.3 K. Lu et al. [105] developed a durable, ultra-broadband multifunctional texture through the modification of silica sol-gel processes (shown in Figure 7a). When this texture is embossed onto the encapsulant polymer backsheet of silicon PV modules, it exhibits an emissivity greater than 0.96 within the atmospheric window. The appropriate incorporation of such radiative cooling effects has been demonstrated to enhance the conversion efficiency of PV systems by 3.13%. It is essential to recognize that while precise optical design is crucial for the development of future novel PV systems, retrofitting existing PV installations to enable radiative cooling is equally important. The development of transparent radiative cooling materials that can be applied over the surface of PV panels presents a promising solution. Lee et al. [106] implemented pyramidal structures into chemically stable and cost-effective PDMS, creating a flexible transparent film. The plain PDMS film with a thickness of 200 μm achieves a high emissivity exceeding 0.9 across the entire IR range of 4–26 μm. Zhou et al. [107] engineered a transparent-colored radiative cooling coating for glass cover plates on silicon PV modules. This coating which enhances both the cooling performance and visual aesthetics, incorporates silicon-core, silica-shell (Si@SiO_2) nanoparticles dispersed within a polymethyl methacrylate (PMMA) matrix. The design achieved a notable balance with a solar transmittance of over 90% and an atmospheric window emissivity greater than 95%, which enabled a cooling potential of approximately 18 to 19 K.

4.3.2. Greenhouse Temperature Regulation

The principles of radiative cooling technologies that are revolutionizing the efficiency of PV systems have found a parallel application in the domain of agriculture, specifically in the design of greenhouses. Greenhouses, serving as agricultural hubs, provide critical protection for crops against inclement weather conditions [108,109]. However, the UV radiation that is transmitted through conventional covering materials of greenhouses can detrimentally affect pollination behaviors, while the substantial heat carried by IR waves constitutes a principal reason for the overheating of these structures during the hot seasons [110]. The application of radiative cooling technologies in greenhouses presents an effective approach to resolve the issue of heat stress in greenhouses during hot seasons without necessitating an increase in water and energy consumption. Distinct from applications in other domains, radiative coolers for greenhouses need to fulfill an additional requirement of allowing the transmission of photosynthetically active radiation (PAR; 400–700 nm), which is vital for supporting the photosynthesis process in the internal plant environment [111]. This dual functionality—merging high solar reflectance and MIR emission with the greenhouse's need for high PAR transmittance—underscores the need for a bespoke design that not only expels thermal energy but also harmonizes with the light requirements essential for plant growth and productivity.

Liu et al. [112] fabricated a monolayer film through co-extrusion molding, incorporating PE with UV stabilizers and metal oxide particles. This innovative material composition realized a temperature reduction of 2.2 °C in an ambient environment of 35 °C. Advancing the practical application of this design principle, Zou et al. [113] engineered a novel transparent radiative cooling film by synthesizing PE terephthalate (PET) with silver (Ag) nanoparticles (shown in Figure 7b). In comparative experiments with traditional polyolefin (PO) films, it was observed that the deployment of the proposed film within greenhouses led to an increase of approximately 20% in the productivity of three crop varieties: Chinese cabbage, pakchoi, and cherry radish. The outcomes of this research provide compelling evidence for the practical efficacy of transparent radiative cooling materials in diminishing thermal stress without sacrificing photosynthetic efficiency, thereby offering a promising solution for augmenting agricultural yield in controlled environments.

4.3.3. Food Preservation and Packaging

The preservation of food is a critical aspect of daily life, with low-temperature cooling being a key method for extending the shelf life of foodstuffs, particularly during long-distance transportation [114]. Passive radiative cooling, as an efficient and environmentally-friendly cooling approach, offers a sustainable pathway for maintaining the freshness of food. Cellulose acetate (CA) is an abundant and eco-friendly cellulose derivative, which is considered as a substrate material for the new generation of sustainable packaging films due to its biodegradable nature, low pollution profile, and non-fossil-fuel-based composition [115–117]. Li et al. [118] developed a flexible, layered packaging film based on CA molecules using the roll-to-roll electrospinning technique. Within this material, interconnected nanofibers create a hierarchy of multi-scale pores ranging from 500 nm to 3 μm, facilitating effective sunlight dispersion and thus granting the material with proficient radiative cooling capabilities (solar reflectance of 0.974, and MIR emissivity of 0.92). When the film is used to package ice cubes or ice cream, it can dramatically reduce the thermal load of chilled outdoor food products from 296 W/m^2 to a mere 1 W/m^2, showcasing its potential to significantly enhance thermal management in food packaging applications. Zhang et al. [119] further advanced the application of CA by merging it with high-index-of-refraction ZnO ($n \approx 2$), engineering a multi-layered porous CA/ZnO film. This composite material exhibits a significantly improved MIR emissivity of 94%, which not only facilitates radiative cooling but also displays exceptional properties such as high hydrophobicity (contact angle of 138°), high tensile strength (2.7 MPa), and low thermal conductivity (0.07 W/mK) benefiting from its multiscale micro-nanostructure. These characteristics render it an ideal material for food preservation packaging applications. Experimental results indicated that utilizing this film for packaging strawberries under direct sunlight can extend their storage time up to 9 days (shown in Figure 7c), demonstrating its potential as a superior solution for food preservation. In further advancements, Chen et al. [120] developed a novel packaging material based on ZnO Nanorods/Cellulose Membrane-Starch Membrane@Bilberry Anthocyanins (ZnO-NRs/CM-SM@BA). This packaging material leverages the rod-like structure of ZnO nanorods, exhibiting a visible light reflectance of 94.4% and an IR emissivity of 98.8%, substantially reducing heat intake and enhancing the dissipation of thermal accumulation. Furthermore, its unidirectional water transport capability, arising from different water contact angles (WCAs) on either side of the membrane (WCA = 100.2°, 34.9°), can maintain a dry environment inside the packaging, thus protecting the packaged food from bacterial spoilage. Experimental tests on the packaging performance with cherry tomatoes and grapes indicated that the packaging material extends the freshness period of the fruits to over five days.

4.3.4. Dew Water Harvesting

Clean freshwater resources are an essential pillar for human survival. However, freshwater scarcity has emerged as a significant threat to the sustainable development of human society, particularly for inhabitants in arid or coastal regions, due to limited freshwater resources on earth, burgeoning populations and escalating environmental pollution. Although around 70% of the world is covered by water, 96.5% is ocean and around 2.5% is freshwater while only ~1% of that freshwater is easily accessible for human usage [121]. Seawater desalination is generally hailed as a panacea to alleviate freshwater shortage [122], based on several mature commercial methodologies—such as reverse osmosis [123,124], multi-stage flash distillation [125–127], and membrane bioreactor systems [128]. However, the considerable economic and energetic inputs required for their operation, constraints relying on large amount of seawater, coupled with the environmental ramifications associated with byproduct pollution, curtail their widespread adoption on a larger and wider scale [129,130]. In addition, there are still challenges and difficulties for freshwater transportation and delivery in remote areas due to insufficient infrastructure, limited facility maintenance and high cost [131]. Thus, decentralized freshwater production is urgently needed, especially for residents living in remote regions and water-deficient countries.

It is noteworthy that atmospheric water is ubiquitous irrespective of geological or hydrological conditions, which is a huge renewable and clean reservoir of water (estimated water volume of around 12,900 km^3), enough to meet the needs of every person on the planet [131,132]. The development of efficient technologies for harvesting atmospheric water represents a promising solution to the global water crisis [133,134]. Passive radiative cooling, offering a propitious avenue for the condensation of water vapor in the air into water, presents a compelling technological trajectory for sustainable water harvesting. Zhou et al. [135] designed and demonstrated a daytime passive radiative condenser, which is composed of a PDMS layer, a silver layer, and an aluminum plate. Their outdoor experiments at UW-Madison in September showed a daily water production of 0.8 mL using a 0.05 m^2 radiative condenser by pumping in humified air (RH = 90–95%) at ambient temperature. These investigations provide compelling evidence of the practicality of utilizing radiative cooling materials for dew water collection; nonetheless, achieving a fully passive and continuous 24-h atmospheric water harvesting system remains an elusive challenge. Haechler et al. [136] designed a water collection system wherein the radiative emitter and the water collecting surface operate independently yet cohesively (shown in Figure 7d). A selective radiative emitter situated at the top, using a conical aluminum structure as a radiative shielding device, intercepts detrimental atmospheric radiative gains, while the bottom condensing collector features a superhydrophobic surface that facilitates the detachment and shedding of droplets. This decoupled and non-inhibitory configuration enables the uninterrupted 24-h harvesting of atmospheric water. Outdoor experiments at ETH (Zurich, Switzerland) in August showed that the mean dew mass flux achieved 50 $g/m^2/hour$ (close to the theoretical limit of such systems) with a RH higher than 90% and a mean solar irradiation of over 200 W/m^2. Additionally, the integration of solar interfacial evaporation with passive radiative cooling represents a viable strategic approach to realizing all-day freshwater collection [137–139]. Xu et al. [140] designed a dual-function film combining daytime solar thermal desalination and nighttime radiative cooling dew collection by applying carbon nanotubes (solar absorption of ~95%) on a flexible PDMS substrate with high IR emissivity (~90% in 8–13 μm range). This film achieves a daytime potable water collection efficiency of 71.1% and a nighttime peak dew water yield of 80 mL/m^2.

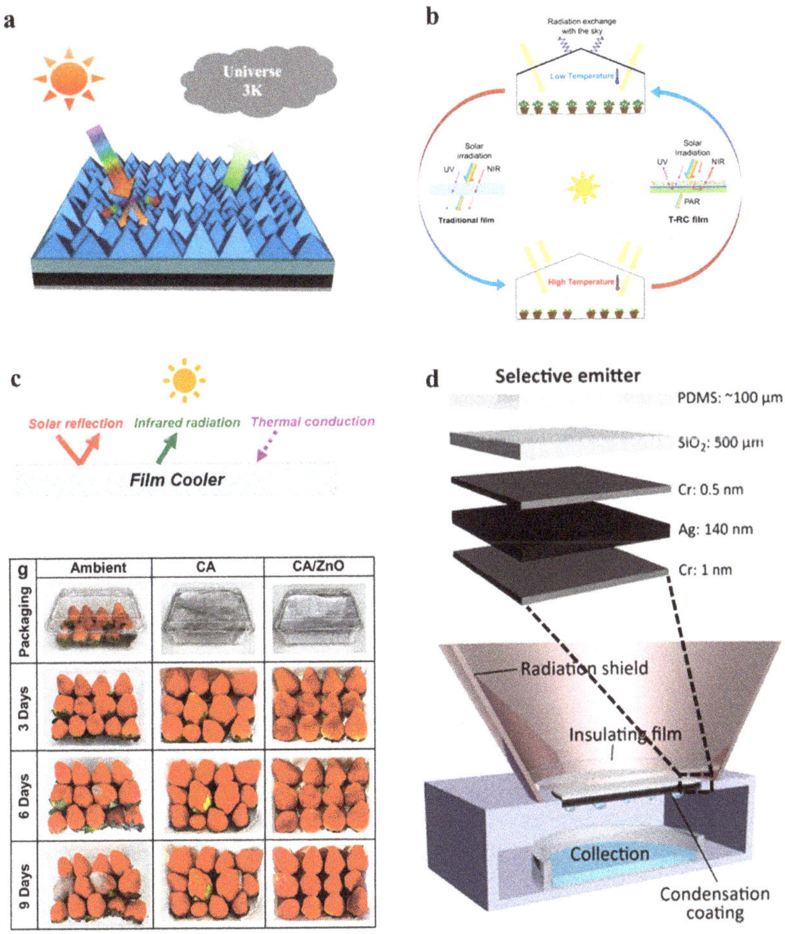

Figure 7. Other applications of radiative cooling. (**a**) Schematic of the imprinted texture glass used in PV modules. Figure by [105] is licensed under CC BY-NC 4.0, with the permission of John Wiley and Sons. (**b**) Schematic diagram for greenhouse temperature regulation using radiative cooling film. Reprinted from [113], with permission from Elsevier [OR APPLICABLE SOCIETY COPYRIGHT OWNER]. (**c**) Cooling of strawberries via the hierarchically designed CA/ZnO film. Reprinted (adapted) with permission from [119]. Copyright 2024 American Chemical Society. (**d**) Design of 24-hour atmospheric water harvesting system driven by radiative cooling. From [136]. Reprinted with permission from AAAS.

A critical factor that merits attention in the field of dew water harvesting is the influence of water condensation and moisture content on the optical characteristics, including infrared transmittance, and the subsequent cooling efficacy of radiative cooling devices. Trosseille et al. [141] provided experimental insights into the correlation between water condensation and the average surface emissivity, revealing that the presence of condensed water on the substrate of radiative coolers can elevate their effective emissivity. Complementarily, Yang et al. [142] employed Monte Carlo ray tracing techniques to model the infrared transmission through water droplets, their simulations confirming the positive impact of water condensation on enhancing the radiative cooling process and highlighting the superiority of superhydrophilic surfaces in maximizing cooling performance. Extending the investigation to porous structured radiative cooling films, Huang et al. [143] refined the diffusion-limited cluster aggregation (DLCA) model to evaluate the role of moisture content in modulating the cooling efficiency of porous radiative cooling films. The results suggested that while moisture presence potentially boosts nighttime cooling efficiency, it conversely compromises the system's performance during daylight hours.

Overall, the convergence of radiative cooling applications across different fields underscores their potential to address the intertwined challenges of energy efficiency and sustainable development. This technology not only addresses current environmental and resource challenges but also offers a pathway towards a more resilient and sustainable future.

5. Outlook

This review systematically encapsulates the current state of passive radiative cooling research, elucidating its potential as a sustainable cooling paradigm. It is evident from the literature that this technology, utilizing the atmospheric window to facilitate heat dissipation into space, introduces a new direction for energy-efficient cooling methods. The breadth of applications, from building envelopes to wearable textiles and from photovoltaic systems, agricultural facilities to dew water collection, demonstrates the vast potential of passive radiative cooling to contribute to energy savings and sustainability across multiple sectors. Nonetheless, the path from research to practical, widespread applications is paved with challenges that demand concerted scholarly attention. Three promising directions for future research are suggested and discussed, which have the potential to substantially influence the utilization and performance of passive radiative cooling solutions across various sectors.

- Enhancement of Durability and Self-Cleaning Properties: The long-term performance and reliability of passive radiative coolers are contingent upon their ability to preserve their optical characteristics—specifically, high solar reflectivity and MIR emissivity—even when subjected to environmental adversities such as pollutant deposition, chemical contamination, and physical weathering. To address these challenges, nano-engineered surfaces with self-cleaning properties emerge as a promising research frontier, which could substantially mitigate degradation and maintain efficiency over prolonged outdoor deployment. Superhydrophobic surfaces, characterized by their exceptional water-repellent properties, are among the most prominent features for self-cleaning applications. Advances in material sciences have enabled the development of innovative approaches to enhance the hydrophobicity of materials. Techniques such as the application of nanoscale particle coatings such as modified silica, the construction of micro and nano-porous structures, and the modification of surface patterning, are all viable strategies that have been explored to repel liquid contaminants effectively. However, the pursuit of extreme superhydrophobicity often leads to delicate structural designs, which can compromise the durability of the material. Therefore, striking a balance between the robustness of the material and its self-cleaning efficiency is crucial. In parallel, photocatalysis-induced self-cleaning technologies offer a complementary mechanism for sustaining the cleanliness of radiative coolers. By employing photocatalytic materials such as ZnO, CdS and ZrO_2, these surfaces leverage the redox potential of photocatalysts to degrade pollutants upon exposure to light. Contrary to the water-repellent approach, these technologies often capitalize on hydrophilic properties to isolate contaminants from cooler surfaces, thus achieving purification. The pursuit of these nanotechnology and advanced material science strategies demands rigorous sustainability assessments to ensure that the environmental footprint of passive radiative cooling technologies is thoroughly evaluated and minimized. Current research on the durability of radiative coolers tends to focus on mechanical resilience (such as impact, tensile, and compressive strength), UV aging, waterproofing, and dust self-cleaning capabilities. However, to propel radiative cooling technology further into mainstream applications, there is a pressing need to extend durability assessments to more extreme conditions. This includes exposure to fire, extreme temperatures, and corrosive agents such as acids and bases, which represent the next frontier for research and development in this field. The integration of self-cleaning properties into radiative coolers, while maintaining their intrinsic durability, represents a sophisticated balance of material science and engineering. As the field advances, the exploration of these novel surfaces must be accompanied by comprehensive environmental impact studies to ensure that the long-term stability and sustainability of radiative cooling technologies are not compromised as we strive for larger-scale implementation.

- Adaptive Cooling Power Modulation: The static nature of traditional passive radiative cooling technologies does not accommodate the dynamic cooling demand imposed by diurnal and seasonal temperature variations, nor does it account for the diverse climatic conditions across different geographical regions. The integration of adaptive mechanisms, capable of modulating radiative cooling power, is therefore a significant research frontier. Materials with tunable emissivity, such as thermochromic, shape-memory, and phase-change materials, offer promising pathways for achieving adaptive cooling capabilities. The synthesis and in-depth characterization of these materials are paramount, with a focus on tailoring their transition temperatures to align with the common temperature ranges encountered in building environments for optimal thermal control. Addressing the phenomenon of hysteresis, which is often observed in temperature-responsive materials, is crucial. This phenomenon, where the transition temperatures for activation and deactivation differ, can lead to inefficiencies in thermal regulation. Minimizing the hysteresis loop is essential for achieving more agile and precise temperature control. Furthermore, the potential of materials responding to other stimuli such as humidity and light intensity should not be overlooked, especially in applications that may benefit from such specific responsiveness. The integration of intelligent control systems is another critical component of adaptive cooling strategies. Optimizing control for standalone radiative coolers may be insufficient for the sophisticated thermal management expectations of modern urban buildings. A holistic approach to intelligent thermal management necessitates the seamless integration of these radiative coolers with conventional heating, ventilation, and air conditioning systems. Such a synergistic system would be responsive not only to ambient temperatures but also aligned with the broader energy and climate control strategies of the structure. The future of intelligent thermal management based on radiative cooling is an interdisciplinary venture that marries the advancements in material science with the complexities of control engineering.
- Scalability and Manufacturing Research: Due to breakthroughs in nanophotonics and materials science, passive radiative cooling technology has made significant strides, offering superior optical performance and cooling effects. However, it is crucial to recognize the accompanying complexity of the manufacturing process and the high production costs involved. Materials commonly used in radiative coolers to enhance infrared emission, such as Al_2O_3, TiO_2, and SiO_2, often require processing techniques that include vacuum etching, photolithography, and magnetron sputtering. These methods not only add substantial manufacturing costs but also present obstacles to scaling up production due to their intricate nature. The key to transitioning these technologies from lab-scale prototypes to widespread commercial applications lies in the innovation of scalable and cost-effective manufacturing processes. In this context, roll-to-roll (R2R) manufacturing emerges as a promising technique. R2R is a continuous production process for thin-film materials that is simpler and more cost-efficient compared to traditional methods, making it an ideal choice for the mass production of radiative coolers. A cost-effective strategy in the long run is the preparation of radiative coolers based on inorganic polymers, which can be inexpensively produced through processes such as painting, dip-coating, or spraying onto various substrates to form direct radiative cooling structures. The incorporation of the phase inversion technique is a successful practice that allows the formation of porous structures, which integrate light-scattering air voids. This technique achieves the requisite high solar reflectance for radiative cooling without introducing metal reflector layers such as

aluminum or silver, thus avoiding the need for micro and nano fabrication technologies. Consequently, this reduces the overall manufacturing cost and complexity and circumvents environmental concerns associated with the use of metal substances. Additionally, developing methodologies for standardized quality control will be vital to ensure the reliability and performance of radiative coolers at scale. Addressing these manufacturing challenges is the key to enabling the commercial adoption of passive radiative cooling technologies.

Author Contributions: Conceptualization, L.L., K.L. and S.C.; formal analysis L.L. and S.B.; investigation, L.L., S.B., S.C. and C.T.K.; writing—original draft preparation, L.L. and S.B.; writing—review and editing, L.L., S.B., K.L., Y.Z. and C.Y.T.; supervision, C.Y.T.; funding acquisition, C.Y.T. All authors have read and agreed to the published version of the manuscript.

Funding: This work was supported by Hong Kong Research Grant Council via General Research Fund (GRF) accounts 11200022 and 11200923, and via the Collaborative Research Fund (CRF) account C1105-20G, and via Strategic Topics Grant (STG) account STG2/E-605/23-N, as well as by the Innovation and Technology Commission via Innovation and Technology Fund (ITF) account ITS/128/22FP.

Institutional Review Board Statement: Not applicable.

Informed Consent Statement: Not applicable.

Data Availability Statement: No new data were created or analyzed in this study. Data sharing is not applicable to this article.

Acknowledgments: The authors wish to thank Lin LU and Jianheng CHEN for the invitation to publish in this Sustainability Special Issue.

Conflicts of Interest: The authors declare no conflicts of interest.

References

1. Shi, J.Y.; Han, D.L.; Li, Z.C.; Yang, L.; Lu, S.G.; Zhong, Z.F.; Chen, J.P.; Zhang, Q.M.; Qian, X.S. Electrocaloric Cooling Materia and Devices for Zero-Global-Warming-Potential, High-Efficiency Refrigeration. *Joule* **2019**, *3*, 1200–1225. [CrossRef]
2. Kumma, N.; Kruthiventi, S.S.H. Current status of refrigerants used in domestic applications: A review. *Renew. Sustain. Energ Rev.* **2024**, *189*, 114073. [CrossRef]
3. Abas, N.; Kalair, A.R.; Khan, N.; Haider, A.; Saleem, Z.; Saleem, M.S. Natural and synthetic refrigerants, global warming: review. *Renew. Sustain. Energy Rev.* **2018**, *90*, 557–569. [CrossRef]
4. Gullo, P.; Elmegaard, B.; Cortella, G. Energy and environmental performance assessment of R744 booster supermarket refrigeratio systems operating in warm climates. *Int. J. Refrig.-Rev. Int. Du Froid* **2016**, *64*, 61–79. [CrossRef]
5. de Paula, C.H.; Duarte, W.M.; Rocha, T.T.M.; de Oliveira, R.N.; de Paoli Mendes, R.; Maia, A.A.T. Thermo-economic an environmental analysis of a small capacity vapor compression refrigeration system using R290, R1234yf, and R600a. *Int. J. Refri* **2020**, *118*, 250–260. [CrossRef]
6. Hossain, M.M.; Gu, M. Radiative Cooling: Principles, Progress, and Potentials. *Adv. Sci.* **2016**, *3*, 1500360. [CrossRef]
7. Catalanotti, S.; Cuomo, V.; Piro, G.; Ruggi, D.; Silvestrini, V.; Troise, G.J.S.E. The radiative cooling of selective surfaces. *Sola Energy* **1975**, *17*, 83–89. [CrossRef]
8. Granqvist, C.G.; Hjortsberg, A. Radiative Cooling to Low-Temperatures—General-Considerations and Application to Selectivel Emitting Sio Films. *J. Appl. Phys.* **1981**, *52*, 4205–4220. [CrossRef]
9. Raman, A.P.; Anoma, M.A.; Zhu, L.; Rephaeli, E.; Fan, S. Passive radiative cooling below ambient air temperature under direc sunlight. *Nature* **2014**, *515*, 540–544. [CrossRef] [PubMed]
10. Zhang, Q.; Wang, S.; Wang, X.; Jiang, Y.; Li, J.; Xu, W.; Zhu, B.; Zhu, J. Recent Progress in Daytime Radiative Cooling: Advance Material Designs and Applications. *Small Methods* **2022**, *6*, e2101379. [CrossRef] [PubMed]
11. Atiganyanun, S.; Plumley, J.B.; Han, S.J.; Hsu, K.; Cytrynbaum, J.; Peng, T.L.; Han, S.M.; Han, S.E. Effective Radiative Cooling b Paint-Format Microsphere-Based Photonic Random Media. *ACS Photonics* **2018**, *5*, 1181–1187. [CrossRef]
12. Chae, D.; Kim, M.; Jung, P.H.; Son, S.; Seo, J.; Liu, Y.; Lee, B.J.; Lee, H. Spectrally Selective Inorganic-Based Multilayer Emitter fo Daytime Radiative Cooling. *ACS Appl. Mater. Interfaces* **2020**, *12*, 8073–8081. [CrossRef] [PubMed]
13. Miyazaki, H.; Okada, K.; Jinno, K.; Ota, T. Fabrication of radiative cooling devices using Si_2N_2O nano-particles. *J. Ceram. Soc. Jp* **2016**, *124*, 1185–1187. [CrossRef]
14. Chen, Z.; Zhu, L.; Raman, A.; Fan, S. Radiative cooling to deep sub-freezing temperatures through a 24-h day-night cycle. *Na Commun.* **2016**, *7*, 13729. [CrossRef]
15. Berdahl, P. Radiative cooling with MgO and/or LiF layers. *Appl. Opt.* **1984**, *23*, 370. [CrossRef]

16. Gao, W.; Lei, Z.Y.; Wu, K.; Chen, Y.P. Reconfigurable and Renewable Nano-Micro-Structured Plastics for Radiative Cooling. *Adv. Funct. Mater.* **2021**, *31*, 2100535. [CrossRef]
17. Aili, A.; Wei, Z.Y.; Chen, Y.Z.; Zhao, D.L.; Yang, R.G.; Yin, X.B. Selection of polymers with functional groups for daytime radiative cooling. *Mater. Today Phys.* **2019**, *10*, 100127. [CrossRef]
18. Zhao, H.; Sun, Q.; Zhou, J.; Deng, X.; Cui, J. Switchable Cavitation in Silicone Coatings for Energy-Saving Cooling and Heating. *Adv. Mater.* **2020**, *32*, e2000870. [CrossRef]
19. Wong, R.Y.M.; Tso, C.Y.; Fu, S.; Chao, C.Y.H. Maxwell-Garnett permittivity optimized micro-porous PVDF/PMMA blend for near unity thermal emission through the atmospheric window. *Sol. Energy Mater. Sol. Cells* **2022**, *248*, 112003. [CrossRef]
20. Yang, X.B.; Geng, J.L.; Tan, X.Y.; Liu, M.; Yao, S.M.; Tu, Y.T.; Li, S.S.; Qiao, Y.L.; Qi, G.G.; Xu, R.Z.; et al. A flexible PDMS@ZrO film for highly efficient passive radiative cooling. *Inorg. Chem. Commun.* **2023**, *151*, 110586. [CrossRef]
21. Zhai, H.T.; Fan, D.S.; Li, Q. Scalable and paint-format colored coatings for passive radiative cooling. *Sol. Energy Mater. Sol. Cells* **2022**, *245*, 111853. [CrossRef]
22. Wang, T.; Wu, Y.; Shi, L.; Hu, X.; Chen, M.; Wu, L. A structural polymer for highly efficient all-day passive radiative cooling. *Nat. Commun.* **2021**, *12*, 365. [CrossRef] [PubMed]
23. Jeong, S.Y.; Tso, C.Y.; Ha, J.; Wong, Y.M.; Chao, C.Y.H.; Huang, B.L.; Qiu, H.H. Field investigation of a photonic multi-layered TiO passive radiative cooler in sub-tropical climate. *Renew. Energy* **2020**, *146*, 44–55. [CrossRef]
24. Harrison, A.W.; Walton, M.R. Radiative Cooling of TiO_2 White Paint. *Sol. Energy* **1978**, *20*, 185–188. [CrossRef]
25. Mandal, J.; Yang, Y.; Yu, N.F.; Raman, A.P. Paints as a Scalable and Effective Radiative Cooling Technology for Buildings. *Joule* **2020**, *4*, 1350–1356. [CrossRef]
26. Chae, D.; Son, S.; Lim, H.; Jung, P.-H.; Ha, J.; Lee, H. Scalable and paint-format microparticle–polymer composite enabling high-performance daytime radiative cooling. *Mater. Today Phys.* **2021**, *18*, 100389. [CrossRef]
27. Tong, Z.; Peoples, J.; Li, X.; Yang, X.; Bao, H.; Ruan, X. Electronic and phononic origins of BaSO as an ultra-efficient radiative cooling paint pigment. *Mater. Today Phys.* **2022**, *24*, 100658. [CrossRef]
28. Li, X.; Peoples, J.; Yao, P.; Ruan, X. Ultrawhite $BaSO_4$ Paints and Films for Remarkable Daytime Subambient Radiative Cooling. *ACS Appl. Mater. Interfaces* **2021**, *13*, 21733–21739. [CrossRef]
29. Farooq, A.S.; Zhang, P.; Gao, Y.F.; Gulfam, R. Emerging radiative materials and prospective applications of radiative sky cooling—A review. *Renew. Sustain. Energy Rev.* **2021**, *144*, 110910. [CrossRef]
30. Nilsson, T.M.; Niklasson, G.A.; Granqvist, C.G. A solar reflecting material for radiative cooling applications: ZnS pigmented polyethylene. *Sol. Energy Mater. Sol. Cells* **1992**, *28*, 175–193. [CrossRef]
31. Huang, J.; Li, M.Z.; Fan, D.S. Core-shell particles for devising high-performance full-day radiative cooling paint. *Appl. Mater. Today* **2021**, *25*, 101209. [CrossRef]
32. Lv, J.P.; Tang, M.Z.; Quan, R.H.; Chai, Z.Y. Synthesis of solar heat-reflective ZnTiO pigments with novel roof cooling effect. *Ceram. Int.* **2019**, *45*, 15768–15771. [CrossRef]
33. Lim, Y.-F. Novel materials and concepts for regulating infra-red radiation: Radiative cooling and cool paint. In *Energy Saving Coating Materials*; Elsevier: Amsterdam, The Netherlands, 2020; pp. 113–131.
34. Page, M.J.; McKenzie, J.E.; Bossuyt, P.M.; Boutron, I.; Hoffmann, T.C.; Mulrow, C.D.; Shamseer, L.; Tetzlaff, J.M.; Akl, E.A.; Brennan, S.E. The PRISMA 2020 statement: An updated guideline for reporting systematic reviews. *Int. J. Surg.* **2021**, *88*, 105906. [CrossRef]
35. Howell, J.R.; Mengüç, M.P.; Daun, K.; Siegel, R. *Thermal Radiation Heat Transfer*; CRC press: Boca Raton, FL, USA, 2020.
36. Zevenhoven, R.; Fält, M. Radiative cooling through the atmospheric window: A third, less intrusive geoengineering approach. *Energy* **2018**, *152*, 27–33. [CrossRef]
37. Fenollosa, R.; Ramiro-Manzano, F.; Garín, M.; Alcubilla, R. Thermal Emission of Silicon at Near-Infrared Frequencies Mediated by Mie Resonances. *ACS Photonics* **2019**, *6*, 3174–3179. [CrossRef]
38. Olivieri, L.; Tenorio, J.A.; Revuelta, D.; Navarro, L.; Cabeza, L.F. Developing a PCM-enhanced mortar for thermally active precast walls. *Constr. Build. Mater.* **2018**, *181*, 638–649. [CrossRef]
39. Lu, X.; Xu, P.; Wang, H.L.; Yang, T.; Hou, J. Cooling potential and applications prospects of passive radiative cooling in buildings: The current state-of-the-art. *Renew. Sustain. Energy Rev.* **2016**, *65*, 1079–1097. [CrossRef]
40. Mooney, C.; Dennis, B. The World Is about to Install 700 Million Air Conditioners. *The Washington Post*, 31 May 2016.
41. Prada, M.; Prada, I.F.; Cristea, M.; Popescu, D.E.; Bungău, C.; Aleya, L.; Bungău, C.C. New solutions to reduce greenhouse gas emissions through energy efficiency of buildings of special importance–Hospitals. *Sci. Total Environ.* **2020**, *718*, 137446. [CrossRef]
42. Staffell, I.; Pfenninger, S.; Johnson, N. A global model of hourly space heating and cooling demand at multiple spatial scales. *Nat. Energy* **2023**, *8*, 1328–1344. [CrossRef]
43. Xu, L.; Shen, Y.; Ding, Y.; Wang, L. Superhydrophobic and Ultraviolet-Blocking Cotton Fabrics Based on TiO_2/SiO_2 Composite Nanoparticles. *J. Nanosci. Nanotechnol.* **2018**, *18*, 6879–6886. [CrossRef] [PubMed]
44. Song, J.; Zhang, W.; Sun, Z.; Pan, M.; Tian, F.; Li, X.; Ye, M.; Deng, X. Durable radiative cooling against environmental aging. *Nat. Commun.* **2022**, *13*, 4805. [CrossRef]
45. Li, M.; Lin, C.; Li, K.; Ma, W.; Dopphoopha, B.; Li, Y.; Huang, B. A UV-Reflective Organic-Inorganic Tandem Structure for Efficient and Durable Daytime Radiative Cooling in Harsh Climates. *Small* **2023**, *19*, e2301159. [CrossRef] [PubMed]

46. Li, T.; Zhai, Y.; He, S.; Gan, W.; Wei, Z.; Heidarinejad, M.; Dalgo, D.; Mi, R.; Zhao, X.; Song, J.; et al. A radiative cooling structural material. *Science* **2019**, *364*, 760–763. [CrossRef] [PubMed]
47. Wang, H.D.; Xue, C.H.; Ji, Z.Y.; Huang, M.C.; Jiang, Z.H.; Liu, B.Y.; Deng, F.Q.; An, Q.F.; Guo, X.J. Superhydrophobic Porous Coating of Polymer Composite for Scalable and Durable Daytime Radiative Cooling. *ACS Appl. Mater. Interfaces* **2022**, *14* 51307–51317. [CrossRef] [PubMed]
48. Meng, X.; Chen, Z.C.; Qian, C.L.; Li, Q.; Chen, X.M. Durable and mechanically robust superhydrophobic radiative cooling coating. *Chem. Eng. J.* **2023**, *478*, 147341. [CrossRef]
49. Chen, Y.; Mandal, J.; Li, W.; Smith-Washington, A.; Tsai, C.C.; Huang, W.; Shrestha, S.; Yu, N.; Han, R.P.S.; Cao, A.; et al. Colored and paintable bilayer coatings with high solar-infrared reflectance for efficient cooling. *Sci. Adv.* **2020**, *6*, eaaz5413. [CrossRef]
50. Tao, S.; Guan, F.Q.; Chen, F.H.; Chen, M.X.; Fang, Z.G.; Lu, C.H.; Xu, Z.Z. Construction of colorful super-omniphobic emitters for high-efficiency passive radiative cooling. *Compos. Commun.* **2021**, *28*, 100975. [CrossRef]
51. Xu, J.; Wan, R.; Xu, W.; Ma, Z.; Cheng, X.; Yang, R.; Yin, X. Colored radiative cooling coatings using phosphor dyes. *Mater. Today Nano* **2022**, *19*, 100239. [CrossRef]
52. Xi, W.; Liu, Y.D.; Zhao, W.X.; Hu, R.; Luo, X.B. Colored radiative cooling: How to balance color display and radiative cooling performance. *Int. J. Therm. Sci.* **2021**, *170*, 107172. [CrossRef]
53. Srinivasarao, M. Nano-Optics in the Biological World: Beetles, Butterflies, Birds, and Moths. *Chem. Rev.* **1999**, *99*, 1935–1962. [CrossRef]
54. Baek, K.; Kim, Y.; Mohd-Noor, S.; Hyun, J.K. Mie Resonant Structural Colors. *ACS Appl. Mater. Interfaces* **2020**, *12*, 5300–5318 [CrossRef] [PubMed]
55. Yalçın, R.A.; Blandre, E.; Joulain, K.; Drévillon, J. Colored radiative cooling coatings with nanoparticles. *ACS Photonics* **2020**, *7*, 1312–1322. [CrossRef]
56. Yu, S.; Zhang, Q.; Wang, Y.; Lv, Y.; Ma, R. Photonic-Structure Colored Radiative Coolers for Daytime Subambient Cooling. *Nano Lett.* **2022**, *22*, 4925–4932. [CrossRef] [PubMed]
57. Zhu, W.; Droguet, B.; Shen, Q.; Zhang, Y.; Parton, T.G.; Shan, X.; Parker, R.M.; De Volder, M.F.L.; Deng, T.; Vignolini, S.; et al Structurally Colored Radiative Cooling Cellulosic Films. *Adv. Sci.* **2022**, *9*, e2202061. [CrossRef] [PubMed]
58. Yang, W.; Xiao, P.; Li, S.; Deng, F.; Ni, F.; Zhang, C.; Gu, J.; Yang, J.; Kuo, S.W.; Geng, F. Engineering Structural Janus MXene-nanofibrils Aerogels for Season-Adaptive Radiative Thermal Regulation. *Small* **2023**, *19*, 2302509. [CrossRef]
59. Li, X.; Sun, B.; Sui, C.; Nandi, A.; Fang, H.; Peng, Y.; Tan, G.; Hsu, P.C. Integration of daytime radiative cooling and solar heating for year-round energy saving in buildings. *Nat. Commun.* **2020**, *11*, 6101. [CrossRef]
60. Ono, M.; Chen, K.; Li, W.; Fan, S. Self-adaptive radiative cooling based on phase change materials. *Opt. Express* **2018**, *26*, A777–A787. [CrossRef]
61. Tang, K.; Dong, K.; Li, J.; Gordon, M.P.; Reichertz, F.G.; Kim, H.; Rho, Y.; Wang, Q.; Lin, C.Y.; Grigoropoulos, C.P.; et al Temperature-adaptive radiative coating for all-season household thermal regulation. *Science* **2021**, *374*, 1504–1509. [CrossRef] [PubMed]
62. Wang, J.; Xie, M.; An, Y.Z.; Tao, Y.J.; Sun, J.Y.; Ji, C. All-season thermal regulation with thermochromic temperature-adaptive radiative cooling coatings. *Sol. Energy Mater. Sol. Cells* **2022**, *246*, 111883. [CrossRef]
63. Wang, T.; Zhang, Y.A.; Chen, M.; Gu, M.; Wu, L.M. Scalable and waterborne titanium-dioxide-free thermochromic coatings for self-adaptive passive radiative cooling and heating. *Cell Rep. Phys. Sci.* **2022**, *3*, 100782. [CrossRef]
64. Mei, X.; Wang, T.; Chen, M.; Wu, L.M. A self-adaptive film for passive radiative cooling and solar heating regulation. *J. Mater. Chem. A* **2022**, *10*, 11092–11100. [CrossRef]
65. Dastidar, S.; Alam, M.M.; Crispin, X.; Zhao, D.; Jonsson, M.P. Janus cellulose for self-adaptive solar heating and evaporative drying. *Cell Rep. Phys. Sci.* **2022**, *3*, 101196. [CrossRef]
66. Du, K.K.; Li, Q.; Lyu, Y.B.; Ding, J.C.; Lu, Y.; Cheng, Z.Y.; Qiu, M. Control over emissivity of zero-static-power thermal emitters based on phase-changing material GST. *Light Sci. Appl.* **2017**, *6*, e16194. [CrossRef]
67. Xu, Z.; Luo, H.; Zhu, H.; Hong, Y.; Shen, W.; Ding, J.; Kaur, S.; Ghosh, P.; Qiu, M.; Li, Q. Nonvolatile Optically Reconfigurable Radiative Metasurface with Visible Tunability for Anticounterfeiting. *Nano Lett.* **2021**, *21*, 5269–5276. [CrossRef]
68. Abdollahramezani, S.; Hemmatyar, O.; Taghinejad, H.; Krasnok, A.; Kiarashinejad, Y.; Zandehshahvar, M.; Alù, A.; Adibi, A. Tunable nanophotonics enabled by chalcogenide phase-change materials. *Nanophotonics* **2020**, *9*, 1189–1241. [CrossRef]
69. Tachikawa, S.; Ohnishi, A.; Okamoto, A.; Nakamura, Y. Development of a variable emittance radiator based on a perovskite manganese oxide. *J. Thermophys. Heat Transf.* **2003**, *17*, 264–268. [CrossRef]
70. Zhao, Y.; Fan, D.S.; Li, Q. Flexible perovskite-based photonic stack device for adaptive thermal management. *Mater. Des.* **2022**, *219*, 110719. [CrossRef]
71. Granqvist, C.G. Recent progress in thermochromics and electrochromics: A brief survey. *Thin Solid Film.* **2016**, *614*, 90–96. [CrossRef]
72. An, Y.D.; Fu, Y.; Dai, J.G.; Yin, X.B.; Lei, D.Y. Switchable radiative cooling technologies for smart thermal management. *Cell Rep. Phys. Sci.* **2022**, *3*. [CrossRef]
73. Zeng, S.; Pian, S.; Su, M.; Wang, Z.; Wu, M.; Liu, X.; Chen, M.; Xiang, Y.; Wu, J.; Zhang, M.; et al. Hierarchical-morphology metafabric for scalable passive daytime radiative cooling. *Science* **2021**, *373*, 692–696. [CrossRef] [PubMed]
74. Zhu, B.; Li, W.; Zhang, Q.; Li, D.; Liu, X.; Wang, Y.; Xu, N.; Wu, Z.; Li, J.; Li, X.; et al. Subambient daytime radiative cooling textile based on nanoprocessed silk. *Nat. Nanotechnol.* **2021**, *16*, 1342–1348. [CrossRef]
75. Hsu, P.C.; Li, X. Photon-engineered radiative cooling textiles. *Science* **2020**, *370*, 784–785. [CrossRef]

76. Steketee, J. Spectral emissivity of skin and pericardium. *Phys. Med. Biol.* **1973**, *18*, 686–694. [CrossRef]
77. Cai, L.; Song, A.Y.; Li, W.; Hsu, P.C.; Lin, D.; Catrysse, P.B.; Liu, Y.; Peng, Y.; Chen, J.; Wang, H.; et al. Spectrally Selective Nanocomposite Textile for Outdoor Personal Cooling. *Adv. Mater.* **2018**, *30*, e1802152. [CrossRef]
78. Cai, L.L.; Peng, Y.C.; Xu, J.W.; Zhou, C.Y.; Zhou, C.X.; Wu, P.L.; Lin, D.C.; Fan, S.H.; Cui, Y. Temperature Regulation in Colored Infrared-Transparent Polyethylene Textiles. *Joule* **2019**, *3*, 1478–1486. [CrossRef]
79. Hsu, P.C.; Song, A.Y.; Catrysse, P.B.; Liu, C.; Peng, Y.; Xie, J.; Fan, S.; Cui, Y. Radiative human body cooling by nanoporous polyethylene textile. *Science* **2016**, *353*, 1019–1023. [CrossRef]
80. Liu, R.; Wang, X.W.; Yu, J.R.; Wang, Y.; Zhu, J.; Hu, Z.M. A Novel Approach to Design Nanoporous Polyethylene/Polyester Composite Fabric via TIPS for Human Body Cooling. *Macromol. Mater. Eng.* **2018**, *303*, 1700456. [CrossRef]
81. Peng, Y.C.; Chen, J.; Song, A.Y.; Catrysse, P.B.; Hsu, P.C.; Cai, L.L.; Liu, B.F.; Zhu, Y.Y.; Zhou, G.M.; Wu, D.S.; et al. Nanoporous polyethylene microfibres for large-scale radiative cooling fabric. *Nat. Sustain.* **2018**, *1*, 105–112. [CrossRef]
82. Song, Y.-N.; Lei, M.-Q.; Lei, J.; Li, Z.-M. Spectrally selective polyvinylidene fluoride textile for passive human body cooling. *Mater. Today Energy* **2020**, *18*, 100504. [CrossRef]
83. Song, Y.N.; Li, Y.; Yan, D.X.; Lei, J.; Li, Z.M. Novel passive cooling composite textile for both outdoor and indoor personal thermal management. *Compos. Part A-Appl. Sci. Manuf.* **2020**, *130*, 105738. [CrossRef]
84. Zhang, X.; Yang, W.; Shao, Z.; Li, Y.; Su, Y.; Zhang, Q.; Hou, C.; Wang, H. A Moisture-Wicking Passive Radiative Cooling Hierarchical Metafabric. *ACS Nano* **2022**, *16*, 2188–2197. [CrossRef] [PubMed]
85. Alberghini, M.; Hong, S.; Lozano, L.M.; Korolovych, V.; Huang, Y.; Signorato, F.; Zandavi, S.H.; Fucetola, C.; Uluturk, I.; Tolstorukov, M.Y.; et al. Sustainable polyethylene fabrics with engineered moisture transport for passive cooling. *Nat. Sustain.* **2021**, *4*, 715–724. [CrossRef]
86. Hu, R.; Wang, N.; Hou, L.; Liu, J.; Cui, Z.; Zhang, C.; Zhao, Y. Bilayer Nanoporous Polyethylene Membrane with Anisotropic Wettability for Rapid Water Transportation/Evaporation and Radiative Cooling. *ACS Appl. Mater. Interfaces* **2022**, *14*, 9833–9843. [CrossRef] [PubMed]
87. Zhong, S.J.; Yi, L.M.; Zhang, J.W.; Xu, T.Q.; Xu, L.; Zhang, X.; Zuo, T.; Cai, Y. Self-cleaning and spectrally selective coating on cotton fabric for passive daytime radiative cooling. *Chem. Eng. J.* **2021**, *407*, 127104. [CrossRef]
88. Cui, C.F.; Lu, J.; Zhang, S.Q.; Su, J.J.; Han, J. Hierarchical-porous coating coupled with textile for passive daytime radiative cooling and self-cleaning. *Sol. Energy Mater. Sol. Cells* **2022**, *247*, 111954. [CrossRef]
89. Hsu, P.C.; Liu, C.; Song, A.Y.; Zhang, Z.; Peng, Y.; Xie, J.; Liu, K.; Wu, C.L.; Catrysse, P.B.; Cai, L.; et al. A dual-mode textile for human body radiative heating and cooling. *Sci. Adv.* **2017**, *3*, e1700895. [CrossRef]
90. Abebe, M.G.; De Corte, A.; Rosolen, G.; Maes, B. Janus-Yarn Fabric for Dual-Mode Radiative Heat Management. *Phys. Rev. Appl.* **2021**, *16*, 054013. [CrossRef]
91. Zhang, X.A.; Yu, S.; Xu, B.; Li, M.; Peng, Z.; Wang, Y.; Deng, S.; Wu, X.; Wu, Z.; Ouyang, M.; et al. Dynamic gating of infrared radiation in a textile. *Science* **2019**, *363*, 619–623. [CrossRef]
92. Catrysse, P.B.; Fan, S. Radiative cooling textiles using industry-standard particle-free nonporous micro-structured fibers. *Nanophotonics* **2024**, *13*, 649–657. [CrossRef]
93. Xu, P.; Xiang, B.; Zhong, W.; Wu, Y.; Zhang, Y.; Chang, Y.; Lei, W.; Zhang, R. Biodegradable, scalable and flexible fiber membrane for green passive radiative cooling. *Sol. Energy Mater. Sol. Cells* **2023**, *253*, 112209. [CrossRef]
94. Chen, C.; Jia, X.; Li, X.; Shi, M.; Hu, J.; Song, M.; Wu, S.; Dai, H.; Wang, X.; Geng, H. Scalable wet-spinning of wearable chitosan-silica textile for all-day radiative cooling. *Chem. Eng. J.* **2023**, *475*, 146307. [CrossRef]
95. Craiut, L.; Bungau, C.; Bungau, T.; Grava, C.; Otrisal, P.; Radu, A.-F. Technology Transfer, Sustainability, and Development, Worldwide and in Romania. *Sustainability* **2022**, *14*, 15728. [CrossRef]
96. Trapani, K.; Millar, D.L.; Smith, H.C.M. Novel offshore application of photovoltaics in comparison to conventional marine renewable energy technologies. *Renew. Energy* **2013**, *50*, 879–888. [CrossRef]
97. Menke, S.M.; Ran, N.A.; Bazan, G.C.; Friend, R.H. Understanding Energy Loss in Organic Solar Cells: Toward a New Efficiency Regime. *Joule* **2018**, *2*, 25–35. [CrossRef]
98. Otth, D.H.; Ross, R. Assessing Photovoltaic Module Degradation and Lifetime from Long Term Environmental Tests. In Proceedings of the 29th Institute of Environmental Sciences Technical Meeting, Los Angeles, CA, USA, 19–21 April 1983; pp. 121–126.
99. Radziemska, E. The effect of temperature on the power drop in crystalline silicon solar cells. *Renew. Energy* **2003**, *28*, 1–12. [CrossRef]
100. Skoplaki, E.; Palyvos, J.A. On the temperature dependence of photovoltaic module electrical performance: A review of efficiency/power correlations. *Solar Energy* **2009**, *83*, 614–624. [CrossRef]
101. Zhao, B.; Hu, M.; Ao, X.; Xuan, Q.; Pei, G. Comprehensive photonic approach for diurnal photovoltaic and nocturnal radiative cooling. *Sol. Energy Mater. Sol. Cells* **2018**, *178*, 266–272. [CrossRef]
102. Smestad, G.P.; Krebs, F.C.; Lampert, C.M.; Granqvist, C.G.; Chopra, K.; Mathew, X.; Takakura, H. *Reporting Solar Cell Efficiencies in Solar Energy Materials and Solar Cells*; Elsevier: Amsterdam, The Netherlands, 2008; Volume 92, pp. 371–373.
103. Gentle, A.; Smith, G. Is enhanced radiative cooling of solar cell modules worth pursuing? *Sol. Energy Mater. Sol. Cells* **2016**, *150*, 39–42. [CrossRef]
104. Zhu, L.X.; Raman, A.; Wang, K.X.; Abou Anoma, M.; Fan, S.H. Radiative cooling of solar cells. *Optica* **2014**, *1*, 32–38. [CrossRef]

105. Lu, Y.H.; Chen, Z.C.; Ai, L.; Zhang, X.P.; Zhang, J.; Li, J.; Wang, W.Y.; Tan, R.Q.; Dai, N.; Song, W.J. A Universal Route to Realiz Radiative Cooling and Light Management in Photovoltaic Modules. *Solar Rrl* **2017**, *1*, 1700084. [CrossRef]
106. Lee, E.; Luo, T.F. Black body-like radiative cooling for flexible thin-film solar cells. *Sol. Energy Mater. Sol. Cells* **2019**, *194*, 222–22 [CrossRef]
107. Zhou, Y.; Jia, C.; Lu, K.; Wang, C.; Ma, L.; Liu, L. Energy-efficient colorful silicon photovoltaic modules driven by transparen colored radiative cooling. *Sol. Energy Mater. Sol. Cells* **2023**, *259*, 112459. [CrossRef]
108. Coelho, J.P.; Oliveira, P.B.D.; Cunha, J.B. Greenhouse air temperature predictive control using the particle swarm optimisatio algorithm. *Comput. Electron. Agric.* **2005**, *49*, 330–344. [CrossRef]
109. Aznar-Sanchez, J.A.; Velasco-Munoz, J.F.; Lopez-Felices, B.; Roman-Sanchez, I.M. An Analysis of Global Research Trends o Greenhouse Technology: Towards a Sustainable Agriculture. *Int. J. Environ. Res. Public Health* **2020**, *17*, 664. [CrossRef] [PubMed
110. Iddio, E.; Wang, L.; Thomas, Y.; McMorrow, G.; Denzer, A. Energy efficient operation and modeling for greenhouses: A literatur review. *Renew. Sustain. Energy Rev.* **2020**, *117*, 109480. [CrossRef]
111. Wang, S.C.; Zhang, J.; Liu, L.; Yang, F.; Zhang, Y.H. Evaluation of cooling property of high density polyethylene (HDPE)/titaniun dioxide (TiO) composites after accelerated ultraviolet (UV) irradiation. *Sol. Energy Mater. Sol. Cells* **2015**, *143*, 120–127. [CrossRef
112. Liu, C.H.; Ay, C.; Tsai, C.Y.; Lee, M.T. The Application of Passive Radiative Cooling in Greenhouses. *Sustainability* **2019**, *11*, 670 [CrossRef]
113. Zou, H.; Wang, C.X.; Yu, J.Q.; Huang, D.F.; Yang, R.G.; Wang, R.Z. Eliminating greenhouse heat stress with transparent radiativ cooling film. *Cell Rep. Phys. Sci.* **2023**, *4*. [CrossRef]
114. Tassou, S.A.; De-Lille, G.; Ge, Y.T. Food transport refrigeration—Approaches to reduce energy consumption and environmenta impacts of road transport. *Appl. Therm. Eng.* **2009**, *29*, 1467–1477. [CrossRef]
115. Gemili, S.; Yemenicioglu, A.; Altinkaya, S.A. Development of cellulose acetate based antimicrobial food packaging materials fo controlled release of lysozyme. *J. Food Eng.* **2009**, *90*, 453–462. [CrossRef]
116. Marrez, D.A.; Abdelhamid, A.E.; Darwesh, O.M. Eco-friendly cellulose acetate green synthesized silver nano-composite a antibacterial packaging system for food safety. *Food Packag. Shelf Life* **2019**, *20*, 100302. [CrossRef]
117. Gouvêa, D.M.; Mendonça, R.C.S.; Soto, M.L.; Cruz, R.S. Acetate cellulose film with bacteriophages for potential antimicrobial us in food packaging. *LWT-Food Sci. Technol.* **2015**, *63*, 85–91. [CrossRef]
118. Li, J.; Liang, Y.; Li, W.; Xu, N.; Zhu, B.; Wu, Z.; Wang, X.; Fan, S.; Wang, M.; Zhu, J. Protecting ice from melting under sunlight vi radiative cooling. *Sci. Adv.* **2022**, *8*, eabj9756. [CrossRef] [PubMed]
119. Zhang, K.; Mo, C.Q.; Tang, X.L.; Lei, X.J. Hierarchically Porous Cellulose-Based Radiative Cooler for Zero-Energy Foo Preservation. *Acs Sustain. Chem. Eng.* **2023**, *11*, 7745–7754. [CrossRef]
120. Chen, Y.N.; Wang, Z.Y.; Dai, Y.T.; Yang, D.Y.; Qiu, F.X.; Li, Y.Q.; Zhang, T. Green Food Packaging with Integrated Functions o High-Efficiency Radiation Cooling and Freshness Monitoring. *ACS Sustain. Chem. Eng.* **2023**, *11*, 15135–15145. [CrossRef]
121. Mishra, R.K. Fresh water availability and its global challenge. *Br. J. Multidiscip. Adv. Stud.* **2023**, *4*, 1–78. [CrossRef]
122. Ng, K.C.; Thu, K.; Kim, Y.; Chakraborty, A.; Amy, G. Adsorption desalination: An emerging low-cost thermal desalinatior method. *Desalination* **2013**, *308*, 161–179. [CrossRef]
123. Linares, R.V.; Li, Z.Y.; Abu-Ghdaib, M.; Wei, C.H.; Amy, G.; Vrouwenvelder, J.S. Water harvesting from municipal wastewater vi osmotic gradient: An evaluation of process performance. *J. Membr. Sci.* **2013**, *447*, 50–56. [CrossRef]
124. Ghermandi, A.; Messalem, R. Solar-driven desalination with reverse osmosis: The state of the art. *Desalin. Water Treat.* **2009**, *7* 285–296. [CrossRef]
125. Tareemi, A.A.; Sharshir, S.W. A state-of-art overview of multi-stage flash desalination and water treatment: Principles, challenges and heat recovery in hybrid systems. *Sol. Energy* **2023**, *266*, 112157. [CrossRef]
126. Khoshrou, I.; Nasr, M.R.J.; Bakhtari, K. New opportunities in mass and energy consumption of the Multi-Stage Flash Distillatior type of brackish water desalination process. *Sol. Energy* **2017**, *153*, 115–125. [CrossRef]
127. El-Dessouky, H.T.; Ettouney, H.M.; Al-Roumi, Y. Multi-stage flash desalination: Present and future outlook. *Chem. Eng. J.* **1999** *73*, 173–190. [CrossRef]
128. Khan, M.A.; Ngo, H.H.; Guo, W.; Liu, Y.; Chang, S.W.; Nguyen, D.D.; Nghiem, L.D.; Liang, H. Can membrane bioreactor be a smart option for water treatment? *Bioresour. Technol. Rep.* **2018**, *4*, 80–87. [CrossRef]
129. Blandin, G.; Verliefde, A.R.; Comas, J.; Rodriguez-Roda, I.; Le-Clech, P. Efficiently combining water reuse and desalinatior through forward osmosis—Reverse osmosis (FO-RO) hybrids: A critical review. *Membranes* **2016**, *6*, 37. [CrossRef]
130. Holloway, R.W.; Achilli, A.; Cath, T.Y. The osmotic membrane bioreactor: A critical review. *Environ. Sci. -Water Res. Technol.* **2015** *1*, 581–605. [CrossRef]
131. Tu, Y.D.; Wang, R.Z.; Zhang, Y.N.; Wang, J.Y. Progress and Expectation of Atmospheric Water Harvesting. *Joule* **2018**, *2*, 1452–1475 [CrossRef]
132. Ejeian, M.; Wang, R.Z. Adsorption-based atmospheric water harvesting. *Joule* **2021**, *5*, 1678–1703. [CrossRef]
133. Bengtsson, L. The global atmospheric water cycle. *Environ. Res. Lett.* **2010**, *5*, 025202. [CrossRef]
134. Wahlgren, R.V. Atmospheric water vapour processor designs for potable water production: A review. *Water Res.* **2001**, *35*, 1–22 [CrossRef] [PubMed]
135. Zhou, M.; Song, H.; Xu, X.; Shahsafi, A.; Qu, Y.; Xia, Z.; Ma, Z.; Kats, M.A.; Zhu, J.; Ooi, B.S.; et al. Vapor condensation with daytime radiative cooling. *Proc.Natl. Acad. Sci. USA* **2021**, *118*, e2019292118. [CrossRef]

136. Haechler, I.; Park, H.; Schnoering, G.; Gulich, T.; Rohner, M.; Tripathy, A.; Milionis, A.; Schutzius, T.M.; Poulikakos, D. Exploiting radiative cooling for uninterrupted 24-hour water harvesting from the atmosphere. *Sci. Adv.* **2021**, *7*, eabf3978. [CrossRef]
137. Xi, Z.; Li, S.; Yu, L.; Yan, H.; Chen, M. All-Day Freshwater Harvesting by Selective Solar Absorption and Radiative Cooling. *ACS Appl. Mater. Interfaces* **2022**, *14*, 26255–26263. [CrossRef]
138. Li, Y.; Lin, C.; Huang, J.; Chi, C.; Huang, B. Spectrally Selective Absorbers/Emitters for Solar Steam Generation and Radiative Cooling-Enabled Atmospheric Water Harvesting. *Glob. Chall.* **2021**, *5*, 2000058. [CrossRef]
139. Chen, Z.; Dong, M.; Wang, C. Passive interfacial photothermal evaporation and sky radiative cooling assisted all-day freshwater harvesting: System design, experiment study, and performance evaluation. *Appl. Energy* **2024**, *355*, 122254. [CrossRef]
140. Xu, J.; Zhang, J.; Fu, B.; Song, C.; Shang, W.; Tao, P.; Deng, T. All-Day Freshwater Harvesting through Combined Solar-Driven Interfacial Desalination and Passive Radiative Cooling. *ACS Appl. Mater. Interfaces* **2020**, *12*, 47612–47622. [CrossRef] [PubMed]
141. Trosseille, J.; Mongruel, A.; Royon, L.; Beysens, D. Effective substrate emissivity during dew water condensation. *Int. J. Heat Mass Transf.* **2022**, *183*, 122078. [CrossRef]
142. Yang, R.; Wang, M.; Du, M.; Wang, X.; Tang, G. Droplet effect on the infrared transmittance of radiative cooler for direct water condensation. *Sol. Energy Mater. Sol. Cells* **2022**, *238*, 111615. [CrossRef]
143. Huang, M.; Tang, G.; Ren, X.; Sun, Q.; Du, M. Effects of microstructure and moisture content on the radiative properties of porous films for radiative cooling. *Sol. Energy* **2023**, *262*, 111855. [CrossRef]

sustainability

MDPI

Review

Nanofluid-Enhanced HVAC&R Systems (2015–2025): Experimental, Numerical, and AI-Driven Insights with a Strategic Roadmap

Aung Myat *, Md Mashiur Rahman and Muhammad Akbar

Department of Mechanical and Manufacturing Engineering, Tennessee State University, Nashville, TN 37209, USA; mrahman6@tnstate.edu (M.M.R.); makbar@tnstate.edu (M.A.)
* Correspondence: amyat@tnstate.edu

Abstract

Heating, ventilation, air conditioning, and refrigeration (HVAC&R) systems account for a significant share of global energy demand, prompting intensive research into advanced thermal enhancement techniques. Among these, nanofluids—colloidal suspensions of nanoparticles in base fluids—have shown promise in boosting heat transfer performance. This review provides a structured and critical evaluation of nanofluid applications in HVAC&R systems, synthesizing research published from 2015 to 2025. A total of 200 peer-reviewed articles were selected from an initial pool of over 900 through a systematic filtering process. The selected literature was thematically categorized into experimental, numerical, hybrid, and AI/ML-based studies, with further classification by fluid type, performance metrics, and system-level relevance. Unlike prior reviews focused narrowly on thermophysical properties or individual components, this work integrates recent advances in artificial intelligence and hybrid modeling to assess both localized and systemic enhancements. Notably, nanofluids have demonstrated up to a 45% improvement in heat transfer coefficients and up to a 51% increase in the coefficient of performance (COP). However, the review reveals persistent gaps, including limited full-system validation, underexplored real-world integration, and minimal use of AI for holistic optimization. By identifying these knowledge gaps and research imbalances, this review proposes a forward-looking, data-driven roadmap to guide future research and facilitate the scalable adoption of nanofluid-enhanced HVAC&R technologies.

Keywords: nanofluids; HVAC & refrigeration; heat transfer enhancement; AI/ML modeling; energy efficiency

1. Introduction

Heating, ventilation, and air conditioning (HVAC) systems account for a substantial portion of global energy consumption, particularly in the residential, commercial, and industrial building sectors. According to the International Energy Agency (IEA) [1], HVAC systems contribute to nearly 40% of total building energy use worldwide, making them a critical focus for energy efficiency improvements. As global temperatures rise and urbanization intensifies, the demand for HVAC systems is projected to increase significantly, further stressing energy infrastructure and contributing to greenhouse gas emissions.

To address these pressing concerns, researchers and engineers have sought to improve the thermal efficiency of HVAC components, particularly through enhancements in heat transfer mechanisms. One of the most effective and widely explored strategies involves upgrading conventional heat transfer fluids—such as water, ethylene glycol, and oils—with advanced working fluids that exhibit superior thermophysical properties [2]. In this context, nanofluids—a class of engineered colloidal suspensions of nanometer-sized particles (e.g., metals, oxides, carbon nanotubes) in a base fluid—have gained significant attention over the past two decades for their potential to revolutionize heat transfer technologies [3].

Nanofluids offer several theoretical advantages over traditional fluids, including anomalously high thermal conductivity (TC) at low nanoparticle concentrations, improved convective heat transfer coefficients, and the potential for microchannel cooling applications [4,5]. These benefits make them attractive candidates for integration into a wide range of HVAC components, such as heat exchangers, radiators, evaporators, and solar collectors. Furthermore, the ability to tailor nanofluid properties through careful selection of the nanoparticle material, size, morphology, and concentration provides unique opportunities for system-level optimization and performance customization [6]. As the field has evolved, a growing body of the literature highlights both the profound promise and the practical challenges—such as long-term stability, increased viscosity, and cost-effectiveness—of using nanofluids in real-world HVAC applications [7]. Despite the growing body of work, existing reviews often lack integration across methodological domains and fall short in addressing the translational gap between laboratory-scale findings and real-world HVAC deployment. Key challenges include the long-term stability of nanofluids under cyclic operation, viscosity-driven pressure penalties, fouling, lifecycle cost, and system compatibility. This review distinguishes itself by not only mapping experimental, numerical, and AI/ML-based methodologies but also by introducing a novel AI-driven system optimization roadmap, offering a strategic synthesis tailored to building energy applications. Unlike prior reviews that focus narrowly on nanofluid thermophysical properties or generalized thermal systems, this work provides an HVAC-specific, application-driven synthesis. It uniquely combines experimental, numerical, and AI/ML-based insights into a unified methodological framework, aligned with building energy performance goals. Additionally, recent studies have employed entropy generation analysis to evaluate the thermodynamic efficiency of nanofluids, especially in hybrid configurations. These approaches offer valuable insight into irreversibility minimization and the optimization of system performance and are discussed in Section 2.

This review aims to critically examine the current state of nanofluid research within HVAC systems by organizing the literature thematically across key application domains. It will synthesize experimental and computational findings, assess performance metrics, and identify future research directions to support the development of high-efficiency, nanofluid-enhanced HVAC technologies.

1.1. Current Gaps in Review of Nanofluids Applications

Over the past five years, several review articles have investigated the intersection of nanofluids and advanced modeling tools, particularly artificial intelligence (AI), machine learning (ML), and numerical simulation techniques. For instance, Riyadi et al. (2024) [8] focused on ML-enhanced porous heat exchangers and sustainable energy systems, offering a computational taxonomy but limited direct applicability to HVAC&R. Basu et al. (2023) [9] and Raza et al. (2024) [10] provided detailed overviews of AI-driven thermophysical property prediction for nanofluids, while Smrity and Yin (2025) [11] emphasized the synergy between experimental data and digital modeling across general heat transfer applications.

Despite their merits, these reviews exhibit several limitations that constrain their practical impact on the HVAC and refrigeration (HVAC&R) sector:

- Lack of HVAC&R-Specific Focus: Prior reviews remain largely disconnected from HVAC&R system architectures. None offers a comprehensive assessment of nanofluid integration into evaporators, condensers, compressors, or performance indicators such as the coefficient of performance (COP), cooling load, or pressure drop under real-world operation.

- Emphasis on Isolated Property Prediction: AI and ML tools have been primarily applied to predict isolated thermophysical properties (e.g., viscosity, thermal conductivity), without bridging the implications of these predictions for full-system modeling, design optimization, or operational energy performance in HVAC settings.
- Absence of Methodological Taxonomy: While some papers classify studies by nanoparticle type or performance metric, a structured methodological synthesis—categorizing experimental, numerical, hybrid, and AI-based investigations—is missing. Such a taxonomy is critical to identify research synergies and implementation pathways in HVAC&R.
- Neglect of Practical and Lifecycle Challenges: Viscosity-induced pressure penalties, long-term nanoparticle stability, fouling, and cost-efficiency trade-offs are acknowledged but not critically analyzed in most prior reviews. For instance, optimization is discussed by Basu et al. (2023) [9], yet lacks connection to lifecycle sustainability or technoeconomic feasibility.
- Missing Integration with Smart HVAC Technologies: None of the cited reviews incorporate nanofluid research within the evolving context of smart HVAC systems, including digital twins, adaptive control, or real-time energy analytics—a direction increasingly relevant to energy-aware building management.

These gaps underscore the need for a domain-focused, methodologically structured, and deployment-oriented review that evaluates nanofluid applications in HVAC&R from both technical and practical lenses. The present review aims to fill this void by offering a layered synthesis that spans material properties, performance metrics, methodological approaches, and AI-driven system integration.

To contextualize the unique contributions of this review, it is essential to benchmark it against the existing literature. While several recent reviews have addressed aspects of nanofluid applications and machine learning techniques, their scopes often remain confined to either predictive modeling or generalized thermal systems without targeted relevance to HVAC and refrigeration. Table 1 presents a comparative assessment of key review articles, highlighting their primary focus areas, methodological breadth, and thematic limitations. This contrast makes clear the distinctive value of the present work, which offers a structured, HVAC-specific synthesis that not only incorporates experimental, numerical, and hybrid studies but also advances the field through integration of lifecycle considerations, technoeconomic insights, and emerging smart HVAC technologies.

Table 1. Comparative scope and contribution of recent review articles on nanofluids vs. this study.

Feature/Focus Area	Riyadi et al. (2024) [8]	Basu et al. (2023) [9]	Smrity & Yin (2025) [11]	This Study (Present Review)
Application Domain	Porous heat exchangers, solar systems	General heat transfer/nanofluid systems	General modeling–experiment synergy	HVAC&R-specific (evaporators, condensers, COP)
AI/ML Integration	Partial (heat exchanger optimization)	Yes (thermophysical property prediction)	Yes (surrogate modeling, data fitting)	Yes (component and system-level modeling + taxonomy)

Table 1. *Cont.*

Feature/Focus Area	Riyadi et al. (2024) [8]	Basu et al. (2023) [9]	Smrity & Yin (2025) [11]	This Study (Present Review)
Methodological Classification	No	No	No	Yes (experimental, numerical, hybrid, AI/ML)
System-Level Analysis	No	No	No	Yes (COP, pressure drop, energy efficiency)
Lifecycle and Technoeconomic Considerations	Brief mention	Not addressed	Not addressed	Discussed (stability, cost, viscosity trade-offs)
Focus on Smart HVAC Integration (e.g., digital twins)	No	No	No	Yes (emerging AI-enabled HVAC technologies)
Research Gaps and Future Roadmap	General	General	General	Detailed and domain-specific

To further contextualize the unique contributions of this review, a conceptual Venn diagram, as shown in Figure 1, is presented to highlight the methodological and thematic intersections explored. While the recent literature often focuses on isolated aspects such as predictive modeling or experimental evaluations, this study deliberately bridges those domains with HVAC&R-specific integration and practical deployment insights.

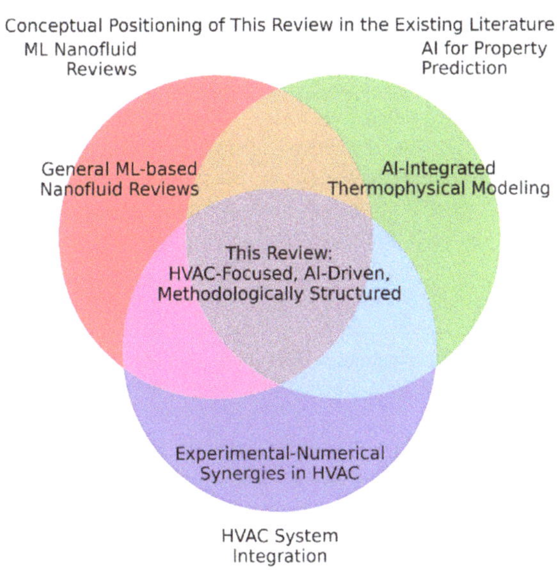

Figure 1. Conceptual positioning of this review. The diagram highlights the intersection of AI-integrated modeling, ML-based nanofluid reviews, and HVAC system-level studies, framing the unique focus of this review.

1.2. Motivation for This Work

This review is motivated by the need for a unified, application-focused understanding of nanofluid technologies specifically tailored to HVAC&R systems. Rather than offering another general survey, this work is purposefully structured to bridge experimental advances, AI-driven modeling, and real-world deployment concerns—domains that have remained siloed in prior studies.

Key features of this review include:

- Component-Specific Relevance: Direct mapping of nanofluid performance metrics to HVAC&R components (e.g., evaporators, condensers) and system-level outputs like COP and thermal load reduction.
- Methodologically Balanced Taxonomy: A structured classification across experimental, numerical, hybrid, and AI/ML-based studies that enables comparative analysis of their strengths, limitations, and evolution.
- Actionable Design Guidance: Synthesized findings are organized by nanoparticle type, base fluid, and operating condition to support engineering design and optimization.
- Deployment-Aware Framing: Particular attention is given to practical challenges—pressure drop penalties, sedimentation, economic feasibility, and integration into smart HVAC systems.

Overall, this review serves not only as a scholarly synthesis but also as a practical reference for researchers, designers, and energy system engineers aiming to implement nanofluid-based enhancements in HVAC&R environments.

1.3. Novelty of This Review Paper

The novelty of this review lies in its application-driven, methodologically structured, and HVAC-specific focus—a combination largely absent in the prior literature. Compared to existing reviews that often generalize nanofluids within broader thermal or materials science contexts, this paper offers the following unique contributions:

- HVAC&R Component-Level Focus: This review uniquely concentrates on the integration of nanofluids into specific HVAC&R components—such as condensers, evaporators, compressors, and heat exchangers—linking them to system-level outcomes like coefficient of performance (COP), heat transfer rates, and energy efficiency.
- Structured Methodological Classification: More than 200 peer-reviewed studies are systematically categorized under a comprehensive framework spanning experimental, numerical, theoretical, hybrid, and AI/ML-based methodologies. This taxonomy allows clear benchmarking of research directions and evidence-based insights for future investigations.
- Comparative and Application-Specific Synthesis: The review presents clustered performance comparisons based on nanoparticle types, base fluids, and operational parameters relevant to HVAC&R environments. This structured synthesis aids researchers and engineers in identifying optimal nanofluid configurations for specific use cases.
- Emphasis on Smart HVAC Integration: Beyond conventional heat transfer evaluation, the paper advances the field by incorporating recent developments in AI, digital twins, and intelligent HVAC controls—areas typically underrepresented in past reviews—thus creating a pathway toward smart, adaptive thermal systems.

By bridging methodological rigor with real-world applicability, this review establishes a research foundation that is not only academically comprehensive but also actionable for next-generation HVAC technologies.

1.4. Organization of This Paper

The remaining parts of this review is organized thematically and methodologically to provide a structured and application-oriented synthesis of nanofluid research relevant to

HVAC and refrigeration systems. The paper is divided into four main sections following the introduction:

- Section 2 presents the core thematic review of nanofluid applications in HVAC systems, categorized into three subdomains:
- Section 2.1 focuses on Thermophysical Properties, evaluating both numerical (Section 2.1.1) and experimental (Section 2.1.2) studies, followed by a Critical Analysis and Research Gaps subsection (Section 2.1.3).
- Section 2.2 reviews advances in Heat Transfer Enhancement, structured into Experimental Studies (Section 2.2.1), Numerical Studies (Section 2.2.2), and AI/ML-Driven Approaches (Section 2.2.3), culminating in a Critical Analysis and Future Directions (Section 2.2.4).
- Section 2.3 synthesizes the broader Research Methodologies and Analytical Trends across the literature, including Experimental Studies (Section 2.3.1), Numerical Investigations (Section 2.3.2), Hybrid Studies (Section 2.3.3), AI/ML-Based Approaches (Section 2.3.4), and Review-Based Papers (Section 2.3.5). It concludes with a Critical Analysis and Future Directions subsection (Section 2.3.6) that integrates methodological gaps and forward-looking insights.
- Section 3 provides a consolidated Conclusion and Future Directions, summarizing the key insights from the review while proposing actionable research trajectories and system-level recommendations for nanofluid deployment in next-generation HVAC&R technologies.

This organization facilitates both a comprehensive understanding of the current research landscape and a practical roadmap for researchers, engineers, and technology developers aiming to translate nanofluid innovations into real-world HVAC applications.

2. Thematic Review of Nanofluid Applications in HVAC Systems

The successful integration of nanofluids into HVAC systems extends beyond their enhanced thermophysical properties. It depends fundamentally on how these improvements translate into measurable performance gains across diverse HVAC components. From compact heat exchangers to district-scale cooling loops, each subsystem presents distinct thermal, hydraulic, and operational challenges that shape the effectiveness of nanofluid utilization. Therefore, understanding nanofluid performance requires a system-specific lens, accounting for variations in flow regime, operating temperature, material compatibility, and long-term stability.

To ensure a focused and high-quality synthesis of the relevant literature, a structured filtering and categorization process was implemented. An initial search across the Scopus database yielded 907 articles published between 2015 and 2025, using compound Boolean keywords that included HVAC, refrigeration, nanofluids, nanoparticles, and heat transfer enhancement. After applying filters to limit the scope to peer-reviewed journal articles and reviews within the engineering and energy domains and written in English, the dataset was reduced to 347 articles.

These remaining articles were further assessed for thematic relevance. Articles that were unrelated to nanofluid applications in thermal systems or represented marginal research clusters (e.g., <10 articles) were excluded. The final selection contained 200 articles, thematically grouped into three core categories:

- Refrigeration, Air Conditioning, and HVAC Systems (main application theme, n = 133),
- Heat Transfer Enhancement (supporting subtheme 1, n = 46), and
- Thermophysical Properties (supporting subtheme 2, n = 21).

The filtering logic and thematic breakdown are illustrated in Figure 2, which provides a visual summary of the literature search strategy, exclusion criteria, and retained themes. Each theme was identified through the literature clustering and reinforced using targeted keyword searches in Scopus. The number of articles selected per theme reflects both the research volume and relevance after exclusion criteria were applied. These categories inform the structure of the following sections, enabling a targeted and critical analysis of nanofluid performance across HVAC&R applications.

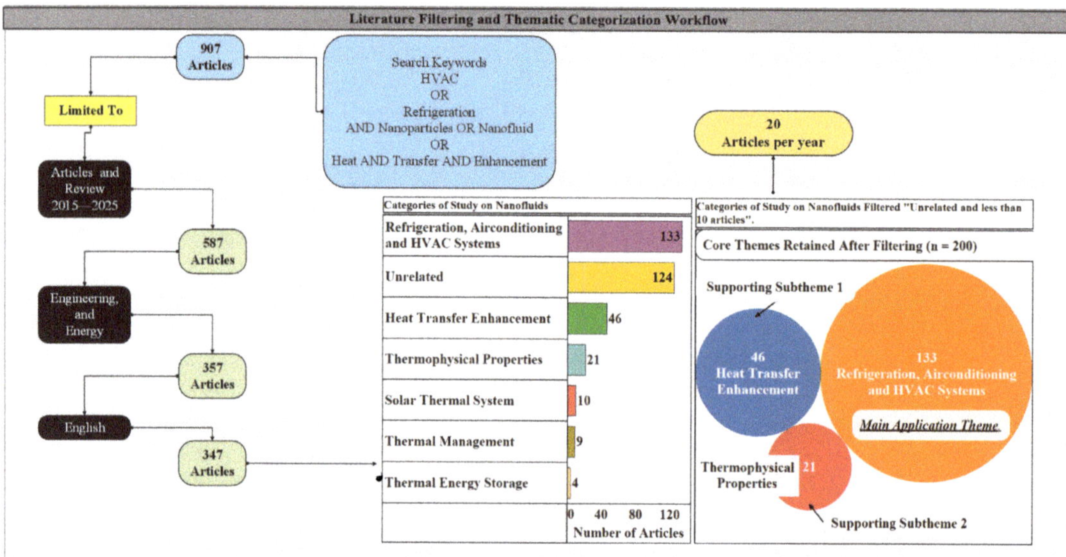

Figure 2. Thematic structure of the review based on application domains in HVAC&R systems.

Within each thematic domain, findings from experimental evaluations, numerical simulations, and hybrid approaches are synthesized to assess performance trends, implementation challenges, and research gaps. This framework provides a structured foundation for understanding the practical relevance and research trajectory of nanofluids across HVAC&R systems.

2.1. Thermophysical Properties

The evaluation of thermophysical properties is fundamental to understanding the suitability and performance of nanofluids in HVAC&R systems. These properties, particularly thermal conductivity, viscosity, density, and specific heat, govern how nanofluids interact with heat exchangers, pumps, and flow circuits [12]. Their accurate characterization under real and simulated operating conditions forms the basis for predicting energy transfer efficiency and system behavior.

2.1.1. Numerical Investigations

Numerical studies serve as the backbone for modeling nanofluid behavior under controlled and parametric variations, providing insight into performance outcomes that would be costly or impractical to capture experimentally. In the context of HVAC&R applications, numerical models are widely used to estimate thermal conductivity, effective viscosity, specific heat capacity, and entropy generation based on nanofluid formulation, temperature, and flow conditions.

Several studies leverage finite volume and finite element approaches to solve coupled momentum and energy equations under laminar and turbulent flow conditions. For instance, Ali et al. (2025) [13] modeled entropy generation in nanofluid-saturated porous channels, incorporating slip velocity and thermodiffusion effects. Their simulations showed that entropy generation could be minimized by optimizing nanoparticle concentration and Darcy number, especially under low Reynolds number conditions typical of HVAC&R circulation loops.

Abbas (2025) [10] used a rotating wavy porous channel geometry to investigate heat transport of Al_2O_3 and hybrid nanofluids. The model revealed that rotational motion and porous structure synergistically enhanced thermal dispersion. Similarly, Gao et al. (2024) [14] conducted a numerical assessment of nanofluid ice slurries, concluding that hybrid formulations could improve cold energy storage density by over 15% without significantly increasing flow resistance.

Other studies incorporated hybrid nanoparticles, including Fe_3O_4-$CoFe_2O_4$ and carbon-based additives. These simulations indicated synergistic thermal benefits, particularly when particle shapes (e.g., cylindrical or platelet-like) and surface treatments were accounted for. Most simulations assume homogeneous dispersion and thermophysical equilibrium, which, while simplifying the problem, introduces limitations when comparing against real-world conditions.

Overall, numerical investigations affirm that thermal conductivity increases nonlinearly with nanoparticle volume fraction, but at the cost of rising viscosity—especially above 1% vol. Several models now include temperature-dependent property correlations, Brownian motion, and thermophoresis to improve predictive accuracy under transient HVAC&R conditions.

2.1.2. Experimental Investigations

Though fewer in number, experimental studies offer indispensable validation for simulation-based predictions and provide real-world insight into nanofluid behavior under operational constraints. Such investigations typically focus on measuring effective thermal conductivity, viscosity, and, less frequently, specific heat capacity and density, using various test setups including guarded hot plate, transient hot wire, and viscometric analyses.

Lakshmi et al. (2025) [15] synthesized Fe_3O_4-$CoFe_2O_4$ nanofluids and characterized them under porous media flow. Their experiments confirmed up to 20% improvement in thermal conductivity at 0.5 vol.% particle loading, with minimal sedimentation over 72 h, attributed to uniform magnetic particle dispersion. However, a corresponding 15–25% increase in viscosity was observed, which could pose challenges for pumping power and flow rate regulation in HVAC&R systems.

Su et al. (2025) [16] performed both experimental and computational assessments of hybrid nanofluids containing TiO_2 and graphene nanoparticles. Their experimental results confirmed a 12–18% enhancement in thermal conductivity, highlighting the importance of nanoparticle morphology and base fluid compatibility. The study also demonstrated that surfactant selection (e.g., SDS, PVP) had a direct impact on long-term stability and the control of aggregation.

A recurring challenge noted across studies is the tendency for nanoparticles to agglomerate or settle over time, especially in systems subject to cyclic heating and cooling, such as HVAC&R loops. This not only affects the thermophysical consistency of the nanofluid but also undermines heat exchanger performance and operational safety. Moreover, accurate measurement of specific heat and dynamic viscosity at low temperature gradients—a typical condition in HVAC&R applications—remains limited due to instrumentation sensitivity and calibration challenges.

2.1.3. Critical Analysis and Research Gaps

A comparative synthesis of the reviewed numerical and experimental studies confirms broad agreement on the trends associated with nanofluid thermophysical behavior. Thermal conductivity enhancement is consistently reported as the primary motivation for nanofluid use, with most studies showing 10–30% improvement, depending on nanoparticle concentration, temperature, and fluid type. However, this enhancement is frequently accompanied by undesirable increases in viscosity, especially beyond 1% vol. loading [17], which imposes a penalty on pumping power and flow dynamics. Figure 3 summarizes the frequency of thermophysical properties mentioned in the reviewed literature.

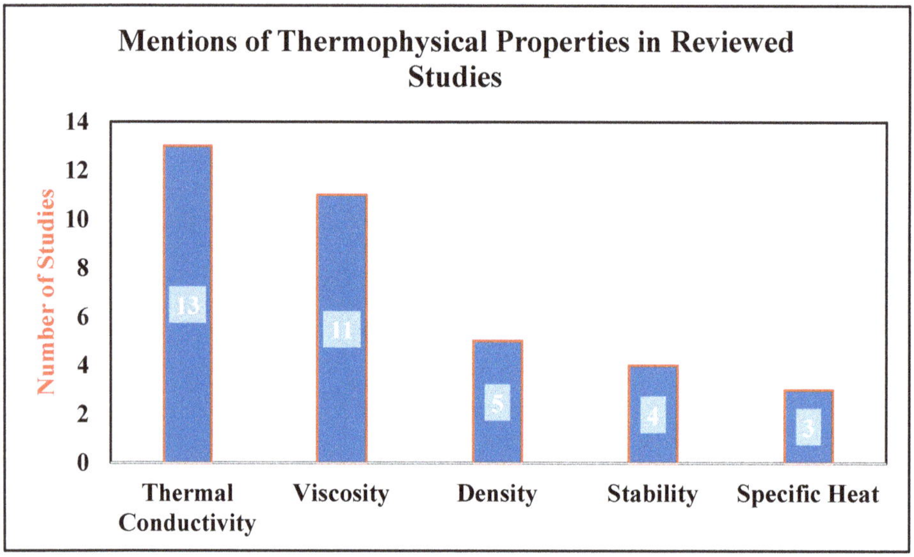

Figure 3. Number of studies of thermophysical properties investigated in the reviewed literature, highlighting the research community's primary focus on thermal conductivity and viscosity.

Despite the progress made, several gaps and inconsistencies persist between numerical predictions and experimental validation:

- Idealized assumptions in simulations (e.g., uniform dispersion, constant properties) often overlook real-world phenomena such as agglomeration, phase separation, and surfactant degradation.
- Long-term operational studies assessing thermal cycling effects, corrosion, and stability under HVAC&R-like conditions remain scarce.
- Viscosity and specific heat are underexplored in terms of their coupled impact on system performance, especially under low-to-moderate temperature gradients typical of HVAC&R loops.

To provide quantitative insight into material preferences in the literature, Table 2 categorizes reviewed studies based on nanoparticle types, the number of publications, their methodological approach (numerical, experimental, or review), and representative findings. It highlights the research emphasis and identifies trends and gaps in nanofluid thermophysical characterization relevant to HVAC&R applications.

Table 2. Summary of nanoparticles studied in thermophysical property investigations, with study types and representative findings.

Ref.	Nanoparticle Type	Study Type Distribution	Representative Findings
[18–22]	Al_2O_3, SiO_2	Numerical	• Density is a less studied property vs. TC and viscosity.
[13,17,23,24]	CuO	Numerical, Review	• TC ↑ by ~57%; COP ↑ by ~27% (with R134a/R152a). • Minor viscosity increase observed.
[15,23,25,26]	Fe_3O_4-$CoFe_2O_4$	Experimental, Numerical	• Investigated MHD effects on heat transfer properties.
[16,27]	Graphene	Experimental, Experimental/Numerical, Numerical	• ↑ Concentration enhances TC and heat transfer. • ↑ Concentration also increases viscosity and friction.
[20,22–24,28]	TiO_2	Experimental	• Studied heating effects on viscosity, TC, and absorbance.
[14,29,30]	Hybrid/Other	Numerical, Review	Improving adsorbent TC and heat exchange area for compact system optimization.
[31,32]	Nanolubricant, Refrigerant	Review	Analyzing thermophysical and hydraulic parameters to enhance refrigeration system COP.

2.2. Heat Transfer Enhancement

Nanofluids have emerged as highly effective thermal carriers for enhancing heat transfer across various components of HVAC&R systems. Their ability to significantly increase thermal conductivity, improve convective heat transfer coefficients, and reduce the size and weight of heat exchangers has prompted extensive investigation. These enhancements arise from several fundamental mechanisms, including enhanced thermal conductivity from high-conductivity nanoparticles, Brownian motion-induced microscale mixing, thermophoresis under strong temperature gradients, and microconvection generated by particle movement. The relative contribution of these mechanisms depends on the nanoparticle type, base fluid properties, concentration, and flow regime, and they often act synergistically to boost performance.

Nanofluids have emerged as highly effective thermal carriers for enhancing heat transfer across various components of HVAC&R systems. Their ability to significantly increase thermal conductivity, improve convective heat transfer coefficients, and reduce the size and weight of heat exchangers has prompted extensive investigation. The improvement is driven by several underlying physical mechanisms:

- Brownian motion—Random motion of nanoparticles at the microscale promotes enhanced mixing in the fluid, disrupting thermal boundary layers and facilitating heat transport.
- Thermophoresis—Migration of nanoparticles from hotter to cooler regions due to temperature gradients, redistributing thermal energy more effectively.
- Microconvection—Localized eddies generated by moving nanoparticles amplify fluid agitation and improve convective heat transfer rates.
- Enhanced thermal conductivity—Nanoparticles with inherently higher conductivity than the base fluid (e.g., Al_2O_3, CuO, graphene) form efficient thermal pathways, accelerating heat transport through the fluid.

- Field-induced effects (optional in specific designs)—Magnetic or electric fields applied to specialized nanofluids can align particles or induce additional motion, further modifying heat transfer behavior.
- Viscous dissipation and porous media effects—Relevant in some configurations where flow resistance or porous structures influence the local energy balance.

Among these, Brownian motion, thermophoresis, microconvection, and enhanced thermal conductivity are most consistently observed in HVAC&R nanofluid applications and form the basis for interpreting experimental and numerical results in this review.

This section provides a comprehensive thematic synthesis of 46 selected studies focusing on heat transfer enhancement using nanofluids. The analysis is organized into five subsections, each representing a dominant methodological perspective or performance trend:

- Section 2.2.1 Experimental Studies examine lab-scale evaluations of nanofluid heat transfer in compact exchangers, solar systems, and augmented flow channels.
- Section 2.2.2 Numerical Investigations delve into Computational Fluid Dynamics (CFD)-based modeling and simulation of nanofluid behavior under varied boundary conditions and channel geometries.
- Section 2.2.3 AI/ML-Driven Approaches highlight recent advances in artificial intelligence for thermal prediction, surrogate modeling, and optimization.
- Section 2.2.4 Critical Analysis and Future Directions discusses limitations, research gaps, and emerging opportunities.

By integrating these perspectives, this section aims to clarify the current landscape of nanofluid heat transfer enhancement and provide direction for next-generation thermal technologies in HVAC&R systems.

2.2.1. Experimental Studies

Experimental investigations form the cornerstone for validating the potential of nanomaterials in enhancing heat transfer performance within HVAC and refrigeration systems. Such studies not only quantify gains in convective or boiling heat transfer but also evaluate critical system-level metrics like COP improvements, pressure drop penalties, and optimal nanoparticle dosages. The following discussion synthesizes insights from selected experimental works, each focusing on different nanomaterial classes—including Al-based, Al_2O_3, TiO_2, ZnO, Ag, and hybrid formulations—applied across diverse thermal systems, ranging from flow boiling loops to passive building composites and phase change storage. Figure 4 furnishes the distribution of experimental studies by nanoparticle type used in nanofluid-based heat transfer enhancement. The figure highlights the prominence of TiO_2 and CuO in laboratory investigations, reflecting their thermal stability, availability, and cost-effectiveness. Less commonly studied nanoparticles such as Cu-based and Single-Wall Carbon Nanotubes (SWCNTs) are emerging but remain underrepresented in experimental validations.

Al-based nanostructures have garnered significant attention for improving fluid-side heat transfer. For instance, Zhang et al. (2025) [33] examined flow boiling of R134a in EDM-textured aluminum tubes enhanced by Al-based nanoparticles, reporting substantial increases in the Nusselt number and heat transfer coefficient, with indirect suggestions of improved COP. Similarly, a separate investigation on R134a nanofluids inside microfin tubes demonstrated impressive local enhancements, with Nusselt numbers improving by up to 210% and heat transfer coefficients increasing by approximately 120%, while also noting modest pressure drop penalties, often below 13%, thus maintaining hydraulic feasibility.

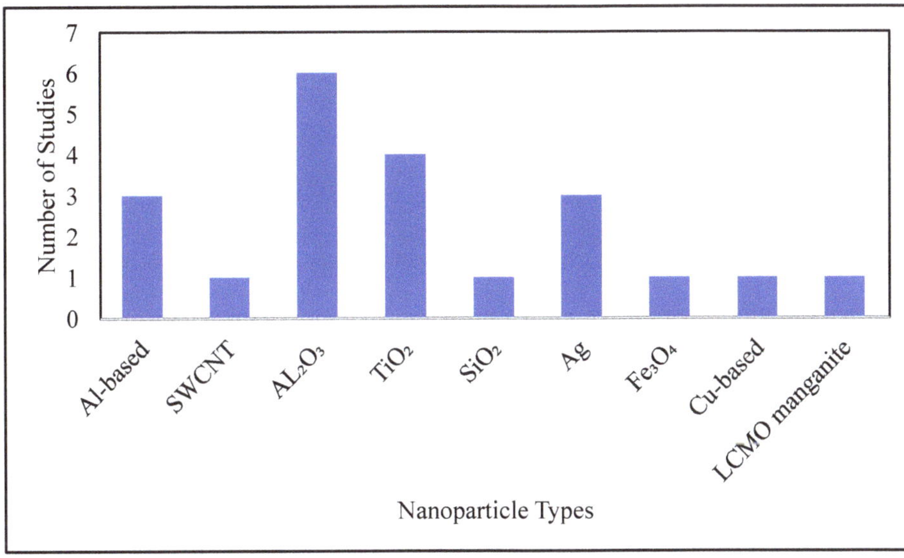

Figure 4. Distribution of experimental studies by nanoparticle type used in nanofluid-based heat transfer enhancement.

Incorporating Al_2O_3 nanoparticles, El Hadi Attia et al. (2025) [34] evaluated honeycomb copper tube arrangements using hybrid nanofluids. Their experiments showed a remarkable range of performance gains—up to 92.9% improvement in the Nusselt number, alongside secondary metrics like evaporator surface absorption and convective effectiveness. This underscores Al_2O_3's continued prominence due to its thermal conductivity stability and chemical inertness in common refrigerants or aqueous systems.

Moving to hybrid configurations, recent work by researchers investigating triply periodic mini-channel structures combined Al-based and Ag nanoparticles, reporting multi-faceted improvements: Nusselt number gains of up to 30%, localized pressure drop increments around 4%, and overall enhancement efficiencies balancing between 12 and 24% across different flow conditions. These studies demonstrate that carefully engineered multi-material nanofluids can exploit synergistic mechanisms—enhanced microconvection from one species and superior thermal conductivity from another—yielding compound benefits.

Silver (Ag) nanoparticles also emerged as a potent agent in phase change thermal management. Abdullah et al. (2024) [35] explored Ag-enhanced PCMs in half-cylindrical solar still configurations, achieving substantial performance jumps, with reported improvements ranging from 55% up to 184% in metrics tied to thermal storage, evaporation rates, and integrated solar efficiency. Such outcomes highlight Ag's unmatched role in accelerating transient heat storage–discharge cycles, pivotal for hybrid HVAC–solar systems.

On the TiO_2 and ZnO front, Jadhav et al. (2024) [36] introduced nano-composite polycarbonate sheets doped with these oxides (and zeolite) for greenhouse-style climate control. They found that doping concentrations around 0.3% led to cabinet temperature reductions up to 21.5%, with TiO_2 exhibiting superior thermal insulation and UV absorption due to its high refractive index. While ZnO offered slightly lower thermal blocking, it still significantly improved optical shielding compared to undoped polycarbonate.

Further extending TiO_2 applications, Liu et al. (2024) [37] designed wood–plastic composites embedded with carbonyl iron microcapsules and TiO_2, achieving thermal management gains up to 89.2% across combined heat storage and regulated release tests.

This underscores how nanomaterials can be embedded in passive construction components to indirectly reduce HVAC loads by flattening thermal swings.

Through key quantitative themes from these diverse experimental efforts, several consistent trends emerge:

- Nusselt number enhancements routinely exceeded 50%, with some specialized microchannel or hybrid geometries pushing improvements beyond 200%.
- COP increases ranged between 5 and 10%, linked to lower compressor work or elevated evaporator effectiveness when using nanofluids.
- Optimal nanoparticle concentrations typically clustered around 0.3–0.4% by weight or volume, balancing enhanced thermal conductivity with manageable viscosity and pressure drop effects.
- In passive systems, cabinet or internal temperature drops of 15–22% were commonly achieved using TiO_2 or ZnO-doped materials.

To further elucidate methodological patterns and material-specific insights, the identified experimental studies were grouped based on their primary nanoparticle type. This classification underscores the diverse thermal enhancement strategies explored, ranging from metal oxides like TiO_2 and Al_2O_3 to metallic nanofluids such as Cu-based suspensions. Table 3 provides a concise summary, cataloging the studies under each nanoparticle category, along with key bibliographic identifiers.

Table 3. Summary of experimental studies on heat transfer enhancement in HVAC&R applications, grouped by nanoparticle type.

Ref.	Nanoparticle Type	Findings
[36–39]	TiO_2	↓ Cabinet temperature up to 21.5% at 125 cm, esp. at 40 °C Optimal at 0.3% doping Provided highest thermal insulation and UV absorption
[34,38,40–45]	Al_2O_3	• ↑ Heat transfer coefficient by 80.5% at 0.4% vol. • ↑ COP by 7.5%, ↓ Power consumption by 5.2% • ↑ Pressure drop by 13.6%, max 25.42 kPa/m
[33,46,47]	Al-based (Hybrid)	• ↑ Nusselt number by 210%, heat transfer coefficient ↑ 120% • Mild pressure drop increase (~13%) • Noted optimal performance around 0.4% concentration
[48]	Cu/CuO	• PEC reached 1.18 (↑ ~18% over baseline) • Most effective at 0.4 wt.% Cu–water nanofluid • ↑ Heat transfer coefficient and pressure drop across flow rates
[36]	ZnO	• Effective UV blocking, better than plain PC • ↓ Refractive index and thermal insulation than TiO_2 • Optimal at 0.3% doping for moderate ↓ temperature
[36,42,47]	Ag	• ↑ PCM storage performance by 55–184% across heat storage and evaporation metrics • Accelerated charge/discharge in solar still hybrid systems
[44]	SWCNT	• ↑ Nusselt number by ~25–35% in nanofluid flows • Slight ↑ viscosity, minimal ΔP rise • Stability achieved with surfactant-assisted dispersion
[49]	LCMO manganite	• ↑ Thermal conductivity and heat transfer by ~15–22% • Optimal at 0.1–0.2% volume fraction • Moderate pressure drop, good long-term dispersion stability

Altogether, these experimental studies provide robust validation of the technical feasibility and scalability of nanomaterial interventions in HVAC and refrigeration. Whether through direct enhancement of fluid-side heat transfer, passive moderation of indoor climates, or integration into phase change storage, the consistent quantitative outcomes—across Nusselt, COP, and temperature metrics—solidify the role of nanoparticles as a key lever for next-generation energy-efficient thermal systems.

2.2.2. Numerical Studies

In parallel with experimental advancements, numerical studies have emerged as a cornerstone for understanding and predicting the complex thermal and flow behaviors associated with nanofluid-enhanced HVAC and refrigeration systems. Computational approaches, primarily employing CFD simulations, finite element methods, or hybrid analytical–numerical frameworks, allow researchers to investigate parameter sensitivities, optimize system configurations, and predict multiscale phenomena that are often impractical or cost prohibitive to test experimentally. Particularly in the domain of nanofluids, numerical models enable a rigorous evaluation of heat transfer enhancements, entropy generation, pressure drop penalties, and magnetic or electric field effects under diverse operating regimes. These insights are invaluable for accelerating the design and deployment of energy-efficient thermal systems. A closer look at recent studies reveals the breadth of computational methodologies and the diversity of nanomaterials investigated.

Khan et al. (2025) [50] conducted a comprehensive CFD study on TiO_2–Au hybrid nanofluids flowing through microchannels under varying Reynolds numbers. Their simulations highlighted an enhancement in Nusselt number of approximately 26–38% compared to pure water, while also noting a moderate increase in friction factor. This underscores the synergy of metal–oxide combinations in augmenting convective heat transfer. Meng et al. (2025) [51] numerically explored thermoelectric cooling systems incorporating SiO_2–Al_2O_3 nanoparticles dispersed in water. Their model revealed that these hybrid suspensions could reduce localized temperatures by up to 12 °C, thereby extending the cooling range and potential application of TEC modules in low-temperature HVAC environments. Raju et al. (2024) [52] investigated GO–Au–Ag hybrid nanofluids in a wedge-shaped duct configuration using finite volume methods. Their results demonstrated a 32% rise in the average heat transfer coefficient, highlighting the pivotal role of graphene oxide in enhancing thermal pathways when combined with noble metal particles.

Balogun et al. (2025) [53] numerically examined gold and silver hybrid nanofluids under diverse parametric variations, showing optimal performance at nanoparticle concentrations near 0.08 vol%, with noted trade-offs between enhanced heat transfer and increased shear stresses. Bompally et al. (2025) [54] modeled EG + MoS_2 + SiO_2 hybrid nanofluids exposed to inclined magnetic fields, capturing the interplay between magnetic intensities and nanoparticle dispersion. Their simulations revealed up to 28% improvements in heat transfer, accompanied by reductions in entropy generation at optimized field angles. Liang et al. (2025) [55] presented a multi-objective optimization study on thermal interface materials with embedded MWCNTs, focusing on cooling modules for active antenna units. They highlighted how incorporating MWCNT nanostructures improved thermal conductivity and reduced hotspots, offering potential extensions to compact HVAC modules.

Kumar et al. (2025) [56] conducted numerical assessments of CNT-based nanofluids for microelectronic thermal management, observing improved local Nusselt profiles and effective temperature suppression across intricate microchannel configurations. Sreelakshmi et al. (2024) [57] analyzed hybrid Cu-based nanofluids in thin liquid films, noting marked enhancements in transient heat transfer behavior due to the inclusion of copper particles, which intensified conduction at the liquid–solid interface. Benkherbache et al. (2024) [58] performed a detailed natural convection study of hybrid multi-nanoparticle dispersions, demonstrating that incorporating Fe_3O_4, Al_2O_3, and Cu nanoparticles led to substantial increases in heat transfer rates while carefully balancing buoyancy-driven flow characteristics. Farooq et al. (2025) [59] investigated a suite of hybrid nanofluid combinations (e.g., $TiO_2 + SiO_2$, $TiO_2 + MoS_2$, TiO_2 + Au) using CFD frameworks, mapping how each hybrid formulation influenced local entropy generation and overall thermal efficiency under solar-assisted flow conditions.

Elboughdiri et al. (2025) [60] examined Cu, TiO_2, and Fe_3O_4 hybrid nanofluid mixtures under variable Stefan blowing and cross-diffusion effects (Soret and Dufour). Their results showcased complex coupled mass-heat transfer enhancements beneficial for evaporative cooling processes. Shinwari et al. (2025) [61] developed a robust computational framework for $Al_2O_3 + TiO_2 + SiO_2$ trihybrid nanofluid flows, incorporating chemical reaction and thermal activation models. They reported notable gains in heat transfer performance coupled with reduced activation energy requirements.

To visually consolidate the diverse hybrid nanoparticle systems investigated across these numerical studies, a schematic representation is presented in Figure 5. This graphical summary clusters the multi-component nanofluids based on their predominant constituents, effectively illustrating the recurrent combinations employed to enhance heat transfer performance. By mapping out these hybrid formulations alongside their corresponding study counts, the figure offers an intuitive overview of the research landscape, underscoring the strategic pairing of metal oxides, carbides, and noble metals to exploit synergistic thermal transport mechanisms. Such visualization not only highlights the methodological breadth of current investigations but also points toward emerging preferences in hybrid nanoparticle engineering within HVAC and refrigeration contexts.

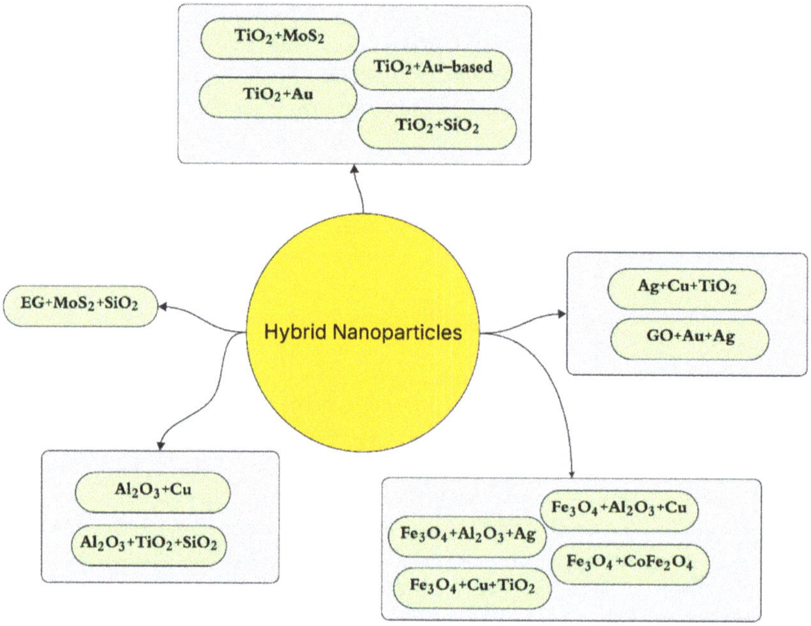

Figure 5. Schematic representation of hybrid nanoparticle systems explored in numerical studies for heat transfer enhancement in HVAC and refrigeration applications.

To further consolidate the diverse numerical investigations into nanoparticle-enhanced heat transfer reviewed in Section 2.2.2, Table 4 summarizes key findings from representative studies. This synthesis highlights the range of nanofluids examined—from simple oxide suspensions to complex multi-metal and carbon-based hybrids—along with the primary enhancements reported in convective heat transfer, entropy reduction, and associated thermophysical behaviors. By considering different nanoparticle configurations and flow phenomena, the table offers a succinct comparative perspective on how numerical methodologies have elucidated the underlying mechanisms and performance trade-offs inherent in these advanced thermal systems.

Table 4. Summary of selected numerical studies on the impact of single and hybrid nanoparticles on heat transfer enhancement.

Ref.	Nanoparticle/Hybrid	Key Findings
[56]	CNT	• Improved microchannel Nusselt and local temperatures
[57]	Al_2O_3 + Cu	• RKF simulation: hybrid more stable vs. mono; HT ↑ with Pr, Eckert; magnetic field raises skin friction
[58]	Fe_3O_4 + Al_2O_3 + Cu/Ag	• Annular cavity: ↑ Ha ↓ temp ↑ velocity ↑ Nusselt; Cu most responsive; fins moderate convection
[59]	TiO_2 + SiO_2, TiO_2 + MoS_2, TiO_2 + Au	• Solar-assisted flows: ↓ entropy, ↑ efficiency under hybrid blends
[60]	Cu, TiO_2, Fe_3O_4	• Stefan blowing and Soret–Dufour: THNF ↑ HT by 22%, hybrid by 17%, mono by 11%; Yamada–Ota > Hamilton–Crosser
[61]	Al_2O_3 + TiO_2 + SiO_2	• Radial magnetic: slip and curvature ↑ temp; activation energy ↑ conc ↓ mass transfer
[62]	Fe_3O_4 + $CoFe_2O_4$	• Magnetic hybrid flows analyzed for entropy and heat transfer
[63]	Ag + Cu + TiO_2	• Simulated multi-metal hybrids under laminar regimes
[64]	Al_2O_3	• Magnetic refrigeration: up to 15% ↑ cooling; small pumping penalty, COP marginal
[65]	ZnO	• Subcooled boiling: ↑ vapor gen, wall Nu with $\phi < 1\%$; bubble dynamics crucial
[66]	CuO (LiBr)	• FEM: CuO ↑ mass flux, rate, coefficient in LiBr absorption
[67,68]	Cu/CuO	• Barocaloric refrigerator: 10% Cu nanofluid ↑ HT by ~30% • Laminar mixed convection: ↑ ϕ ↑ Nu, rib AR and Ri interplay key for buoyancy vs. forced
[69]	Diamond + Cu	• ANFIS + PSO predicted pool boiling HT of refrigerant–oil with nanoparticles
[55]	MWCNT	• In interface cooling modules ↓ hotspots, ↑ thermal uniformity
[54]	EG + MoS_2 + SiO_2	• Inclined magnetic ↑ HT by ~28%, ↓ entropy
[52]	GO + Au + Ag	• Wedge boundary layer ↑ HT
[53]	Hybrid (Au–Ag)	• Optimal ~0.08 vol% ↑ HT by ~28–41%, balanced vs. shear
[51]	SiO_2, Al_2O_3	• In TEC, hybrid ratios ↓ local temps, extended cooling
[50]	TiO_2 + Au	• Magnetized non-Newtonian flow ↑ Nusselt, sensitive to field and ϕ

2.2.3. AI/ML-Driven Approaches

The adoption of artificial intelligence (AI) and machine learning (ML) methodologies in the study of nanofluid-enhanced heat transfer has expanded significantly, driven by the need to model and optimize highly nonlinear, multi-parameter thermal systems. These approaches offer robust predictive frameworks that surpass traditional analytical and purely numerical methods, especially in handling the complex interplay of nanoparticle concentration, flow dynamics, thermal conductivity, and entropy generation. This is critically important in HVAC&R applications, where balancing heat transfer enhancement with pressure drop and entropy minimization remains a central engineering challenge.

Among single-component systems, Zolghadri et al. (2023) [70] investigated the application of artificial neural networks (ANNs) with Levenberg–Marquardt (LM) optimization to predict thermal parameters of Al_2O_3 nanofluids. Their study reported high correlation coefficients and very low prediction errors, demonstrating the capability of ANN-LM to effectively capture thermal behavior under varying operational conditions. Furthermore, by employing genetic algorithms (GAs), they identified optimal nanoparticle volume fractions that maximized thermal effectiveness while mitigating additional pumping power demands.

In hybrid systems, Kanti et al. (2025) [71] focused on Al_2O_3 + TiO_2 nanofluids, utilizing Gaussian Process Regression (GPR) with a 5/2 Matern kernel to construct surrogate models that predicted Nusselt numbers and entropy generation rates in microchannel flows. Their GPR approach significantly reduced computational time compared to full CFD simulations while maintaining prediction errors under 3%, illustrating its utility for rapid design iterations.

For more intricate multi-component systems, Ullah et al. (2025) [72] examined nanofluids composed of Al_2O_3, CuO, Al, and Cu nanoparticles. They applied recurrent neural networks (RNNs) enhanced with long short-term memory (LSTM) architectures to capture the temporal evolution of convective heat transfer and entropy generation. Their models effectively identified operating regimes that balanced improved heat transfer with acceptable increases in frictional and entropic losses, providing a nuanced tool for thermal system optimization.

Expanding to hybrid nanoparticle systems that incorporate carbon nanostructures, Almohamadi et al. (2024) [73] analyzed Al_2O_3, SWCNT, and Au nanofluids using quantum-like machine learning (QLM) integrated with bidirectional recurrent neural networks (BRNNs). This advanced hybrid approach successfully predicted enhancements in effective thermal conductivity and convective heat transfer coefficients, demonstrating the synergistic benefits of multi-nanoparticle systems, which are often poorly captured by classical regression or single-output models.

Finally, a particularly rigorous investigation by Qureshi et al. (2024) [74] focused on a ternary nanofluid system comprising Al_2O_3, CuO, and Cu, flowing over a stretchable wavy cylinder. By integrating a Local Mean Field Approximation Neural Network (LMFA NN) within their hybrid computational framework, they achieved exceptionally low mean square errors—on the order of 10^{-10} to 10^{-13}—across training, testing, and validation datasets. Their study showcased how ML frameworks can precisely capture complex velocity, temperature, and nanoparticle concentration fields, outperforming conventional numerical solvers, particularly for geometrically intricate domains.

Collectively, these diverse investigations underscore the pivotal role of AI/ML techniques in advancing the understanding and optimization of nanofluid-based heat transfer systems. By leveraging data-driven algorithms tailored to capture multi-parameter dependencies, researchers have demonstrated substantial improvements in predictive accuracy, design optimization speed, and the identification of operational envelopes that maximize thermal performance with controlled hydrodynamic penalties. As the field continues to mature, the integration of AI/ML models with experimental validation and high-fidelity simulations is poised to drive the next generation of energy-efficient HVAC&R systems.

To further elucidate the integration of artificial intelligence and machine learning (AI/ML) techniques with nanofluid-based heat transfer applications, a concept map has been developed, as shown in Figure 6. This graphical representation systematically links the various AI/ML models deployed in the reviewed studies to the types of nanoparticles investigated, thereby highlighting both the methodological diversity and the material scope of recent research. By visually mapping these associations, readers can quickly discern predominant algorithmic strategies, frequently studied nanofluid systems, and overlapping areas where multiple models have been employed to predict or optimize thermal performance. This holistic view facilitates a clearer understanding of how advanced data-driven approaches are progressively shaping the field of nanofluid-enhanced HVAC&R systems.

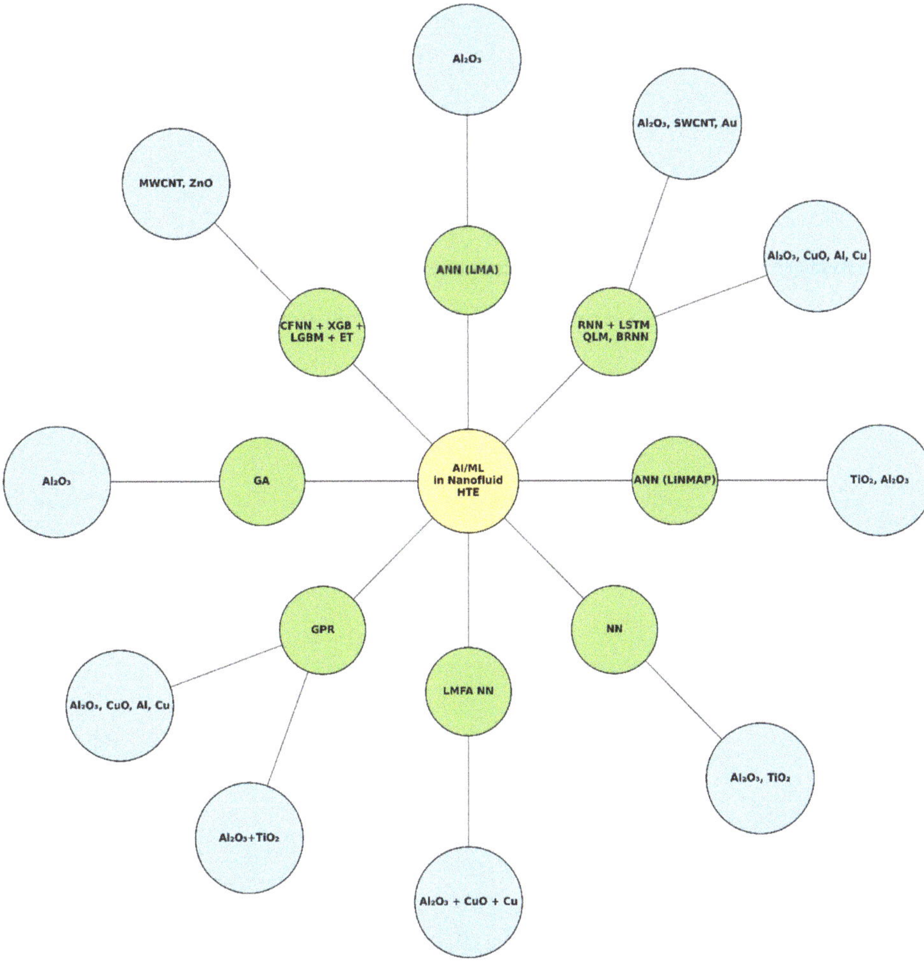

Figure 6. Mapping of AI/ML algorithms to nanoparticle materials used in nanofluid heat transfer enhancement studies for HVAC&R applications. The central node represents the application domain, green nodes indicate the AI/ML techniques applied, and blue nodes show the corresponding nanoparticle materials. While nanoparticle size information is not explicitly shown to maintain clarity, it is provided in the cited studies.

To provide a consolidated overview of the methodological advancements and key insights discussed in this subsection, Table 5 summarizes the specific types of nanoparticles investigated, the AI/ML algorithms employed, and the principal findings reported across the reviewed studies. This tabular synthesis facilitates a rapid comparative assessment of recent contributions, highlighting both the breadth of data-driven techniques applied and the diverse nanofluid systems explored within the context of heat transfer enhancement for HVAC&R applications.

Table 5. Summary of recent studies applying AI/ML techniques to nanofluid-based heat transfer enhancement.

Ref.	Nanoparticle Type	AI/ML Algorithms	Key Findings
[70,75,76]	Al_2O_3	ANN, LM, GA	• ANN models predicted nanofluid heat exchanger parameters accurately. • GA optimized volume fraction for max thermal effectiveness.
[71,77]	$Al_2O_3 + TiO_2$	GPR	• GPR predicted Nusselt number and entropy rates in microchannels. • Achieved surrogate models reducing CFD computation time.
[72,78]	Al_2O_3, CuO, Al, Cu	RNN + LSTM	• LSTM captured time-dependent heat transfer dynamics. • Predicted optimal conditions balancing heat transfer and pressure drop.
[73,79]	Al_2O_3, SWCNT, SWCNT, Au, ZnO	CFNN, ET, QLM, BRNN	• Hybrid AI approaches forecasted thermal conductivity enhancements. • Showed multi-nanoparticle blends raised convective HT coefficients.

To complement the thematic synthesis presented in this review, a high-level decision-flow schematic (Figure 7) has been developed to guide readers in the practical selection of nanofluids and associated system design parameters for HVAC&R applications. The flow begins with identifying the application type (e.g., chiller, heat pump, secondary loop), followed by selecting a suitable nanoparticle–base fluid combination based on thermal conductivity, stability, and compatibility criteria. Subsequent decision points address the optimum particle volume fraction, recommended flow velocities for enhanced heat transfer without excessive pressure drop, and suitable channel sizes/geometries to balance performance with manufacturability. This schematic is distilled from the collective insights of experimental, numerical, and AI-driven studies reviewed in Sections 2.1–2.3 and is intended as a practical visual guide rather than a rigid design formula.

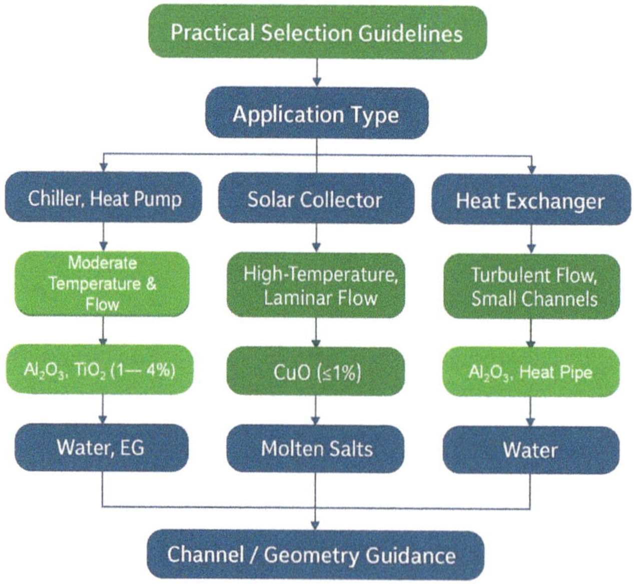

Figure 7. Selection framework for nanofluid-based thermal enhancement in HVAC&R systems, illustrating sequential decision points from material choice to system-level deployment.

2.2.4. Critical Analysis and Future Directions

A comprehensive review of experimental, numerical, and AI/ML-driven investigations into nanofluid-based heat transfer enhancement reveals both significant promise and persistent challenges that shape the future research landscape. While results across studies consistently demonstrate meaningful gains in thermal performance—frequently exceeding 30–50% in Nusselt number or local heat transfer coefficients—such improvements often come with nuanced trade-offs that necessitate deeper scrutiny.

From an experimental standpoint (Section 2.2.1), investigations on Al_2O_3, TiO_2, ZnO, and Ag nanofluids, along with emerging SWCNT and hybrid systems, have validated the capacity of nanoparticles to boost convective and boiling heat transfer in compact exchangers and passive thermal control structures. However, these studies also underscore recurring issues: viscosity escalation leading to higher pressure drops; challenges in sustaining long-term nanoparticle dispersion; and practical limits on concentration (~0.3–0.4 wt.%), beyond which benefits taper or even reverse due to excessive ΔP or sedimentation risks. Passive building materials doped with TiO_2 or ZnO have shown encouraging internal temperature reductions (15–22%), but such applications are still largely prototype-scale, with questions remaining about durability under cyclic thermal loads.

Numerical analyses (Section 2.2.2) have played a pivotal role in dissecting parametric influences, offering detailed mappings of how hybrid nanoparticle blends—like GO + Au + Ag or Fe_3O_4 + Al_2O_3 + Cu—impact entropy generation, frictional losses, and heat transfer rates under controlled simulations. Yet, many models rely on steady-state assumptions, ideal dispersion, or simplified thermophysical correlations. As a result, real-world complexities such as particle clustering, transient fouling, or multi-phase slip effects are often underrepresented. Even sophisticated multi-objective frameworks that balance entropy minimization with thermal gains rarely incorporate lifecycle cost or maintenance aspects, limiting their immediate translatability to HVAC&R scale systems.

The recent influx of AI/ML-driven studies (Section 2.2.3) signals a transformative methodological shift. Surrogate models like GPR, advanced neural architectures (RNN + LSTM, BRNN), and hybrid algorithms (QLM + CFNN) have achieved high-fidelity predictions with mean errors typically below 3%, dramatically reducing the computational burden compared to exhaustive CFD. These models excel at capturing multi-parametric nonlinearities, as evident in works examining Al_2O_3, TiO_2, SWCNT, and multi-metal hybrids. However, their effectiveness remains fundamentally tied to the depth and quality of training data. Most models are validated on narrowly defined datasets under controlled lab or simulation conditions, raising concerns about generalizability under fluctuating load scenarios, contaminant exposure, or long-duration HVAC operation.

Looking forward, several critical pathways can enhance the robustness and applicability of nanofluid technologies in HVAC&R:

- Integrated multi-objective frameworks: Future studies should unify AI/ML optimization with CFD or experimental data streams to jointly minimize entropy, maintain low pressure drops, and maximize Nusselt gains—bridging energy efficiency with operational reliability.
- Stability and fouling resilience: There is a pressing need for dynamic investigations tracking nanoparticle agglomeration, surface fouling, and re-dispersion under cyclic on–off operation typical in HVAC systems. Coupled magnetic or ultrasonic field techniques may offer novel avenues for in situ stability management.
- Next-generation nanomaterials: Graphene derivatives, MXenes, or bio-inspired hybrid composites hold promise for achieving high thermal conductivities at lower concentrations, potentially decoupling the viscosity trade-off that has constrained many oxide-based systems.

- Lifecycle and techno-economic perspectives: Very few studies holistically assess long-term cost-effectiveness, environmental footprint, or regulatory considerations tied to nanoparticle deployment and disposal. These analyses are vital to guide commercial translation.
- Integration into digital twins and adaptive controls: As smart HVAC infrastructures proliferate, embedding AI-driven nanofluid models into real-time digital twin systems could enable predictive adjustments to flow rates, nanoparticle dosing, or auxiliary magnetic fields—dynamically optimizing thermal performance and prolonging system lifespan.

Collectively, while nanofluids remain a highly promising strategy for elevating heat transfer capabilities in HVAC&R, unlocking their full potential demands a concerted push toward more integrative, multidisciplinary investigations. By coupling rigorous experimental validation, high-fidelity simulations, advanced machine learning, and lifecycle assessments, next-generation systems can move closer to practical, scalable deployments that deliver robust energy savings with engineered thermal precision.

2.3. Research Methodologies and Analytical Trends in Nanofluid-Enhanced HVAC&R Systems

Integrating nanofluids into HVAC&R and thermal management systems has prompted the adoption of diverse research methodologies over the past decade. This section synthesizes the methodological landscape of nanofluid research from 2015 to 2025, based on a thematic categorization of 134 peer-reviewed studies, as visualized in Figure 7.

A detailed alluvial diagram, Figure 7 maps the relationship between research methods—including experimental, numerical, theoretical, hybrid, and AI/ML-based approaches—and the targeted application domains such as refrigeration systems and general HVAC&R applications. The aim is to reveal not only the methodological diversity but also the alignment (or mismatch) between methods and specific application areas.

From this mapping, several dominant trends emerge:

- Experimental studies (n = 69) overwhelmingly dominate the literature and are most commonly applied to refrigeration systems (n = 105) and HVAC systems (n = 23). These studies often focus on evaluating nanofluid-enhanced heat exchangers, condensers, and evaporators under lab or bench-scale conditions.
- Numerical (n = 14) and theoretical investigations (n = 11) are primarily used to model heat transfer, fluid flow, and entropy generation using simulation tools or closed-form analytical methods. These studies complement experimental findings and explore parameter sensitivity under controlled assumptions.
- Hybrid methodologies (n = 5), although less common, serve a vital role in validating numerical models against experimental data. These studies demonstrate the highest methodological robustness but remain underutilized.
- AI/ML-driven research (n = 20) is an emerging area, primarily focused on predictive modeling and performance optimization. These studies show increasing potential for integration into real-time HVAC&R system control and design optimization.
- Review papers (n = 14) synthesize findings across various applications and methodologies, often identifying knowledge gaps or proposing research directions.

To maintain focus and clarity, certain categories were excluded from the detailed thematic analysis due to their limited representation in the literature:

- Heat Exchanger studies (n = 1);
- Thermal management applications (n = 2);
- Magnetic nanofluid studies (n = 1).

While these areas offer promising avenues for nanofluid application, the current volume of work is insufficient to support meaningful synthesis or trend evaluation. However, their presence indicates niche interest and underscores the need for expanded research in these subdomains.

To better visualize the distribution of research efforts, an alluvial diagram (Figure 8) maps the linkage between methodological approaches and their corresponding application domains in the nanofluid-based HVAC literature from 2015 to 2025. This representation highlights dominant pairings, such as experimental studies with refrigeration systems, and reveals underexplored intersections that represent potential research opportunities.

Figure 8. Alluvial diagram mapping research methodologies to application domains in the nanofluid HVAC literature (2015–2025), highlighting dominant research patterns and underexplored intersections.

2.3.1. Experimental Studies

This section synthesizes the experimental investigations that explore the integration of nanofluids into HVAC&R systems. These studies employ laboratory-scale prototypes, component-level setups (e.g., chillers, condensers), and thermal loops to quantify improvements in heat transfer, energy efficiency, and system reliability. To structure the discussion, experimental studies have been grouped into three thematic categories:

- Experimental Fluid Characterization for System Integration,
- System-Level Performance Enhancement, and
- Comparative and Application-Specific Analysis.

Each theme captures a distinct experimental focus, ranging from nanofluid preparation and thermophysical profiling to performance validation and benchmarking under real or simulated operating conditions.

Figure 9 provides a visual distribution of experimental studies across these thematic categories and their corresponding application domains (HVAC vs. Refrigeration systems). Studies are grouped by (i) Experimental Fluid Characterization for System Integration, (ii) System-Level Performance Enhancement, and (iii) Comparative and Application-Specific Analysis. The diagram highlights that the majority of studies focus on refrigeration systems, particularly within system-level performance evaluations and nanofluid characterization. The figure reveals several key patterns:

- Experimental Fluid Characterization for System Integration accounts for 26 studies, again with a strong concentration in refrigeration systems (24 articles). These works primarily investigate nanofluid formulation, dispersion stability, and thermophysical properties (e.g., thermal conductivity, viscosity) in application-ready contexts, often under flow or heat load conditions.
- System-Level Performance Enhancement is the most extensively studied category, comprising 31 experimental studies, with the majority (25 articles) focusing on refrigeration systems. These studies commonly aim to validate gains in Coefficient of Performance (COP), energy reduction, or thermal efficiency when nanofluids are used in actual system components, such as evaporators, compressors, or heat exchangers.
- Comparative and Application-Specific Analysis includes 10 studies, split between HVAC&R systems. These studies test multiple nanofluids or operating parameters to evaluate performance differences under controlled comparative conditions.

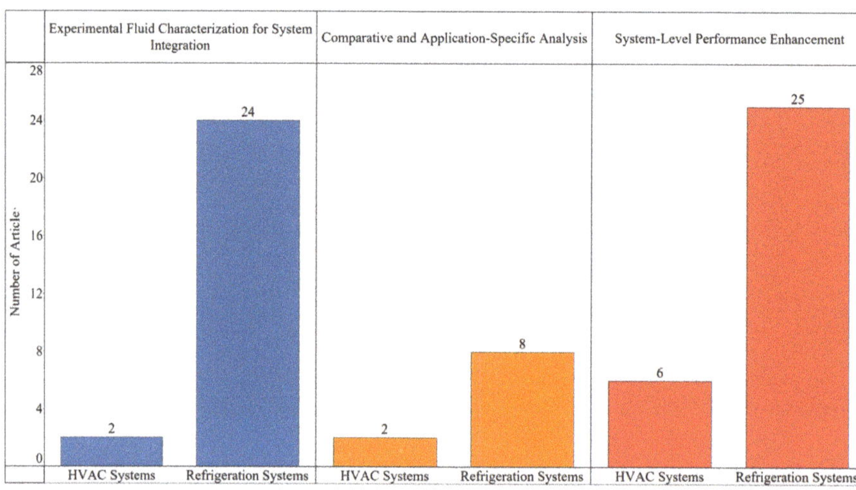

Figure 9. Distribution of experimental studies in nanofluid HVAC&R systems categorized by thematic focus and application domain.

The clear dominance of refrigeration systems across all three categories emphasizes their continued relevance and practical feasibility in nanofluid research. In contrast, HVAC&R systems, while still represented, have seen fewer experimental evaluations—highlighting a potential research gap and opportunity for future exploration.

This thematic organization ensures that each subgroup of studies can be analyzed in detail, with attention to methodological consistency, performance metrics, and application-specific trends. The following subsections will explore these categories in depth, identifying key findings, technical contributions, and unresolved challenges in each domain.

Experimental Fluid Characterization for System Integration

Experimental fluid characterization is a foundational stage in the deployment of nanofluids in HVAC&R systems. These studies primarily focus on understanding how nanofluid properties—such as thermal conductivity, viscosity, density, and stability—respond to nanoparticle type, concentration, base fluid selection, and flow conditions. A consistent observation across this category is the central role of nanolubricants and hybrid nanofluids in enhancing refrigerant performance, particularly in vapor compression cycles.

Silica- and oxide-based nanolubricants have emerged as promising candidates. Khan and ul Haque (2025) [80] performed a detailed thermophysical evaluation of silica-based nanolubricants in refrigeration systems. Their results demonstrated that the incorporation of 0.1 wt% silica nanoparticles in polyolester (POE) oil significantly improved thermal conductivity and reduced dynamic viscosity, offering better lubrication and energy transfer. This improvement was sustained across various temperature gradients and was accompanied by good dispersion stability, reinforcing its viability for system integration.

Similarly, Kumar et al. (2023) [81] proposed a hybrid nanofluid formulation involving TiO_2 and SiO_2 nanoparticles suspended in POE lubricant. Their study aimed at reducing the environmental footprint of conventional refrigerants. The experimental data showed that the hybrid nanofluid reduced total hydrofluorocarbon (HFC) emissions by improving heat transfer efficiency and reducing compressor workload, demonstrating environmental and performance benefits simultaneously.

Carbon nanotube-enhanced nanofluids also exhibit notable performance traits. Cassim, Hairuddin, and Cut (2024) [82] examined a multiwalled carbon nanotube (MWCNT)–CuO nanofluid blend in a test loop mimicking HVAC coil operation. They found that the hybrid nanofluid enhanced overall system energy absorption while maintaining thermal stability and dispersion. Importantly, their study utilized both empirical testing and visual sedimentation analysis to confirm fluid reliability over extended operation times.

Other investigations explored frictional losses and stability trade-offs. Dağıdır and Bilen (2023) [83] studied the use of Al_2O_3 and TiO_2 nanoparticles blended into POE lubricants in R134a refrigeration systems. They noted a complex interaction between nanoparticle loading and fluid pressure drop—where small concentrations improved thermal response, but higher doses led to increased flow resistance. These findings stress the importance of optimizing particle size and concentration to balance thermal gains against system pressure losses.

From a system optimization standpoint, comparative evaluations also fall into this group when focused on fluid behavior. Ahsan et.al (2025) [84] conducted an experimental comparison of nanolubricants in refrigeration systems using different refrigerant–nanofluid pairs. Though their primary objective was COP benchmarking, their work involved extensive viscosity and thermal conductivity profiling, underscoring the practical importance of fluid characterization in real operating scenarios.

In general, studies in this category confirm that the nanoparticle material, morphology, and concentration, along with surfactant use and mixing technique, play decisive roles in nanofluid performance. However, despite promising short-term gains, few studies explore long-term reliability, biofouling risks, or compatibility with legacy components, suggesting areas for future research.

To facilitate system-level integration, understanding the fluid performance characteristics under varying nanoparticle compositions and refrigerant pairings is crucial. Table 6 presents a consolidated summary of key experimental studies that explored a range of nanofluids—including metal oxides, carbon-based nanomaterials, and hybrid blends—dispersed in conventional compressor oils or refrigerants. The table delineates the specific performance parameters evaluated (e.g., COP, cooling capacity, power reduction), alongside the working fluids and nanoparticles involved. Notably, consistent improvements in COP (ranging from 2.4% to over 45%) and reductions in power consumption were observed, with hybrid and carbon-based nanomaterials often yielding the highest performance gains. These findings underscore the potential of nanofluid-enhanced systems in improving the energy efficiency and operational reliability of refrigeration cycles.

Table 6. Summary of nanoparticle-enhanced nanolubricants and nanorefrigerants used in experimental refrigeration systems, highlighting the evaluated performance parameters, refrigerants employed, and observed enhancements in system metrics such as COP, power consumption, and cooling capacity.

Ref.	Nanoparticle Involved	Evaluated Parameter	Refrigerant	Findings
[85–88]	Al_2O_3	COP	R134a, R600a	↑ COP (14 to 45%) Power Reduction (5–9%)
[80,84,88]	SiO_2	COP	R600a	↑ COP (15–40%)
[80,81,84,88–92]	(25%) TiO_2-(75%) SiO_2, ZnO	COP, Compressor Power, Cooling Capacity	POE oil, R600a	↑ COP (14–51%) ↓ Power (8–13%) ↑ Refrigerated Capacity (13%)
[93–100]	CuO, Ag, Cu	COP, Compressor Power, Refrigerating Capacity	R600a, R134a, R290	↑ COP (12.2–35%) ↓ Power (1.39–24%) ↑ Refrigerated Capacity (13%) ↓ Friction 9.9%
[82,83,101]	CNT, MWCNT, Graphene	COP, Compressor Power, Cooling Capacity	POE oil	↑ COP (12.2–35%) ↓ Power (1.39–6.8%)
[102–106]	Ag, Diamond,	COP, Cooling Capacity	R600a, R134a, R410A, R32	↑ COP (2.4–39.5%) ↓ Power (1.39–6.8%) ↑ Refrigeration Capacity (2.4%)

System-Level Performance Enhancement

While previous section emphasized fluid-level thermophysical characterization, such as thermal conductivity, viscosity, and nanoparticle concentration effects, the actual impact of nanofluids on system performance must be validated under integrated operating conditions. These system-level studies account for component interactions, dynamic load variations, and long-term operational stability—factors essential to real-world HVAC&R applications. Accordingly, this subsection focuses on experimental research that evaluates how nanofluid incorporation affects the overall energy performance, cooling capacity, compressor behavior, and environmental efficiency of complete systems.

Ikpe and Udofia (2024) [107] investigated the application of a hybrid nanofluid blend of Al_2O_3 and CuO in a domestic vapor compression refrigeration system. Their study highlighted significant improvements in thermal conductivity and heat absorption characteristics, which translated into measurable reductions in compressor work and improved coefficient of performance (COP). The hybrid formulation contributed to enhanced dispersion stability and synergistic thermal performance, outperforming single-particle nanofluids. These findings are particularly relevant for systems requiring robust part-load performance and extended duty cycles.

Jiang et al. (2023) [108] explored the role of TiO_2-based nanolubricants in an R134a-based refrigeration cycle. The experimental setup revealed tangible improvements in both cooling capacity and power consumption, owing to improved lubricant film strength and more effective heat transfer within the compressor. Lower compressor discharge temperatures were also noted, suggesting not only enhanced performance but also a likely extension of component life due to reduced mechanical stress.

In HVAC systems, Micali et al. (2025) [109] and Milanese et al. (2022) [110] provided valuable insights from full-scale implementations of nanofluid-enhanced air conditioning units. Micali et al. focused on optimizing CO_2 emissions and energy consumption in data center applications, showing that nanofluid-enhanced secondary loops can cut CO_2 emissions by over 20%, while improving thermal control and reducing compressor cycling. Milanese et al. emphasized the operational resilience of nanofluids, confirming improved load-following performance and control precision under variable ambient conditions. These studies underscore the broader applicability of nanofluids beyond component optimization, highlighting their role in achieving whole-system energy goals.

Raghavulu and Rasu (2025) [111] adopted a sustainability-driven approach, evaluating nanorefrigerants in environmentally friendly R600a systems. Their results demonstrated up to 35% COP enhancement and considerable energy savings. The use of nanoparticles in low Global Warming Potential (GWP) refrigerant environments offers a path forward for balancing energy performance with climate impact mitigation—an increasingly critical metric for future-ready HVAC&R technologies.

Taken together, these experimental investigations validate that nanofluids can play a transformative role in system-level optimization. The improvements are not merely confined to static test benches but extend to integrated operating conditions, influencing compressor behavior, refrigerant flow regulation, and thermal stability across various subsystems. These findings reinforce the need for future work to explore long-term reliability, optimal nanoparticle loading, and compatibility with newer refrigerants and lubricants. Ultimately, nanofluid adoption at the system scale offers a scalable and practical route to achieve next-generation energy efficiency and decarbonization targets in thermal systems.

To better understand the performance characteristics and comparative benefits of various nanofluids applied in HVAC systems, Table 7 synthesizes key experimental findings from the recent literature. Studies are organized according to the type of nanoparticle used, base fluid or refrigerant, and the evaluated parameters such as COP enhancement, temperature reduction, pressure drop, or energy efficiency. This grouping not only reveals the breadth of experimental conditions but also underscores the recurring trends and dominant nanofluid combinations explored for thermal performance improvements.

Table 7. Grouped summary of experimental studies on nanofluid-enhanced HVAC systems, categorized by nanoparticle type, base fluid or refrigerant, and performance parameters evaluated.

Ref.	Nanoparticle Type	Base Fluid/Refrigerant	Key Findings
[87,112–115]	Hybrid $Al_2O_3/CuO/TiO_2$	Zeotropic refrigerant blend (23% R32/25% R125/52% R134a)	↑ COP 3.10% (at blend 011) ↑ Compressor power coefficient 13.51% ↑ Cooling capacity by 5.78% ↑ TEGWI 1.06 kg/s CO_2 ↑ Optimal heat transfer at 0.0075 Vol.% nanoparticle concentration
[100,101,107–111,115–120]	Al_2O_3/TiO_2	Chiller fluid (not explicitly stated, inferred to be water-based), Distilled Water, H_2O–LiBr, POE (polyol ester) oil, R134a, Water–glycol mixture; Ammonia–water solution, NH_3–H_2O, NH_3–H_2O–LiBr, Water	↑ COP (9.1% to 14%) ↓ Energy consumption about 32%, ↓ 50.9 tons/year CO_2 emission; net LCA benefit of 36.6 tons/year Winter: Avg. ↑ 9.8% Summer: Avg. ↑ 8.9% Al_2O_3/R134a compared to pure refrigerant COP ↑ to 6.5 (about 15% improvement) ↑ COP 17.27% Absorption efficiency ↑ up to 85% Improved Thermal Performance

Table 7. *Cont.*

Ref.	Nanoparticle Type	Base Fluid/Refrigerant	Key Findings
[88]	Graphene-Based	Polyester Oil (POE)	↑ COP by ~29%
[112,121–123]	Binary Nanofluids (e.g., Iron Oxide + LiBr-H_2O)	LiBr-H_2O; Water–ethylene glycol; H_2O–LiBr	↑ Absorption rate 4.9–7.6% without magnetic field ↑ TC by up to 37.7% over baseline fluid, ↑ Max desorption rate by 7.9% (CNT); ↑ Max SCOP by 14%; ↑ cooling effect up to 6%; Optimal desorber modules: <50, cost cap: EUR 35/module
[124–127]	Noble/Carbon-Based (Ag, MWCNT)	Water; Various compressor oils	↑ COP with nanofluid absorbent ↑ Overall efficiency from 77.3% to 81% with Ag nanofluid Exergy efficiency of the system: 29% At 50 °C and 0.1% mass conc., ↑ viscosity by 40–90%, depending on base oil;
[128–131]	Hybrid Experimental (Al_2O_3, CuO, TiO_2)	Deionized Water; Water	↑ Max COP: 24.2% (vs. water) Al_2O_3/H_2O had 6.2% higher COP vs. hybrid TiO_2 had 5.6% lower COP vs. hybrid ↓ Compression ratio up to 6.9%
[130,132]	Copper-Based (Cu, CuO)	SAE50 Oil; Water	↓ Viscosity by up to 15% at low vol. fractions; ↑ by 14% at 1% vol.; ↑ COP by 23% (Al_2O_3) and 72% (Cu) at 5% volume fraction Higher TC of Cu yielded better performance than Al_2O_3
[133]	Mixed Metal Oxides (MgO, TiO_2)	Water	↑ COP by 16.7% (MgO) and 11.5% (TiO_2) at 1% wt. Max performance at PVC fill, 12 mm, 90° spray angle
[134–137]	TiO_2 with Surfactants (SDBS)	NH_3–H_2O	↑ COP up to 27% overall ↑ 14–26.2% under challenging conditions (e.g., low evap temp, high cooling water temp); best performance with 0.5% TiO_2 + 0.02% SDBS

To consolidate the diverse experimental investigations discussed in this section, a summary table was developed by systematically grouping studies based on the nanoparticle types and their respective base fluids or refrigerants. The table captures essential parameters such as system configuration, performance metrics evaluated, and the key outcomes. This clustering not only highlights the predominance of Al_2O_3, TiO_2, and copper-based nanofluids across HVAC subsystems but also facilitates cross-comparison of their thermal enhancement trends. It becomes evident that while certain nanoparticles exhibit superior heat transfer or system efficiency, others offer advantages in terms of stability, economic feasibility, or compatibility with renewable integrations. The insights drawn here provide a foundational reference for selecting optimal nanofluid formulations in future HVAC&R applications. Building upon these consolidated findings, the next subsection delves deeper into comparative and application-specific analyses. By examining performance outcomes across similar HVAC&R configurations and operating conditions, we aim to uncover nuanced trends and evaluate the practical viability of each nanofluid formulation. This approach not only clarifies relative advantages but also aligns material choices with targeted HVAC&R applications and performance goals.

Comparative and Application-Specific Analysis

A nuanced understanding of nanofluid integration into HVAC&R systems requires comparative assessment across specific application domains and working conditions. The current body of experimental literature has shown that nanofluid-enhanced systems do not offer uniform improvements across all components and configurations; instead, performance gains are context dependent—shaped by factors such as nanoparticle type, concentration, dispersion medium (refrigerant or lubricant), and the targeted subsystem (e.g., compressor, condenser, evaporator).

In the realm of refrigeration systems, Yousif, Al-Obaidi, and Al-Muhsen (2024) [138] investigated the use of CuO nanoparticles dispersed in R134a refrigerant in a vapor–compression refrigeration cycle. Their experimental outcomes showed a consistent improvement in the coefficient of performance (COP), with a peak increase of approximately 19% when compared to the baseline. The authors noted reductions in compressor work and energy consumption, emphasizing that nanorefrigerants can substantially enhance system efficiency when nanoparticle stability and uniform dispersion are ensured.

Echoing similar findings, Shewale et al. (2024) [139] focused on the application of TiO_2 nanorefrigerants in R134a-based vapor compression systems. Their comparative analysis revealed that the introduction of nanofluids not only elevated the COP but also shortened the cooling time and reduced power consumption. They emphasized that these enhancements are attributable to increased thermal conductivity and improved refrigerant flow characteristics induced by the nanoparticles. However, their study also highlighted challenges regarding long-term dispersion stability and potential clogging effects, particularly at higher concentrations.

In contrast, Mohammed et al. (2024) [140] concentrated on HVAC systems, where Al_2O_3 nanoparticles were integrated into R600a refrigerants. Their work aimed to analyze both thermodynamic and exergetic performance across a range of nanoparticle concentrations. Results indicated a peak improvement of 24.56% in cooling capacity and COP at a 0.1% mass fraction. Notably, the exergy destruction in the compressor and evaporator components was significantly reduced, underscoring how nanoparticle integration can mitigate entropy generation in HVAC applications.

Further supporting this trend in HVAC-focused applications, Patel et al. (2024) [141] explored hybrid Al_2O_3–TiO_2 nanoparticles dispersed in R600a in both experimental and simulation settings. Their results demonstrated a substantial enhancement of 16.5% in COP at an optimal nanoparticle volume fraction of 0.2%. Additionally, the hybrid nanofluid contributed to improved compressor work reduction and system responsiveness under fluctuating thermal loads. Their study also noted that hybrid nanoparticles can offer synergistic benefits—combining high thermal conductivity from one material, such as Al_2O_3–TiO_2, with good dispersion and stability from another—thus making them especially effective in HVAC systems with variable operating conditions [142,143].

A distinct application-specific innovation is presented by Narayanasarma and Kuzhiveli (2020) [144], who focused not on the refrigerant, but on the lubricant used in refrigeration compressors. They experimentally evaluated the performance of SiO_2 nanoparticles added to polyolester (POE) and mineral oils. Their analysis revealed enhanced thermal conductivity and superior lubricity properties, which collectively led to improved compressor reliability and efficiency. Importantly, the study highlighted that nanolubricants could also reduce wear and friction losses, demonstrating a valuable yet often overlooked application route for nanotechnology in refrigeration systems.

From a comparative perspective, these studies collectively demonstrate that the influence of nanofluids varies across systems and target components. In refrigeration applications, most gains are reported in COP improvement, compressor power reduction, and shortened cooling cycles—especially when using CuO or TiO_2 in refrigerants such as R134a. However, system stability is highly dependent on the nanoparticle concentration and surface treatment, which affects dispersion behavior. Meanwhile, HVAC&R systems tend to benefit from hybrid nanoparticles and lower GWP refrigerants like R600a, with pronounced improvements in heat transfer and exergy efficiency. Studies using Al_2O_3 and hybrid Al_2O_3–TiO_2 (Patel et al., 2024 [141]; Mohammed et al., 2024 [140]; Babarinde et al., 2020 [145]) show the importance of nanoparticle design and refrigerant compatibility in dictating system responsiveness and thermal load handling.

Additionally, the application of nanofluids to lubricants—as opposed to refrigerants—offers a complementary pathway to system performance enhancement. While this approach may not directly alter thermodynamic cycles, it can extend equipment lifespan and reduce energy consumption through mechanical efficiency improvements.

Quantitatively, the reported studies collectively reveal that nanofluid integration can yield COP enhancements ranging from 16.5% to 24.6%, with compressor work reductions up to 15% depending on the nanoparticle–refrigerant combination and concentration (Devendra et al.,2019 [146], Lin et al., 2017 [147]). Lubricant-based modifications with SiO_2 nanoparticles demonstrated thermal conductivity improvements of around 18%, contributing to smoother mechanical operation and improved compressor performance. Most optimal concentrations for achieving these gains fall within the 0.05% to 0.2% range by volume or mass, beyond which performance often plateaus or deteriorates due to viscosity increase or sedimentation. These figures underscore that nanofluid application in HVAC&R systems is a quantitatively validated strategy, though success depends on careful alignment of formulation, system design, and operational stability.

While experimental studies provide concrete evidence of nanofluid-enhanced performance in HVAC&R systems, they often face constraints related to cost, scalability, and real-time control over boundary conditions. To address these limitations and explore a broader design space, researchers have increasingly turned to numerical modeling and simulation. Section 2.3.2 delves into these numerical investigations, highlighting how computational tools such as CFD and thermodynamic modeling frameworks are used to predict nanofluid behavior, optimize system parameters, and validate experimental findings under controlled and hypothetical scenarios.

2.3.2. Numerical Investigations

The growing reliance on numerical approaches in nanofluid research for HVAC&R systems reflects the need for scalable, cost-effective, and repeatable methods to probe complex thermal phenomena. Section 2.3.2 synthesizes findings from 25 numerical studies, categorized into two principal methodological clusters: Computational Fluid Dynamics (CFD)-based studies (n = 14) and theoretical or analytical modeling approaches (n = 11).

These numerical methods serve two essential roles: first, they provide detailed insight into local heat transfer behavior, phase interactions, and pressure distributions that are difficult to measure experimentally; second, they allow rapid prototyping of design configurations and parametric sensitivity analysis under idealized boundary conditions. Notably, CFD simulations have been widely adopted for their ability to model nanofluid flow in complex geometries, such as microchannels, condenser tubes, and evaporators. Meanwhile, theoretical studies often rely on thermodynamic cycle modeling, heat balance equations, and exergy analyses to assess the macroscopic performance impact of nanofluid integration.

In terms of application domains, refrigeration systems account for the majority of numerical investigations (17 studies), compared to 7 studies on HVAC&R systems, and a single study focused on thermal management. This reflects a continued interest in optimizing the vapor compression cycle, compressor performance, and refrigerant behavior, particularly using nanorefrigerants and nanolubricants. HVAC-oriented models, on the other hand, emphasize energy efficiency at the system scale, often incorporating external variables such as ambient temperature, airflow dynamics, and hybrid heat exchanger designs.

This section aims to distill methodological insights, model structures, boundary condition considerations, and validation strategies employed across these studies. By doing so, it bridges the gap between conceptual modeling and practical implementation, highlighting how numerical tools contribute to both hypothesis testing and performance forecasting in nanofluid research.

Among the numerical approaches, Computational Fluid Dynamics (CFD) has emerged as a particularly powerful and versatile tool for evaluating nanofluid behavior in HVAC&R applications. By solving governing equations for fluid flow and heat transfer at both micro and macro scales, CFD studies enable detailed visualization of temperature distribution, velocity fields, pressure drops, and thermal boundary layer characteristics. The following subsection synthesizes insights from 14 CFD-based investigations, emphasizing their modeling frameworks, simulation parameters, and how these studies validate or extend experimental findings across various application domains.

Computational Fluid Dynamics (CFD) Studies

Computational Fluid Dynamics (CFD) plays a vital role in exploring the thermal and hydrodynamic behavior of nanofluids in HVAC&R systems. These simulation-based investigations provide detailed insight into convective mechanisms, pressure gradients, entropy generation, and system optimization—parameters often challenging to capture through experimental approaches alone. Fourteen studies in this category demonstrate the versatility and predictive power of CFD in modeling nanofluid-enhanced systems across various application domains (Table 8).

In the HVAC sector, Aziz et al. (2025) [148] developed a hybrid modeling approach combining CFD with advanced neural network optimization to assess the energy performance of a PV-integrated heat pump system. Their simulation employed nanofluids to improve thermal conductivity within the condenser and evaporator loops. Results showed that system efficiency improved significantly, and the integration of Levenberg–Marquardt backpropagation models allowed for real-time optimization of operational parameters, such as flow rate and temperature setpoints. Benazzouz et al. (2025) [149] examined the thermal behavior of a domestic-scale HVAC system using nanorefrigerants through 3D CFD modeling. The simulation framework incorporated variable nanoparticle concentrations of Al_2O_3 dispersed in R134a refrigerant. Their results indicated that optimal nanoparticle dispersion enhanced heat transfer efficiency by up to 21%, particularly under transient load conditions. They also demonstrated that the nanoparticle-enhanced refrigerant suppressed thermal stratification and enabled quicker attainment of setpoint temperatures, reducing overall energy demand.

In refrigeration systems, Boldoo et al. (2022) [150] performed a numerical investigation on the thermal performance of nanolubricants in a vapor–compression refrigeration system. The study utilized Al_2O_3 and TiO_2 nanoparticles suspended in mineral oil as the lubricant base. CFD simulations showed that Al_2O_3 nanolubricants enhanced thermal conductivity by approximately 12% and improved compressor work efficiency by 8%, while TiO_2 showed better flow stability across narrow piping configurations. Their comparative modeling confirmed the suitability of these nanolubricants for miniaturized systems operating under fluctuating thermal loads.

Khan et al. (2023) [151] advanced the field with a three-dimensional mathematical model of a solar-assisted HVAC&R system integrated with nanofluid-based thermal storage. While the full set of simulation parameters and nanoparticle types was not explicitly detailed, the model focused on exergy analysis and entropy reduction across system components. The incorporation of CFD allowed the authors to validate the energy gain potential of solar–HVAC coupling and demonstrate the utility of nanofluids in dynamic thermal storage control, although their conclusion was not explicitly provided.

Another significant contribution to HVAC&R system modeling comes from Colangelo et al. (2021) [152], who numerically assessed a complex HVAC network using hybrid nanofluids—specifically Al_2O_3-Cu/water suspensions—in multiple loop configurations. Their CFD model revealed up to 28% improvement in heat transfer coefficient compared to conventional water-based fluids, particularly in spiral and wavy tube geometries. This study underscored the effectiveness of multi-component nanoparticles in enhancing both thermal and flow performance, offering strong implications for future HVAC system design strategies (Colangelo et al., 2021) [152].

Collectively, these CFD-based studies illustrate that nanoparticle-enhanced working fluids—whether used as refrigerants, lubricants, or coolants—offer significant thermo-hydraulic benefits. Simulation frameworks commonly apply steady-state or transient Navier–Stokes solvers coupled with energy conservation equations, often incorporating turbulence models (e.g., k–ε or RNG) and conjugate heat transfer formulations. The studies differ in complexity and scope, from system-level simulations of solar-assisted HVAC loops to microchannel and heat exchanger-focused investigations.

Moreover, nanoparticle type and base fluid pairing strongly influence simulation outcomes. Al_2O_3 remains the most commonly modeled nanoparticle due to its well-documented thermal properties and availability. However, hybrid combinations and metal oxide variants (e.g., CuO, TiO_2) are increasingly simulated to capture synergistic heat transfer effects. Base fluids range from water and ethylene glycol in cooling applications to R134a and R410A in refrigeration, each selected based on the intended system's thermal load characteristics and environmental footprint.

In summary, CFD studies bridge the gap between conceptual design and experimental validation by offering a cost-effective, scalable, and insightful tool for evaluating nanofluid integration in thermal systems. These models not only confirm the potential performance benefits observed in lab-scale tests but also offer parametric sensitivity insights that guide optimal system configurations for real-world deployment.

Table 8. Types of nanoparticles, base fluid/refrigerant, and key findings.

Ref.	Nanoparticle Type	Base Fluid/Refrigerant	Key Findings
[151,153–156]	Al_2O_3, SiO_2, Fe	R134a, Water	↑ HVAC system efficiency by approximately 10% ↓ Compressor power at 0–5% vol fraction ↑ 10% in thermal performance using nanofluids Maximum COP with 0.2% volume fraction of nanoparticles ↑ 7.6% in effectiveness of a plate heat exchanger ↓ 11.6% compressor power 14.6% improvement in COP ↑ cold chain efficiency by reducing energy consumption ↑ Thermal characteristics by inlet velocity, but a higher pressure drop
[150]	MWCNT	R1234yf	Analyze the specific heats of the various MWCNT nanoparticle-enhanced [HMIM] cation-based ionic liquids Effect of concentration (0–1%) on COP at assorted Generator temperatures (303–383 K) were obtained experimentally at a temperature and a concentration of 303–383 K and 0–1 wt%, respectively, and were subsequently used for modeling the solubility and absorption refrigeration cycle
[152]	Hybrid	Ethyl glycol	Impact of activation parameters and Schmit number on dimensionless concentration for different nanofluids
[156–160]	CuO, Cu-based nanofluids	Water, R134a, ethylene-glycol	Optimal performance at 4% nanofluid concentration ↑ cold chain efficiency by ↓ energy consumption ↑ COP 20% for CuO/Water, 25% for CuO/water with insulation ↑ Cooling power (about 5%) Simulation results show that, a 9 m^2 solar collector, a 0.3 m^3 storage tank, and 0.05 m thick polystyrene insulation I The active barocaloric regenerative refrigeration cycle is supposed to work as a domestic refrigerator in temperature range of 255 ÷ 290 K

Table 8. *Cont.*

Ref.	Nanoparticle Type	Base Fluid/Refrigerant	Key Findings
[153,156,158,160,161]	TiO_2, MO	R134a, POE Oil, LPG	0.5 to 1 g TiO_2 addiction to the POE oil performs better than that of 1.5 g TiO_2. Predicting irreversibility using ANFIS-SC model: RSME = 0.998, MAPE = 0.078% ↑ cold chain efficiency by ↓ energy consumption The comparison of variance, root mean square error (RMSE), mean absolute percentage error (MAPE) were 0.996–0.999, 0.0296–0.1726 W, and 0.108–0.176% marginal variability values

Theoretical Modeling and Analytical Studies

Theoretical and analytical modeling plays a critical role in evaluating the performance of nanofluids in HVAC&R systems by allowing researchers to predict system behavior under idealized or parametric conditions, often without the cost or complexity of experimental setups. These models provide insight into energy and exergy performance, thermoeconomic viability, and the thermophysical impact of nanoparticle inclusion within various refrigerant cycles.

Li and Lu (2022) [162] conducted a comprehensive thermodynamic modeling of vapor–compression refrigeration cycles using Al_2O_3-based nanorefrigerants across several environmentally friendly refrigerants, including R600a, R134a, R1234yf, and R1233zd(E). Their theoretical framework employed both energetic and exergetic analyses, revealing that the coefficient of performance (COP) improved by more than 20% with Al_2O_3 nanofluids. Moreover, the highest exergetic efficiency was observed with R600a, whereas R1233zd(E) yielded the lowest environmental impact, making it the most sustainable option despite lower thermodynamic output.

Akhtar and Rajput (2023) [163] explored the energy and exergy dynamics of a vapor–compression refrigeration system using nanolubricants, specifically MWCNTs and graphene blended with POE oil. Their theoretical investigation showed a potential reduction in power consumption by up to 16%, along with lower pressure ratios and improved compressor performance. This highlights the efficacy of carbon-based nanoparticles in enhancing thermal conductivity and lubrication properties in traditional systems.

In a similar context, Kumar and Rajput (2023) [164] assessed a computerized refrigeration cycle operating with CuO-Al_2O_3-based mineral oil nanolubricants and low-GWP refrigerants such as R600a, R134a, R1234yf, and R1233zd(E). Their results indicated up to a 23% reduction in power consumption and substantial improvement in exergy efficiency, especially when hybrid nanoparticles were used. This study also emphasized the role of computerized control systems in optimizing nanofluid-based cycles through dynamic parameter tuning.

The study by Yilmaz et al. (2024) [165] focused on the theoretical analysis of a cascade vapor–compression refrigeration system utilizing CNT, CuO, and TiO_2 nanofluids with R290 and R1233zd(E) refrigerants. Their thermodynamic analysis demonstrated an 11% reduction in compressor power consumption and significant improvement in environmental metrics, such as CO_2 emissions. Among the refrigerants tested, R290 consistently exhibited better energetic performance, while the hybrid nanofluids led to enhanced heat transfer across cascade stages.

Gürbüz, Keçebaş, and Sözen (2022) [166] applied exergy and thermoeconomic modeling to evaluate nanofluid integration in NH_3–H_2O absorption refrigeration systems. Utilizing various binary nanofluids such as MgO–Al_2O_3, ZnO–Al_2O_3, and TiO_2, their analysis revealed TiO_2 as the most effective additive in terms of both exergy efficiency and cost–performance balance. The levelized cost of cooling for the system ranged between USD 0.110 and USD 0.180 per kWh, suggesting that thermoeconomic modeling provides essential decision-making tools for large-scale deployment.

Collectively, these theoretical studies underscore the utility of nanofluids in enhancing both energy efficiency and sustainability in vapor–compression and absorption systems. Across various refrigerants and base oils, nanoparticle inclusion—whether metallic oxides like Al_2O_3 or carbon-based like CNTs and graphene—proved to offer substantial thermodynamic benefits, including up to 23% power reduction and over 20% improvement in COP. Furthermore, the incorporation of exergetic and economic modeling provides a more holistic understanding of system feasibility, especially in the context of emerging low-GWP refrigerants and renewable-integrated HVAC architectures.

To consolidate the diverse theoretical investigations discussed in this subsection, a summary table, Table 9, has been developed that captures the key modeling characteristics and performance outcomes. The table categorizes each study by nanoparticle type, base fluid or refrigerant, modeling scope, and reported thermodynamic or economic benefits. This structured synthesis facilitates cross-comparison of findings and highlights recurring modeling trends—such as the predominance of Al_2O_3 and hybrid nanofluids, frequent use of R600a and R1233zd(E) as sustainable refrigerants, and consistent improvements in COP, exergy efficiency, or power savings. These insights provide a valuable reference point for researchers and practitioners seeking to integrate nanofluid-based enhancements into advanced HVAC&R models.

Table 9. Summary of theoretical and analytical modeling studies on nanofluid-enhanced HVAC&R systems, highlighting nanoparticle types, base fluids or refrigerants, modeling approaches, and key quantitative findings.

Ref.	Nanoparticle Type	Base Fluid/Refrigerant	Key Findings
[162,164,167–169]	AL_2O_3, CuO-Al_2O_3/MO	LiBr-H_2O, R600a, R134a, R1234yf, R1233ZDE	↑ COP more than 20%, ↑ max exergy efficiency: 38.48% ↓ Power consumption: up to 23%, Exergy destruction up to 49%, ↑ COP by 29%, ↑ Second law efficiency to 28% ↑ COP of the solar cycle by 4.51% ↑ COP range: 16 to 25%
[164,165,167,168]	Cu, CuO	R600a, R134a, R1234yf, R1233ZDE	↓ Power consumption: up to 23%, ↑ COP by 29%, ↑ Second law efficiency up to 28% ↓ Power consumption by 11%, ↓ CO_2 emission: 12.5%

Table 9. *Cont.*

Ref.	Nanoparticle Type	Base Fluid/Refrigerant	Key Findings
[166,170,171]	TiO_2, $FeOTiO_2$	POE Oil, NH_3–H_2O	TiO_2 is most effective, Levelized cost ranges USD 0.773/kWh to USD 0.875/kWh ↑ COP, circulation ratio, and internal temperature reached 17%, 57%, and 20% ↓ Irreversibility up to 18%
[163,165]	CNT, MWCNT	POE Oil, R290/R1233ZDE	↓ Power consumption: up to 16%, ↓ Pressure ratio up to 5.6%, ↑ COP by 11%
[163]	Graphene	POE Oil	↓ Power consumption reduced: up to 16%, ↓ Pressure ratio up to 5.6%, ↑ COP by 11%
[172,173]	Ag, Fe_2O_3	Ethylene glycol/DI-water	↑ convective HTC increased up to 11%

2.3.3. Hybrid Studies

Hybrid studies that integrate both experimental testing and numerical modeling provide a robust approach for evaluating nanofluid performance in HVAC&R systems. By complementing empirical data with theoretical and computational insights, these studies offer both the realism of physical systems and the flexibility of simulation to explore design alternatives, parameter optimization, and scale-up feasibility.

Akhayere et al. (2023) [174] performed a comprehensive assessment of nanorefrigerants in a modified vapor compression system, examining their effects on system efficiency and environmental impact under post-Kyoto Protocol climate objectives. The hybrid methodology involved comparative simulation and real-world testing to evaluate thermal and energetic performance improvements using advanced refrigerant formulations.

Khetib et al. (2022) [175] explored the impact of nanofluid-enhanced ejector refrigeration systems for thermal management in lithium-ion battery storage. Their hybrid approach combined CFD simulation with experimental evaluation using Al_2O_3 nanoparticles. Results demonstrated that a 0.3% volume fraction of nanofluids improved the system's coefficient of performance by 20% and reduced the evaporator outlet temperature by 5.3 °C, thereby enhancing cooling uniformity and reducing the temperature gradient within battery cells. The study highlighted nanofluids as an effective strategy for improving both energy efficiency and thermal safety in advanced energy storage applications.

In a different line of inquiry, Gürbüz et al. (2022) [166] conducted a hybrid investigation involving zinc oxide (ZnO) nanofluids in diffusion absorption refrigeration systems (DARS). The dual-method approach enabled validation of heat transfer coefficients, flow stability, and refrigerant absorption rates under varying nanoparticle concentrations. Their study focused on solar-integrated refrigeration systems and emphasized the influence of nanoparticle morphology on both heat and mass transfer behaviors.

Wang et al. (2018) [176] examined the mass transfer performance of nanosolution-based falling film absorption systems. Their hybrid analysis involved formulating nanosolutions using LiBr-H_2O as the base pair with metallic nanoparticles. The experimental portion confirmed enhanced absorption rates, while the numerical model evaluated surface tension effects, diffusion rates, and film stability. This study demonstrated that adding nanoparticles reduced absorption time and increased film efficiency, thereby improving refrigeration system responsiveness under partial load conditions.

Finally, Li et al. (2019) [177] performed one of the most rigorous comparative hybrid investigations using H_2O-based silica nanofluids as secondary refrigerants. Both experimental measurements and modeling demonstrated that using nanofluids significantly improved heat exchanger performance. Key thermophysical properties such as thermal conductivity and viscosity were characterized and validated, showing a consistent heat transfer enhancement of up to 18% and a reduction in thermal response time by over 15%. This dual-method study confirmed nanofluids as viable candidates for retrofitting secondary fluid loops in refrigeration systems.

These hybrid studies collectively validate the tangible benefits of nanofluid application across diverse refrigeration contexts—from absorption and ejector cycles to battery cooling and secondary loop retrofits. The blend of modeling and experimentation provides more reliable assessments of nanoparticle behavior in real-world conditions, accounting for non-idealities such as agglomeration, pressure drops, and thermal degradation. Across all studies, nanofluids based on Al_2O_3, ZnO, and silica consistently improved heat transfer and system performance, while also contributing to better environmental outcomes in terms of reduced energy consumption and enhanced thermal efficiency.

2.3.4. AI/ML-Based Approaches

The integration of Artificial Intelligence (AI) and Machine Learning (ML) in nanofluid research has shown transformative potential, particularly in accelerating predictions of thermophysical properties and performance metrics like coefficient of performance (COP), heat transfer coefficients, and cooling capacity. While there remains a notable gap in direct applications of AI-ML to full-scale HVAC&R systems incorporating nanofluids, a significant number of studies have focused on foundational thermophysical modeling, parameter optimization, and component-level system enhancements. These studies collectively offer a trajectory for future work that may bridge data-driven modeling and system-level optimization.

Artificial intelligence (AI) and machine learning (ML) techniques have emerged as powerful tools for analyzing complex nonlinear behaviors and optimizing performance in thermal systems. While direct applications of AI/ML to full-cycle HVAC or refrigeration systems containing nanofluids remain limited, a rapidly expanding body of the literature has focused on predicting thermophysical properties, key system parameters such as coefficient of performance (COP), heat transfer coefficient, and thermal conductivity—parameters that underpin system-level behavior. These studies serve as foundational work for integrating AI/ML tools into larger-scale predictive frameworks, control systems, and optimization modules for HVAC&R systems enhanced by nanotechnology.

In refrigeration systems, hybrid models combining support vector regression (SVR), genetic algorithms, and extreme learning machines have demonstrated substantial improvement in the prediction of relative cooling power (RCP) of magnetocaloric materials (Shamsah, 2024) [178]. Similarly, Gill et al. (2020) [179] employed adaptive neuro–fuzzy inference systems (ANFIS) to analyze exergetic performance of LPG-refrigerated systems using TiO_2-based nanolubricants, reporting predictive accuracies with variations under 1% from experimental observations.

Several studies have advanced the use of Gaussian Process Regression (GPR), Support Vector Regression (SVR), and Artificial Neural Networks (ANNs) for optimizing nanofluid performance in HVAC systems. Sharma et al. (2021) [180] utilized GPR and SVR to predict COP and cooling capacity for TiO_2-POE nanolubricants in a domestic refrigeration cycle, achieving R^2 values exceeding 0.99. Meanwhile, Aziz et al. (2025) [148] developed an ANN model integrated with the Levenberg–Marquardt algorithm to predict thermal performance in photovoltaic-based HVAC systems using tri-hybrid nanofluids, achieving extremely low error rates (10^{-7} to 10^{-5}).

Optimization studies combining multi-objective evolutionary algorithms with AI models are also notable. Boyaghchi et al. (2016) [181] employed Non-dominated Sorting Genetic Algorithm II (NSGA-II) to enhance both exergetic efficiency and economic cost for cascade refrigeration systems using CuO nanofluids and eco-friendly refrigerants. Their study achieved thermal and exergy COP improvements of 10.13% and 0.63, respectively.

In lubrication and material synthesis domains, AI models have demonstrated precise predictive capabilities. For instance, a cascaded forward neural network (CFNN) developed by a recent study modeled viscosity and thermal conductivity of Al_2O_3–CuO nanofluids, yielding correlation coefficients above 0.99 (MAPE < 1%)—highlighting its suitability for predictive integration in solar thermal systems. In another innovative study, an ANN-GA hybrid was successfully used to predict the optimal synthesis conditions of ZnO nanoparticles for photocatalytic use, predicting sizes as small as 5.3 nm with high agreement to experimental outcomes.

The scope of ML deployment has extended even to nanoscale simulations. A study combining molecular dynamics with decision tree and K-nearest neighbors models predicted thermophysical properties of iron-oxide nanofluids in CO_2 media, with the decision tree achieving a 99% accuracy rate across temperature and concentration ranges.

These works collectively show the potential of AI/ML models in reducing experimental loads, improving predictive capabilities, and enabling real-time optimization of nanofluid-based systems. However, most models are currently constrained to material or subsystem level. Future work should focus on:

- Developing system-level digital twins for HVAC&R applications that integrate nanofluid property predictions into full-cycle simulations.
- Applying reinforcement learning or advanced control-based AI models for real-time system optimization, considering dynamic environmental loads.
- Creating large-scale datasets through experimental or simulated studies to train generalizable AI models across multiple refrigerants and nanofluids.

This transition from property-level modeling to system-level deployment represents the next frontier in intelligent HVAC&R systems research.

To facilitate a rapid understanding of the methodological and performance-related trends across AI/ML-based investigations on nanofluids, a condensed summary is provided Table 10. The table categorizes studies based on nanoparticle type and highlights the key modeling techniques, performance parameters evaluated, and the principal findings. Abbreviations are employed to maintain brevity without compromising technical clarity. This synthesized overview enables researchers and practitioners to identify promising material–method combinations and guide future investigations toward system-level HVAC&R integration.

While Table 10 contains details study and AI methodology, to better illustrate the chronological development of artificial intelligence and machine learning applications in nanofluid heat transfer studies, a timeline diagram is presented in Figure 10. The diagram highlights the progressive adoption of various models—from NN and ANFIS to advanced hybrid techniques like QLM and BRNN—across prediction tasks such as thermal conductivity, irreversibility, viscosity, and performance metrics. The increasing model complexity reflects growing computational integration within thermal-fluid research.

Table 10. Summary of AI/ML-based studies grouped by nanoparticle type, including employed techniques, evaluated parameters, and condensed key findings.

Ref.	Type of Nanoparticle	Method/Technique	Parameter Evaluated	Key Findings
[148,177,179]	RE2TM2Y ternary intermetallic compounds (e.g., Gd, Tb, Dy, Ho, Er, Tm with Ni, Cu, Co and Sn, In, Cd, Ga, A	MOO using (NSGA-II), (GPR) (SVR) with RBF kernel, (SELM), (GSVR) with Gaussian and Polynomial kernels	Thermal and exergy COP, total product cost rate, Relative Cooling Power (RCP), applied magnetic field, ionic radii, RMSE, MAE, correlation coefficient (CC)	GU-GSVR vs. SELM (1–7 T field): RMSE ↓ 27.97% MAE ↓ 76.01% CC ↑ 10.55% R134a Refrigerant: Thermal COP: 10.13% Exergy COP: 0.6530% ↑ 8.5%/12.3% over base case R1234ze Refrigerant: Annual cost reduced to USD 6148 (↓ 2.4%) Optimal nanoparticle fraction: 0.0149–0.041 GPR-RBF Model: Highest predictive accuracy COP improved with 0.5–1.0 g TiO_2 Performance declined at 1.5 g TiO_2
[182–184]	Halloysite nanotubes (HNTs) in SAE 5W40	RSM, NSGA-II, MLP, ML algorithms, Cascaded Forward Neural Network (CFNN), GPR, ANN, SVM, DT	Dynamic viscosity, TC, density, specific heat capacity	Property Enhancement: Max TC increase: 23.24% at 1 vol.% and 60 °C Max viscosity increase: 21.4% at 1 vol.% and 30 °C Trend: TC and viscosity increase with nanoparticle concentration
[183,185–192]	Hybrid nanofluid (Al_2O_3–Cu–Water) in porous media	(ANN) with (PSO), (ANN) + (GA), (SVR), (RF); CFD-based simulation, SVR, MLP, RF (regressor models)	Temperature, time, NaOH concentration → Nanoparticle size, pressure, substrate thermal effusivity, Nusselt number, heat transfer efficiency, thermal performance factor, Reynolds number, permeability, nanoparticle volume fraction, entropy generation, shear stress	ML Prediction of CHF Trends: MLP: Best performance (mean R^2 ~87%) RF: High single-run R^2 (90%) but lower mean R^2 SVR: Lowest performance Top features: Pressure, nanoparticle size Substrate thermal effusivity: Negligible impact ML models outperformed Kandlikar's correlation Heat Transfer Enhancement: CNi inserts: 12.5–23.9% improvement CDNi inserts: Up to 32.6% improvement ML trained on 220 CFD + experimental data showed high prediction accuracy ZnO Nanoparticle Synthesis and Efficiency: Predicted size: 5.67 nm at 59 °C, 56 min, 0.08 M NaOH Experimental: 5.3 ± 0.4 nm Photocatalytic efficiency: 74% (vs. 58% for commercial ZnO) Flow and Entropy Analysis: Heat transfer ↑ with Re, permeability, and volume fraction Entropy shows nonmonotonic trend ANN-PSO provided accurate correlations for thermal parameters

Table 10. *Cont.*

Ref.	Type of Nanoparticle	Method/Technique	Parameter Evaluated	Key Findings
[148,189,193,194]	Tri-hybrid nanofluid (specific composition not explicitly stated); used as coolant under PV panels	(ANN) trained with Levenberg–(LMA-TNN); combined with Lobatto IIIa numerical solver, Bayesian (BRNN), (QLM), k (ANN) with hyperparameter optimization, combined with CFD using Keller Box Method (KBM)	Temperature, velocity profile, thermal radiation, magnetic field, Casson parameter, ANN regression accuracy, Reynolds number, Deborah number, MSE, regression, ST index, heat transfer rate, shear stress, magnetic field	High model accuracy with errors ranging from 10^{-7} to 10^{-11} ANN models reduced computational time and effectively captured flow/temperature behavior ANN and KBM achieved near-perfect regression ($R^2 \approx 1.0$) PG–water nanofluid showed comparable heat transfer to EG-based nanofluid Curvature and suction increased velocity and decreased temperature in flow regime BRNN addressed uncertainty and reduced overfitting efficiently Magnetic field and curvature enhanced overall heat transfer
[179,195,196]	TiO_2 nanoparticles in MO oil; LPG refrigerant	ANFIS with GP, SC, FCM clustering; WS optimization, Gene Expression Programming (GEP), and Adaptive Neuro–Fuzzy Inference System (ANFIS), Adaptive Neuro–Fuzzy Inference System (ANFIS) using Grid Partitioning and Subtractive Clustering	Thermal conductivity, viscosity, density, specific heat, temperature (20–60 °C), concentration (0–0.3%), specific heat capacity (SHC), Second-law efficiency, total irreversibility, RMSE, MAPE, variance	ANFIS (subtractive clustering): High accuracy (Var: 0.996–0.999, RMSE: 0.0296–0.1726 W, MAPE: 0.108–0.176%); outperformed ANN GP-ANFIS: Best for SHC (R = 0.99992, MAPE = 0.036%) and TC (R = 0.99833, MAPE = 0.218%) SC-ANFIS: Best for density and viscosity ($R \approx 0.9989$) Key sensitivities: Density_np (0.53%), VF (1.9%), SHC_np (1.6%), TC_np (1.7%) GEP vs. ANFIS: GEP more accurate (R > 0.9825, RMSE = 0.79, MAPE < 2.16%) than ANFIS (R > 0.96, RMSE = 1.50, MAPE < 2.93%)

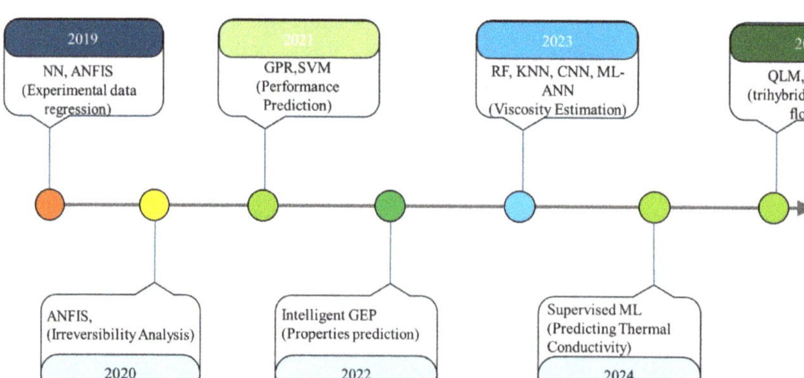

Figure 10. Evolutionary timeline of AI and ML applications in nanofluid-enhanced heat transfer (2019–2025).

2.3.5. Review-Based Papers

A growing body of the literature has systematically reviewed the role of nanofluids and nanolubricants in enhancing the performance of HVAC&R systems. These review studies provide consolidated insights into the thermophysical mechanisms, experimental advancements, modeling strategies, and emerging trends associated with the integration of nanoparticles into conventional systems. Unlike primary studies that focus on specific nanofluid formulations or experimental validations, review-based investigations offer a broader outlook across various nanomaterials, system configurations, and operating conditions.

Several reviews have focused on vapor compression refrigeration (VCR) systems using nanorefrigerants. Bhattad et al. (2018) [197] provided one of the most comprehensive evaluations of nanofluids in refrigeration, identifying mechanisms such as reduced viscosity, improved thermal conductivity, and nanoparticle dispersion stability as key drivers of performance gains. Bilen et al. (2022) [198] and Kumar et al. (2023) [199] expanded upon this by evaluating system-level impacts, such as enhancement in COP, compressor work reduction, and energy savings across different refrigerant–nanoparticle combinations.

In the domain of vapor absorption systems, Pisal et al. (2024) [200], Jiang et al. (2022) [201], and Talpada and Ramana (2021) [202] examined the effects of adding nanoparticles to absorbents and working fluids. These studies underscore that metal oxide-based nanofluids (e.g., Al_2O_3, TiO_2, ZnO) significantly enhance heat and mass transfer in absorber and generator sections, while stability and sedimentation remain limiting factors.

Another critical theme is the use of nanolubricants in compressors, as discussed by Marcucci Pico et al. (2023) [203]. Their review highlights the synergistic effect of nanoparticles (e.g., CuO, MoS2) in reducing friction, enhancing lubrication, and mitigating compressor wear. Azmi et al. (2017) [204] and Patil et al. (2015) [205] similarly noted the dual benefit of nanolubricants in both improving energy efficiency and extending component life.

From a methodological standpoint, reviews by Ponticorvo et al. (2022) [206] and R et al. (2021) [207] provide crucial insights into nanoparticle dispersion stability, agglomeration prevention techniques, and the tribological behavior of nanorefrigerants. Their analyses are pivotal in identifying research gaps around surfactant compatibility, long-term stability, and nano-additive lifecycle effects.

Further, Du et al. (2018) [208] offered a broader scope by evaluating phase change materials (PCMs) integrated with nanostructures for cooling, heating, and power generation. Although not confined to refrigeration cycles alone, their findings have relevance in cold storage and building thermal management systems.

A few reviews, such as those by Kundan and Singh (2020) [209] and Valiyandi and Thampi (2023) [210], explored hybrid systems and alternative working fluids. Their discussions provide a forward-looking perspective on multi-component working pairs, hybrid nanofluids, and eco-friendly refrigerants for sustainable development.

Overall, the review-based literature plays an indispensable role in consolidating evidence, identifying optimal nanoparticle–base fluid combinations, addressing stability concerns, and shaping future experimental and computational research directions in nanofluid-enhanced HVAC&R systems.

2.3.6. Critical Analysis and Future Directions

The body of research conducted between 2015 and 2025 has successfully established nanofluids as a promising technology for enhancing the performance of HVAC&R systems. The reported improvements in heat transfer and the coefficient of performance (COP) are substantial. However, a critical evaluation of the existing literature reveals significant imbalances and unresolved challenges that currently impede the transition from laboratory-scale validation to widespread commercial deployment. This section provides a critical analysis of the current research landscape and proposes a strategic roadmap for future investigations.

A Critical Appraisal of the Current Research Landscape

A systematic review of the research methodologies, application scope, and analytical depth highlights five key areas of concern:

- Methodological Imbalance and Limited Synergy: The field is disproportionately reliant on experimental investigations. While empirical data is indispensable, the pronounced scarcity of hybrid studies—which integrate experimental work with validated numerical modeling—suggests a prevailing trial-and-error approach. This underutilization of synergistic methodologies means the predictive and optimization power of computational tools is not being fully leveraged to accelerate discovery and reduce research overhead.
- Asymmetrical Application Focus: There is a distinct and substantial focus on refrigeration systems at the expense of broader HVAC applications. This neglects the significant energy-saving potential in areas such as commercial chillers, large-scale heat pumps, and data center cooling, which constitute a major portion of global energy consumption for thermal management. This represents a strategic misalignment of research effort with the sectors promising the greatest potential impact.
- The Unresolved Challenge of Long-Term Viability: A critical deficiency in the current literature is the predominant focus on short-term performance metrics. Issues of paramount importance for practical implementation—such as long-duration fluid stability, nanoparticle agglomeration and sedimentation, potential for erosion and corrosion of system components, and material compatibility—remain largely unaddressed. Without robust, long-term operational data, the remarkable performance enhancements reported in laboratory settings remain academically compelling but commercially unproven.
- Deficiency in Economic and Lifecycle Analysis: The research convincingly demonstrates that nanofluids can improve thermodynamic efficiency, but it largely fails to address whether this improvement is economically viable or environmentally beneficial over the full lifecycle. The high initial cost of nanomaterials and the energy intensity of fluid preparation processes may offset the economic and environmental gains from operational energy savings. A comprehensive technoeconomic analysis (TEA) and lifecycle assessment (LCA) are conspicuously absent from most studies.
- Nascent Integration of AI and Machine Learning: While the application of AI/ML is a promising development, its current use is rudimentary. Existing models are almost exclusively focused on predicting the thermophysical properties of the nanofluids themselves or the performance of isolated components. A critical gap exists between this foundational, property-level prediction and the development of holistic, system-level models capable of dynamic optimization and control.

3. Conclusions and Future Directions

This review has systematically synthesized and critically evaluated the body of research on nanofluid applications in HVAC&R systems published between 2015 and 2025. The findings unequivocally confirm that nanofluids offer significant potential to enhance thermal performance, with numerous studies demonstrating substantial improvements in heat transfer and system efficiency. However, despite two decades of research, the transition from academic promise to widespread commercial adoption is impeded by several persistent and critical challenges. The current research landscape is characterized by a methodological imbalance favoring experimental studies; a narrow application focus on refrigeration systems; and a general deficiency in the long-term, economic, and system-level analyses required for industrial confidence.

To accelerate progress and mature this technology, future research must adopt a more strategic and integrated approach. Based on the critical analysis conducted in this review, the following directions are proposed as a roadmap for the field:

1. Prioritize Long-Term, System-Level Validation: The foremost priority must be to shift from short-term, component-based experiments to long-duration studies on integrated, pilot-scale systems. For example, this involves running a nanofluid-enhanced chiller for a full cooling season (>2000 h) while continuously monitoring for performance degradation, changes in pressure drop, filter clogging, and signs of component erosion. This is the only way to generate conclusive data on fluid stability, material compatibility, and sustained performance under realistic operational cycling.
2. Broaden the Application Scope to High-Impact HVAC Systems: Research efforts should be deliberately expanded into under-investigated, high-impact areas where the potential for energy savings is greatest. Specific examples include investigating nanorefrigerants in large-scale commercial chillers for office buildings, employing nanolubricants in the compressors of multi-zone heat pumps, and using nanofluids in the secondary loops of data center cooling systems.

While the thermodynamic benefits of these applications are promising, comprehensive evaluations of their cost-effectiveness are often lacking. Specifically, few studies provide quantifiable comparisons between the upfront cost premium of nanofluids and the achievable long-term energy savings. As a result, direct assessments of return on investment or payback periods remain limited. Future work should aim to bridge this gap by integrating performance testing with technoeconomic metrics to better support real-world decision-making and adoption.

3. Promote Synergistic Hybrid Methodologies: To move beyond trial-and-error, hybrid studies that couple experimental work with validated numerical simulations (CFD) and theoretical models must become the standard. A practical workflow would involve using CFD to computationally screen and optimize the geometry of a microchannel heat exchanger for a specific nanofluid before fabricating and experimentally testing only the most promising designs. This synergy will enable more rapid and cost-effective optimization.
4. Advance from Property Prediction to System-Level AI/ML: The research community must leverage AI and machine learning to develop dynamic "digital twins" of entire HVAC&R systems. For instance, an AI model could be trained to predict the onset of nanoparticle agglomeration using real-time sensor data (temperature, pressure, flow rate) and automatically adjust system operation to maintain stability, or forecast the remaining useful life of the fluid for predictive maintenance.

However, while data-driven AI/ML models have shown high predictive accuracy, their performance is highly dependent on the quality and representativeness of training data. These models may be vulnerable to extrapolation errors, especially under dynamic or extreme operating conditions such as low-temperature gradients or variable loads. To address this, future research should integrate physics-informed machine learning (PIML) frameworks that embed governing equations, boundary conditions, and thermodynamic constraints into model architectures. Such hybrid approaches can enhance generalizability, improve interpretability, and enable robust predictions with limited data—marking a critical step toward trustworthy AI-assisted nanofluid system design.

5. Mandate Technoeconomic and Lifecycle Assessments: To establish commercial viability, rigorous technoeconomic analysis (TEA) and lifecycle assessment (LCA) must be integrated into research projects. This means quantifying whether the operational energy savings from a 15% COP improvement can justify the initial cost of graphene-based nanoparticles and the energy footprint of their synthesis over a 10-year system lifespan.
6. Foster Innovation in Material Science for Inherent Stability: Foundational research into novel nanomaterials is critical to overcoming the core challenge of stability. This includes developing advanced materials like core-shell nanoparticles, where a highly conductive core (e.g., copper) is coated with a chemically inert shell (e.g., silica) to prevent oxidation and improve long-term dispersion in water–glycol mixtures without relying on potentially degradable surfactant additives.

By systematically addressing these strategic imperatives, the scientific community can guide the evolution of nanofluid technology from a promising concept into a validated, economically sound, and impactful solution for enhancing energy efficiency in thermal management systems worldwide.

Funding: This research is funded by The National Science Foundation, grant Number 2200515 and Tennessee State University, MSIPP Seed Grant Awards 5-1-25.

Conflicts of Interest: The authors declare no conflict of interest.

Abbreviations

The following abbreviations are used in this manuscript:

AI	Artificial Intelligence
ANFIS	Adaptive Neuro–Fuzzy Inference System
ANN	Artificial Neural Network
BRNN	Bayesian Regression Neural Network
CFNN	Cascaded Forward Neural Network
CNN	Convolutional Neural Network
COP	Coefficient of Performance
conc	Concentration
Da	Darcy Number
DT	Decision Tree
Ec	Eckert Number
EG	Ethylene Glycol
ET	Extra Trees Algorithm
Fr	Forchheimer Number
GA	Genetic Algorithm
GEP	Gene Expression Programming
GPR	Gaussian Process Regression
Ha	Hartmann number
HMIM	1-hexyl-3-methylimidazolium
HT	Heat Transfer
HTC	Heat Transfer Coefficient
HVAC&R	Heating, Ventilation, Air Conditioning, and Refrigeration
KBM	Keller Box Method
LCA	Lifecycle Analysis
LGBM	Light Gradient-Boosting Machine
LINMAP	Linear Programming Technique for Multidimensional Analysis of Preference
LMA	Levenberg–Marquardt Algorithm

LMA-TNN	Levenberg–Marquardt Algorithm Trained Neural Network
LMFA NN	Local Mean Field Approximation Neural Network
LSTM	Long Short-Term Memory
MAPE	Mean Absolute Percentage Error
ML	Machine Learning
MLP	Multi-Layer Perceptron
MO	Mineral Oil
MOD	Mean of Deviation
np	Nanoparticle
NSGA-II	Non-dominated Sorting Genetic Algorithm II
PCM	Phase Change Material
Pe	Peclet Number
PG	Propylene Glycol
PVP	Polyvinylpyrrolidone
QLM	Quasi-Linearization Method
RCP	Relative Cooling Power
Rd	Radiation parameter
RF	Random Forest
Ri	Richardson Number
RMSE	Root Mean Square Error
RNN	Recurrent Neural Network
RSM	Response Surface Methodology
Sc	Schmidt Number
SC	Subtractive Clustering
SDS	Sodium Dodecyl Sulfate
SHC	Specific Heat Capacity
SVM	Support Vector Machine
SWCNT	Single-Wall Carbon Nanotube
TC	Thermal Conductivity
TDMA	Tri-Diagonal Matrix Algorithm
THNF	Trihybrid nanofluid
VF	Volume Fraction
VST	Dynamic Viscosity
WF	Weight Fraction
XGB	Extreme Gradient Boosting
ZnO	Zinc Oxide
↑	Increase
↓	Decrease
>	Greater or outperform
<	Less than

References

1. International Energy Agency. Cooling. IEA. 2023. Available online: https://www.iea.org/energy-system/buildings/cooling (accessed on 15 May 2025).
2. Kasaeian, A.; Nasrin, R.; Vatan, S. A review on solar-assisted heat pump systems using nanofluid. *Renew. Sustain. Energy Rev.* **2015**, *46*, 113–124. [CrossRef]
3. Saidur, R.; Leong, K.Y.; Mohammad, H.A. A review on applications and challenges of nanofluids. *Renew. Sustain. Energy Rev.* **2011**, *15*, 1646–1668. [CrossRef]
4. Wang, Y.; Yu, J.; Qi, C.; Zhang, W. Review on Coupled Thermo-Hydraulic Performance of Nanofluids and Microchannels. *Nanomaterials* **2022**, *12*, 3979. [CrossRef]
5. Das, S.K.; Choi, S.U.; Yu, W.; Pradeep, T. *Nanofluids: Science and Technology*; John Wiley & Sons: Hoboken, NJ, USA, 2007.
6. Mahian, O.; Kianifar, A.; Sahin, A.Z.; Wongwises, S. A review of the applications of nanofluids in solar energy. *Int. J. Heat Mass Transf.* **2019**, *139*, 1205–1224. [CrossRef]

7. Zhang, A.Q.; Wang, L.M.; Li, H.B. Critical review on the stability and thermal conductivity of water-based hybrid nanofluids fo heat transfer applications. *RSC Adv.* **2025**, *15*, 14088–14125. [CrossRef]
8. Riyadi, T.; Herawan, S.; Tirta, A.; Ee, Y.; Hananto, A.; Paristiawan, P.; Yusuf, A.; Venu, H.; Veza, I. Nanofluid heat transfer anc machine learning: Insightful review of machine learning for nanofluid heat transfer enhancement in porous media and hea exchangers as sustainable and renewable energy solutions. *Results Eng.* **2024**, *24*, 103002. [CrossRef]
9. Basu, A.; Saha, A.; Banerjee, S.; Roy, P.; Kundu, B. A Review of Artificial Intelligence Methods in Predicting Thermophysica Properties of Nanofluids for Heat Transfer Applications. *Energies* **2024**, *17*, 1351. [CrossRef]
10. Raza, M.; Faiz, M.; Hassan, W.; Abbas, M.; Raza, J.; Kumail, Z.; Nawaz, T.; Shabir, S.; Jan, A.; Li, F. AI-powered optimization anc numerical techniques for nanofluid heat transfer systems-a review. *Multiscale Multidiscip. Model. Exp. Des.* **2025**, *8*, 316. [CrossRef
11. Smrity, A.; Yin, P. Bridging computational and experimental boundaries: A review of theoretical modeling and experimenta validation of hybrid nanofluids in heat transfer applications. *Int. J. Heat Fluid Flow* **2025**, *115*, 109873. [CrossRef]
12. Khalatbari, S.; Jalili, P.; Jalili, B.; Ganji, D. Investigating the improvement of heat transfer and flow characteristics of hybric nanofluids: A comprehensive review. *Proc. Inst. Mech. Eng. Part E J. Process Mech. Eng.* **2025**. [CrossRef]
13. Ali, D.; Ullah, H.; Fiza, M.; Jan, A.; Akgül, A.; Hendy, A.; Islam, S. Investigating slip velocity effects on thermal and mass transpor in magnetized nanoparticle squeeze flow via numerical scheme. *Proc. Inst. Mech. Eng. Part N J. Nanomater. Nanoeng. Nanosyst* **2025**. [CrossRef]
14. Gao, Y.; Jin, Y.; Xu, M.; Liaw, K.L.; Zhang, K.; Mohit, M.; Kurnia, J.; Sasmito, A. Thermal characteristics of nanofluid ice slurry flowing through a spiral tube: A computational study. *Case Stud. Therm. Eng.* **2024**, *61*, 104882. [CrossRef]
15. Lakshmi, B.; Dharmaiah, G.; Anjum, A.; Samdani; Naheed, M. Influence of Darcy–Forchheimer Fe_3O_4—$CoFe_2O_4$—H_2O hybrid nanofluid flow with magnetohydrodynamic and viscous dissipation effects past a permeable stretching sheet: A numerical contribution. *Multiscale Multidiscip. Model. Exp. Des.* **2024**, *8*, 115. [CrossRef]
16. Su, Z.; Zhao, N.; Li, X.; Liu, J.; Yang, L. Experimental and numerical investigation on heat transfer mechanism and multifunctional thermodynamic properties of supercritical carbon dioxide/amino-functionalized graphene nanofluids. *Int. J. Heat Mass Transf* **2025**, *241*, 126702. [CrossRef]
17. Bibin, B.S.; Gundabattini, E. Investigation on transport properties, heat transfer characteristics and pressure drop of CuO enhanced R1234yf based refrigerant. *Case Stud. Therm. Eng.* **2023**, *49*, 103229. [CrossRef]
18. Satheeshkumar, P.; Selwynraj, A.I. Pathways to enhance the performance of adsorption cooling system: An overview. *Results Eng* **2024**, *23*, 102855. [CrossRef]
19. Alzahrani, J. A three-dimensional unsteady flow of Casson nanofluid with suspension of microorganisms with variable thermal conductivity: A modified Fourier theory approach. *Case Stud. Therm. Eng.* **2024**, *63*, 105218. [CrossRef]
20. Bibin, B.S.; Gundabattini, E. Investigation on the density of Al_2O_3/R1234yf, TiO_2/R1234yf and CuO/R1234yf nano-refrigerants *Proc. Inst. Mech. Eng. Part N J. Nanomater. Nanoeng. Nanosyst.* **2022**, *237*, 121–129. [CrossRef]
21. Ahmed, Z.; Bhargav, A. Effect of Aggregation Morphology on Thermal Conductivity and Viscosity of Al_2O_3-CO_2 Nanofluid: A Molecular Dynamics Approach. *Nanosci. Technol. Int. J.* **2021**, *12*, 19–37. [CrossRef]
22. Zawawi, N.; Azmi, W.; Redhwan, A.; Sharif, M.; Samykano, M. Experimental investigation on thermo-physical properties of metal oxide composite nanolubricants. *Int. J. Refrig.* **2018**, *89*, 11–21. [CrossRef]
23. Elboughdiri, N.; Fatima, N.; El-Rahman, M.; Abbas, M.; Rashid, R.; Shomurotova, S.; Said, M.; Mahariq, I.; SHendy, A.; MGalal, A Characteristics of unsteady thermo-bioconvection chemical reactive stagnation point flow of trihybrid nanofluid around rotating sphere with Oxytactic Microorganisms. *Case Stud. Therm. Eng.* **2024**, *61*, 104981. [CrossRef]
24. Yıldız, G.; Ağbulut, Ü.; Gürel, A. A review of stability, thermophysical properties and impact of using nanofluids on the performance of refrigeration systems. *Int. J. Refrig.* **2021**, *129*, 342–364. [CrossRef]
25. Sarac, B.; Soprunyuk, V.; Herwig, G.; Gümrükçü, S.; Kaplan, E.; Yüce, E.; Schranz, W.; Eckert, J.; Boesel, L.; Sarac, A. Thermomechanical properties of confined magnetic nanoparticles in electrospun polyacrylonitrile nanofiber matrix exposed to a magnetic environment: Structure, morphology, and stabilization (cyclization). *Nanoscale Adv.* **2024**, *6*, 6184–6195. [CrossRef]
26. Mehyo, M.; Özcan, H. Thermophysical Properties of Nanoferrofluid (Fe_3O_4–Acetone/$Znbr_2$) as a Working Fluid for Use in Absorption Refrigeration Applications. *Int. J. Thermodyn.* **2021**, *24*, 103–109. [CrossRef]
27. Sica, L.; Contreras, E.; Bandarra Filho, E.; Parise, J. An experimental viscosity investigation on the use of non-Newtonian graphene heat transfer nanofluids at below-ambient temperatures. *Int. J. Energy Res.* **2021**, *45*, 14530–14546. [CrossRef]
28. Jiang, W.; Du, K.; Li, Y.; Yang, L. Experimental investigation on the influence of high temperature on viscosity, thermal conductivity and absorbance of ammonia–water nanofluids. *Int. J. Refrig.* **2017**, *82*, 189–198. [CrossRef]
29. Abbas, T. Numerical simulation of heat transfer in williamson hybrid nanofluid (MgO−Ag/water) flow with viscous dissipation effects. *Multiscale Multidiscip. Model. Exp. Des.* **2025**, *8*, 293. [CrossRef]
30. Mohammed, H.; Giddings, D.; Walker, G.; Power, H. CFD assessment of the effect of nanoparticles on the heat transfer properties of acetone/ZnBr2 solution. *Appl. Therm. Eng.* **2018**, *128*, 264–273. [CrossRef]

31. Said, Z.; Rahman, S.; Sohail, M.; Bibin, B.S. Analysis of thermophysical properties and performance of nanorefrigerants and nanolubricant-refrigerant mixtures in refrigeration systems. *Case Stud. Therm. Eng.* **2023**, *49*, 103274. [CrossRef]
32. Said, Z.; Rahman, S.; Sohail, M.; Bahman, A.; Alim, M.; Shaik, S.; Radwan, A.; El-Sharkawy, I. Nano-refrigerants and nano-lubricants in refrigeration: Synthesis, mechanisms, applications, and challenges. *Appl. Therm. Eng.* **2023**, *233*, 121211. [CrossRef]
33. Zhang, X.; Yang, Z.; Wang, Y. Enhanced flow boiling heat transfer study on EDM-modified aluminum surface. *Appl. Therm. Eng.* **2025**, *274*, 126642. [CrossRef]
34. El Hadi Attia, M.; Harby, K.; Amin, M.; Abdelgaied, M. Evaluating honeycomb-arranged hollow copper tubes and nanoparticle-coated flax fibers for improved hemispherical solar still performance. *Process Saf. Environ. Prot.* **2025**, *197*, 106986. [CrossRef]
35. Abdullah, A.; Alqsair, U.; Alotaibi, F.; Alrwuais, F.; Omara, Z.; Essa, F. Improving thermal performance of half-cylindrical solar stills with convex/corrugated absorbers, wick materials, reflector, and nano-enhanced phase change materials. *J. Energy Storage* **2024**, *100*, 113462. [CrossRef]
36. Jadhav, A.; Chavan, S.; Sutar, K.; Chendake, Y.; Shaikh, I.; Sihna, M. Experimental investigation on climate-control nano-composite polycarbonate sheets enhanced with TiO_2, ZnO, and zeolite nanoparticles for polyhouse use. *Eng. Res. Express* **2024**, *6*, 045553. [CrossRef]
37. Liu, H.; Dong, C.; Xiong, P.; Nan, H.; Yan, D.; Li, R.; An, Y.; Lei, J.; Li, Z. Triple Thermal Management Wood–Plastic Composite Plate with Excellent Waterproof and Mechanical Properties. *Ind. Eng. Chem. Res.* **2024**, *64*, 453–465. [CrossRef]
38. Cuce, P.; Cuce, E.; Guclu, T.; Shaik, S.; Alshahrani, S.; Saleel, C. Effect of using hybrid nanofluids as a coolant on the thermal performance of portable thermoelectric refrigerators. *Sustain. Energy Technol. Assess.* **2022**, *53*, 102685. [CrossRef]
39. Wiriyasart, S.; Suksusron, P.; Hommalee, C.; Siricharoenpanich, A.; Naphon, P. Heat transfer enhancement of thermoelectric cooling module with nanofluid and ferrofluid as base fluids. *Case Stud. Therm. Eng.* **2021**, *24*, 100877. [CrossRef]
40. Ali, A.; Salam, B. A review on nanofluid: Preparation, stability, thermophysical properties, heat transfer characteristics and application. *SN Appl. Sci.* **2020**, *2*, 1636. [CrossRef]
41. Sanukrishna, S.S.; Murukan, M.; Reshmi Krishnan, S. Effect of nanoparticles on the flow boiling heat transfer characteristics of refrigerant and performance of refrigeration systems: An experimental investigation. *Int. J. Thermofluids* **2025**, *27*, 101263. [CrossRef]
42. Gorfie, E.; Moltot, A.; Zergaw, G.; Dessie, H. Influence of Joule Heating and Nonlinear Thermal Radiation on the Electrical Conductivity of Second-Grade Hybrid Nanofluid Flow Over a Stretching Cylinder. *Eng. Rep.* **2025**, *7*, e70129. [CrossRef]
43. Muhammad, U.; Bhattacharyya, D.; Endrino, J.; Fereres, S. The effects of ejector adiabatic absorber on heat and mass transfer of binary nanofluid with heat transfer additives. *Emergent Mater.* **2021**, *4*, 1665–1678. [CrossRef]
44. Mahbubul, I.; Khan, M.; Ibrahim, N.; Ali, H.; Al-Sulaiman, F.; Saidur, R. Carbon nanotube nanofluid in enhancing the efficiency of evacuated tube solar collector. *Renew. Energy* **2018**, *121*, 36–44. [CrossRef]
45. Mahbubul, I.; Saadah, A.; Saidur, R.; Khairul, M.; Kamyar, A. Thermal performance analysis of Al_2O_3/R-134a nanorefrigerant. *Int. J. Heat Mass Transf.* **2015**, *85*, 1034–1040. [CrossRef]
46. Ye, H.; Lum, L.; Kandasamy, R.; Zhao, H.; Ho, J. Flow boiling heat transfer enhancement of R134a in additively manufactured minichannels with microengineered surfaces. *Appl. Therm. Eng.* **2024**, *256*, 124150. [CrossRef]
47. Kilic, G. Performance Evaluation of Triply Periodic Minimal Surface Heat Exchangers Using Nanofluids at High Flow Rates for Enhanced Energy Efficiency. *Appl. Sci.* **2025**, *15*, 4140. [CrossRef]
48. Khoshvaght-Aliabadi, M.; Alizadeh, A. An experimental study of Cu–water nanofluid flow inside serpentine tubes with variable straight-section lengths. *Exp. Therm. Fluid Sci.* **2015**, *61*, 1–11. [CrossRef]
49. Bagum, M.; Hossen, M.; Chowdhury, F.; Hoque, S. Influence of lightly doped Ca on the structural, magnetic and magnetocaloric properties of $LaCaMnO_3$ nanoparticles. *Mater. Charact.* **2025**, *225*, 115153. [CrossRef]
50. Khan, M.; Lu, D.; Imran, M.; Anwar, M. Numerical insights into enhanced heat transfer mechanisms in TiO_2-Au ethylene glycol nanofluids within Darcy porous media using fractional calculus. *Case Stud. Therm. Eng.* **2025**, *70*, 106063. [CrossRef]
51. Meng, F.; Zhou, L.; Song, C.; Zhang, J. Numerical study of thermoelectric coolers for cooled infrared detectors using microchannel heat sinks and nanofluids. *Appl. Therm. Eng.* **2025**, *266*, 125457. [CrossRef]
52. Raju, S.; Durgaprasad, P.; Díaz Palencia, J.; Wakif, A.; Raju, C.; Dinesh kumar, M.; Muneerah, A.; Ali, J. Contour Analysis for Heat Transfer Rate in a Wedge Geometry with Non-Uniform Shapes Nanofluid: Gradient Descent Machine Learning Technique. *Results Eng.* **2024**, *23*, 102714. [CrossRef]
53. Balogun, O.; Aslam, M.; Iqbal, M.; Arshad, M. Investigation of gold and silver based hybrid nanofluid with effects of thermal radiation over a stretching sheet. *Alex. Eng. J.* **2025**, *117*, 352–363. [CrossRef]
54. Bompally, R.; Gurijala, R.; Reddy, R.; Thumma, T. Effect of inclined magnetic field, non-uniform heat source on hybrid EG-MoS_2-SiO_2 radiative nanofluid flow with viscous and Joule dissipation over convectively heated elongating surface. *Multiscale Multidiscip. Model. Exp. Des.* **2025**, *8*, 292. [CrossRef]
55. Liang, Y.; Lin, Z.; Luo, X.; Chen, J.; Yang, Z.; Wu, R.; Chen, Y. Bi-objective optimization of thermal interface material-microchannel-nanofluid-based thermal management system design for microchip array. *Appl. Therm. Eng.* **2025**, *274*, 126727. [CrossRef]

56. Kumar, M.; Chandel, S.; Shukla, P.; Sinha-Ray, S. Thermal Management of Microelectronics Using Microchannel Heat Sink wit Inclined Geometry and Cnt-Based Nanofluid. *J. Enhanc. Heat Transf.* **2025**, *32*, 37–60. [CrossRef]
57. Sreelakshmi, K.; Mary, G.; Alqsair, U.; Elsemary, I.; Alsayegh, R.; Khan, S.; Kolsi, L. Evaluation of heat transfer for unstead thin film flow of mono and hybrid nanomaterials with five different shape features. *Case Stud. Therm. Eng.* **2024**, *62*, 10516 [CrossRef]
58. Benkherbache, S.; Amroune, S.; Belaadi, A.; Zergane, S.; Farsi, C. Numerical Analysis of Natural Convection in an Annular Cavit Filled with Hybrid Nanofluids under Magnetic Field. *Energies* **2024**, *17*, 4671. [CrossRef]
59. Farooq, U.; Imran, M.; Noreen, S.; Fatima, N.; Muhammad, T. Heat Transfer Performance of Hybrid Nanofluid Radiative Flow vi a Rotating Disk With Heat Source–Sink Effects and Response Surface Methodology. *Math. Methods Appl. Sci.* **2025**. [CrossRef]
60. Elboughdiri, N.; Abbas, M.; Ayed, H.; Mouldi, A.; Fatima, N.; Abduvalieva, D.; Said, M.; Mahariq, I.; Hendy, A.; Galal, A Numerical simulation of Stephan blowing impacts on thermally laminated 3D flow of MHD trihybrid nanofluid with Soret an Dufour effects. *Case Stud. Therm. Eng.* **2025**, *66*, 105460. [CrossRef]
61. Shinwari, W.; Hayat, T.; Abbas, Z.; Momani, S. Trihybrid fluid flow with Arrhenius activation energy and slip conditions i porous space: A numerical analysis. *Alex. Eng. J.* **2025**, *117*, 534–544. [CrossRef]
62. Shree, S.; Hanumagowda, B.; Saini, G.; Singh, K.; Kulshreshta, A.; Varma, S.; Punith Gowda, R. Heat and mass transfer i electrically conducting hybrid nanofluid flow between two rotating parallel stretching surfaces: An entropy analysis. *Multiscal Multidiscip. Model. Exp. Des.* **2024**, *8*, 18. [CrossRef]
63. Hema, S.; Venkatesh, P.; Gireesha, B.; Pavithra, C. Flow and heat transfer analysis of MHD ternary hybrid nanofluid flow throug a vertical porous microchannel with slip boundary conditions. *Multiscale Multidiscip. Model. Exp. Des.* **2025**, *8*, 183. [CrossRef]
64. Scarpa, F.; Bianco, V. Improving the performance of room temperature magnetic regenerators using Al_2O_3-water nanofluid. *App Therm. Eng.* **2024**, *236*, 121711. [CrossRef]
65. Mohammed, H.; Giddings, D.; Walker, G. CFD simulation of a concentrated salt nanofluid flow boiling in a rectangular tube. *In J. Heat Mass Transf.* **2018**, *125*, 218–228. [CrossRef]
66. Gao, H.; Mao, F.; Song, Y.; Hong, J.; Yan, Y. Effect of adding copper oxide nanoparticles on the mass/heat transfer in falling film absorption. *Appl. Therm. Eng.* **2020**, *181*, 115937. [CrossRef]
67. Aprea, C.; Greco, A.; Maiorino, A.; Masselli, C. Enhancing the Heat Transfer in an Active Barocaloric Cooling System Usin Ethylene-Glycol Based Nanofluids as Secondary Medium. *Energies* **2019**, *12*, 2902. [CrossRef]
68. Hassan, M.; Teamah, M.; El-Maghlany, W. Numerical investigation for heat transfer enhancement using nanofluids over ribbe confined one-end closed flat-plate. *Alex. Eng. J.* **2017**, *56*, 333–343. [CrossRef]
69. Saee, A.; Baghban, A.; Zarei, F.; Zhang, Z.; Habibzadeh, S. ANFIS based evolutionary concept for estimating nucleate pool boilin heat transfer of refrigerant-ester oil containing nanoparticles. *Int. J. Refrig.* **2018**, *96*, 38–49. [CrossRef]
70. Zolghadri, A.; Maddah, H.; Ahmadi, M.; Sharifpur, M. Predicting Parameters of Heat Transfer in a Shell and Tube Heat Exchange Using Aluminum Oxide Nanofluid with Artificial Neural Network (ANN) and Self-Organizing Map (SOM). *Sustainability* **202** *13*, 8824. [CrossRef]
71. Kanti, P.; Wanatasanappan, V.; Said, N.; Saini, S.; Mishra, V.; Paramasivam, P.; Yusuf, M. Thermal performance, entropy generatio and machine learning insights of Al_2O_3-TiO_2 hybrid nanofluids in turbulent flow. *Sci. Rep.* **2025**, *15*, 16035. [CrossRef]
72. Ullah, B.; Bashir, H.; Ahmed, N.; Khan, U.; Kumar, A.; Garayev, M. Enhancing heat transfer in nanofluids: Exploring nanoparticl aggregation using deep learning optimization. *Mod. Phys. Lett. B* **2025**, *39*, 2550195. [CrossRef]
73. AlMohamadi, H.; Rubbab, Q.; Garalleh, H.; Atta, G.; Amjad, M.; Jamshed, W.; ElSeabee, F.; Bayram, M. Artificial intelligence an numerical simulation based assessment of trihybrid structured flow over a curved geometry: Thermalized case analysis. *Result Eng.* **2025**, *25*, 103829. [CrossRef]
74. Qureshi, H.; Pasha, A.; Shah, Z.; Zahoor Raja, M.; Algarni, S.; Alqahtani, T.; Irshad, K.; Khan, W. Application of machine learnin for thermal exchange of dissipative ternary nanofluid over a stretchable wavy cylinder with thermal slip. *Case Stud. Therm. Eng* **2024**, *60*, 104599. [CrossRef]
75. Hojjat, M. Nanofluids as coolant in a shell and tube heat exchanger: ANN modeling and multi-objective optimization. *App Math. Comput.* **2020**, *365*, 124710. [CrossRef]
76. Vahdat Azad, A.; Vahdat Azad, N. Application of nanofluids for the optimal design of shell and tube heat exchangers usin genetic algorithm. *Case Stud. Therm. Eng.* **2016**, *8*, 198–206. [CrossRef]
77. Jyothi, K.; Venkateswarlu, B.; Reddy, P.; Kodi, R.; Annapureddy, D. Neural network-driven analysis of MHD boundary layer flow and heat transfer in Sisko nanofluids. *Multiscale Multidiscip. Model. Exp. Des.* **2025**, *8*, 291. [CrossRef]
78. Zhang, Y.; Xu, X. Predicting the thermal conductivity enhancement of nanofluids using computational intelligence. *Phys. Lett. A* **2020**, *384*, 126500. [CrossRef]
79. Alipour Bonab, S.; Yazdani-Asrami, M. Machine learning-based model for the intelligent estimation of critical heat flux i nanofluids. *Nano Express* **2024**, *5*, 025012. [CrossRef]

80. Khan, A.; ul Haque, M. Thermophysical Evaluation of Silica Nanoparticles in Refrigerant Compressor Oil for Enhanced Refrigeration Performance. *Int. J. Heat Technol.* **2025**, *43*, 172–180. [CrossRef]
81. Kumar, N.; Kumar, P.; Goyal, K. A novel approach of using TiO_2 and SiO_2 nanoparticles and hydrocarbon refrigerant R600a in retrofitted vapour compression refrigeration system for environment protection and system performance enhancement. *Appl. Nanosci.* **2023**, *13*, 6299–6317. [CrossRef]
82. Cassim, S.; Hairuddin, A.; Cut, V. The effects of Multiwalled Carbon Nanotube to the performance of a refrigeration system. *Proc. Inst. Mech. Eng. Part C J. Mech. Eng. Sci.* **2024**, *238*, 6691–6705. [CrossRef]
83. Dağıdır, K.; Bilen, K. Experimental investigation of usage of POE lubricants with Al_2O_3, graphene or CNT nanoparticles in a refrigeration compressor. *Beilstein J. Nanotechnol.* **2023**, *14*, 1041–1058. [CrossRef] [PubMed]
84. Ahsan, M.; Chand, P.; Namrata, K. Comparative Experimental Investigation to Analyze the Use of SiO_2 Nanoparticles with R600a and Pure R600a in Vapour Compression Refrigeration System. *Recent Pat. Mech. Eng.* **2025**, *18*. [CrossRef]
85. Hariram, V.; Al Riyami, H.; Nadanakumar, V.; Godwin John, J.; Christu Paul, R.; Nakandhrakumar, R.; Rameshkumar, P. Microstructure and Structural Comparison between Additively Manufactured Metal and Ceramics. *Int. J. Veh. Struct. Syst.* **2022**, *14*. [CrossRef]
86. Karthick, M.; Karuppiah, S.; Kanthan, V. Performance investigation and exergy analysis of vapor compression refrigeration system operated using R600a refrigerant and nanoadditive compressor oil. *Therm. Sci.* **2020**, *24*, 2977–2989. [CrossRef]
87. Yusof, T.; Arshad, A.; Suziyana, M.; Chui, L.; Basrawi, M. Experimental Study of a Domestic Refrigerator with POE-Al_2O_3 Nanolubricant. *Int. J. Automot. Mech. Eng.* **2015**, *11*, 2243–2252. [CrossRef]
88. Ohunakin, O.; Adelekan, D.; Babarinde, T.; Leramo, R.; Abam, F.; Diarra, C. Experimental investigation of TiO_2-, SiO_2- and Al_2O_3-lubricants for a domestic refrigerator system using LPG as working fluid. *Appl. Therm. Eng.* **2017**, *127*, 1469–1477. [CrossRef]
89. Govindasamy, S.; Kalibangan, M.; Sadhasivam, S.; Kadasari, R. Experimental analysis of domestic refrigeration system using nanorefrigerant [CeO_2+ZnO+R134a]. *Therm. Sci.* **2022**, *26*, 969–974. [CrossRef]
90. Zhang, J.; Zhu, Y.; Yao, S.; Sun, Q. Highly efficient grooved NiTi tube refrigerants for compressive elastocaloric cooling. *Appl. Therm. Eng.* **2023**, *228*, 120439. [CrossRef]
91. Akkaya, M.; Menlik, T.; Sözen, A.; Gürü, M. The Effects of Triton X-100 and Tween 80 Surfactants on the Thermal Performance of a Nano-Lubricant: An Experimental Study. *Int. J. Precis. Eng. Manuf. Technol.* **2020**, *8*, 955–967. [CrossRef]
92. Dhamneya, A.; Rajput, S.; Singh, A. Comparative performance analysis of ice plant test rig with TiO_2-R-134a nano refrigerant and evaporative cooled condenser. *Case Stud. Therm. Eng.* **2018**, *11*, 55–61. [CrossRef]
93. Senthilkumar, A.; Anderson, A.; Sekar, M. Performance analysis of R600a vapour compression refrigeration system using CuO/Al_2O_3 hybrid nanolubricants. *Appl. Nanosci.* **2021**, *13*, 899–915. [CrossRef]
94. Mohamed, H.; Camdali, U.; Biyikoglu, A.; Aktas, M. Performance analysis of R134a vapor compression refrigeration system based on CuO/CeO_2 mixture nanorefrigerant. *J. Braz. Soc. Mech. Sci. Eng.* **2022**, *44*, 220. [CrossRef]
95. Jeyakumar, N.; Uthranarayan, C.; Narayanasamy, B. Energy conservation in the refrigeration system through improvement of Coefficient of Performance and power consumption reduction using Nanofluids. *Int. J. Ambient Energy* **2019**, *43*, 1120–1126. [CrossRef]
96. Mishra, S.; Harge, P. Effect of Addition of Different Mass Fractions of CuO Nano Particles in Compressor oil upon COP of the Refrigeration System. *Int. J. Recent Technol. Eng.* **2019**, *8*, 3732–3736. [CrossRef]
97. Yilmaz, A. Performance evaluation of a refrigeration system using nanolubricant. *Appl. Nanosci.* **2020**, *10*, 1667–1678. [CrossRef]
98. Kaushik, R.; Kundan, L.; Sharma, R. Investigating the Performance of Nanorefrigerant (R134a + CuO)-Based Vapor Compression Cycle: A New Scope. *Heat Transf. Res.* **2021**, *52*, 33–53. [CrossRef]
99. Manikanden, V.; Avinash, A. An experimental insight into the effects of silver-doped cupric oxide nanoparticles on the performance of hydrocarbon refrigeration system. *SN Appl. Sci.* **2019**, *1*, 1462. [CrossRef]
100. Sanukrishna, S.S.; Vishnu, A.S.; Jose Prakash, M. Nanorefrigerants for energy efficient refrigeration systems. *J. Mech. Sci. Technol.* **2017**, *31*, 3993–4001. [CrossRef]
101. Vasconcelos, A.; Cárdenas Gómez, A.; Bandarra Filho, E.; Parise, J. Experimental evaluation of SWCNT-water nanofluid as a secondary fluid in a refrigeration system. *Appl. Therm. Eng.* **2017**, *111*, 1487–1492. [CrossRef]
102. Afolalu, S.; Ikumapayi, O.; Ogedengbe, T.; Adegbenjo, A.; Jen, T. Evaluation and Analysis of an Agro-Based Nano Refrigerant to Improve the Performance of a Domestic Refrigeration System. *Int. J. Heat Technol.* **2022**, *40*, 1305–1310. [CrossRef]
103. Selimefendigil, F. Experimental Investigation of Nano Compressor Oil Effect on the Cooling Performance of a Vapor-Compression Refrigeration System. *J. Therm. Eng.* **2018**, *5*, 100–104. [CrossRef]
104. Marcucci Pico, D.; da Silva, L.; Schneider, P.; Bandarra Filho, E. Performance evaluation of diamond nanolubricants applied to a refrigeration system. *Int. J. Refrig.* **2019**, *100*, 104–112. [CrossRef]

105. Marcucci Pico, D.; da Silva, L.; Hernandez Mendoza, O.; Bandarra Filho, E. Experimental study on thermal and tribological performance of diamond nanolubricants applied to a refrigeration system using R32. *Int. J. Heat Mass Transf.* **2020**, *152*, 119493. [CrossRef]
106. Wang, J.; Li, X.; Wang, J.; Zhu, T.; Bao, Y. Thermal performance evaluation of a thermoelectric cooler coupled with corona wind. *Appl. Therm. Eng.* **2020**, *179*, 115753. [CrossRef]
107. Ikpe, A.; Udofia Ekpenyong, A. Buhar sıkıştırmalı soğutma sistemindeki hibrit-nanoakışkan zeotropik karışımların termo-fiziksel değerlendirilmesi. *Gümüşhane Üniversitesi Fen Bilim. Enstitüsü Derg.* **2024**, *14*, 1021–1038. [CrossRef]
108. Jiang, W.; Tang, X.; Song, J.; Jia, T.; Zhang, Y.; Li, Y.; Yang, L.; Li, S.; Su, W. Experimental study on the influence of adding TiO_2 nanoparticles on practical ammonia-water absorption refrigeration system-the generation and rectification processes. *Appl. Therm. Eng.* **2023**, *230*, 120763. [CrossRef]
109. Micali, F.; Milanese, M.; Colangelo, G.; de Risi, A. Reducing CO_2 emissions by improving HVAC system efficiency of data centers through nanofluids: A case study. *Appl. Therm. Eng.* **2025**, *275*, 126889. [CrossRef]
110. Milanese, M.; Micali, F.; Colangelo, G.; de Risi, A. Experimental Evaluation of a Full-Scale HVAC System Working with Nanofluid. *Energies* **2022**, *15*, 2902. [CrossRef]
111. Raghavulu, K.; Rasu, N. An Experimental Study on the Improvement of Coefficient of Performance in Vapor Compression Refrigeration System Using Graphene Lubricant Additives. *Energy Sources Part A Recover. Util. Environ. Eff.* **2021**, *47*, 6449–6464. [CrossRef]
112. Ahmed, M.; Elsaid, A. Effect of hybrid and single nanofluids on the performance characteristics of chilled water air conditioning system. *Appl. Therm. Eng.* **2019**, *163*, 114398. [CrossRef]
113. Venegas, M.; García-Hernando, N.; Zacarías, A.; de Vega, M. Performance of a Solar Absorption Cooling System Using Nanofluids and a Membrane-Based Microchannel Desorber. *Appl. Sci.* **2020**, *10*, 2761. [CrossRef]
114. Ahmed, F.; Ahmed Khan, W.; Nayfeh, J. Experimental Investigation on the Performance of Heat Pump Operating with Copper and Alumina Nanofluids. *Comput. Mater. Contin.* **2021**, *66*, 2843–2856. [CrossRef]
115. Nourafkan, E.; Asachi, M.; Jin, H.; Wen, D.; Ahmed, W. Stability and photo-thermal conversion performance of binary nanofluids for solar absorption refrigeration systems. *Renew. Energy* **2019**, *140*, 264–273. [CrossRef]
116. Elibol, E.; Yilmazoglu, M.; Aksoy, A.; Aktas, F. Thermal analysis of a refrigeration system integrated with a thermoelectric couple and microchannels using TiO_2-water nanofluid. *Int. J. Heat Fluid Flow* **2024**, *108*, 109491. [CrossRef]
117. Anand, R.; Jawahar, C.; Brusly Solomon, A.; Bellos, E.; Ajay Vasanth, X. Experimental investigation of a two-phase closed thermosyphon with Al_2O_3/R134a nanorefrigerant. *Proc. Inst. Mech. Eng. Part E J. Process Mech. Eng.* **2022**, *238*, 56–66. [CrossRef]
118. Jin, Z.; Li, S.; Zhou, R.; Xu, M.; Jiang, W.; Du, K. Experimental investigation on the effect of TiO_2 nanoparticles on the performance of NH_3—H_2O—LiBr absorption refrigeration system. *Int. J. Refrig.* **2021**, *131*, 826–833. [CrossRef]
119. Laplane, C.; Ren, P.; Roberts, R.; Lu, Y.; Volz, T. Inert Shell Coating for Enhanced Laser Refrigeration of Nanoparticles: Application in Levitated Optomechanics. *ACS Photonics* **2024**, *11*, 963–968. [CrossRef]
120. Cuce, E.; Guclu, T.; Cuce, P. Improving thermal performance of thermoelectric coolers (TECs) through a nanofluid driven water to air heat exchanger design: An experimental research. *Energy Convers. Manag.* **2020**, *214*, 112893. [CrossRef]
121. Zhou, R.; Wang, M.; Jin, Z.; Li, S. Modeling and experimental verification of the enhancement of TiO_2 nanofluid on ammonia falling film absorption process. *Int. J. Therm. Sci.* **2022**, *184*, 107917. [CrossRef]
122. Jin, Z.; Li, S.; Zhou, R.; Xu, M.; Jiang, W.; Du, K. Effects of absorption pressure and temperature on NH_3-H_2O-LiBr-TiO_2 nanofluid absorption performance and system COP. *Appl. Therm. Eng.* **2023**, *219*, 119353. [CrossRef]
123. Sözen, A.; Özbaş, E.; Menlik, T.; Skender, N.; Kılınç, C.; Çakır, M. Performance investigation of a diffusion absorption refrigeration system using nano-size alumina particles in the refrigerant. *Int. J. Exergy* **2015**, *18*, 443. [CrossRef]
124. Ahmed, F. Experimental investigation of Al_2O_3-water nanofluid as a secondary fluid in a refrigeration system. *Case Stud. Therm. Eng.* **2021**, *26*, 101024. [CrossRef]
125. Safaei, M.; Ranjbarzadeh, R.; Hajizadeh, A.; Bahiraei, M.; Afrand, M.; Karimipour, A. Effects of cobalt ferrite coated with silica nanocomposite on the thermal conductivity of an antifreeze: New nanofluid for refrigeration condensers. *Int. J. Refrig.* **2019**, *102*, 86–95. [CrossRef]
126. Pourfayaz, F.; Imani, M.; Mehrpooya, M.; Shirmohammadi, R. Process development and exergy analysis of a novel hybrid fuel cell-absorption refrigeration system utilizing nanofluid as the absorbent liquid. *Int. J. Refrig.* **2019**, *97*, 31–41. [CrossRef]
127. Dalkilic, A.; Mahian, O.; Kucukyildirim, B.; Eker, A.; Ozturk, T.; Jumpholkul, C.; Wongwises, S. Experimental Study on the Stability and Viscosity for the Blends of Functionalized MWCNTs with Refrigeration Compressor Oils. *Curr. Nanosci.* **2018**, *14*, 216–226. [CrossRef]
128. Wu, S.; Rincon Ortiz, C. Experimental investigation of the effect of magnetic field on vapour absorption with LiBr–H_2O nanofluid. *Energy* **2020**, *193*, 116640. [CrossRef]
129. Bondre, D.; Joshi, A.; Shinde, T.; Deshmukh, A.; Dhanawade, K. Experimental Performance and Analysis of Domestic Refrigeration System Using Nano-Refrigerants. In *Lecture Notes in Mechanical Engineering*; Springer: Singapore, 2018; pp. 389–399. [CrossRef]

30. Talpada, J.; Ramana, P. Experimental Analysis of H_2O–LiBr Absorption Refrigeration System Using Al_2O_3 Nanoparticles. *Int. J. Air-Cond. Refrig.* **2020**, *28*, 2050010. [CrossRef]
31. Hatami, M.; Domairry, G.; Mirzababaei, S. Experimental investigation of preparing and using the H_2O based nanofluids in the heating process of HVAC system model. *Int. J. Hydrogen Energy* **2017**, *42*, 7820–7825. [CrossRef]
32. Elsaid, A. A novel approach for energy and mass transfer characteristics in wet cooling towers associated with vapor-compression air conditioning system by using MgO and TiO_2 based H_2O nanofluids. *Energy Convers. Manag.* **2020**, *204*, 112289. [CrossRef]
33. Hussain, T.; Ahmad, W.; Quadri, I. Performance Evaluation of a Vapor Compression Refrigeration System Using Oxide Nanoparticles: Experimental Investigation. *Nanosci. Nanotechnol.-Asia* **2021**, *11*, 189–196. [CrossRef]
34. Esfe, M.; Dalir, R.; Bakhtiari, R.; Afrand, M. Simultaneous effects of multi-walled carbon nanotubes and copper oxide nanoparticles on the rheological behavior of cooling oil: Application for refrigeration systems. *Int. J. Refrig.* **2019**, *104*, 123–133. [CrossRef]
35. Jiang, W.; Li, S.; Yang, L.; Du, K. Experimental investigation on performance of ammonia absorption refrigeration system with TiO_2 nanofluid. *Int. J. Refrig.* **2019**, *98*, 80–88. [CrossRef]
36. Han, H.; Kim, C.; Kim, N. Steam Condensation on Copper-Enhanced Tubes Used in an Absorption Refrigeration System. *J. Enhanc. Heat Transf.* **2021**, *28*, 1–16. [CrossRef]
37. Aramesh, M.; Pourfayaz, F.; Haghir, M.; Kasaeian, A.; Ahmadi, M. Investigating the effect of using nanofluids on the performance of a double-effect absorption refrigeration cycle combined with a solar collector. *Proc. Inst. Mech. Eng. Part A J. Power Energy* **2019**, *234*, 981–993. [CrossRef]
38. Yousif, S.; Al-Obaidi, M.; Al-Muhsen, N. Towards More Efficient Refrigeration: A Study on the Use of TiO_2 and Al_2O_3 Nanoparticles. *Int. J. Heat Technol.* **2024**, *42*, 1251–1256. [CrossRef]
39. Shewale, V.; Kapse, A.; Sonawane, V. Experimental analysis of vapour compression refrigeration system using nano lubricant with refrigerant R-134a. *Therm. Sci.* **2024**, *28*, 3687–3697. [CrossRef]
40. Mohammed, K.; Al-Mirani, A.; Ali, B.; Alomar, O. Impacts of Using AlO Nano Particle to Compressor Oil on Performance of Automobile Air Conditioning System. *Front. Heat Mass Transf.* **2024**, *22*, 839–854. [CrossRef]
41. Patel, A.; Rajput, S.; Kumar, R.; Sharma, A.; Saxena, K.; Agrawal, M.; Kadhim, I. Experimental and numerical investigation of metal oxide base nano particles for VCRS test rig. *Int. J. Interact. Des. Manuf. (IJIDeM)* **2023**, *18*, 5909–5916. [CrossRef]
42. Pasha, K. Controlling the Nusselt Number in a TiO_2/R134a Nano-refrigerant System. *Int. J. Heat Technol.* **2019**, *37*, 179–187. [CrossRef]
43. Nair, V.; Parekh, A.; Tailor, P. Experimental investigation of a vapour compression refrigeration system using R134a/Nano-oil mixture. *Int. J. Refrig.* **2020**, *112*, 21–36. [CrossRef]
44. Narayanasarma, S.; Kuzhiveli, B. The effect of silica nanoparticle on thermal, chemical, corrosive, and the nature-friendly properties of refrigerant compressor lubricants—A comparative study. *Asia-Pac. J. Chem. Eng.* **2020**, *15*, e2551. [CrossRef]
45. Babarinde, T.; Akinlabi, S.; Madyira, D.; Ekundayo, F. Comparative Study of Energy Performance of R600a/TiO_2 and R600a/MWCNT Nanolubricants in a Vapor Compression Refrigeration System. *Int. J. Energy A Clean Environ.* **2020**, *21*, 317–332. [CrossRef]
46. Devendra; RaviPrakash, M.; Kapilan, N.; Yadav, S.P.S. Performance of Vapor Compression Refrigeration System using Nanoparticles—Based Mineral Oil Lubricant. *Int. J. Recent Technol. Eng.* **2019**, *8*, 730–734. [CrossRef]
47. Lin, L.; Peng, H.; Chang, Z.; Ding, G. Experimental investigation on TiO_2 nanoparticle migration from refrigerant–oil mixture to lubricating oil during refrigerant dryout. *Int. J. Refrig.* **2017**, *77*, 75–86. [CrossRef]
48. Aziz, A.; Shah, S.; Bahaidarah, H.; Zamir, T.; Aziz, T. Advanced neural network modeling with Levenberg–Marquardt algorithm for optimizing tri-hybrid nanofluid dynamics in solar HVAC systems. *Case Stud. Therm. Eng.* **2025**, *65*, 105609. [CrossRef]
49. Benazzouz, A.; Ammar, M.; Settou, B.; Marif, Y.; Djemoui, K. Thermal performance analysis of a domestic-size absorption cooling system incorporating nanofluid with thermal insulation in hot climate of Algeria. *Int. J. Refrig.* **2025**, *170*, 192–213. [CrossRef]
50. Boldoo, T.; Lee, M.; Cho, H. Numerical investigation on thermal performance of absorption refrigeration system using MWCNT nanoparticle-enhanced 1-hexyl-3-methylimidazolium cation-based ionic liquids. *Appl. Therm. Eng.* **2022**, *206*, 118093. [CrossRef]
51. Khan, W.; Nabwey, H.; EL-Kabeir, S.; Rashad, A.; AbdElnaem, S.; Seddek, L. Solar Energy Encouragement in Solar HVAC Using Eyring-Powell Ternary-Hybrid Nanofluid Flow in Porous Medium with Cattaneo-Christov Heat and Mass Fluxes. *Spec. Top. Rev. Porous Media Int. J.* **2023**, *14*, 71–91. [CrossRef]
52. Colangelo, G.; Raho, B.; Milanese, M.; de Risi, A. Numerical Evaluation of a HVAC System Based on a High-Performance Heat Transfer Fluid. *Energies* **2021**, *14*, 3298. [CrossRef]
53. Prabakaran, R.; Gomathi, B.; Saravanan, A.L.; Jeyalakshmi, P.; Lal, D.M.; Kim, S.C. Machine learning-based performance and optimal refrigerant charge prediction for a split air conditioning system. *Appl. Therm. Eng.* **2025**, *274*, 126764. [CrossRef]
54. Babaei, S.; Razmi, A.; Soltani, M.; Nathwani, J. Quantifying the effect of nanoparticles addition to a hybrid absorption/recompression refrigeration cycle. *J. Clean. Prod.* **2020**, *260*, 121084. [CrossRef]

155. Shankar Prasad, U.; Mishra, R.; Das, R. Experimental studies of vapour compression refrigeration system with eco-Friendl primary refrigerant and brine mixed with nano particles as secondary refrigerant. *Mater. Today Proc.* **2021**, *45*, 3857–385 [CrossRef]
156. Hernández Jaramillo, D.; Nieto Londoño, C.; Zapata-Benabithe, Z. Analysis of working nanofluids for a refrigeration syster *DYNA* **2016**, *83*, 176–183. [CrossRef]
157. Alqaed, S.; Mustafa, J.; Sharifpur, M.; Alharthi, M.A. Numerical simulation and artificial neural network modeling of exergy an energy of parabolic trough solar collectors equipped with innovative turbulators containing hybrid nanofluids. *J. Therm. Ana Calorim.* **2023**, *148*, 8611–8626. [CrossRef]
158. Aprea, C.; Greco, A.; Maiorino, A.; Masselli, C. The use of barocaloric effect for energy saving in a domestic refrigerator wit ethylene-glycol based nanofluids: A numerical analysis and a comparison with a vapor compression cooler. *Energy* **2020**, *19* 116404. [CrossRef]
159. Talpada, J.S.; Ramana, P.V. The Theoretical Analysis of H_2O-LiBr Absorption Refrigeration System using Al_2O_3 Nanoparticle *Int. J. Mech. Prod. Eng. Res. Dev.* **2019**, *9*, 303–322. [CrossRef]
160. Ndoye, F.; Schalbart, P.; Leducq, D.; Alvarez, G. Numerical study of energy performance of nanofluids used in secondary loops c refrigeration systems. *Int. J. Refrig.* **2015**, *52*, 122–132. [CrossRef]
161. Gill, J.; Singh, J.; Ohunakin, O.S.; Adelekan, D.S. Artificial neural network approach for irreversibility performance analysis c domestic refrigerator by utilizing LPG with TiO2–lubricant as replacement of R134a. *Int. J. Refrig.* **2018**, *89*, 159–176. [CrossRef
162. Li, S.; Lu, J. A Theoretical Comparative Study of Vapor-Compression Refrigeration Cycle using Al_2O_3 Nanoparticle wit Low-GWP Refrigerants. *Entropy* **2022**, *24*, 1820. [CrossRef] [PubMed]
163. Akhtar, M.; Rajput, S. Energy and exergy analysis of vapour compression test rig using R134a blended with GN- MWCNT/PO hybrid nano-lubricants. *Energy Sources Part A Recover. Util. Environ. Eff.* **2023**, *46*, 188–208. [CrossRef]
164. Kumar, A.; Rajput, S. Performance of a computerized refrigeration cycle test rig for varying concentrations of (CuO-Al_2O_3 /MC hybrid nano-lubricants and R600a refrigerant charges: An energetic and exergetic approach. *Energy Sources Part A Recover. Uti Environ. Eff.* **2023**, *45*, 9975–9992. [CrossRef]
165. Yilmaz, M.; Cimsit, C.; Keven, A.; Karaali, R. Analysis of cascade vapor compression refrigeration system using nanorefrigerant Energy, exergy, and environmental (3E). *Case Stud. Therm. Eng.* **2024**, *57*, 104373. [CrossRef]
166. Gürbüz, E.; Keçebaş, A.; Sözen, A. Exergy and thermoeconomic analyses of the diffusion absorption refrigeration system wit various nanoparticles and their different ratios as work fluid. *Energy* **2022**, *248*, 123579. [CrossRef]
167. Asgharian, H.; Baniasadi, E.; Colpan, C. Energy, exergy and exergoeconomic analyses of a solar refrigeration cycle using nanoflui *Int. J. Exergy* **2019**, *30*, 63–85. [CrossRef]
168. Sarkar, J. Performance improvement of double-tube gas cooler in CO_2 refrigeration system using nanofluids. *Therm. Sci.* **2015**, *1* 109–118. [CrossRef]
169. Gürbüz, E.; Sözen, A.; Keçebaş, A.; Özbaş, E. Experimental and numerical investigation of diffusion absorption refrigeratio system working with $ZnOAl_2O_3$ and TiO_2 nanoparticles added ammonia/water nanofluid. *Exp. Heat Transf.* **2020**, *35*, 197–22 [CrossRef]
170. Özbaş, E. Energy and Exergy Analysis of Using $FeOTiO_2$ Nanofluid in Diffusion Absorption Refrigeration Systems. *Heat Trans Res.* **2021**, *52*, 11–24. [CrossRef]
171. Selimefendigil, F.; Oztop, H. Exergetic performance of vapor-compression refrigeration system with TiO_2-nanoadditive in th compressor oil. *Therm. Sci.* **2021**, *25*, 637–642. [CrossRef]
172. Fragelli, R.; Sanchez, L.; Ingraci Neto, R.; Scalon, V. Refrigeration capacity of silver nanofluids under electrohydrodynamic effec oriented to heat removal in machining process. *Exp. Therm. Fluid Sci.* **2018**, *96*, 11–19. [CrossRef]
173. Fahar, M.; Otanicar, T. Experimental and theoretical description of a technique for the concentration measurement of binar liquids containing nanoparticles. *Int. J. Heat Mass Transf.* **2015**, *85*, 561–567. [CrossRef]
174. Akhayere, E.; Adebayo, V.; Adedeji, M.; Abid, M.; Kavaz, D.; Dagbasi, M. Investigating the effects of nanorefrigerants in cascaded vapor compression refrigeration cycle. *Int. J. Energy Environ. Eng.* **2022**, *14*, 601–612. [CrossRef]
175. Khetib, Y.; Sait, H.; Habeebullah, B.; Hussain, A. Improving the performance of a lithium-ion battery connected to a solar syster using a nano-refrigerant in an ejector refrigeration cycle. *J. Energy Storage* **2022**, *45*, 103601. [CrossRef]
176. Wang, G.; Zhang, Q.; Zeng, M.; Xu, R.; Xie, G.; Chu, W. Investigation on mass transfer characteristics of the falling film absorptio of LiBr aqueous solution added with nanoparticles. *Int. J. Refrig.* **2018**, *89*, 149–158. [CrossRef]
177. Li, Z.; Renault, F.; Gómez, A.; Sarafraz, M.; Khan, H.; Safaei, M.; Filho, E. Nanofluids as secondary fluid in the refrigeratior system: Experimental data, regression, ANFIS, and NN modeling. *Int. J. Heat Mass Transf.* **2019**, *144*, 118635. [CrossRef]
178. Shamsah, S. Relative cooling power modeling of RE2TM2Y ternary intermetallic rare-earth-based magnetocaloric compound for magnetic refrigeration application using extreme learning machine and hybrid intelligent method. *Int. J. Refrig.* **2024**, *16* 122–134. [CrossRef]

179. Gill, J.; Singh, J.; Ohunakin, O.; Adelekan, D.; Atiba, O.; Nkiko, M.; Atayero, A. Adaptive neuro-fuzzy inference system (ANFIS) approach for the irreversibility analysis of a domestic refrigerator system using LPG/TiO_2 nanolubricant. *Energy Rep.* **2020**, *6*, 1405–1417. [CrossRef]
180. Sharma, H.; Kumari, S.; Dutt, A.; Kumar, P.; Makhatha, M. A Machine Learning Approach to Predict the Performance of Refrigerator and Air Conditioning Using Gaussian Process Regression and Support Vector Methods. *Recent Pat. Mech. Eng.* **2021**, *14*, 550–557. [CrossRef]
181. Boyaghchi, F.; Mahmoodnezhad, M.; Sabeti, V. Exergoeconomic analysis and optimization of a solar driven dual-evaporator vapor compression-absorption cascade refrigeration system using water/CuO nanofluid. *J. Clean. Prod.* **2016**, *139*, 970–985. [CrossRef]
182. Wadi, V.; Özmen, Ö.; Caliskan, A.; Karamış, M. Experimental study and computational intelligence on dynamic viscosity and thermal conductivity of HNTs based nanolubricant. *Ind. Lubr. Tribol.* **2022**, *74*, 102–110. [CrossRef]
183. Kiannejad Amiri, M.; Ghorbanzade Zaferani, S.; Sarmasti Emami, M.; Zahmatkesh, S.; Pourhanasa, R.; Sadeghi Namaghi, S.; Klemeš, J.; Bokhari, A.; Hajiaghaei-Keshteli, M. Multi-objective optimization of thermophysical properties GO powders-DW/EG Nf by RSM, NSGA-II, ANN, MLP and ML. *Energy* **2023**, *280*, 128176. [CrossRef]
184. Marulasiddeshi, H.; Kanti, P.; Jamei, M.; Prakash, S.; Sridhara, S.; Said, Z. Experimental study on the thermal properties of Al_2O_3-CuO/water hybrid nanofluids: Development of an artificial intelligence model. *Int. J. Energy Res.* **2022**, *46*, 21066–21083. [CrossRef]
185. Alizadeh, R.; Abad, J.; Fattahi, A.; Mohebbi, M.; Doranehgard, M.; Li, L.; Alhajri, E.; Karimi, N. A Machine Learning Approach to Predicting the Heat Convection and Thermodynamics of an External Flow of Hybrid Nanofluid. *J. Energy Resour. Technol.* **2021**, *143*, 070908. [CrossRef]
186. Sharma, P.; Ramesh, K.; Parameshwaran, R.; Deshmukh, S. Thermal conductivity prediction of titania-water nanofluid: A case study using different machine learning algorithms. *Case Stud. Therm. Eng.* **2022**, *30*, 101658. [CrossRef]
187. Zhu, T.; Mei, X.; Zhang, J.; Li, C. Prediction of Thermal Conductivity of EG–Al_2O_3 Nanofluids Using Six Supervised Machine Learning Models. *Appl. Sci.* **2024**, *14*, 6264. [CrossRef]
188. Bemani, A.; Madani, M.; Kazemi, A. Machine learning-based estimation of nano-lubricants viscosity in different operating conditions. *Fuel* **2023**, *352*, 129102. [CrossRef]
189. Ahmad, S.; Alsalmah, H.A. Artificial intelligence based evaluation of the tri-hybridized (WS_2+Fe_3O_4+CuS) flow subjected to the thermal performance in porous systems. *Results Eng.* **2025**, *26*, 104717. [CrossRef]
190. Serrao, B.; Kim, K.; Duarte, J. Analysis of the Effects of Different Nanofluids on Critical Heat Flux Using Artificial Intelligence. *Energies* **2023**, *16*, 4762. [CrossRef]
191. Syed, J. Enhancement of Heat Transfer Using Water/Graphene Nanofluid and the Impact of Passive Techniques—Experimental, Numerical, and ML Approaches. *Energies* **2024**, *18*, 77. [CrossRef]
192. Ajila, F.; Manokaran, S.; Ramaswamy, K.; Thiyagarajan, D.; Pappula, P.; Ali, S.; Dillibabu, S.; Kasi, U.; Selvaraju, M. Prediction of nanofluid thermal conductivity and viscosity with machine learning and molecular dynamics. *Therm. Sci.* **2024**, *28*, 717–729. [CrossRef]
193. Hussain, M.; Khan, M.; Waqas, H.; Al-Mdallal, Q. Optimizing solar collector efficiency and safety: A comparative thermal analysis of non-toxic hybrid nanofluid mixtures using machine learning. *Case Stud. Therm. Eng.* **2025**, *72*, 106221. [CrossRef]
194. Martínez-Vargas, B.; González Rodríguez, L.; Pacheco-Álvarez, M.; Peralta-Hernández, J.; Picos, A. Use of artificial intelligence models for the reduction of nanoparticle size in the synthesis of ZnO. *J. Chem. Technol. Biotechnol.* **2023**, *98*, 1868–1877. [CrossRef]
195. Sharma, P.; Said, Z.; Memon, S.; Elavarasan, R.; Khalid, M.; Nguyen, X.; Arıcı, M.; Hoang, A.; Nguyen, L. Comparative evaluation of AI-based intelligent GEP and ANFIS models in prediction of thermophysical properties of Fe_3O_4-coated MWCNT hybrid nanofluids for potential application in energy systems. *Int. J. Energy Res.* **2022**, *46*, 19242–19257. [CrossRef]
196. Zhang, Z.; Al-Bahrani, M.; Ruhani, B.; Heybatian Ghalehsalimi, H.; Zandy Ilghani, N.; Maleki, H.; Ahmad, N.; Nasajpour-Esfahani, N.; Toghraie, D. Optimized ANFIS models based on grid partitioning, subtractive clustering, and fuzzy C-means to precise prediction of thermophysical properties of hybrid nanofluids. *Chem. Eng. J.* **2023**, *471*, 144362. [CrossRef]
197. Bhattad, A.; Sarkar, J.; Ghosh, P. Improving the performance of refrigeration systems by using nanofluids: A comprehensive review. *Renew. Sustain. Energy Rev.* **2018**, *82*, 3656–3669. [CrossRef]
198. Bilen, K.; Dağidir, K.; Arcaklioğlu, E. The effect of nanorefrigerants on performance of the vapor compression refrigeration system: A comprehensive review. *Energy Sources Part A Recover. Util. Environ. Eff.* **2022**, *44*, 3178–3204. [CrossRef]
199. Kumar, A.; Narendran, G.; Perumal, D. Effects of Nanorefrigerants for Refrigeration System: A review. *Nanosci. Technol. Int. J.* **2023**, *14*, 17–36. [CrossRef]
200. Pisal, H.; Deulgaonkar, V.; Tamkhade, P.; Desale, A.; Patil, R.; Raut, S. Vapour Absorption Refrigeration System and Nanofluids: Recent Research, Development and Applications. *Int. J. Veh. Struct. Syst.* **2024**, *16*. [CrossRef]
201. Jiang, W.; Song, J.; Jia, T.; Yang, L.; Li, S.; Li, Y.; Du, K. A comprehensive review on the pre-research of nanofluids in absorption refrigeration systems. *Energy Rep.* **2022**, *8*, 3437–3464. [CrossRef]

202. Talpada, J.; Ramana, P. A review on performance of absorption refrigeration system using new working pairs and nano-particles. *Int. J. Ambient Energy* **2021**, *43*, 5654–5672. [CrossRef]
203. Marcucci Pico, D.; Parise, J.; Bandarra Filho, E. Nanolubricants in refrigeration systems: A state-of-the-art review and latest developments. *J. Braz. Soc. Mech. Sci. Eng.* **2023**, *45*, 88. [CrossRef]
204. Azmi, W.; Sharif, M.; Yusof, T.; Mamat, R.; Redhwan, A. Potential of nanorefrigerant and nanolubricant on energy saving in refrigeration system—A review. *Renew. Sustain. Energy Rev.* **2017**, *69*, 415–428. [CrossRef]
205. Patil, M.; Kim, S.; Seo, J.; Lee, M. Review of the Thermo-Physical Properties and Performance Characteristics of a Refrigeration System Using Refrigerant-Based Nanofluids. *Energies* **2015**, *9*, 22. [CrossRef]
206. Ponticorvo, E.; Iuliano, M.; Cirillo, C.; Maiorino, A.; Aprea, C.; Sarno, M. Fouling Behavior and Dispersion Stability of Nanoparticle-Based Refrigeration Fluid. *Energies* **2022**, *15*, 3059. [CrossRef]
207. Praveen Bharathwaj, R.; Shaik, J.; Raju, J.J.; Padmanathan, P.; Satheesh, A. A critical review on nanorefrigerants: Boiling, condensation and tribological properties. *Int. J. Refrig.* **2021**, *128*, 139–152. [CrossRef]
208. Du, K.; Calautit, J.; Wang, Z.; Wu, Y.; Liu, H. A review of the applications of phase change materials in cooling, heating and power generation in different temperature ranges. *Appl. Energy* **2018**, *220*, 242–273. [CrossRef]
209. Kundan, L.; Singh, K. Improved performance of a nanorefrigerant-based vapor compression refrigeration system: A new alternative. *Proc. Inst. Mech. Eng. Part A J. Power Energy* **2020**, *235*, 106–123. [CrossRef]
210. Valiyandi, S.; Thampi, G. Review on performance analysis in diffusion absorption refrigeration system (DARS) using different working fluids. *Int. J. Ambient Energy* **2023**, *44*, 2185–2199. [CrossRef]

sustainability

MDPI

.rticle

?easibility Analysis of Indirect Evaporative Cooling System \ssisted by Liquid Desiccant for Data Centers in Hot-Humid Regions

Venchao Shi [1], Xiaochen Ma [1], Yunran Min [2,*] and Hongxing Yang [1]

1 Renewable Energy Research Group (RERG), Department of Building Environment and Energy Engineering, The Hong Kong Polytechnic University, Hong Kong, China
2 College of Marine Equipment and Mechanical Engineering, Jimei University, Xiamen 361021, China
* Correspondence: minyunran@jmu.edu.cn

Abstract: The rapid development of data centers (DCs) has led to a marked increase in energy consumption in recent years, which poses a direct challenge to global efforts aimed at reducing carbon emissions. In regions with hot and humid climates, the energy demand is largely driven by air conditioning systems necessarily to maintain appropriate operational temperatures. This study proposes a novel multi-stage indirect evaporative cooling (IEC) system, incorporating a liquid desiccant in the primary air channel to address the cooling demands of such DCs. Our approach involves a two-stage process where the first stage uses a liquid desiccant-based IEC (LD-IEC) for air dehumidification and the second stage utilizes the treated air from the first stage as the secondary air to enhance the cooling effect. A simulation model of the proposed system is established with validation, and the performance of the multi-stage system was also discussed based on different operation modes. Furthermore, a case study was conducted to investigate the feasibility of using this system in the DC under a typical hot and humid zone. The findings reveal that the first-stage LD-IEC is capable of diminishing the wet-bulb temperature of the ambient air. Furthermore, the case study demonstrates that the proposed system can greatly improve the temperature drop by 72.7% compared to the single IEC, which noticeably reduces the operation time of energy-intensive supplementary cooling equipment from 5092 h to 31 h given the supply air temperature threshold of 25 °C. In summary, the proposed system could substantially decrease reliance on traditional cooling systems, which demonstrates a promising avenue to fully use this passive cooling technology for cooling DCs.

Keywords: air conditioning; indirect evaporative cooling; data center; liquid desiccant

1. Introduction

As stated by the most recent Sixth Assessment Report from the United Nations Intergovernmental Panel on Climate Change (IPCC), capping global warming to 1.5 °C will require a significant shift towards zero- or low-carbon sources for nearly all electricity generation by 2050 [1]. This transition would involve embracing renewables, fossil fuels with carbon capture and storage technology, and increased electrification of energy demand. In addition to transforming existing energy systems, reducing energy use in economies is important as well to relieve the pressure on the supply side, which is conducive to carbon peaking and neutrality [2].

As one of the major energy consumers, buildings have received significant attention in the context of dual carbon goals [3,4]. In China, the urbanization rate has shifted from rapid growth to steady growth, resulting in a gradual slowdown in the development pace of new residential and commercial buildings. However, data centers (DC), which are more important in the digital age, become a rapidly growing type of building. They link a large number of computing resources and storage resources together and are the center of massive data computing, exchange, and storage [5,6]. Nowadays, various activities, online communications, entertainment, and knowledge generation and spread increasingly rely on the internet. The increasing demand for timely interaction through the Internet has led to a sharp increase in human demand for DCs [7]. Therefore, the scale of DCs is becoming larger, the computational complexity is constantly increasing, and the structure is becoming more and more precise. However, the expansion of DCs is associated with significant energy demands, with the cooling systems alone accounting for approximately 40% of the total electricity consumption [8]. As devices progress, electronic devices are capable of delivering enhanced computational performance while also generating increased heat dissipation. Therefore, it is imperative to give due attention to both cooling and thermal management aspects [9,10], while a high percentage of cooling energy consumption obviously runs counter to dual-carbon goals.

To lessen the energy consumption of cooling systems in data rooms, researchers have focused on how to take full advantage of natural cooling, extend the operating time of natural cooling approaches, and reduce the working time of the mechanical vapor compression system (MVCS). Indirect evaporative cooling (IEC) is one of the emerging air conditioning (AC) approaches [11–13]. Figure 1 illustrates a cross-flow IEC system wherein the primary air stream to be cooled flows perpendicular to the secondary air stream that facilitates evaporation. In this configuration, the secondary air flows vertically and is in direct contact with water within the heat exchanger. As the air flows through the secondary air channel, due to the gradient in moisture content between the water film and air, water evaporation occurs and removes the latent heat. Then, the sensible heat from the primary air transfers to the secondary air stream, which is then vented externally. This approach enables the air cooling process to maintain the same humidity level and effectively prevents cross-contamination [14–16]. Furthermore, this system circumvents energy-intensive compressors or environmentally hazardous refrigerants. The air conditioning process relies on a natural phenomenon, namely the evaporation of water, with the system's main energy consumption stemming from the operation of fans and the pump. As a result, the COP and energy consumption of the IEC system are lower than those of traditional air conditioning systems, which also correlates with reduced carbon emissions [17,18].

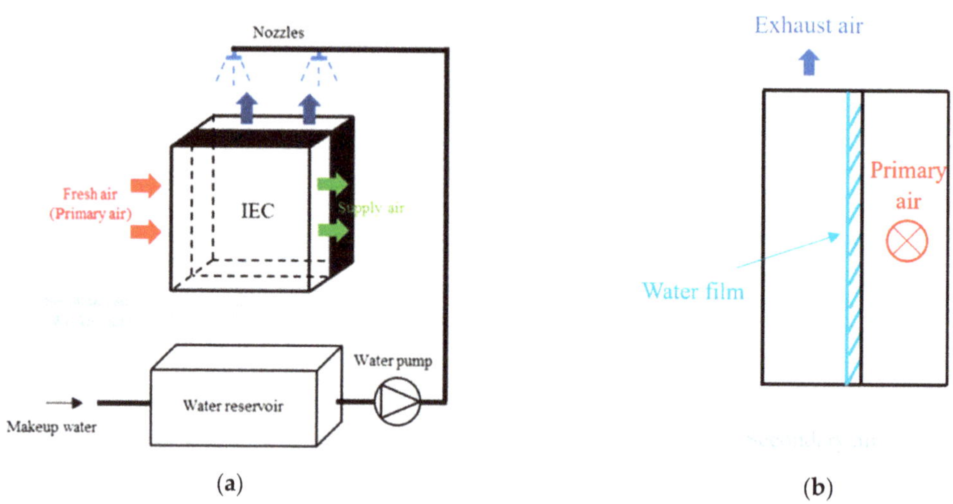

Figure 1. (**a**) Configuration of an IEC system, (**b**) side view of the IEC.

2. Literature Review

Over the past decades, fruitful research has been conducted. At the heat exchanger level, advancements have been made in both novel structures and optimal water spraying.

For instance, De Antonellis et al. evaluate the effect of spraying nozzle locations and airflow orientations on the cooling capacity of an IEC system [12]. Baffles were incorporated into the channel surface to create a small internal vortex, thereby enhancing the heat transfer coefficient [15]. Sun et al. developed a ceramic-based tubular IEC model and fabricated a prototype to test its performance under intermittent spraying [19]. Shi et al. proposed a novel IEC with porous secondary air channel surfaces to enhance the water absorbent ability and implemented periodic spraying modes to minimize operational energy consumption [20]. Corrugated channels were designed and produced to increase the contact area in comparison to traditional IECs [21]. Ahmed et al. assessed the external, internal, and mixed spraying modes for the same IEC under varying inlet conditions, identifying the internal spraying plan as a reliable method to improve mass transfer in the hexagonal IEC [22]. After evaluating six configurations, they recommended a top or a horizontal spraying scheme for optimal water distribution.

At the system level, hybrid IEC systems have been developed to meet cooling requirements and decrease energy consumption. Wan et al. evaluated the IEC system integrated with a paraffin-based phase change material (PCM) under commercial operating conditions. The use of PCMs shifts the peak load and enables a higher temperature of the chiller water [23]. Nemati et al. proposed to precool the air by the ground source heat exchanger and further cool the supply air through the IEC, which resulted in a 45% reduction in water consumption [24]. Shi et al. performed a techno-economic evaluation of a dedicated outdoor air system using an IEC and an auxiliary cooling coil in an office building. Their findings underscore the viability of extending IEC applications in regions characterized by hot and humid climates [25]. Chen et al. incorporated an IEC into a mechanical vapor compression system to partially undertake the cooling load, reducing the energy consumption by at least 19% [26]. Yan et al. proposed a DXAC-IEC system with an adaptive controller and evaluated the performance under hot and arid climates, which saved 27.7% of energy during the transition season compared to the standalone IEC [27]. Tripathi and Kumar et al. investigated a tubular IEC system powered by solar photovoltaic panels under various inlet conditions in a hot and arid city in India [28]. Hu et al. proposed a novel synergy between IEC and radiative cooling, which obtained an 11.9% improvement in dew-point effectiveness and superior cooling performance in dry climates [29].

Since the load in the DC is mainly from equipment heat dissipation, the proportion of sensible heat is extremely high. The handling of sensible heat loads is a key feature and advantage of IECs. Consequently, in recent years, projects in hot and arid areas have begun to incorporate direct evaporative cooling (DEC) and IEC in the DC's AC system, which have received noticeable energy saving benefits [30–32]. The implementation of the DEC for fresh air at the Facebook DC achieves significant energy savings in Plinville, a dry region in the USA [33]. Kao Data in London yielded an average PUE of 1.21 at a full IT load by incorporating an IEC in its AC infrastructure [34]. Yan et al. performed multi-objective optimizations of a counterflow IEC for DCs under hot and dry climate conditions based on response surface methodology [35]. Cui et al. analyzed the climatic applicability of a single-stage IEC with mechanical cooling for DCs in China. Results show sufficient cooling in western arid regions, while supplementary cooling is necessary with a prolonged operation period in hot and humid regions [36].

From the literature above, despite these successes in the application of IEC technology in DCs of arid regions, some research gaps still remain to be addressed, especially in hot and humid regions. Firstly, due to the high humidity of ambient air, evaporation is constrained so that the cooling effect will be noticeably reduced, and this is the main reason why IEC technology is yet to be adopted in DCs of such areas. Secondly, the cooling performance of an IEC system through the addition of a liquid desiccant to overcome geographical limitations in hot and humid areas has not been quantitatively detailed. Thirdly, the system feasibility remains to be analyzed based on the local weather profile of a typical hot-humid region.

In response to these gaps, this study proposed a novel multi-stage IEC system tailored for a DC situated in hot-humid regions. A simulation model was established with validation, and the cooling performance of the system was investigated under a variety of conditions through parametric analysis. Then, a case study was performed to assess the viability of the hybrid system in a typical hot-humid area (Hong Kong), indicating a potential approach to lessen the reliance on traditional AC systems and fully use passive cooling technology for DC cooling in similar climatic conditions. The limitations and future works are summarized as well to guide the further research steps.

3. Methodology

3.1. System Configuration

As illustrated in the Section 1, the evaporative cooling process is inherently constrained by climatic conditions. Particularly in the summer months, the ambient outdoor relative humidity tends to be substantially elevated when the cooling requirement intensifies. Typically, both the primary air and secondary air are derived from outdoor air in an IEC. The secondary air is intended to facilitate evaporation upon contact with the water through the humidity difference with the saturated liquid surface. Nonetheless, in hot and humid regions, the relative humidity during summer can be so high that it sometimes reaches 100%, at which point the evaporation process in the IEC becomes very difficult. This impediment can lead to a cooling output that fails to satisfy the requisite standards for thermal management. To mitigate this limitation and procure drier air for an improved cooling effect, this study proposes a multi-stage evaporative cooling system tailored for DC applications. The proposed system encompasses a liquid-desiccant-based (LD) IEC, a conventional IEC, and an auxiliary cooling section to fulfill the cooling demand. In DCs, there is typically no need for fresh air, or a separate AC system is responsible for handling outdoor fresh air and delivering it to specific personnel areas. The hot air produced by the electronic devices is handled using AC equipment and circulated. The schematic diagram of the airflow configuration used in a DC is shown in Figure 2. The LD-IEC and the normal IEC are made of a metal material with good anti-corrosion. In summer (Figure 2a), liquid desiccant dehumidification is needed to reduce the humidity of outdoor air so that evaporation can happen under a larger gradient in the secondary air channel. The outdoor fresh air passes the LD-IEC as primary air, and the liquid solution is sprayed into the primary air channel to dehumidify moist fresh air. The primary outlet air from LD-IEC is divided into two air streams. Air stream 1 is delivered to the second-stage IEC as secondary air, while air stream 2 is introduced to the secondary air channel of the first-stage LD-IEC. As air stream 2 has been dehumidified, it can result in a greater humidity differential between it and the water film on the secondary channel, which enhances the evaporation process. This, in turn, contributes to a decrease in the temperature of the total air stream in the primary air channel compared to the situation without dehumidification and removes some latent heat released during the dehumidification process. After passing through the secondary air channel and aiding in the evaporation process, air stream 2 is discharged into the ambient environment at the exit. Then, the circulated hot air from the DC can be cooled when it is passing the primary air channel of the second-stage IEC. To ensure that the supply air temperature is able to cool the electronic device in the DC, a supplementary cooling section is positioned after the second-stage IEC. In spring and autumn (Figure 2b), only the second-stage IEC needs to operate, and the on/off status of the water spraying depends on the supply air temperature and weather conditions.

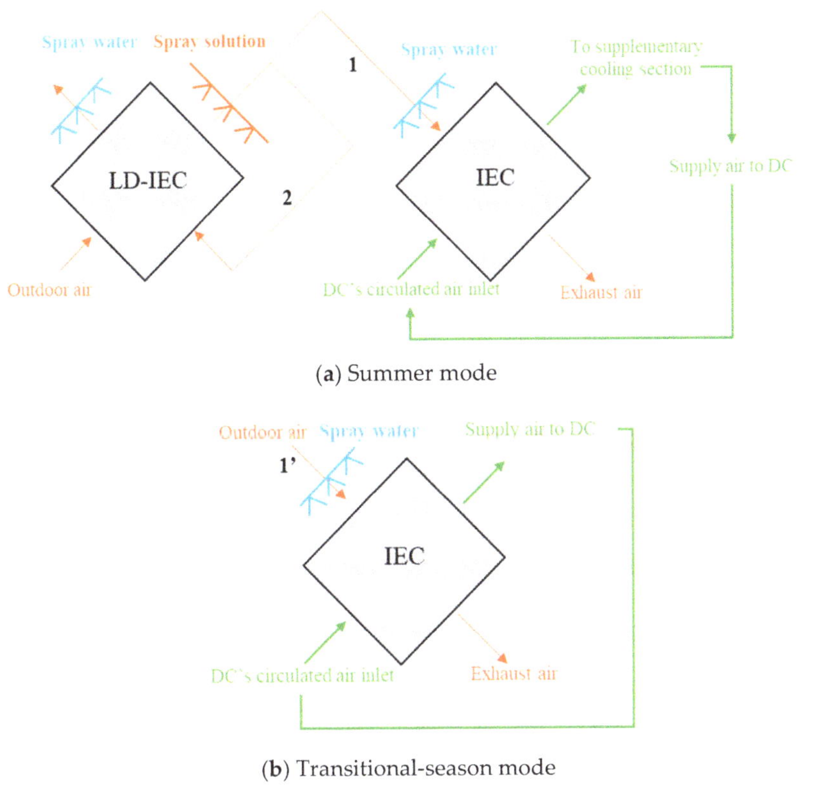

Figure 2. Schematic diagram of the multi-stage evaporative cooling system used for a DC. 1: First-stage LD-IEC outlet primary air as second-stage IEC secondary air. 2: First-stage LD-IEC outlet primary air as its secondary air. 1′: Outdoor air as secondary air of IEC.

3.2. Model Establishment

To facilitate the modeling of the aforementioned air conditioning (AC) devices, it is imperative to delineate certain assumptions, which are also widely endorsed in existing literature [37,38] as follows: (1) The external environment is considered to be isolated from the LD-IEC/IEC, negating any heat or mass exchange with the surroundings; (2) Both air and water are treated as incompressible fluids in steady state; (3) Due to their small *Re* number, the states of the fluids can be considered as laminar flow; (4) Thermal resistance of the thin channel sheet and the liquid membrane is negligible because their tiny thickness; (5) The solution and water are distributed uniformly over the whole of the passages; (6) The desiccant solution is maintained at a consistent concentration, with the potential for regeneration via a heat source such as solar energy; (7) The heat and mass transfer occur solely along the fluid's flow direction.

3.2.1. IEC Model

The 2D numerical model of the IEC has been developed according to the general assumptions outlined in our previous study, and the essential equations are briefly summarized as follows. The core equations that govern the behavior of the secondary air and its interaction with the water membrane are expressed from Equations (1)–(3):

$$h_s(t_w - t_s)\cdot dxdy + h_{fg}h_{ms}(\omega_{t_w} - \omega_s)dxdy = \dot{m}_s\frac{\partial i_s}{\partial y}dy \tag{1}$$

$$h_{ms}(\omega_{s,\,sat} - \omega_s)\cdot dxdy = m_s \frac{\partial \omega_s}{\partial y}\cdot dy \quad (2)$$

$$\dot{m_s} = \frac{dx}{L} m_s \quad (3)$$

The heat transfer equation for the primary air is:

$$h_p(t_p - t_w)\cdot dxdy = c_{pa}\dot{m_p}\frac{\partial t_p}{\partial y}dx \quad (4)$$

$$\dot{m_p} = \frac{dy}{H} m_p \quad (5)$$

The energy balance can be written as:

$$\dot{m_s}\frac{\partial i_s}{\partial y} - c_{pa}\dot{m_p}\frac{\partial t_p}{\partial x} = c_{pw}t_{ew}\frac{\partial m_e}{\partial y} \quad (6)$$

$$\frac{\partial m_e}{\partial y} = \dot{m_s}\frac{\partial w_s}{\partial y} \quad (7)$$

The convective heat and mass transfer coefficients are connected through the Lewis number, as shown in Equation (8) [39], while the convective heat transfer coefficient is correlated with the Nusselt number, as seen in Equation (9).

$$Le^{\frac{2}{3}} = \frac{h_s}{h_{ms}c_{pa}} \quad (8)$$

$$h = \frac{Nu\cdot\lambda}{d_e} \quad (9)$$

$$Nu = 0.664Re^{\frac{1}{2}}Pr^{\frac{1}{3}} \quad (10)$$

where the Lewis number is determined as 1 [40]; c_{pa} = 1.005 kJ/(kg·°C).

The saturated humidity ratio of the interface of the liquid film and secondary air is formulated by Equation (12) [18]:

$$ln(P_{sat}) = \frac{e_1}{t_w} + e_2 + e_3t_w + e_4{t_w}^2 + e_5{t_w}^3 + e_6\ln(t_w) \quad (11)$$

$$w_{s,\,sat} = 0.622\frac{P_{sat}}{P_{atm} - P_{sat}} \quad (12)$$

where: e_1 = 5800.2206; e_2 = 1.3914993; e_3 = −0.04860239; e_4 = 4.1764769 × 10^{-5}; e_5 = −1.4452093 × 10^{-5}; e_6 = 6.5459673; P_{atm} = 101.325 kPa; and t_w is the water film temperature, °C.

3.2.2. LD-IEC Model

For the LD-IEC, the liquid desiccant is dispersed into the primary air channel to extract water vapor from the humid air. Thus, Equations (13) and (14) are added to characterize the gradient of the moisture content, and the new energy balance formula is written as Equation (15). The equations to illustrate the heat and mass transfer in the secondary air channel are identical to that shown in the IEC model in Section 3.2.1. A lithium chloride solution, as a common desiccant material, is utilized in this study. The dehumidification of the moist air is driven by the vapor pressure difference between the air–solution interface and the main moist air stream. The equilibrium vapor pressure on the desiccant solution film surface is a function of both its temperature and concentration. Based on the thermodynamic properties acquired from published studies [41], the equilibrium vapor pressure and corresponding moisture content at the liquid desiccant interface within the primary air channel can be obtained through Equations (16) and (17).

$$h_{mp}(\omega_{sol}-\omega_p)\cdot dxdy = \dot{m_p}\frac{\partial \omega_p}{\partial x}\cdot dx \tag{13}$$

$$-\dot{m_p}\frac{\partial \omega_p}{\partial x} = \frac{\partial m_d}{\partial y} \tag{14}$$

$$\dot{m_s}\frac{\partial i_s}{\partial y} - \dot{m_p}\frac{\partial i_p}{\partial x} = c_{pw}t_{ew}\frac{\partial (m_e)}{\partial y} + c_{pw}t_{dw}\frac{\partial (m_d)}{\partial y} \tag{15}$$

$$P_{sol} = C_1 + C_2 t_{sol} + C_3\sigma + C_4 {t_{sol}}^2 + C_5\sigma^2 + C_6 t_{sol}\sigma \tag{16}$$

where $C_1 = -1.004$; $C_2 = 0.1265$; $C_3 = 5.982$; $C_4 = 0.003644$; $C_5 = 11.13$; $C_6 = -0.7975$; t_{sol} is the solution film temperature, °C; σ is solution concentration; and $c_{pw} = 4.18$ kJ/(kg·°C).

$$w_{sol} = 0.622\frac{P_{sol}}{P_{atm} - P_{sol}} \tag{17}$$

where t_{sol} is the solution temperature within the range from 20 to 50 °C in Equation (13), and the solution concentration (σ) is limited within the range from 0.20 to 0.40 to avoid crystallization [42].

Referring to the system configuration in Figure 2, the boundary conditions of the first-stage LD-IEC and the second-stage IEC models are listed as follows. For the LD-IEC: $t_{p,in,LD-IEC} = t_{amb}$; $w_{p,in,LD-IEC} = w_{amb}$; $m_{p,in,LD-IEC} = m_{in,\ total}$; $t_{s,in,LD-IEC} = t_{p,out,LD-IEC}$; $w_{s,in,LD-IEC} = w_{p,out,LD-IEC}$; $m_{s,in,LD-IEC} = m_{p,in,LD-IEC}(1-r)$. For the IEC: $t_{p,in,IEC} = t_{circ,DC}$; $w_{p,in,IEC} = w_{circ,DC}$; $m_{p,in,IEC} = m_{p,LD,in}r$; $t_{s,in,IEC} = t_{p,out,LD}$; $w_{s,in,LD} = w_{p,out,LD}$; $m_{s,in,IEC} = m_{p,LD,in}r$; $r = \frac{m_{p,LD,in} - m_{s,LD,in}}{m_{p,LD,in}}$.

3.3. Model Solution and Grid Independence Monitoring

The multi-stage IEC system consists of the first-stage LD-IEC and the second-stage IEC. For the first-stage LD-IEC, we use Equations (1)–(3) and (8)–(17). For the second-stage IEC, we use Equations (1)–(12) for calculation. The finite difference method was utilized to discretize the aforementioned sets of partial differential equations into algebraic equations. Combined with the boundary conditions, MATLAB 2018b software was employed for code programing and solving. The numerical solution procedures are illustrated as follows: (1) Input air inlet conditions and geometry data (and concentration ratio and mass flow ratio for LD-IEC); (2) Calculate heat and mass transfer coefficients; (3) Assume the channel plate temperature and calculate saturated humidity; (4) Solve the governing equations of LD-IEC or IEC; (5) Figure out the new channel plate temperature, compare it with the assumed one, and judge the convergence; (6) Output the results or adjust the channel plate temperature if it cannot meet the convergence criteria. The grid number is checked to achieve a balance between accuracy and computational efficiency. As depicted in Figure 3, the variations of the results are within 0.05% when the number exceeds 140, indicating that the effect of increasing the grid number is tiny and negligible. Therefore, 140 × 140 is selected for further calculations. A threshold of 0.001 is determined as the convergence criterion.

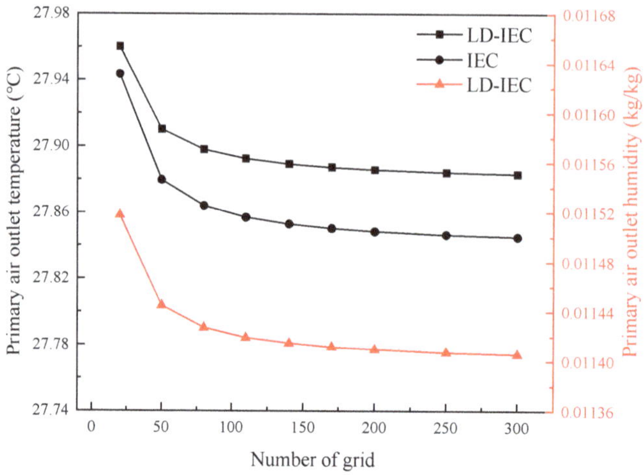

Figure 3. Outlet parameters under grid independence monitoring.

3.4. Model Validation

With respect to the model validation, the multi-stage IEC system for DC cooling has not been explicitly proposed in the existing studies so it may not be straightforwardly validated. Nonetheless, the two components in the system can be verified individually using the published results [43,44]. Comparisons between the results from the models and published studies are exhibited in Figure 4. The maximum relative deviation of the IEC primary air outlet temperature is 3.3%, while the greatest relative discrepancies of the primary air outlet temperature and humidity of the LD-IEC are −6.4% and 2.6%, respectively. These values underscore the acceptable predictive precision of the established models, thereby indicating their applicability for subsequent analysis of the proposed system for DC cooling.

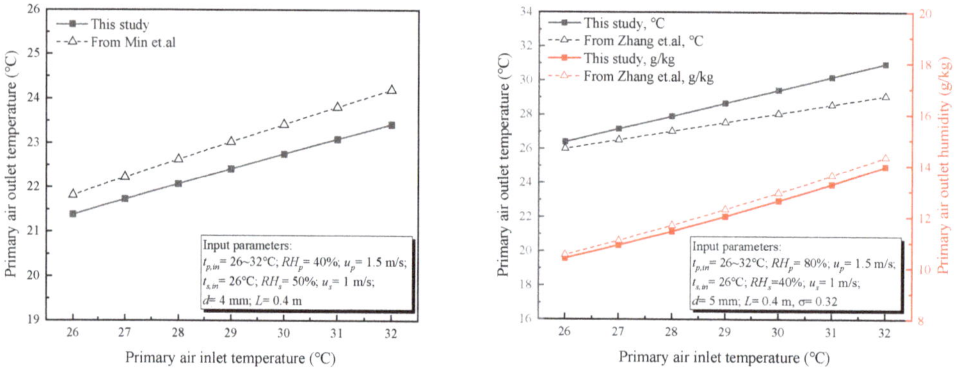

(**a**) (**b**)

Figure 4. Comparisons between the results from the present models and published research on (**a**) IEC [44], (**b**) LD-IEC [43].

4. Results and Discussions

To assess the thermal performance of the proposed multi-stage IEC system for DC cooling, the effects of essential parameters are presented. The outlet air temperature and humidity of the LD-IEC, the outlet air temperature (supply air) of the IEC, the primary air temperature drop of the IEC, and wet-bulb efficiency are illustrated in various conditions in this section. The ranges of analyzed parameters are listed in Table 1.

Table 1. Ranges of essential parameters.

Symbol (Unit)	Range	Basic Value
$t_{p,in,LD\text{-}IEC}$ (°C)	[26:2:34]	30
$RH_{p,in,LD\text{-}IEC}$ (%)	[65:5:95]	80
σ (-)	[0.2:0.05:0.4]	0.3
r (-)	[0.5:0.1:0.9]	0.7
$t_{p,in,IEC}$ (°C)	[32:2:40]	34
$u_{p,in,IEC}$ (m/s)	[2:0.5:4]	3
$m_{p,in,LD}$ = 0.003 kg/s; L = 400 mm; one channel pair for both IEC and LD-IEC		

Note: [a:b:c]: a is the start value; c is the end value; b is the interval.

4.1. Effect of Primary Air Temperature

Figure 5 delineates the impact of outdoor ambient air temperature (primary air temperature of the LD-IEC) on the thermal performance of the proposed system. A direct correlation is noticed in Figure 5a between the outdoor air temperature and the system's outlet conditions. As the outdoor air temperature rises, both the outlet temperature and humidity of the first-stage LD-IEC experience an increase. Specifically, when the outdoor air temperature at the LD-IEC inlet increases from 26 °C to 34 °C, the outlet primary air temperature and humidity of the LD-IEC increase from 23.3 °C to 30.3 °C and from 10.79 g/kg to 16.64 g/kg, respectively. In each case, there is a significant decrease in the wet-bulb temperature of the LD-IEC outlet air compared to the inlet air, decreasing from 23.3–30.9 °C to 18.1–24.3 °C.

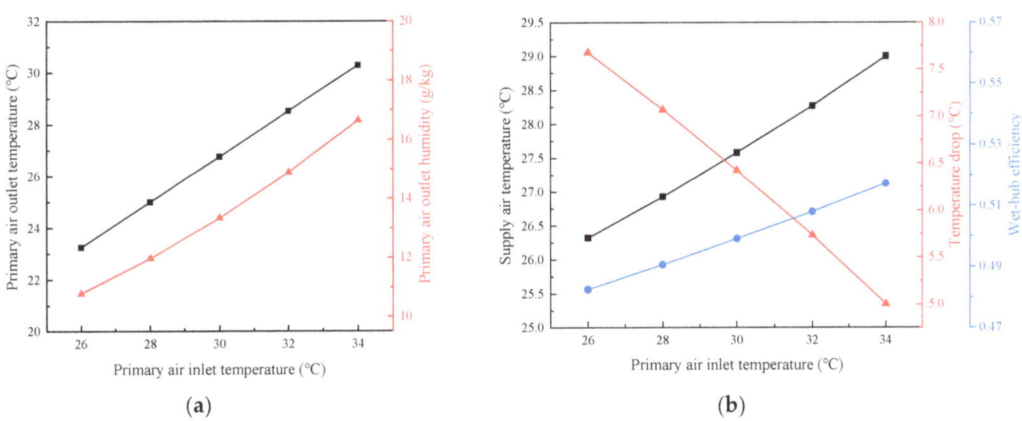

Figure 5. Effect of primary air temperature on the thermal performance of (**a**) first-stage LD-IEC, (**b**) second-stage IEC.

Due to the utilization of a fraction of the LD-IEC primary air as the secondary air for the second-stage IEC, the supply air temperature (primary air outlet temperature of the IEC) increases from 26.3 °C to 29 °C. The associated temperature decrement achieved by the IEC is observed to decrease from 7.7 °C to 5.0 °C. DCs typically require the supply temperature to be controlled within the range of 18 °C to 27 °C. Consequently, when the inlet air temperature of the LD-IEC exceeds 28 °C, the proposed system fails to meet the specified supply temperature under this operating condition. An auxiliary cooling mechanism is necessitated to maintain the supply air within the prescribed thermal range. In addition, the wet-bulb efficiency of the IEC is influenced by both the primary air temperature drop of the IEC and the secondary air (i.e., outlet air status of the LD-IEC). As depicted in Figure 4b, a marginal enhancement in wet-bulb efficiency is noted with increasing inlet temperatures, with the wet-bulb efficiency ascending from 0.482 to 0.517.

4.2. *Effect of Primary Air Relative Humidity*

Figure 6 presents the effect of variations in the relative humidity (RH) at the inlet of the primary air on the thermal performance of the multi-stage IEC system. As the RH of the primary air increases, although the liquid desiccant can remove moisture content from the air, the overall humidity ratio increases, and the air's capacity to evaporate water diminishes, leading to a higher wet-bulb temperature and thus elevated outlet temperature and humidity after passing through the first-stage LD-IEC. As shown in Figure 6a, an escalation in the RH of the primary air from 65% to 95% at the LD-IEC inlet leads to an increase in both the temperature and humidity at the primary air outlet, with values ranging from 26.3 °C to 27.3 °C and 17.39 g/kg to 25.75 g/kg, respectively. The LD-IEC contributes to a significantly lower wet-bulb temperature compared to directly utilizing high-humidity outdoor air as the secondary air for the IEC. The wet-bulb temperature exhibits a considerable decrease from a span of 24.7 °C to 29.3 °C down to a range of 19.6 °C to 22.7 °C, indicating a more favorable potential for air cooling.

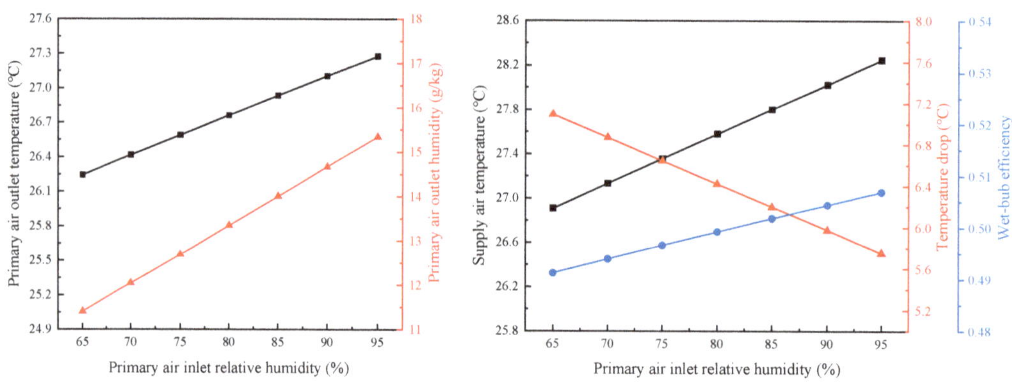

(**a**) (**b**)

Figure 6. Effect of primary air relative humidity on the thermal performance of (**a**) first-stage LD-IEC, (**b**) second-stage IEC.

Furthermore, the increased temperature and humidity of the primary air emerging from the LD-IEC, which subsequently functions as the secondary air for the second-stage IEC, causes an upsurge in the supply air temperature from 26.9 °C to 28.3 °C. Consequently, the cooling ability of the second-stage IEC for the primary air decreases from 7.1 °C to 5.8 °C. Notably, even at an RH of 65%, the supply air temperature nears the upper threshold of 27 °C, prompting the need for an auxiliary cooling section to maintain the supply air temperature within the desired parameters. Additionally, the wet-bulb temperature of the IEC is influenced by both the temperature difference of the primary air and the secondary air conditions (i.e., outlet air temperature of the primary air from the LD-IEC in this study). As observed in Figure 6b, an increase in the RH of the air at the LD-IEC inlet results in a slight improvement in the wet-bulb efficiency, rising from 0.491 to 0.507.

4.3. *Effect of Mass Flow Ratio*

Figure 7 illustrates the effect of variation in the mass flow ratio (r) on the thermal performance of the proposed system. When the r value grows, the outlet temperature and humidity of the LD-IEC rise, with a steeper slope. As presented in Figure 7a, it is noticed that the higher r value from 0.5 to 0.9 corresponds to a decrease in the mass flow rate of the secondary air, resulting in greater outlet temperature and humidity of the primary air of the LD-IEC from 26.2 to 28.0 °C and from 13.13 to 13.93 g/kg, respectively. The wet-bulb temperature of the air processed by the LD-IEC ranges from 20.8 to 21.9 °C, significantly lower than the inlet wet-bulb temperature (27.1 °C), indicating a greater potential for air cooling as the secondary air for the IEC.

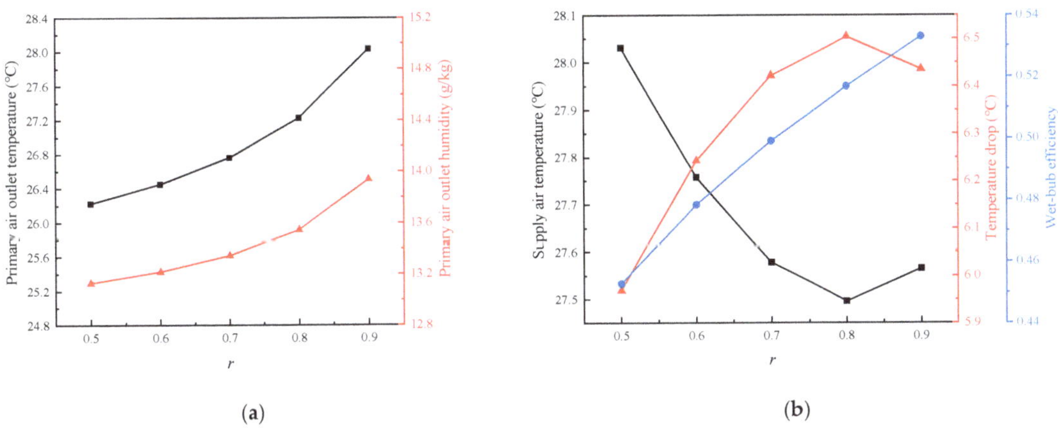

Figure 7. Effect of primary air to secondary air ratio on the thermal performance of (**a**) first-stage LD-IEC, (**b**) second-stage IEC.

In contrast to the consistent trends observed in the previous parameters, Figure 7b presents a more complex relationship when it comes to the supply air temperature as r increases. It is seen that as the r value increases from 0.5 to 0.8, the supply air temperature gradually decreases, albeit at a reduced rate. However, the supply air temperature tends to grow when r is greater than 0.8. Correspondingly, the temperature drop generated by the second-stage IEC initially increases from approximately 6.0 °C to 6.5 °C, and then slightly decreases to 6.4 °C. This phenomenon can be explained by the fact that although the outlet temperature of the first-stage LD-IEC increases with an increase in the r value, the primary air from the LD-IEC serves as the secondary air for the second-stage IEC. An increase in the r value implies an increase in the flow rate entering the secondary channel of the IEC. With a constant channel spacing, a larger flow rate results in a higher convective heat and mass transfer, allowing the supply air temperature of the IEC to continue decreasing until an r value of 0.8. However, as the wet-bulb temperature also increases, its influence on the supply air temperature gradually becomes greater than the effect of the increased convective heat transfer coefficient, ultimately leading to an increase in the supply air temperature. An essential insight from this analysis is that optimizing the r value is critical to maintaining the cooling performance of the proposed system. A value that is too high may lead to less effective cooling, as demonstrated by the supply air temperature increasing beyond the optimal range for r values greater than 0.8.

In the analysis of this parameter, the supply air temperature exceeds 27 °C, indicating the need for supplementary cooling to ensure the desired supply air temperature. Regarding the wet-bulb temperature of the IEC, Figure 7b shows a significant increase in wet-bulb efficiency from 0.453 to 0.533 as the r value increases.

4.4. Effect of the Solution Concentration

Figure 8 presents the influence of variations in the concentration of the desiccant solution (σ) sprayed in the primary air channel of the LD-IEC on the system's thermal performance. As the concentration becomes stronger, the difference in vapor pressure between the air–liquid interface and the main moist airstream increases, creating a larger gradient and facilitating moisture migration in the humid air. Consequently, the outlet temperature and humidity of the LD-IEC decrease as the concentration of the desiccant solution increases. In Figure 8a, it is observed that an increase in σ from 0.2 to 0.4 corresponds to a decrease in the outlet temperature and humidity of the primary air of the LD-IEC from 27.7 °C to 26.2 °C and from 17.24 g/kg to 11.30 g/kg, respectively. Moreover, the wet-bulb temperature of the air processed by the LD-IEC decreases from the inlet temperature of 27.1 °C to a range of 19.5–24 °C. As the temperature and humidity of the primary air from the LD-IEC (which serves as the secondary air for the second-stage IEC) decrease, the supply air temperature of the second-stage IEC increases from 28.9 °C to 26.9 °C, and the cooling ability of the IEC for the primary air increases from 5.1 °C to 7.1 °C. When σ reaches 0.4, the IEC's supply air temperature falls below 27 °C, indicating that supplementary cooling can be suspended at this point. This implies that at higher σ values, the multi-stage IEC system could potentially meet the cooling requirements without the need for additional cooling methods. It is important to mention that the regeneration of the desiccant solution, which involves restoring the diluted solution that has absorbed moisture from the humid air, requires a dedicated regeneration device. This ensures a continuous and stable humidity gradient between the primary air and the solution surface. The regeneration process can utilize waste heat or solar thermal energy, and further investigation will be conducted in future research to optimize this aspect. The discussion of solution regeneration is beyond the scope of this paper. Regarding the wet-bulb temperature of the IEC, it is seen from Figure 8b that as the concentration becomes stronger, the wet-bulb efficiency slightly decreases from 0.513 to 0.491.

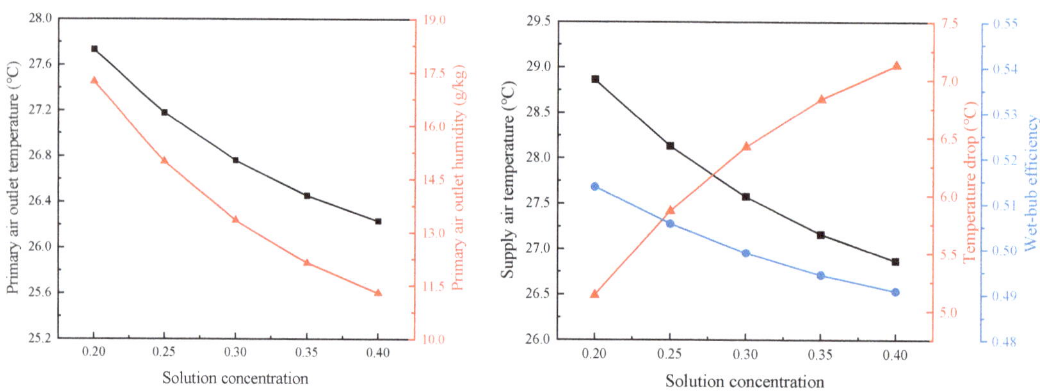

(a) (b)

Figure 8. Effect of solution concentration on the thermal performance of (**a**) first-stage LD-IEC, (**b**) second-stage IEC.

4.5. *Effect of Temperature of Circulated Air from the DC*

Figure 9 presents the impact of variations in the inlet air temperature of the second-stage IEC's primary air channel on the system's thermal performance. The impetus for this investigation stems from the fluctuating air temperatures in the DC, attributable to heat emissions from electronic components, which are influenced by factors such as the DC size, energy-saving requirements, and density of the installed equipment. Initially, the first-stage LD-IEC reduces the wet-bulb temperature of the humid air from 27.1 °C to 21.1 °C, which is then utilized as the secondary air for the second-stage IEC, significantly enhancing the IEC's cooling capacity. As the inlet temperature of the circulating air gradually increases, the outlet air temperature of the IEC also rises. Figure 9 demonstrates that when the inlet temperature increases from 32 °C to 40 °C, the IEC's outlet air temperature increases from

26.6 °C to 30.4 °C, resulting in an increase in the IEC's cooling capacity for the primary air from 5.4 °C to 9.7 °C. Upon the circulating air temperature surpassing 32 °C, the outlet air temperature of the second-stage IEC exceeds 27 °C, signifying the requisition of auxiliary cooling to attain the prescribed supply air temperature for DCs. Regarding the wet-bulb temperature of the IEC, Figure 9 shows a slight increase in wet-bulb efficiency from 0.495 to 0.510 as the temperature of the circulating air rises.

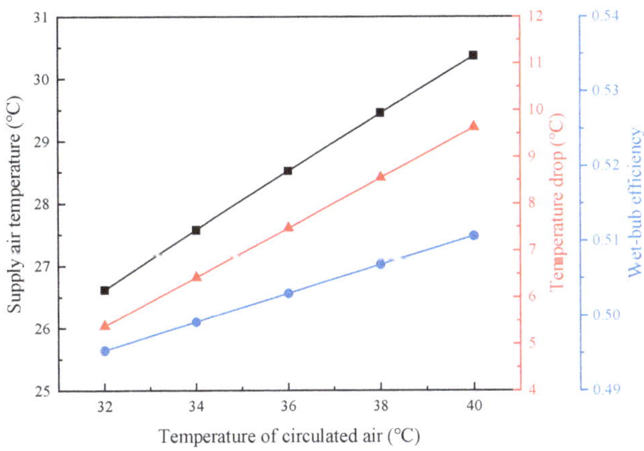

Figure 9. Influence of temperature of circulated air from the DC on the thermal performance of second-stage IEC.

4.6. Effect of Velocity of Circulated Air from the DC

Figure 10 illustrates the effect of variations in the air velocity of the second-stage IEC's primary air channel on the system's thermal performance. Since the velocity of the second-stage IEC does not impact the air treatment process of the first-stage LD-IEC, as discussed in the previous section, the first-stage LD-IEC significantly reduces the wet-bulb temperature of the inlet humid air to 21.1 °C, which serves as the secondary air for the second-stage IEC. It is observed that as the velocity of the primary channel of the second-stage IEC increases, the outlet air temperature of the IEC also grows. If the air is accelerated from 2 m/s to 4 m/s, the IEC's outlet air temperature rises from 25.5 °C to 27.4 °C. This inversely affects the IEC's temperature drop of the primary air from 6.5 °C to 4.6 °C. Although an increase in velocity enhances convective heat transfer, it also raises the mass flow rate, and the latter has a more significant effect, leading to an increase in the temperature of the supply air as the circulated air velocity increases. When the air velocity in the channel exceeds 3.5 m/s, the IEC's outlet air temperature surpasses 27 °C, indicating the need to activate supplementary cooling to meet the desired supply air temperature for data centers. Regarding the wet-bulb temperature of the IEC, Figure 10 demonstrates a rapid increase in wet-bulb efficiency from 0.602 to 0.424 as the air velocity rises. This efficiency upsurge is predominantly ascribed to the intensified air flow, which, despite enhancing convective heat transfer to a certain degree, primarily leads to a notable temperature increment. Thus, the primary air velocity in the IEC should be carefully considered in practical engineering projects.

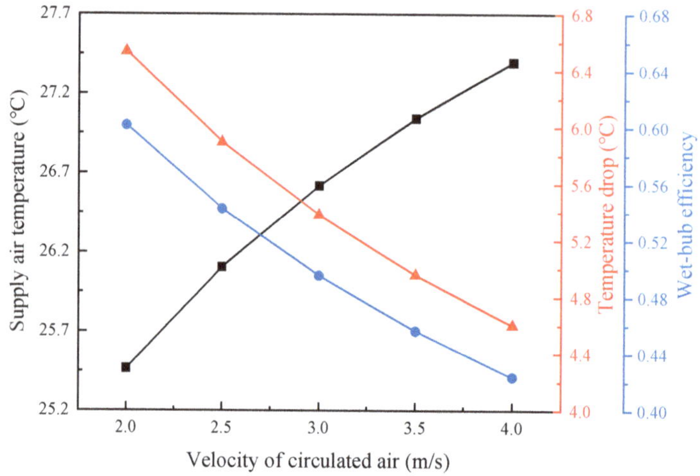

Figure 10. Effect of velocity of circulated air from the DC on the thermal performance of second-stage IEC.

4.7. Case Study for the Application in a Hot and Humid Region

As mentioned in the previous section, the integration of the liquid desiccant in the IEC system mitigates the detrimental effects of high ambient humidity levels on the evaporation process, particularly within hot and humid climates. To present the annual feasibility of employing the multi-stage IEC system, Hong Kong's weather conditions in the typical meteorological year (TMY) are input as the primary air of the LD-IEC. According to the existing literature on the AC operation status of the typical DC, the circulated air after removing the heat from the IT equipment can be determined as 34 °C. In addition, the recommended range of supply air temperature to the electronic equipment zone is from 18 °C to 27 °C.

Figure 11 presents the supply air temperature of a single IEC and the proposed IEC system given the outdoor air conditions in Hong Kong. When the multi-stage IECs or the single IEC cannot handle the circulated air to the required temperature (27 °C), the rest of the cooling load will be treated by the supplementary cooling section, and the operating hour is counted. As shown in Figure 11, both the standalone IEC system and the multi-stage IEC system demonstrate effective cooling of circulating air when outdoor temperatures and humidity are relatively low, such as in early spring, late autumn, and winter. However, the cooling capacity of the single IEC diminishes greatly during the summer months due to the severe constraints imposed by high outdoor temperatures and humidity on water evaporation. The average temperature drop is only 5.5 °C in summer, which is much lower than the annual average temperature drop of 8.7 °C. The number of hours exceeding the upper temperature limit (27 °C) reaches as high as 3550 h, necessitating prolonged operation of supplementary cooling equipment, particularly in the summer. Furthermore, for engineering projects where the supply air temperature is designed to be 25 °C, this would significantly increase the operational time of supplementary cooling equipment to 5092 h annually. In comparison to a single IEC, this system benefits from the incorporation of an LD-IEC, which provides a lower wet-bulb temperature for the secondary air used in the IEC. Consequently, this enables the IEC to consistently reduce the supply air temperature to below 27 °C. During the summer months, the system achieves an average temperature difference of 9.5 °C, which is improved by 72.7% compared to that of the standalone IEC. The annual average temperature drop is 11.5 °C, both of which are significantly superior to a standalone IEC system. Given that the supplementary cooling module is activated if the supply air exceeds 25 °C, this multi-stage IEC system reduces the operational hours of the supplementary cooling module to a mere 31 h per year, substantially cutting down the runtime of the supplementary cooling system. This advancement represents a significant stride toward the goal of relying solely on passive cooling methods for DC cooling.

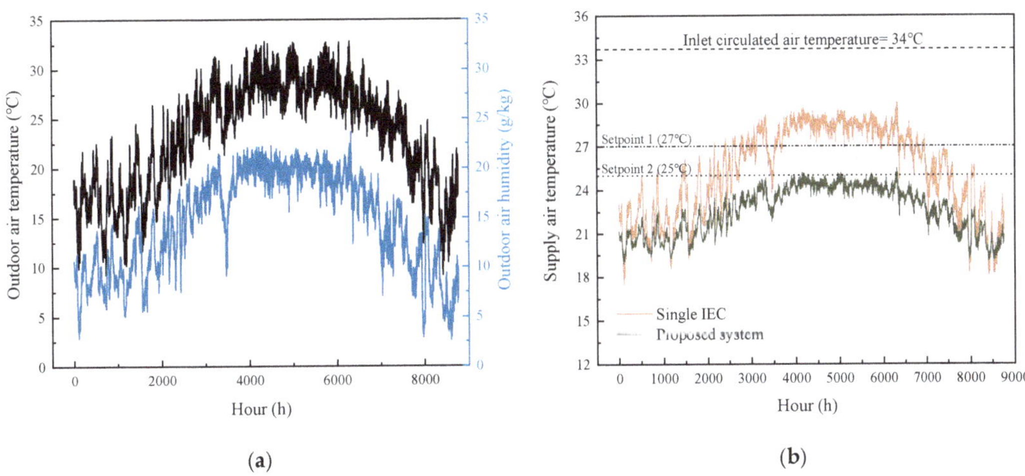

Figure 11. Profile of (**a**) outdoor air status in Hong Kong, (**b**) supply air temperature of the proposed system.

4.8. Discussions

In hot and humid regions, the interaction between the primary air temperature and humidity with the system's outlet conditions underscores the reliance of evaporative cooling processes on ambient air properties. The increased temperature and RH challenge the cooling performance due to the diminished humidity gradient between the air and the wet surfaces within the IEC. The LD-IEC's role in pre-treating the air before it enters the second-stage IEC is instrumental in enhancing the overall thermal performance. By reducing the humidity, the first-stage LD-IEC effectively lowers the wet-bulb temperature of the air, which is an essential factor of the evaporative cooling potential. The mass flow ratio (r) emerges as a critical parameter, with an optimal range necessary to balance the cooling efficiency against the system's cooling performance. The desiccant solution concentration (σ) dictates the dehumidification strength and the subsequent cooling potential as well. Regarding the second-stage IEC, inlet circulated air temperature from the DC and the channel velocity influence the cooling capacity, where increased air temperatures can improve the IEC's temperature drop and wet-bulb efficiency. The elevated air velocity can improve the convective heat transfer, but the increased mass flow rate is the predominant factor that ultimately leads to higher outlet temperatures. Lastly, the case study involving Hong Kong's climatic conditions indicates that the multi-stage IEC system's ability to maintain the supply air temperature within the desired threshold for the majority of the year marks a substantial improvement over conventional single-stage IEC systems, which offers insights into the system's annual operational feasibility and demonstrates the system's capability in reducing the reliance on the energy-intensive traditional cooling system for DCs in hot and humid regions.

Table 2 presents a comparison between recent studies on hybrid IEC systems and this study. Currently, a few composite systems related to DCs are reported, especially in hot and humid regions. In hot and arid regions, both the primary and secondary air for IEC can be sourced from outdoors, as the lower wet-bulb temperature outside makes it possible for a single-stage IEC to achieve the required supply air temperature. However, in hot and humid regions, the high outdoor humidity significantly limits the effectiveness of evaporative cooling. Therefore, some researchers propose to utilize exhaust air from air-conditioned spaces as the secondary air to enhance the cooling effect of the IEC [23,25]. However, for buildings with specialized uses such as DCs which typically involve air recirculating between the indoor space and air handling units without fresh air [30], the source of secondary air can only be from outdoors, and this outdoor air needs to be pre-treated to reduce its humidity and wet-bulb temperature before it can be used as the secondary air in the second-stage IEC. Hence, this study has proposed the LD-IEC to achieve this purpose. Compared to existing studies, the achieved temperature drop is satisfactory, demonstrating the potential of using this system in hot and humid regions.

Table 2. Comparison with the published studies.

Research	Nemati et al. [24]	Hu et al. [29]	Wan et al. [23]	Present Study
Type of IEC	Counter flow	Counter flow	Counter flow	Cross flow
IEC size (L × H, m^2)	0.5 × 0.5	1 × 1	0.6 × 0.15	0.6 × 0.6
System configuration	Earth-air heat exchanger + IEC	RC + IEC	IEC + cooling coil	LD-IEC + IEC
Building type	Residential building	Residential building	Commercial building	Data center
Climate region	Hot and arid	Hot and arid	Hot and humid	Hot and humid
Working air source	Outdoor space	Outdoor space	Indoor AC space	Outdoor space
Temperature drop (°C)	4.2	8.8	10	9.5

In addition to being a technical solution to DC cooling in hot-humid regions, this study may be possible to motivate the update of some policies related to IECs and DCs. Firstly, IEC systems are originally suitable for use in arid and hot regions. However, the results mentioned above have demonstrated that with the assistance of a desiccant solution, a composite IEC system can be effectively applied in hot and humid areas. Consequently, policies can encourage the adoption of IEC systems in DCs within hot and humid regions, supplemented by the liquid desiccant described in this paper (i.e., spraying the desiccant solution in the primary air channel in the first stage). In addition, DCs are typically low-rise buildings with ample roof area. Policies can suggest making full use of the rooftop space to install solar energy equipment for regeneration of the desiccant solution or even power supply to such IEC systems.

5. Limitations and Future Works

This study has conceptually proposed a multi-stage system and analyzed its thermal performance for data center (DC) cooling in hot and humid climates. Nevertheless, the system's integration remains unrealized in the laboratory in reality. The detailed approach for regenerating the liquid desiccant—a critical component of the system—has not been included in this research. The strategies to incorporate renewable energy such as solar energy are not presented as well. Furthermore, the integrated system is not optimized to contribute a better cooling performance. Additionally, a comprehensive economic analysis, which would include the determination of potential economic benefits and payback periods, is requisite for practical engineering applications. Therefore, future works will be carried out based on the above limitations: (1) To assemble the multi-stage IEC on the laboratory scale or in a pilot-scale setting; (2) To incorporate the liquid desiccant regeneration system for further investigation; (3) To involve a renewable energy system to enhance the low-carbon and sustainability aspects; (4) To optimize the proposed system for optimal operating conditions; (5) To carry out an economic analysis to estimate the system's cost effectiveness, operational expenses, and return on investment.

6. Conclusions

In this study, a multi-stage IEC system was developed for the cooling of data centers (DCs) situated in hot and humid regions. Liquid desiccant (LD) was incorporated into the first-stage IEC to mitigate the influence of high-humidity ambient air on the cooling performance and expand the application range of the IEC technology. The model of the proposed system was constructed with validation from the published literature, and the parametric analysis was carried out to investigate the thermal performance under various inlet conditions. Furthermore, a case study was conducted based on a typical hot-humid climatic zone (Hong Kong) to demonstrate the feasibility of the proposed system for achieving the required supply air temperature for a DC. The main findings are summarized as follows.

(1) The first-stage LD-IEC effectively reduces the wet-bulb temperature of the ambient air, thereby enhancing the cooling performance of the second-stage IEC. However, a higher ambient air temperature, relative humidity, and circulated air velocity may diminish the temperature drop, while an increased solution concentration and circulated air temperature can expand the temperature drop.
(2) The cooling performance of the system improves with an increasing mass flow ratio up to 0.8, but experiences a reduction as the mass flow ratio continues to grow to 0.9. This highlights the importance of determining the optimal value during the design stage.
(3) The case study illustrates that the multi-stage IEC system can alleviate the constraints of high-humidity ambient air on the cooling effect compared to a single IEC, which greatly improves the temperature drop by 72.7%. This results in a significant reduction in the operation time of energy-intensive supplementary cooling equipment from 5092 h to 31 h given the supply air temperature threshold of 25 °C. This advancement brings us closer to the goal of utilizing fully natural passive cooling technology for the cooling of DCs in hot and humid regions.
(4) Future works should include the integration of the multi-stage IEC in real practice, the combination of the liquid desiccant regeneration system and a renewable energy system, optimization, and economic analysis, which should demonstrate the reliable cooling, energy saving, and cost-friendliness of the proposed system in DCs of hot-humid regions.

Author Contributions: Conceptualization, methodology, programming, formal analysis, writing—original draft: W.S.; validation, data curation, resources: X.M.; conceptualization, methodology, writing—review & editing: Y.M.; supervision, funding acquisition, writing—review & editing: H.Y. All authors have read and agreed to the published version of the manuscript.

Funding: This research work was supported by the National Natural Science Foundation of China (No. 52208111). This work was also supported by the PolyU Distinguished Postdoctoral Fellowship Scheme (1-YWDD).

Institutional Review Board Statement: Not applicable.

Informed Consent Statement: Not applicable.

Data Availability Statement: The data presented in this study are available on request from the corresponding author.

Conflicts of Interest: The authors declare that they have no known competing financial interests or personal relationships that could have appeared to influence the work reported in this paper.

Nomenclature

c_{pa}	Specific heat of air, J/(kg·°C)
d	Channel height, mm
d_e	Hydraulic diameter of channel, m
h	Heat transfer coefficient, W/(m^2·°C)
h_m	Mass transfer coefficient, kg/(m^2·s)
h_{fg}	Heat capacity of vaporization, 2501 kJ/kg
H	Height, m
i	Specific enthalpy, kJ/kg
L	Length, m
Le	Lewis number
m	Mass flow rate of air, kg/s
Nu	Nusselt number
P	Pressure, kPa
Pr	Prandtl number
r	Mass flow ratio
Re	Reynolds number
t	Temperature, °C
u	Air velocity, m/s
w	Humidity ratio, g/kg
Greek symbols	
σ	Solution concentration
λ	Thermal conductivity, W/(m·°C)
Abbreviation	
AC	Air conditioning
DC	Data center
LD-IEC	Liquid desiccant-based indirect evaporative cooling
IEC	Indirect evaporative cooling
RH	Relative humidity
Subscripts	
amb	Ambient air
atm	Atmospheric
circ	Circulated air
d	Dehumidification
e	Evaporation
p	Primary air
s	Secondary air
sat	Saturated status
sol	Solution
w	Water film
in	Inlet
out	Outlet

References

1. Fan, J.L.; Li, Z.; Huang, X.; Li, K.; Zhang, X.; Lu, X.; Wu, J.; Hubacek, K.; Shen, B. A net-zero emissions strategy for China's power sector using carbon-capture utilization and storage. *Nat. Commun.* **2023**, *14*, 5972. [CrossRef]
2. Lai, L.; Wang, X.; Hu, E.; Choon Ng, K. A vision of dew point evaporative cooling: Opportunities and challenges. *Appl. Therm. Eng.* **2024**, *244*, 122683. [CrossRef]
3. Liu, X.; Liu, X.; Jiang, Y.; Zhang, T.; Hao, B. Photovoltaics and Energy Storage Integrated Flexible Direct Current Distribution Systems of Buildings: Definition, Technology Review, and Application. *CSEE J. Power Energy Syst.* **2023**, *9*, 829–845. [CrossRef]
4. Chen, J.; Gong, Q.; Lu, L. Evaluation of passive envelope systems with radiative sky cooling and thermally insulated glazing materials for cooling. *J. Clean. Prod.* **2023**, *398*, 136607. [CrossRef]
5. Ma, X.; Zeng, C.; Zhu, Z.; Zhao, X.; Xiao, X.; Akhlaghi, Y.G.; Shittu, S. Real life test of a novel super performance dew point cooling system in operational live data centre. *Appl. Energy* **2023**, *348*, 121483. [CrossRef]
6. Zhang, Y.; Shan, K.; Li, X.; Li, H.; Wang, S. Research and Technologies for next-generation high-temperature data centers—State-of-the-arts and future perspectives. *Renew. Sustain. Energy Rev.* **2023**, *171*, 112991. [CrossRef]

7. Liu, Y.; Yang, X.; Li, J.; Zhao, X. Energy savings of hybrid dew-point evaporative cooler and micro-channel separated heat pipe cooling systems for computer data centers. *Energy* **2018**, *163*, 629–640. [CrossRef]
8. Zhang, Y.; Zhao, Y.; Dai, S.; Nie, B.; Ma, H.; Li, J.; Miao, Q.; Jin, Y.; Tan, L.; Ding, Y. Cooling technologies for data centres and telecommunication base stations—A comprehensive review. *J. Clean. Prod.* **2022**, *334*, 130280. [CrossRef]
9. Faraji, H.; Alami, M.E.; Arshad, A.; Hariti, Y. Numerical Survey on Performance of Hybrid NePCM for Cooling of Electronics: Effect of Heat Source Position and Heat Sink Inclination. *J. Therm. Sci. Eng. Appl.* **2021**, *13*, 4049431. [CrossRef]
10. Faraji, H.; Yıldız, Ç.; Arshad, A.; Arıcı, M.; Choukairy, K.; El Alami, M. Passive thermal management strategy for cooling multiple portable electronic components: Hybrid nanoparticles enhanced phase change materials as an innovative solution. *J. Energy Storage* **2023**, *70*, 108087. [CrossRef]
11. Kabeel, A.E.; Bassuoni, M.M.; Abdelgaied, M. Experimental study of a novel integrated system of indirect evaporative cooler with internal baffles and evaporative condenser. *Energy Convers. Manag.* **2017**, *138*, 518–525. [CrossRef]
12. De Antonellis, S.; Joppolo, C.M.; Liberati, P. Performance measurement of a cross-flow indirect evaporative cooler: Effect of water nozzles and airflows arrangement. *Energy Build.* **2019**, *184*, 114–121. [CrossRef]
13. Kim, H.-J.; Ham, S.-W.; Yoon, D.-S.; Jeong, J.-W. Cooling performance measurement of two cross-flow indirect evaporative coolers in general and regenerative operation modes. *Appl. Energy* **2017**, *195*, 268–277. [CrossRef]
14. Shi, W.; Yang, H.; Ma, X.; Liu, X. Performance prediction and optimization of cross-flow indirect evaporative cooler by regression model based on response surface methodology. *Energy* **2023**, *283*, 128636. [CrossRef]
15. Kabeel, A.E.; Abdelgaied, M. Numerical and experimental investigation of a novel configuration of indirect evaporative cooler with internal baffles. *Energy Convers. Manag.* **2016**, *126*, 526–536. [CrossRef]
16. Cui, X.; Yang, C.; Yan, W.; Zhang, L.; Wan, Y.; Chua, K.J. Experimental study on a moisture-conducting fiber-assisted tubular indirect evaporative cooler. *Energy* **2023**, *278*, 128014. [CrossRef]
17. Xiao, X.; Liu, J. A state-of-art review of dew point evaporative cooling technology and integrated applications. *Renew. Sustain. Energy Rev.* **2024**, *191*, 114142. [CrossRef]
18. Ma, X.; Shi, W.; Yang, H. Spray parameter analysis and performance optimization of indirect evaporative cooler considering surface wettability. *J. Build. Eng.* **2024**, *82*, 108175. [CrossRef]
19. Sun, T.; Huang, X.; Chen, Y.; Zhang, H. Experimental investigation of water spraying in an indirect evaporative cooler from nozzle type and spray strategy perspectives. *Energy Build.* **2020**, *214*, 109871. [CrossRef]
20. Shi, W.; Yang, H.; Ma, X.; Liu, X. A novel indirect evaporative cooler with porous media under dual spraying modes: A comparative analysis from energy, exergy, and environmental perspectives. *J. Build. Eng.* **2023**, *2023*, 106874. [CrossRef]
21. Golizadeh Akhlaghi, Y.; Badiei, A.; Zhao, X.; Aslansefat, K.; Xiao, X.; Shittu, S.; Ma, X. A constraint multi-objective evolutionary optimization of a state-of-the-art dew point cooler using digital twins. *Energy Convers. Manag.* **2020**, *211*, 112772. [CrossRef]
22. Al-Zubaydi, A.Y.T.; Hong, G. Experimental study of a novel water-spraying configuration in indirect evaporative cooling. *Appl. Therm. Eng.* **2019**, *151*, 283–293. [CrossRef]
23. Wan, Y.; Huang, Z.; Soh, A.; Jon Chua, K. On the performance study of a hybrid indirect evaporative cooling and latent-heat thermal energy storage system under commercial operating conditions. *Appl. Therm. Eng.* **2023**, *221*, 119902. [CrossRef]
24. Nemati, N.; Omidvar, A.; Rosti, B. Performance evaluation of a novel hybrid cooling system combining indirect evaporative cooler and earth-air heat exchanger. *Energy* **2021**, *215*, 119216. [CrossRef]
25. Shi, W.; Yang, H.; Ma, X.; Liu, X. Techno-economic evaluation and environmental benefit of hybrid evaporative cooling system in hot-humid regions. *Sustain. Cities Soc.* **2023**, *2023*, 104735. [CrossRef]
26. Chen, Q.; Kum Ja, M.; Burhan, M.; Akhtar, F.H.; Shahzad, M.W.; Ybyraiymkul, D.; Ng, K.C. A hybrid indirect evaporative cooling-mechanical vapor compression process for energy-efficient air conditioning. *Energy Convers. Manag.* **2021**, *248*, 114798. [CrossRef]
27. Yan, H.; Chen, Y.; Min, Y.; Pan, Y. An adaptive controller based dynamic simulation of household air-conditioner with indirect evaporative cooler as dedicated outdoor air system. *Energy Build.* **2022**, *274*, 112454. [CrossRef]
28. Tripathi, R.J.; Kumar, D. Performance assessment of solar-driven indirect evaporative cooling with a novel wet channel: An experimental study. *J. Build. Eng.* **2023**, *78*, 107674. [CrossRef]
29. Hu, M.; Seng Lee, P. Performance evaluation of a passive air conditioning module integrating radiative sky cooling and indirect evaporative cooling. *Appl. Therm. Eng.* **2024**, *2024*, 122608. [CrossRef]
30. Chu, J.; Huang, X. Research status and development trends of evaporative cooling air-conditioning technology in data centers. *Energy Built Environ.* **2023**, *4*, 86–110. [CrossRef]
31. Li, C.; Mao, R.; Wang, Y.; Zhang, J.; Lan, J.; Zhang, Z. Experimental study on direct evaporative cooling for free cooling of data centers. *Energy* **2024**, *288*, 129889. [CrossRef]
32. Mohammed, R.H.; El-Morsi, M.; Abdelaziz, O. Indirect evaporative cooling for buildings: A comprehensive patents review. *J. Build. Eng.* **2022**, *50*, 104158. [CrossRef]
33. Heiliger, J. Building efficient data centers with the open compute project. In *Facebook Engineering Notes*; 2011. Available online: https://engineering.fb.com/2011/04/07/data-center-engineering/building-efficient-data-centers-with-the-open-compute-project/ (accessed on 25 February 2024).
34. Finch, P.; ENG, C. Indirect Evaporative Cooling Drives Data Center Efficiency. *ASHRAE J.* **2020**, *62*, 40–47.

35. Yan, W.; Cui, X.; Zhao, M.; Meng, X.; Yang, C.; Zhang, Y.; Liu, Y.; Jin, L. Multi-objective optimization of dew point indirec evaporative coolers for data centers. *Appl. Therm. Eng.* **2024**, *241*, 122425. [CrossRef]
36. Cui, X.; Yang, C.; Yan, W.; Wang, X.; Liu, Y.; Jin, L. Climatic applicability of indirect evaporative cooling strategies for data centers in China. *J. Build. Eng.* **2024**, *83*, 108431. [CrossRef]
37. Mumtaz, M.; Pamintuan, B.C.; Fix, A.J.; Braun, J.E.; Warsinger, D.M. Hybrid membrane dehumidification and dewpoin evaporative cooling for sustainable air conditioning. *Energy Convers. Manag.* **2023**, *294*, 117547. [CrossRef]
38. Anisimov, S.; Pandelidis, D.; Danielewicz, J. Numerical analysis of selected evaporative exchangers with the Maisotsenko cycle *Energy Convers. Manag.* **2014**, *88*, 426–441. [CrossRef]
39. Adam, A.; Han, D.; He, W.; Amidpour, M.; Zhong, H. Numerical investigation of the heat and mass transfer process within a cross-flow indirect evaporative cooling system for hot and humid climates. *J. Build. Eng.* **2022**, *45*, 103499. [CrossRef]
40. Zheng, B.; Guo, C.; Chen, T.; Shi, Q.; Lv, J.; You, Y. Development of an experimental validated model of cross-flow indirec evaporative cooler with condensation. *Appl. Energy* **2019**, *252*, 113438. [CrossRef]
41. Cui, X.; Islam, M.R.; Mohan, B.; Chua, K.J. Theoretical analysis of a liquid desiccant based indirect evaporative cooling system *Energy* **2016**, *95*, 303–312. [CrossRef]
42. Zhang, H.; Ma, H.; Ma, S.; Yang, M. Investigation on the performance of an indirect evaporative cooling system integrated with liquid dehumidification. *Energy Build.* **2021**, *251*, 111356. [CrossRef]
43. Zhang, Y.; Zhang, H.; Yang, H.; Chen, Y.; Leung, C.W. Counter-crossflow indirect evaporative cooling-assisted liquid desiccant dehumidifier: Model development and parameter analysis. *Appl. Therm. Eng.* **2022**, *217*, 119231. [CrossRef]
44. Min, Y.; Chen, Y.; Yang, H. Numerical study on indirect evaporative coolers considering condensation: A thorough comparison between cross flow and counter flow. *Int. J. Heat Mass Transf.* **2019**, *131*, 472–486. [CrossRef]

sustainability

MDPI

Article

Photovoltaic Strings on Large, Flat Roofs: Experimental Wind Loads on Representative Configurations

Giacomo Scrinzi [1,*], Enrico Sergio Mazzucchelli [1] and Sara Muggiasca [2]

1 Department of Architecture, Built Environment and Construction Engineering, Politecnico di Milano, 20133 Milan, Italy
2 Department of Mechanical Engineering, Politecnico di Milano, 20156 Milan, Italy
* Correspondence: giacomo.scrinzi@polimi.it; Tel.: +39-02-2399-6019

Abstract

The integration of tilted photovoltaic strings on large, flat roofs, typical of industrial and commercial buildings, raises complex design challenges, particularly regarding wind-induced loads. This study presents a comprehensive wind tunnel investigation aimed at evaluating the aerodynamic effects on rooftop PV strings under various representative configurations and the correlation between characteristic geometric parameters such as tilt angle, bottom clearance, row spacing, and wind direction. Following a literature review, a detailed 1:10 scaled model with geometric adjustment capabilities was developed and eventually tested in a boundary-layer wind tunnel. High-resolution pressure measurements were processed to derive force and moment resultants normalised by reference wind pressure. Envelopes of force/moment resultants are presented for each representative geometric configuration and for each wind exposure angle. The results present severe variations in local wind actions, particularly significant at the strings' free ends and for oblique wind angles. The severe underestimation of local wind loads by standard codes is discussed. The findings underline the importance of detailed wind-load assessment for both new constructions and retrofits, suggesting that reliance solely on code provisions might result in unsafe designs.

Keywords: photovoltaics; flat roof; renewable; wind loads

1. Introduction

In order to meet nowadays' increasing energy efficiency needs, the integration of photovoltaic (PV) systems on building roofs is a topical aspect, and especially for those building such as factories, warehouses, or exhibition halls, notoriously energy-intensive and usually featuring large, flat roofs which can accommodate high-capacity power plants.

In order to maximise the productivity, solar arrays should be installed with optimal orientation (azimuth) and tilt angle, depending on the site location, where the best orientation is generally 0° to the south with tilt angles rising up to about 30°.

However, greater tilt angles increase the structural stresses transferred from the PV modules to their support frames and, consequently, to the building structure [1]. As a result, flat or co-planar configurations are more common.

While the azimuth can often be selected freely, on large flat roofs, which are usually rectangular in plan, PV strings are often aligned with the building's edges. This typically results in a simplification of the design of support frames, cable routing, and maintenance procedures. Moreover, standing seam roofs, which are extremely common for such buildings, require a free thermal elongation capability that, in the presence of roof-attached PV strings, may lead to severe damage to the roof metal panels [2].

In this framework, designers should be provided with adequate information in order to correctly calculate and ensure the reliability of rooftop PV arrays, both in new constructions and revamping interventions, considering that actions on flat and tilted strings (see Figure 1) can be significantly different.

(a) (b)

Figure 1. PV modules on a large, flat, metal roof, arrayed: (**a**) in groups parallel to the slope; (**b**) in tilted strings parallel to the building edges.

In fact, calculations should aim to verify both the global resistance of the roof structure and the resistance of PV string components (i.e., PV modules, frames, brackets, fixings, etc.). For the latter, local wind effects are particularly important as they can be significantly greater than mean actions, as better described in the following, and may result in failures even under loads related to ordinary wind events. In this respect, accurate and detailed pressure coefficient distributions are strongly necessary.

In this respect, the national and European standards used to derive wind actions only provide instructions about the design loads on different configurations of bare roofs, not considering the specificity of roof-mounted, tilted photovoltaic strings. As stated in the standards, the aerodynamic coefficients (pressure or friction coefficients) to be used for the design of buildings or building parts with a particular shape should be derived from appropriate scientific documentation or specific wind tunnel tests, which can provide more precise and accurate results to be applied to specific layouts while also investigating local pressure effects.

Several studies investigated the wind-induced load and effects on solar panels or arrays of multiple panels in order to optimise the design and ultimately the cost of the photovoltaic panels, the array support structure, and its foundation or anchoring. The literature provides a wide set of approaches, including both small- and large-scale wind tunnel experiments as well as numerical simulations made by computational fluid dynamics modelling.

In the late 1970s and early 1980s, several commercial industrial projects aimed to initiate an effort towards reducing the cost of solar arrays were promoted. Among them, the most relevant are those by the Bechtel National Inc. (San Francisco, CA, USA) and Miller and Zimmerman. Bechtel National Inc. (San Francisco, CA, USA) [3] reported the wind loads on a ground-mounted solar array using strain gauges in a boundary-layer wind tunnel, while Miller & Zimmerman [4–6] used scaled models provided with pressure taps in a boundary-layer wind tunnel to measure wind loads on very-high-aspect-ratio (span/chord) flat plate photovoltaic arrays located on or in close proximity to the ground, in close proximity to each other, and exposed to the wind, providing pressure distributions on the solar panel surface. The aim of these studies was to determine wind-loading criteria in relation to chord lengths, array spacings, the height of the arrays from the ground, wind directions, and array angles of attack.

The model scale issue was also stressed by Aly & Bitsuamlak [7], who experimentally and numerically investigated the sensitivity of wind loads on solar panels, although ground-mounted, at different geometric scales. This study showed that the mean surface pressures are not significantly affected by the scale and then by the turbulence intensity, while the standard deviation and peak load coefficients can vary significantly according to the model size. Another similar study by Aly [8] in fact confirmed that the geometric scale is a primary reason for the discrepancies in peak wind loads, which can however be avoided by adapting the inflow turbulence, overcoming the lack of large-scale turbulence.

Since wind loads on low-rise buildings and their roof-attached elements also depend on many parameters relative to the building geometry and the wind characteristics, different values were collected through boundary-layer wind tunnel tests on small-scale [1,9] or large-scale models [10–12]. Among these, Kopp [1] particularly focused on different aerody namic mechanisms to be assessed, i.e., the equalisation of the building-generated pressures and array-generated turbulence. Particularly, Kopp et al. [13] investigated through wind tunnel tests the effect of buildings on roof-mounted solar arrays, compared to ground-mounted arrays, showing that the aerodynamics of roof-mounted arrays have a complexity beyond that of building aerodynamics, due their presence in a flow field containing vortices generated by flow separations at the roof edges, turbulence induced by the array, and turbulence in the atmospheric boundary layer. The difference in wind peak loads on the bare roof and on roof-mounted solar arrays was also studied through experimental tests on various scales models by Banks [14] and Abiola-Ogedengbe et al. [15].

However, although the response of different roof-mounted arrays layouts was experimentally evaluated on large scale models [11], previous studies produced results for several different although hardly comparable configurations. In this regard, the present research aims to provide a comprehensive and definitive investigation of the influence of typical geometric parameters that generally characterise rooftop tilted PV strings in representative configurations on strings' wind-loading pattern. Experimental results had been intended to eventually confirm hypotheses derived by the literature and highlight any correlations between said parameters.

In order to achieve the most accurate and reliable design coefficients for representative configurations, wind tunnel tests were carried out at the Politecnico di Milano Wind Tunnel, which has a ~14 m wide, ~4 m high boundary-layer section equipped with a circular turning table which allows for an easy investigation of all wind directions.

2. Materials and Methods

2.1. Experimental Setup Definition

Reference scaled-model essential features were firstly defined based on the following:

- The reference building has a rectangular footprint, which is standard for the most of buildings provided with large, flat roofs;
- The roof slope is assumed to be perfectly horizontal (typical flat roof pitches down to 0.5/1.0% do not significantly affect wind-pressure coefficients);
- Photovoltaic modules are assumed to have standard dimensions of 1600 × 1000 mm;
- Photovoltaic modules are assumed to be installed in a tilted position with an angle ranging from 5° to 30° and their long side parallel to the roof plane, with their short sides against each other in order to reduce the overall drag;

- Photovoltaic strings are assumed to be parallel to the building edges, which is a configuration commonly adopted to simplify and standardise support frames and cableways, provide a cleaner layout for maintenance, and also, in the case of metal roofs, avoid issues related to the differential thermal expansions of adjacent metal panels onto which string frames are attached.

Concerning the configuration of rooftop PV strings, the literature provided useful contributions to the present research case [1,9,10], which suggest that:

- For straight wind, the magnitude of force coefficients decreases across rows as a result of a sheltering effect from the upwind solar panels, as the first row experiences the maximum wind loads (drag and lift) for all wind directions;
- The largest reduction in wind-load coefficients is on the second row, but the amount of reduction drops quickly afterwards, becoming minimal after the fourth or fifth row;
- Modules located in the outer string ends could be subjected to higher wind loads compared to modules located in the middle of the strings, mainly due to edge vortices that are also particularly relevant for oblique winds;
- Increasing the side spacing between adjacent modules has a minimal effect on the force coefficients, while the wind-load coefficients increase as the rows spacing is increased;
- Furthermore, for roof-mounted arrays, the wind-load coefficients appeared to decrease with increasing perimeter gap from the building edge.

According to the previous findings, the experimental setup characteristics were defined as follows:

- A number of 8 rows was chosen so that the length is enough to appropriately highlight the downstream dissipation of sheltering effects and to investigate possible peaks of pressure coefficients at the downwind flow reattachment point;
- A 12-module, 1920 mm (scaled size) long string was considered adequate to avoid the affection of pressure coefficients by corner vortexes;
- The model strings were designed without providing any gap between adjacent modules;
- A fixed 1.5 m distance from the building edge was chosen as it can be considered the minimum to allow for the integration of a perimetral gutter and a maintenance walkway.

The roof height was ultimately set to 10 m (1 m in model scale), which can be assumed as representative for typical large, flat-roofed commercial/industrial buildings in suburban areas.

In the end, the main parameters to be investigated through wind tunnel tests were:

- Wind direction (α);
- Photovoltaic panels tilt angle (β);
- Rows spacing (d);
- Clearance between the panel's bottom and the roof plane (c).

Concerning the tilted modules, it has to be noted that increasing the β angle results in a shading over the upcoming strings. For this reason, the net row spacing was assumed equal, for each configuration, to the maximum value among the standard gap, to allow for passing and maintenance purposes (taken as 60 cm), and the distance needed to avoid the shadowing given by the adjacent string was calculated (conservatively, for any latitude) as presented by Aronescu & Appelbaum [16], referring to parameters in Figure 2:

$$L \cdot sin\beta = d \cdot tan\delta + d \cdot tan\gamma + L \cdot cos\beta \cdot tan\gamma \quad (1)$$

where δ is the sun declination at the winter solstice (23.45°) and γ is the slope pitch. For ease of experimental investigating, the latter is assumed equal to zero, whereas it is consistent with general reference buildings' "flat" roofs' pitch (1% to 3%), and such a minimal difference will not affect wind-loading pattern.

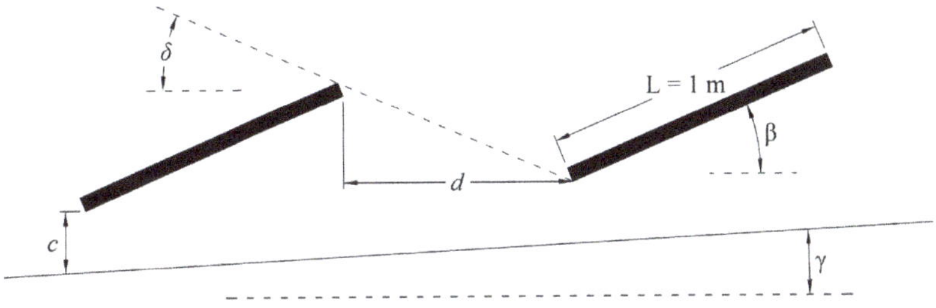

Figure 2. Geometric parameters for strings' spacing calculation.

For the considered interval of tilt angles (5° to 30°), the resulting spacing values versus tilt angle are shown in Figure 3.

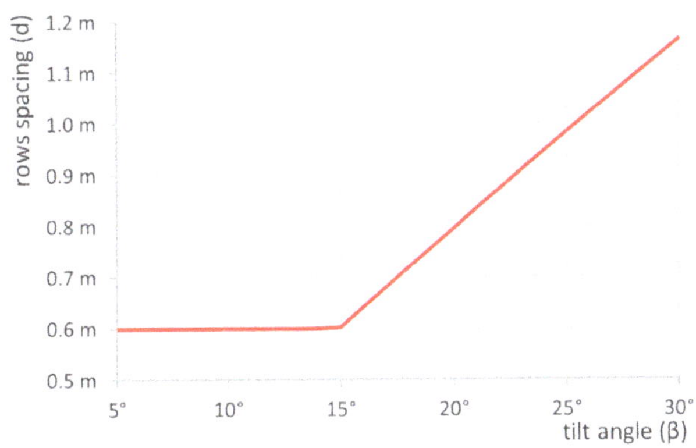

Figure 3. Net row spacing (d) versus tilt angle (β).

While most of previous studies were performed in boundary-layer wind tunnels not large enough to fit large-scale models, Politecnico di Milano Wind Tunnel's dimensions make it suitable to accommodate larger-scale models.

In order to achieve a better results accuracy, the reference building model scale was defined as 1:10, also according to Aly [8], who showed the similarity between mean pressure distribution on 1:1 and 1:10 solar panels obtained by CFD simulations and experimental results. Specifically, higher scales allow for a better flow similarity between the model and the real scale due to a higher Reynolds number with no need for very high tunnel speed.

Moreover, Politecnico di Milano's wind tunnel section size also allowed for the integration of boundary-layer simulation equipment in order to provide further accuracy, as better described in the following paragraph.

2.2. Scaled-Model Design and Creation

Based upon the characteristics mentioned above, the design of the 1:10 scale model [2] was first made using 3D CAD software, in order to create a rectangular envelope with a flat top, provided with 8 strings with sixteen modules each and a system to adjust their tilt angle, the clearance from the roof plane, and the row spacing (see Figure 4).

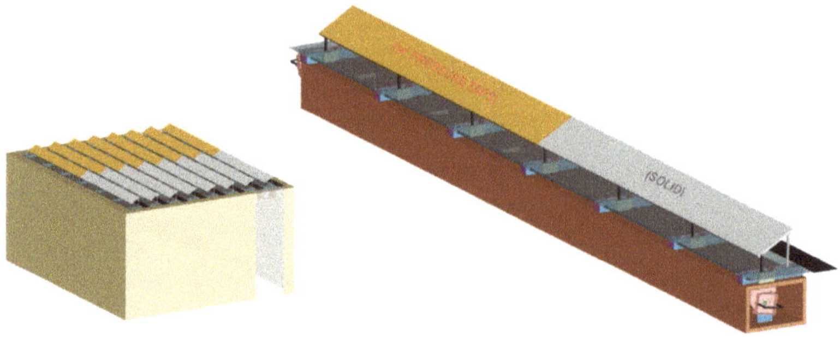

Figure 4. A CAD representation of the scaled-model and adjustable PV strings.

This meant having 8 independent strings, each provided with its own tilt adjustment and capable to slide away from each other to increase/decrease the row spacing, integrated into an outer box capable of extending its length to comply with larger row spacing and higher tilt-angle configurations.

Each string was then provided with NEMA17 stepper motors controlled by an Arduino shield in order to quickly adjust tilt angle and bottom clearance, allowing for easy transitions between different string configurations.

Due to the model symmetry, a half of each string was made of scaled modules provided with pressure taps, while the other half was dummy (in Figure 4, respectively, orange- and grey-coloured). Each instrumented semi-string featured 56 pressure taps, one on each side (needed to derived net pressure coefficients) in 28 different positions (see Figure 5), giving a total of 448 pressure taps.

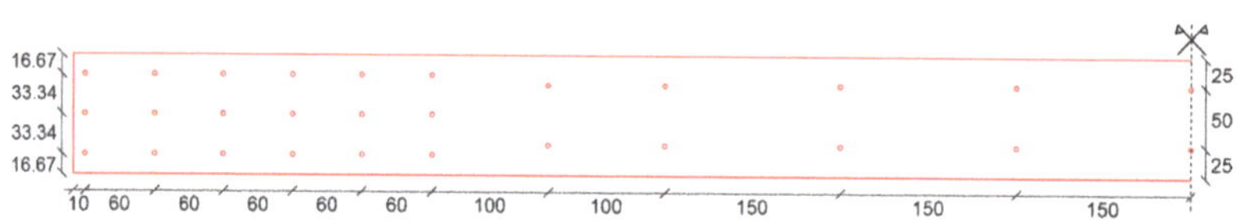

Figure 5. Pressure-tap position on the half string, where each position corresponds to two taps, one on each side of the string (dimensions in millimeters).

The 3D printing of scaled-model components and their preparation were presented by Scrinzi et al. in [17]. The full model is depicted in Figures 6 and 7.

Figure 6. A closeup of adjustable strings and pressure tubes distribution.

Figure 7. The assembled scaled model inside the boundary-layer wind tunnel at Politecnico di Milano. In the background, spikes and bricks are used to create a suburban velocity/turbulence profile.

The model was therefore provided with 16 high-speed pressure scanners with 32 channels each, in turn connected to a high-speed data acquisition controller.

2.3. Boundary-Layer Wind Tunnel Tests

Tests were conducted in the boundary-layer chamber, the dimensions of which (13.84 m wide, 3.84 m height) allowed us to avoid freestream flow blockage issues even in the least favourably oriented wind directions, with a blockage ratio always lower than 5% for a scaled-model maximum cross-section of ~2.7 × 1.0 m. Different orientations (wind exposure angle) were investigated by rotating the model around its middle point on the turning table.

Appropriate vertical spikes installed at the chamber inlet and a brick pattern all along the floor upwind of the scaled model (see Figure 7) were used to generate a wind turbulence/velocity profile approximating a Class II (mild roughness) profile by Eurocode [18], as depicted in Figure 8. Experimental profiles were measured 1 m upwind of the scaled model, at a distance of 30+ m from the boundary-layer chamber inlet.

In order to reduce the number of different configurations to be tested, the tilt angle, bottom clearance, and wind-direction angle values had to be discretized, as shown in the following Tables 1 and 2, with references to the depiction of parameters in Figures 2 and 9.

All-round wind directions were covered by 45° steps, only reducing the step to 15° for the first eighth of a turn (instrumented modules upwind), as previous studies [13] showed the highest corner vortexes effects on upwind rows edges for oblique winds with angles up to 45°.

Two sizes of bottom clearance were selected: 15 and 30 cm (on a real scale). Usually, when installing photovoltaic modules on a flat roof, an air gap remains between the bottom panels' edge and the roof. This particularly happens in case of metal roofing systems due to the presence of the standing seam and the anchor clamps, which would always allow a certain air flow. A 15 cm clearance is assumed to be the minimum actually achievable.

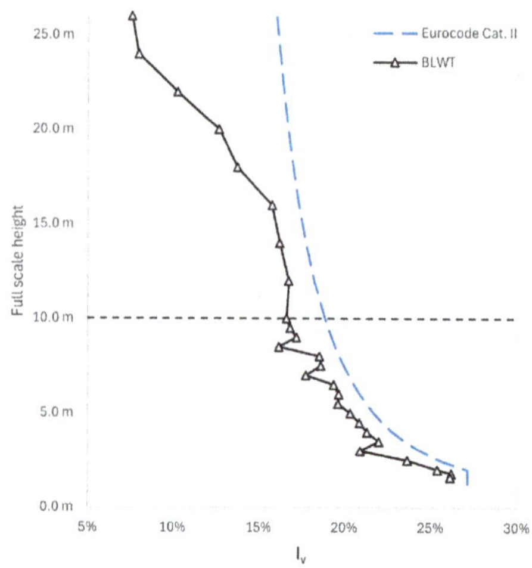

(a)

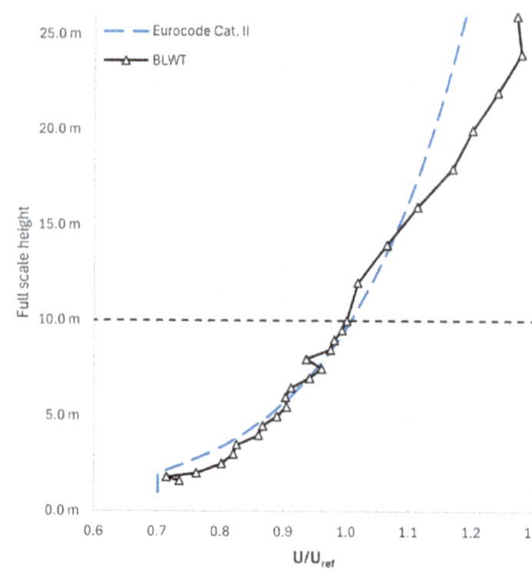

(b)

Figure 8. Boundary-layer wind tunnel (BLWT) turbulence (**a**) and velocity (**b**) profiles compared to Eurocode Class II terrain profiles.

Table 1. Summary of scaled-model geometric parameters for each configuration.

c Bottom Clearance	*β* Tilt Angle	*d* Net Row Spacing	Configuration ID
15 cm (real scale)	5°	60 cm (real scale)	**C 15 T 05**
	10°	60 cm (real scale)	**C 15 T 10**
	20°	80 cm (real scale)	**C 15 T 20**
	30°	120 cm (real scale)	**C 15 T 30**
30 cm (real scale)	5°	60 cm (real scale)	**C 30 T 05**
	10°	60 cm (real scale)	**C 30 T 10**
	20°	80 cm (real scale)	**C 30 T 20**
	30°	120 cm (real scale)	**C 30 T 30**

Table 2. Summary of wind-direction testing angles.

α Wind Direction									
−135°	−90°	−45°	+0°	+15°	+30°	+45°	+90°	+135°	+180°

Tilt angles were assumed 5° to 30°, which is an optimal angle range for local latitudes, efficiency wise.

Each testing cycle lasted 300 s in order to provide statistical relevance in terms of peak values.

In order to validate the experimental setup and the measurements' accuracy, multiple preliminary tests were run in specific conditions/configurations and the consistency of the results was checked before starting the actual tests.

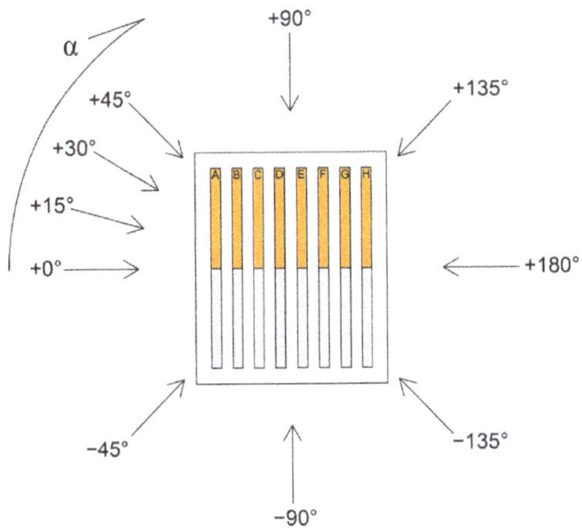

Figure 9. Scale-model plan with evidence of wind-direction test angles, where orange half strings are instrumented (see Figure 4), and strings identified as A to H.

2.4. Data Processing

Differential (with reference to the inner wind tunnel static pressure) local pressures (p) were measured for each tap at 500 Hz over a 5 min time span and divided by the kinetic wind pressure (q_w) calculated with the speed (v) and air density (ρ) at the mock-up roof height (10 m in real scale), in order to obtain instant local pressure coefficients ($c_{p,\pm}$):

$$c_{p,\pm} = \frac{p_\pm}{q_w} = \frac{p_\pm}{\frac{1}{2}\rho v^2} \tag{2}$$

For each couple of pressure taps (top and bottom side) on the PV strings, an instant net pressure coefficient (c_p) is obtained by subtracting the value of the bottom-side coefficient from the value of its respective top-side coefficient. Conventionally, c_p is positive for pressures directed towards the top surface of the string.

A fully probabilistic assessment method by Cook & Mayne [19,20] was applied in order to neglect instant peaks. A 78% fractile (namely a 22% risk of exceedance) of extreme wind-load distribution was extracted as the experimental result. The indicated fractile has been selected to comply with serviceability requirements, providing an adequately conservative estimation of typical loads without focusing on extreme action, as PV strings are actually non-structural elements which may instead require very higher fractiles. It is also to be considered that, in such an application, occasional exceedance poses relatively little risk, as common metal support structures are generally capable of redistributing and/or absorbing extra loads in case of extreme events.

Eventually, net pressure coefficients' time histories (i.e., c_p versus time) of each wind tunnel test case were then processed to derive the time histories of force (C_F) and moment (C_M) resultants normalised by q_w pressure, for each section, which are obtained by appropriately combining c_p values multiplied by their reference section width (see Figure 10). C_F and C_M are dimensionally represented, respectively, as [m] and [m^2]. Once multiplied by q_w pressure [N/m^2], the resulting forces [N/m] and moments [Nm/m] are intended per string length unit.

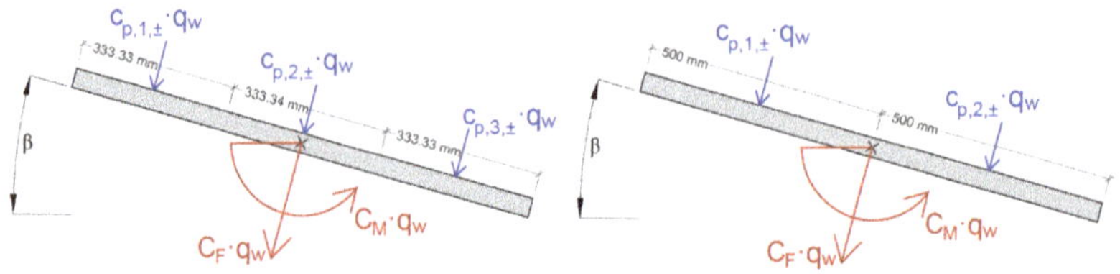

Figure 10. Normalised force (C_F) and moment (C_M) resultants definition for 3-tap (**left**) and 2-tap (**right**) sections of each PV string (dimensions in millimeters, real scale).

3. Results

The experimental results support the initial hypotheses quite well based on the literature review above, concerning the influence of different geometric parameters. Such good consistency further proves the measurements' accuracy.

The envelopes of corrected positive and negative C_F (normalised force resultants) and C_M (normalised moment resultants) are plotted in Figures A1–A16 of Appendix A, for each configuration ID, where strings are depicted with reference to the identification in Figure 9, and the dotted line at the bottom of each picture represents the mid-string symmetry line.

Actual forces and moments may be derived for each row and each section by multiplying plotted values by the reference wind pressure provided by standards [18] or, if available, environmental data logs.

Referring to the sign convention in Figure 10, C_F positive peak values at strings' free ends decrease, moving from the front row to the back rows (referring to the first string as the one upwind for a 0° wind exposure), presenting maximum peaks in proximity to the free end of each string which increase with the tilt angle and reach up to 4.90 and 5.77, respectively, for the 0.15 m and 0.30 m clearance configurations.

Instead, negative C_F values at the strings' extreme edge increase from the first row to the last ones, especially with higher tilt angles. Negative peaks are (in terms of absolute value) 30% lower, at least, than the positive ones in the same section, reaching −4.10 for 0.15 m and −3.83 for 0.30 m clearance.

In this respect, a higher clearance results in a reduction in upward forces, while presenting higher downward peak actions.

Both positive and negative C_F peaks at the very ends dissipate and tend to semi-constant values within a few metres towards the string midpoint. Such values are generally greater by those provided by Eurocode 1 (see Figure 11) by 50 to 300% for positive (downward) forces and by 10 to 50% for negative (upward) forces.

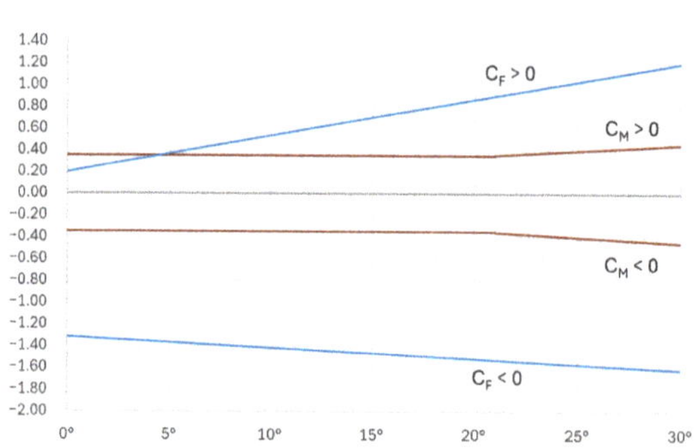

Figure 11. Conservative (regardless of blockage ratio) positive and negative values of C_F and C_M for monopitch canopies (which can be compared to rooftop tilted PV strings) derived by Eurocode 1 §7.3 [18], with reference to the sign convention in Figure 10.

Concerning C_M values, as shown in the charts, the variation trend in different configurations is less noticeable. The high variability is mainly due to the turbulent part of the stream and the vortexes generated by flow deviations that make it difficult to describe. In general, effects are particularly evident for the free ends of each string and for the first and the last rows over their whole length, which are more exposed to the stream turbulence. However, the maximum magnitude of C_M values is fairly small and generally does not exceed 0.70, although slightly exceeding values provided by Eurocode 1 for monopitch canopies (which can be compared to rooftop tilted PV strings), as shown in Figure 11.

The lower stream blockage on account of the larger bottom clearance results, in case of higher tilt angles, in a reduction and homogenization of the force resultants in the inner part of the strings towards the middle, but seems not to compensate for the greater positive pressure experienced, particularly by the first few rows near the strings' extremity, while it just provides a slight reduction in the upward negative forces.

Recalling that the results plotted in Figures A1–A16 represent, for each configuration, the envelope of C_F and C_M values of all tested wind exposures, fluctuations over the string length are, in general, mainly due to specific wind exposure angles.

In order to better visualise the influence of wind exposure angle, Appendix B provides plots of maximum positive/negative C_F and C_M coefficients for each direction angle (α).

Concerning positive values, as expected [13], oblique wind-direction angles result in the most severe force resultants.

In particular, +15°, +30°, and +45° exposure angles are the most critical (see Figures A25, A27 and A29), where the highest loads in correspondence of the first strings' (A to D) free ends are given by +30° exposure and the most evident vortex effects are visible on the third and fourth rows for +45° exposure.

Similarly, for negative values, the most critical exposure appears to be an angle of +135°, resulting in maximum negative forces on the last strings (E to H).

4. Discussion and Conclusions

The envelope of wind-load distribution obtained by wind tunnel tests, depicted in Appendix A, shows a significant difference between mean and peak pressure coefficients over PV strings, reaching up to a 1:4 ratio, and several fluctuations over the strings' length for certain configurations.

Such a condition suggests the strict need of performing accurate local verifications of PV module brackets, string frames, and frame anchors/ballast which are usually calculated to resist against lower upward and downward forces derived by reference standards.

Alternatively, especially for interventions on existing buildings, PV string support structures shall be appropriately installed in order to distribute and transfer the local overloads to a greater surface area and to a higher number of anchors/ballasts.

It will therefore be the designer's duty to evaluate the most appropriate configuration to adopt, depending on the PV support structure robustness and how this distributes and transfers the forces to the roof anchors, in order to avoid failures or detachments of PV modules or other roof components. This is particularly needed in the case of underperforming roofing systems (e.g., thin standing seam roofs, lighter support materials), especially in the light of increasingly severe wind events nowadays.

Moreover, it is reasonable to assume that for most roofing systems, the upward forces are the most critical in terms of safety, and therefore should be adopted as a design driver. In this regard, although configurations presenting a higher clearance generally result in higher downward (positive) forces, they are also characterised by lower upward (negative) forces, which is possibly the preferable condition, as it is also safer to avoid airborne debris. This is particularly true for metal roofing systems, very commonly used on large flat roofs, as their anchors generally offer limited pull-off resistance, which is just provided by a thin metal sheet, and a local failure may lead to the whole roof detachment with disastrous outcomes.

The loading patterns discussed above also proved the importance for designers of evaluating different configurations and different wind exposure angles in order to possibly adopt, only where needed, adequate means of wind-load reduction, such as porous fences, flaps, etc.

In conclusion, for the integration of tilted PV strings on large, flat roofs, particularly in sites presenting significant wind speeds, designers are recommended to refer to wind tunnel tests results to derive suitable pressure coefficients or force/moment coefficients, both for the global and local verification of string components and roof anchors.

Author Contributions: Conceptualization, G.S. and E.S.M.; methodology, G.S.; resources, G.S. and S.M.; investigation, G.S.; formal analysis, G.S. and S.M.; validation, G.S.; visualisation, G.S.; writing—original draft preparation, G.S.; writing—review and editing, G.S., E.S.M. and S.M.; supervision, E.S.M. and S.M.; project administration, E.S.M. All authors have read and agreed to the published version of the manuscript.

Funding: This research received no external funding.

Institutional Review Board Statement: Not applicable.

Informed Consent Statement: Not applicable.

Data Availability Statement: Data is contained within the article and its appendices.

Conflicts of Interest: The authors declare no conflicts of interest.

Abbreviations

The following abbreviations are used in this manuscript:

PV Photovoltaic

Appendix A

The present appendix contains the plots, for each configuration ID, of the envelopes of corrected positive and negative C_F (normalised force resultants) and C_M (normalised moment resultants) over all wind exposure angles. Strings are depicted with reference to their identification in Figure 9. The thick dotted line at the bottom of each picture represents the mid-string symmetry line.

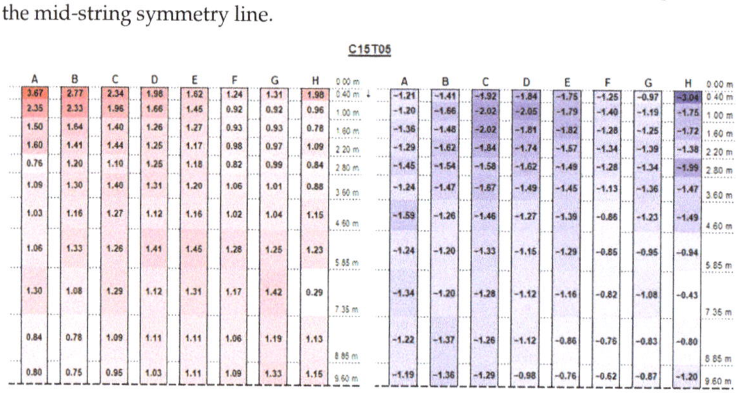

Figure A1. The envelope (all wind exposure angles) of maximum positive and negative values of force resultants normalised by wind pressure (C_F) for the C15T05 configuration.

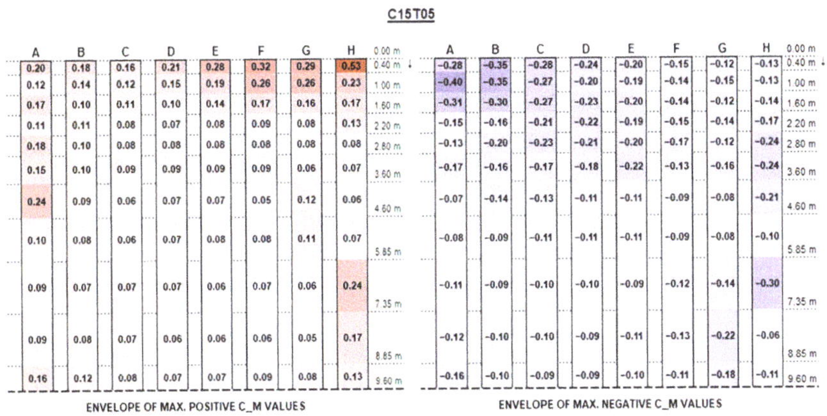

Figure A2. The envelope (all wind exposure angles) of maximum positive and negative values of moment resultants normalised by wind pressure (C_M) for the C15T05 configuration.

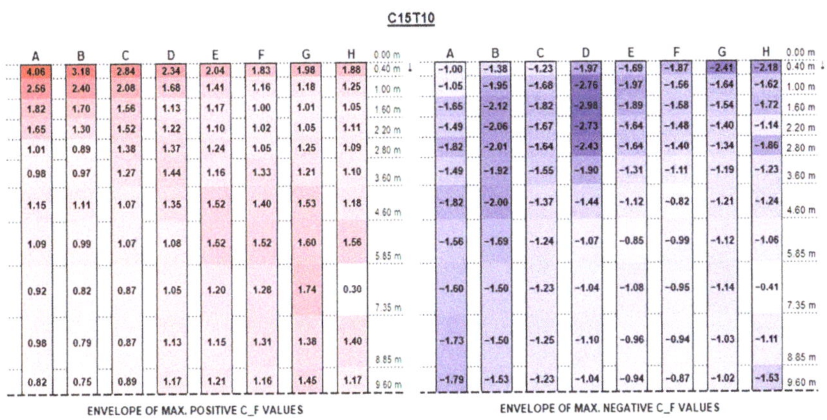

Figure A3. The envelope (all wind exposure angles) of maximum positive and negative values of force resultants normalised by wind pressure (C_F) for the C15T10 configuration.

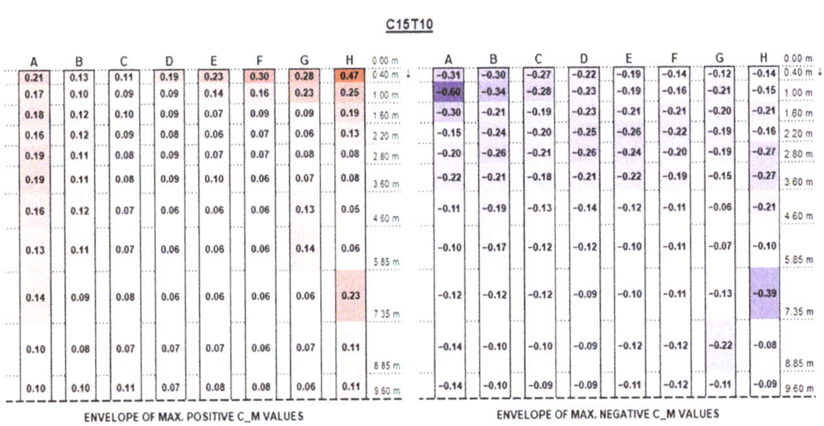

Figure A4. The envelope (all wind exposure angles) of maximum positive and negative values of moment resultants normalised by wind pressure (C_M) for the C15T10 configuration.

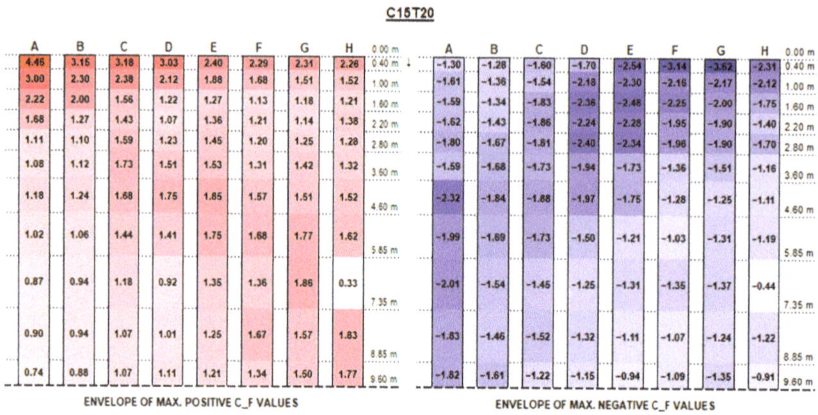

Figure A5. The envelope (all wind exposure angles) of maximum positive and negative values of force resultants normalised by wind pressure (C_F) for the C15T20 configuration.

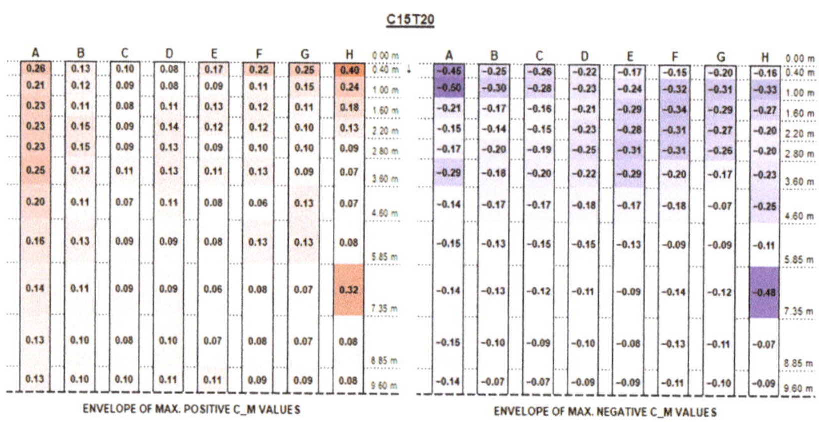

Figure A6. The envelope (all wind exposure angles) of maximum positive and negative values of moment resultants normalised by wind pressure (C_M) for the C15T20 configuration.

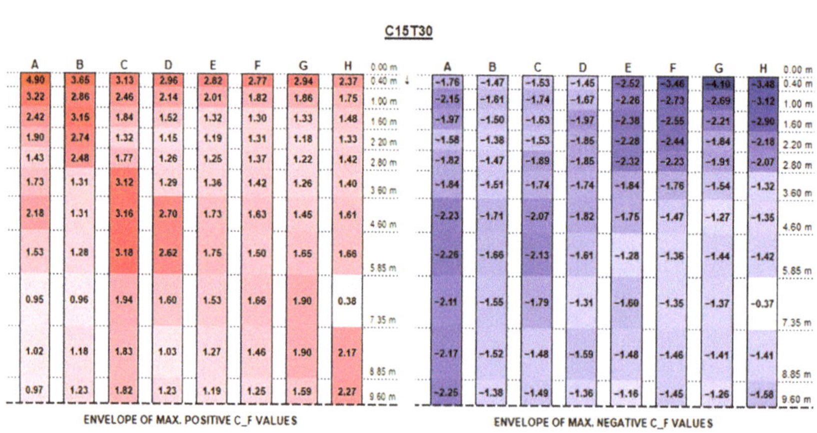

Figure A7. The envelope (all wind exposure angles) of maximum positive and negative values of force resultants normalised by wind pressure (C_F) for the C15T30 configuration.

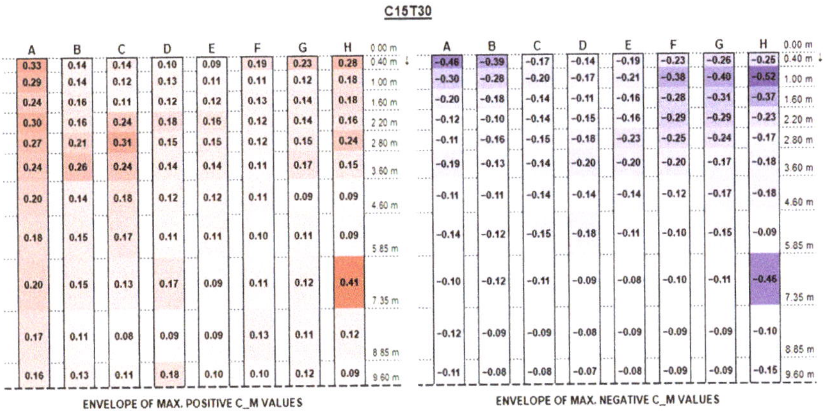

Figure A8. The envelope (all wind exposure angles) of maximum positive and negative values of moment resultants normalised by wind pressure (C_M) for the C15T30 configuration.

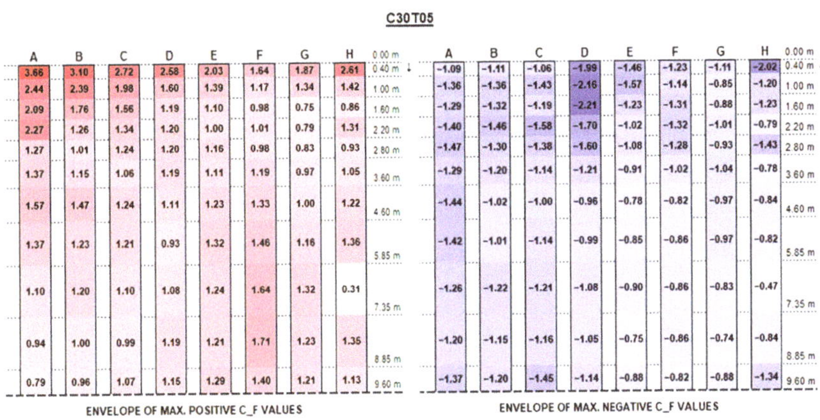

Figure A9. The envelope (all wind exposure angles) of maximum positive and negative values of force resultants normalised by wind pressure (C_F) for the C30T05 configuration.

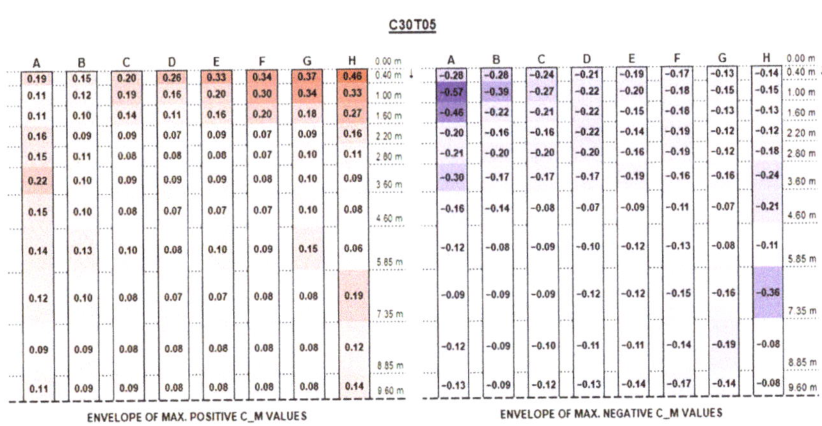

Figure A10. The envelope (all wind exposure angles) of maximum positive and negative values of moment resultants normalised by wind pressure (C_M) for the C30T05 configuration.

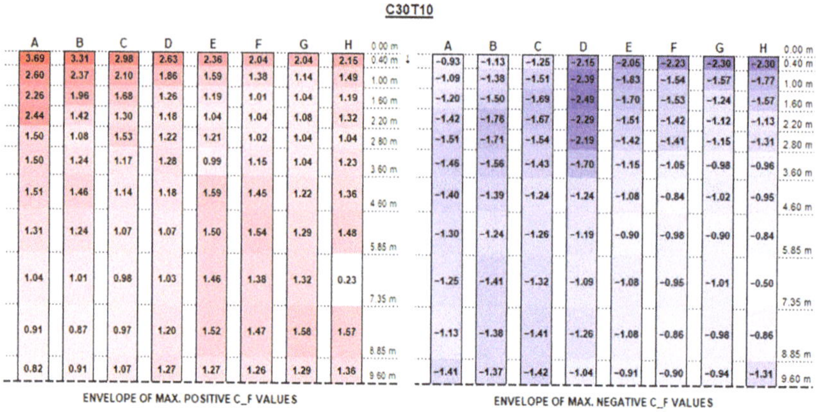

Figure A11. The envelope (all wind exposure angles) of maximum positive and negative values of force resultants normalised by wind pressure (C_F) for the C30T10 configuration.

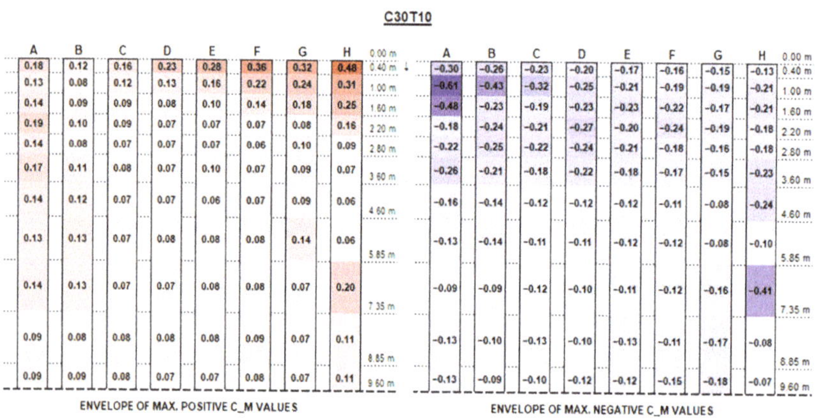

Figure A12. The envelope (all wind exposure angles) of maximum positive and negative values of moment resultants normalised by wind pressure (C_M) for the C30T10 configuration.

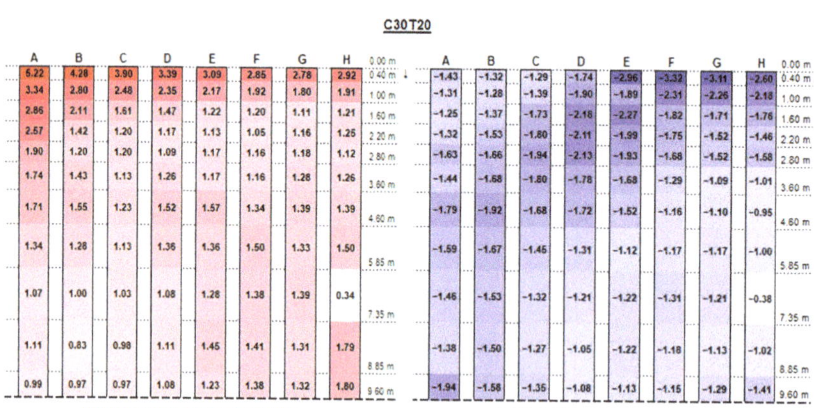

Figure A13. The envelope (all wind exposure angles) of maximum positive and negative values of force resultants normalised by wind pressure (C_F) for the C30T20 configuration.

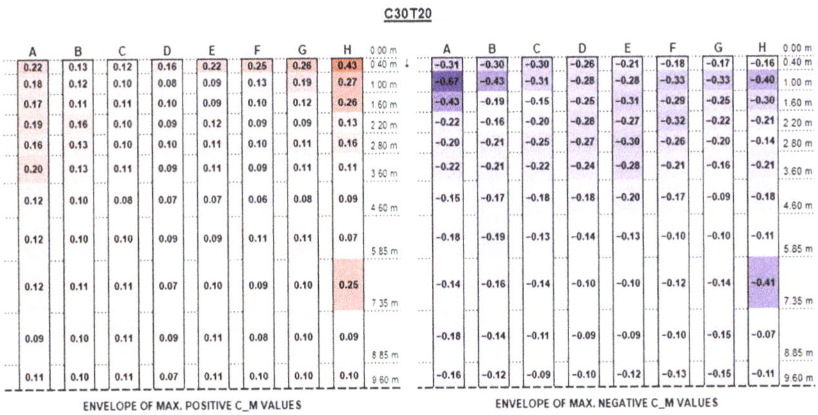

Figure A14. The envelope (all wind exposure angles) of maximum positive and negative values of moment resultants normalised by wind pressure (C_M) for the C30T20 configuration.

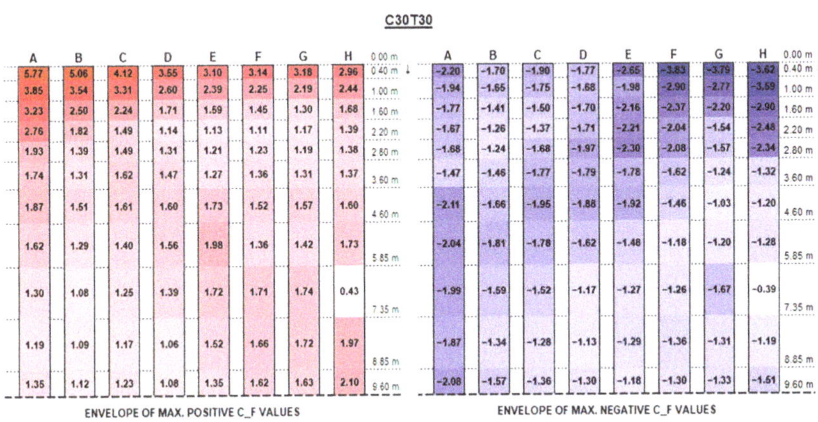

Figure A15. The envelope (all wind exposure angles) of maximum positive and negative values of force resultants normalised by wind pressure (C_F) for the C30T30 configuration.

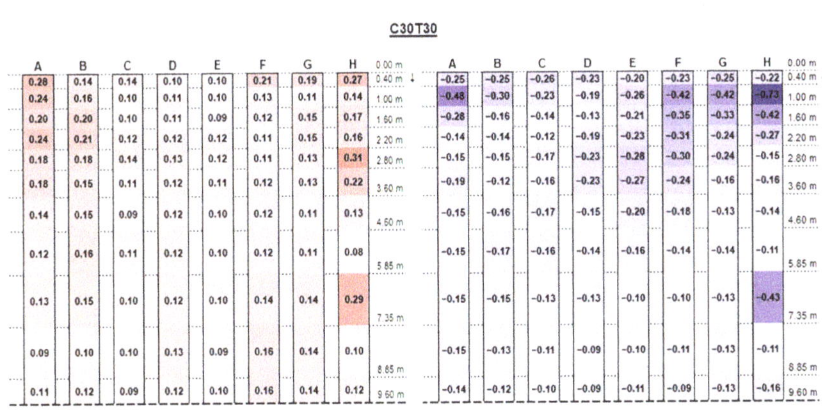

Figure A16. The envelope (all wind exposure angles) of maximum positive and negative values of moment resultants normalised by wind pressure (C_M) for the C30T30 configuration.

Appendix B

This appendix contains the plots, for each wind exposure angle, of corrected positive and negative C_F (normalised force resultants) and C_M (normalised moment resultants) over all PV string configurations. Strings are depicted with reference to their identification in Figure 9. The thick dotted line at the bottom of each picture represents the mid-string symmetry line.

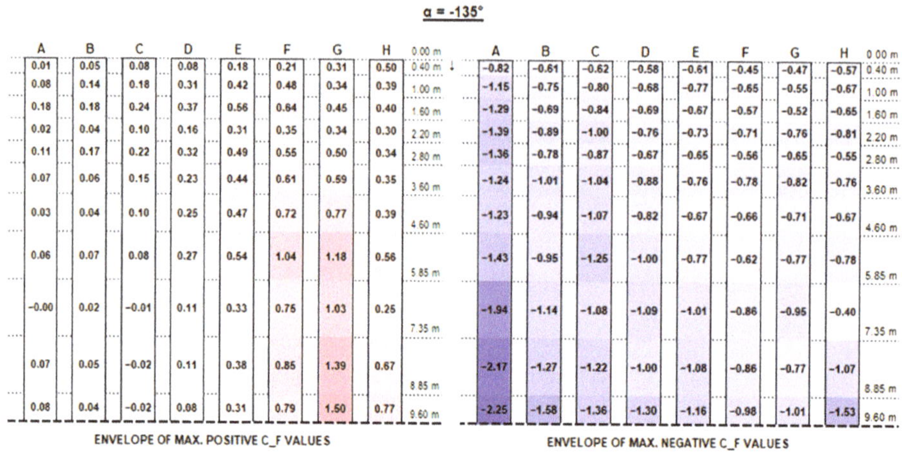

Figure A17. Envelope (all configurations) of maximum positive and negative values of force resultants normalised by wind pressure (C_F) for $-135°$ wind exposure angle (see Figure 9).

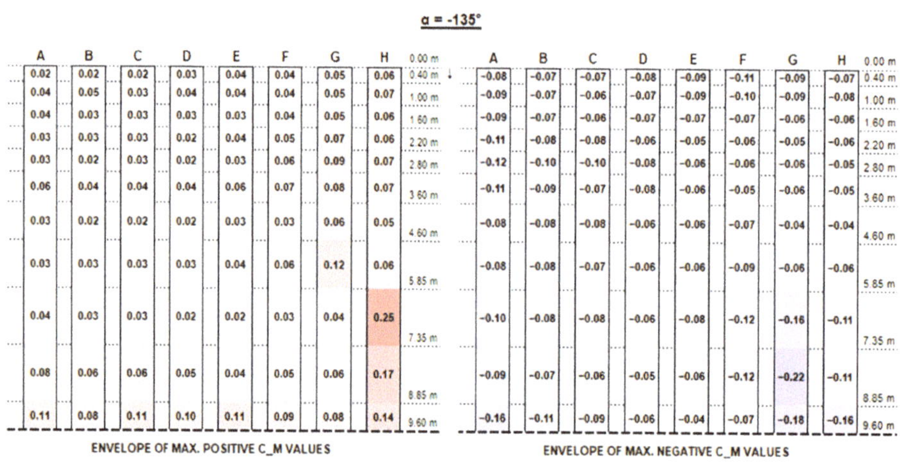

Figure A18. Envelope (all configurations) of maximum positive and negative values of moment resultants normalised by wind pressure (C_M) for $-135°$ wind exposure angle (see Figure 9).

Figure A19. Envelope (all configurations) of maximum positive and negative values of force resultants normalised by wind pressure (C_F) for −90° wind exposure angle (see Figure 9).

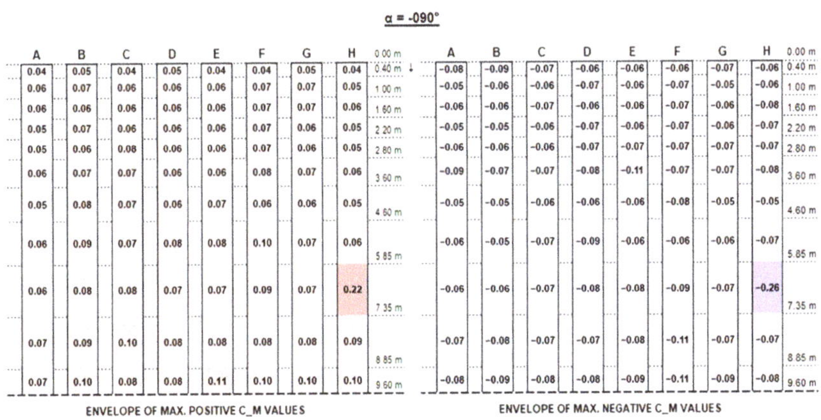

Figure A20. Envelope (all configurations) of maximum positive and negative values of moment resultants normalised by wind pressure (C_M) for −90° wind exposure angle (see Figure 9).

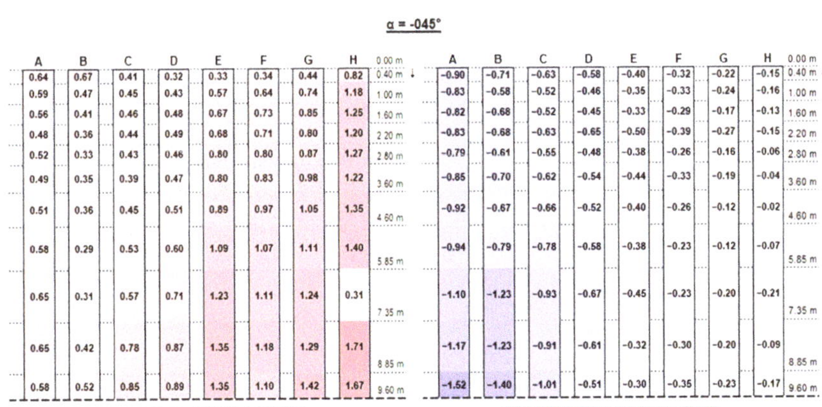

Figure A21. Envelope (all configurations) of maximum positive and negative values of force resultants normalised by wind pressure (C_F) for −45° wind exposure angle (see Figure 9).

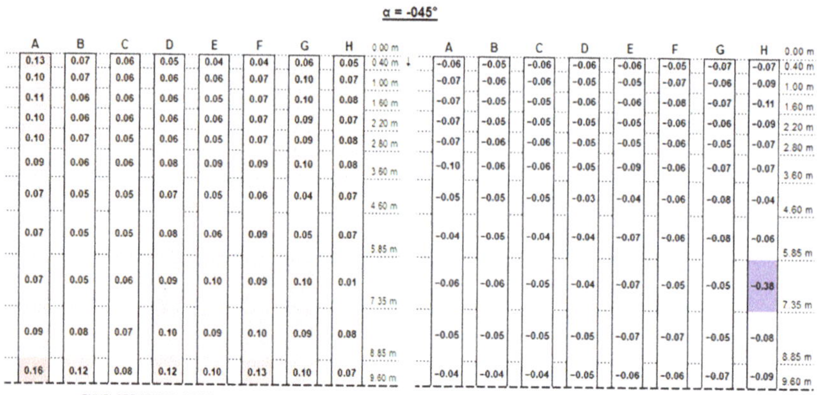

Figure A22. Envelope (all configurations) of maximum positive and negative values of moment resultants normalised by wind pressure (C_M) for −45° wind exposure angle (see Figure 9).

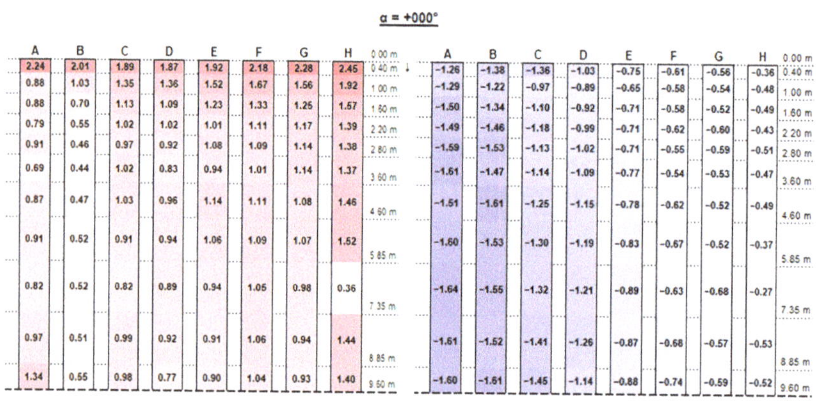

Figure A23. Envelope (all configurations) of maximum positive and negative values of force resultants normalised by wind pressure (C_F) for 0° wind exposure angle (see Figure 9).

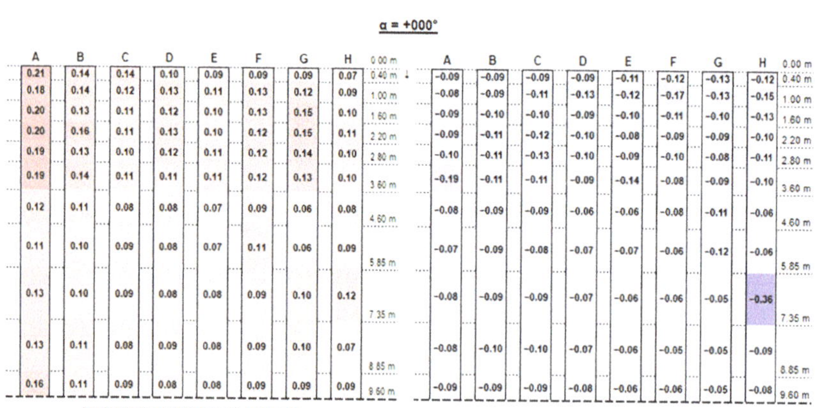

Figure A24. Envelope (all configurations) of maximum positive and negative values of moment resultants normalised by wind pressure (C_M) for 0° wind exposure angle (see Figure 9).

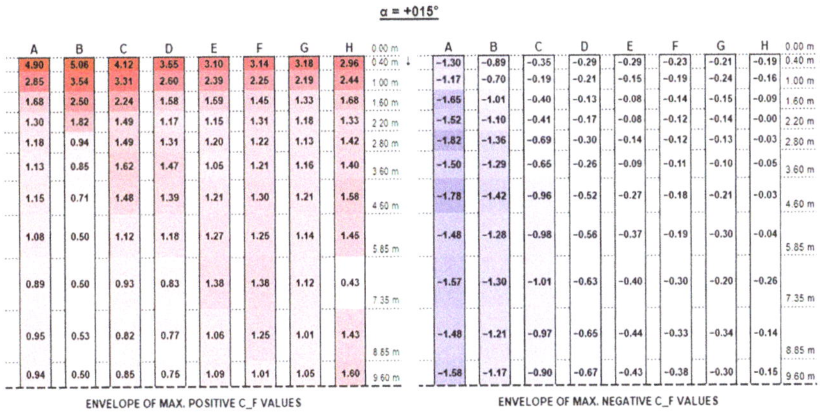

Figure A25. Envelope (all configurations) of maximum positive and negative values of force resultants normalised by wind pressure (C_F) for +15° wind exposure angle (see Figure 9).

Figure A26. Envelope (all configurations) of maximum positive and negative values of moment resultants normalised by wind pressure (C_M) for +15° wind exposure angle (see Figure 9).

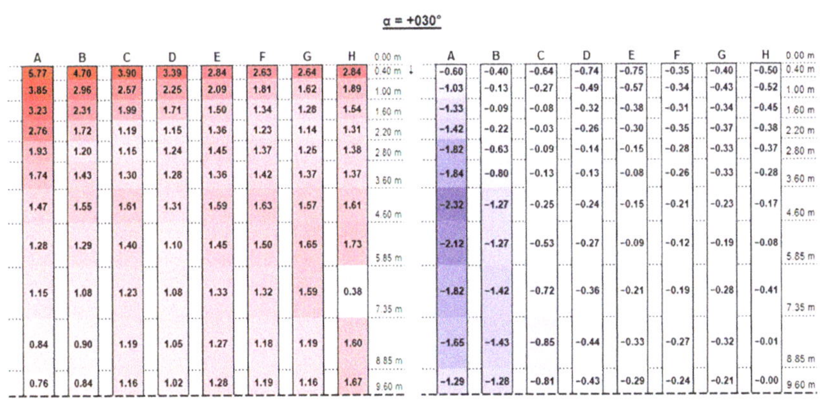

Figure A27. Envelope (all configurations) of maximum positive and negative values of force resultants normalised by wind pressure (C_F) for +30° wind exposure angle (see Figure 9).

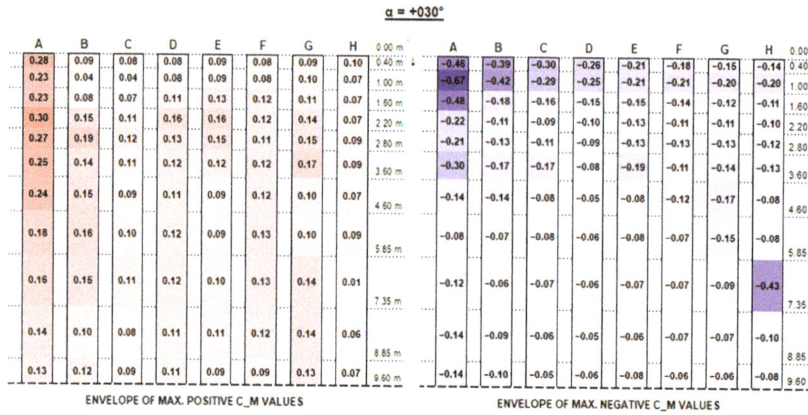

Figure A28. Envelope (all configurations) of maximum positive and negative values of moment resultants normalised by wind pressure (C_M) for +30° wind exposure angle (see Figure 9).

Figure A29. Envelope (all configurations) of maximum positive and negative values of force resultants normalised by wind pressure (C_F) for +45° wind exposure angle (see Figure 9).

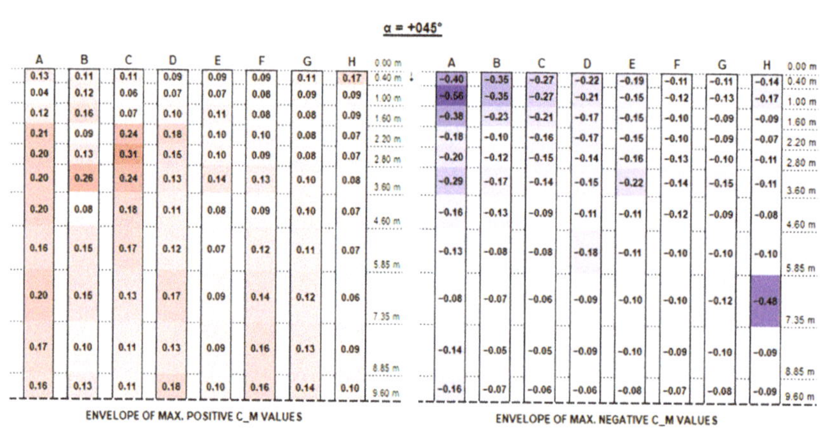

Figure A30. Envelope (all configurations) of maximum positive and negative values of moment resultants normalised by wind pressure (C_M) for +45° wind exposure angle (see Figure 9).

Figure A31. Envelope (all configurations) of maximum positive and negative values of force resultants normalised by wind pressure (C_F) for +90° wind exposure angle (see Figure 9).

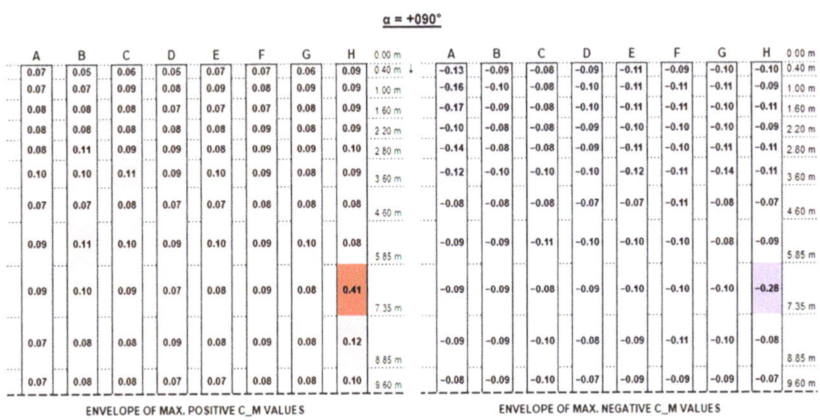

Figure A32. Envelope (all configurations) of maximum positive and negative values of moment resultants normalised by wind pressure (C_M) for +90° wind exposure angle (see Figure 9).

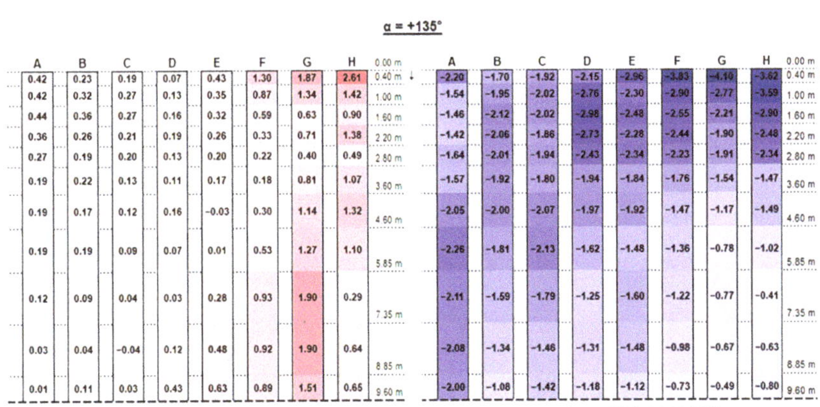

Figure A33. Envelope (all configurations) of maximum positive and negative values of force resultants normalised by wind pressure (C_F) for +135° wind exposure angle (see Figure 9).

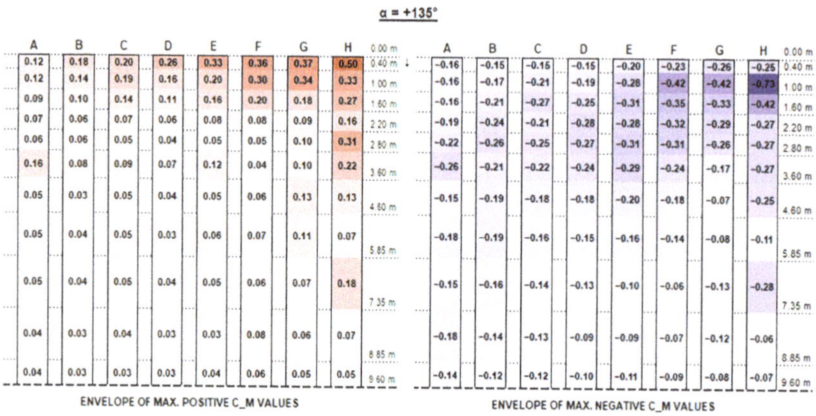

Figure A34. Envelope (all configurations) of maximum positive and negative values of moment resultants normalised by wind pressure (C_M) for +135° wind exposure angle (see Figure 9).

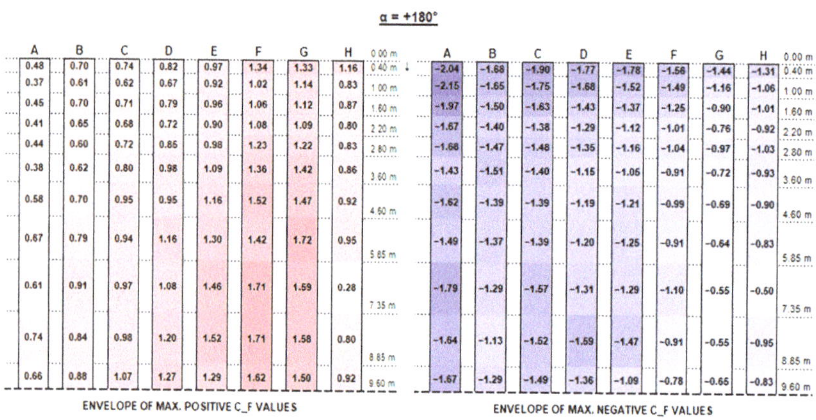

Figure A35. Envelope (all configurations) of maximum positive and negative values of force resultants normalised by wind pressure (C_F) for +180° wind exposure angle (see Figure 9).

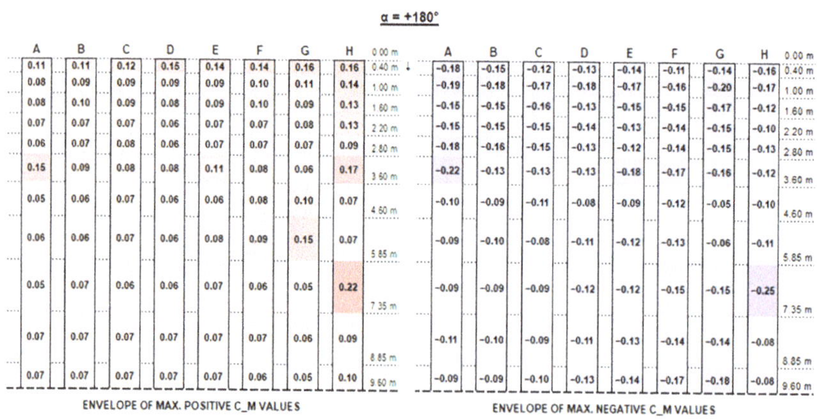

Figure A36. Envelope (all configurations) of maximum positive and negative values of moment resultants normalised by wind pressure (C_M) for +180° wind exposure angle (see Figure 9).

References

1. Kopp, G.A. Wind Loads on Low-Profile, Tilted, Solar Arrays Placed on Large, Flat, Low-Rise Building Roofs. *J. Struct. Eng.* **2014**, *140*, 04013057. [CrossRef]
2. Scrinzi, G. Experimental Testing Methods for the Performance Control of Standing Seam Roofing Systems for Large Roof Applications and Photovoltaic Systems Integration. Ph.D. Thesis, Politecnico di Milano (Department of Architecture, Built Environment and Construction Engineering), Milan, Italy, 2023.
3. Bechtel National Inc. *Wind Design of Flat Panel Photovoltaic Array Structures*; Sandia National Laboratories: Albuquerque, Mexico, 1980.
4. Miller, R.D.; Zimmerman, D.K. *Wind Loads on Flat Plate Photovoltaic Array Fields—Phase II Final Report*; Boeing Engineering and Construction Company: Seattle, DC, USA, 1979.
5. Miller, R.D.; Zimmerman, D.K. *Wind Loads on Flat Plate Photovoltaic Array Fields—Phase III Final Report*; Boeing Engineering and Construction Company: Seattle, DC, USA, 1981.
6. Miller, R.D.; Zimmerman, D.K. *Wind Loads on Flat Plate Photovoltaic Array Fields (Nonsteady Winds)—Phase IV Final Report*; Boeing Engineering and Construction Company: Seattle, DC, USA, 1981.
7. Aly, A.M.; Bitsuamlak, G. Aerodynamics of ground-mounted solar panels: Test model scale effects. *J. Wind Eng. Ind. Aerodyn.* **2013**, *123*, 250–260. [CrossRef]
8. Aly, A.M. On the evaluation of wind loads on solar panels: The scale issue. *Sol. Energy* **2016**, *135*, 423–434. [CrossRef]
9. Warsido, W.P.; Bitsuamlak, G.T.; Barata, J.; Gan Chowdhury, A. Influence of spacing parameters on the wind loading of solar array. *J. Fluids Struct.* **2014**, *48*, 295–315. [CrossRef]
10. Cao, J.; Yoshida, A.; Saha, P.K.; Tamura, Y. Wind loading characteristics of solar arrays mounted on flat roofs. *J. Wind Eng. Ind. Aerodyn.* **2013**, *123*, 214–225. [CrossRef]
11. Somekawa, D.; Taniguchi, T.; Taniike, Y. Wind loads acting on PV panels and support structures with various layouts. In Proceedings of the 8th Asia-Pacific Conference on Wind Engineering, Chennai, India, 10–14 December 2013.
12. Stathopoulos, T.; Zisis, I.; Xypnitou, E. Local and overall wind pressure and force coefficients for solar panels. *J. Wind Eng. Ind. Aerodyn.* **2014**, *125*, 195–206. [CrossRef]
13. Kopp, G.A.; Farquhar, S.; Morrison, M.J. Aerodynamic mechanisms for wind loads on tilted, roof-mounted, solar arrays. *J. Wind Eng. Ind. Aerodyn.* **2012**, *111*, 40–52. [CrossRef]
14. Banks, D. The role of corner vortices in dictating peak wind loads on tilted flat solar panels mounted on large, flat roofs. *J. Wind Eng. Ind. Aerodyn.* **2013**, *123*, 192–201. [CrossRef]
15. Abiola-Ogedengbe, A.; Hangan, H.; Siddiqui, K. Experimental investigation of wind effects on a standalone photovoltaic (PV) module. *Renew. Energy* **2015**, *78*, 657–665. [CrossRef]
16. Aronescu, A.; Appelbaum, J. Shading on Photovoltaic Collectors on Rooftops. *Appl. Sci.* **2020**, *10*, 2977. [CrossRef]
17. Scrinzi, G.; Mazzucchelli, E.S. 3D-printing high precision components of building scaled-models for wind tunnel testing. *AM Perspect.* 2025; *accepted for publication*.
18. *EN 1991-1-4*; Eurocode 1: Actions on Structures—Part 1–4: General Actions—Wind Actions. Available online: https://www.phd.eng.br/wp-content/uploads/2015/12/en.1991.1.4.2005.pdf (accessed on 14 April 2025).
19. Cook, N.J.; Mayne, J.R. A novel working approach to the assessment of wind loads for equivalent static design. *J. Wind Eng. Ind. Aerodyn.* **1979**, *4*, 149–164. [CrossRef]
20. Cook, N.J.; Mayne, J.R. A refined working approach to the assessment of wind loads for equivalent static design. *J. Wind Eng. Ind. Aerodyn.* **1980**, *6*, 125–137. [CrossRef]

sustainability

MDPI

Review

Overview of Sensing and Data Processing Technologies for Smart Building Services and Applications

Hamza Elkhoukhi [1,*], Abdellatif Elmouatamid [2,*], Achraf Haibi [3], Mohamed Bakhouya [4] and Driss El Ouadghiri [3]

[1] Department of Mathematics and Computer Science, University Toulouse 2 Jean Jaurès, 31100 Toulouse, France
[2] Department of Sciences and Technologies, UMR Espace-Dev, University of French Guiana, 97300 Cayenne, France
[3] IA Laboratory, Faculty of Sciences, Moulay Ismail University, Meknes 11201, Morocco; a.haibi@umi.ac.ma (A.H.); d.elouadghiri@umi.ac.ma (D.E.O.)
[4] LERMA Lab, College of Engineering and Architecture, International University of Rabat, Sala El Jadida 11103, Morocco; mohamed.bakhouya@uir.ac.ma
* Correspondence: hamza.elkhoukhi@gmail.com or hamza.el-khoukhi@univ-tlse2.fr (H.E.); abdellatif.elmouatamid@univ-guyane.fr (A.E.)

Abstract: Internet of things (IoT) and big data technologies are increasingly gaining significance in the implementation of various services and applications. Consequently, much of the research focused on energy efficiency and building management revolves around integrating IoT and big data technologies for data collection and processing. Occupancy detection, comfort, and energy management are the most important services for optimizing building energy consumption in smart buildings, and environmental data play a key role in improving these services. Furthermore, the integration of advanced and recent techniques, such as IoT, big data, and machine learning, is progressively becoming more vital for both researchers and industries. This paper presents and discusses various emerging technologies that will contribute to designing novel IoT-based architectures to improve smart building services. These technologies offer innovative solutions to address the challenges of interoperability, scalability, and real-time data processing within intelligent environments, paving the way for more efficient, adaptive, and user-centric smart building systems. The main aim of this research is to help researchers define an optimal architecture that presents all layers, from sensing to big data stream processing. We established comparative criteria between the most popular data processing techniques to select the appropriate framework for developing intelligent platforms for managing building services, such as occupancy detection systems and occupants' comfort management, and further, to enhance the deployment of digital twins for critical environment monitoring and anomaly detection. The proposed architecture uses Apache Kafka, Apache Storm, and Apache SAMOA as its core components, creating a comprehensive platform for efficient data collection, monitoring, and processing with high performance in terms of fault tolerance and low latency.

Keywords: internet of things; big data; occupancy detection; smart building; energy efficiency in building

1. Introduction

Energy security is considered an important challenge in numerous countries to maintaining sustainable economic development in the world. The construction and building sectors consume the most energy globally. Moreover, recent public statistics related to energy consumption around the world cite that 22% of total energy consumption is consumed by residential buildings distributed in different residential end uses (i.e., HVAC, lighting,

https://doi.org/10.3390/su17094029

water heating, cooking, and other appliances) [1]. In fact, the growth of energy demand gives rise to power degradation. Furthermore, the excessive use of fossil fuels can lead to extensive environmental impact. Therefore, developing new strategies to manage energy use and integrating renewable energy sources are a major focus for researchers worldwide.

Most governments around the world investigate the development of new strategies for optimizing energy use and protecting the environment. For example, the Moroccan government aims to achieve 20% of energy-saving potential by 2030 [2]. Accordingly, modern residential and commercial buildings integrate building energy management systems (BEMSs) as advanced methods for controlling different building services, with the aim of optimizing energy consumption and maintaining the occupants' comfort. BEMSs integrate different services, such as lighting, shading, and HVAC. Moreover, advanced controllers, such as MPC and GPC, can provide significant indoor air quality with potential energy savings in a specific area [3]. Several control methods can merge thermal data to efficiently control HVAC and ventilation systems [4]. Moreover, the integration of renewable sources into the existing grid poses a big challenge to controlling approaches and storage instrumentation. Therefore, many investigators have proposed different smart grid ideas with the intention of securing electricity for consumers and exploiting generated electricity efficiently [5]. In fact, badly managing building's active systems may considerably increase the wasted energy. Therefore, BEMSs have been proposed by numerous studies in order to facilitate the control and management of energy use in buildings, where weather information and occupant impact are considered key parameters [6–9]. In addition, building information models play an important role in BEMSs, where different information could be integrated and considered (i.e., such as geometric information, physical properties, and component information) for optimal computing and calculating essential building services such as thermal efficiency and lighting intensity. The main aim of smart buildings is to integrate physical and computational components to create an environment where the energy is well exploited, and the occupants' comfort is preserved. Each building has a unique characteristic. Therefore, emerging real-time and adaptive approaches to environmental settings are a big challenge enabled by the development of smart buildings. Dealing with this issue requires the integration of new adaptive techniques and deep insight into control modeling. With recent advances in data processing sensing networks, many industries and researchers have confirmed the potential of IoT as an enabler in the development of intelligent and context-aware services and applications [10,11]. Most studies demonstrate that building services require the integration of recent IoT and big-data technologies due to the complexity of large amounts of data that can be monitored and exploited [12–15]. These services can dynamically react to the changes in the environment and users' preferences. Their main objective is to create a comfortable environment for occupants depending on their location and ongoing activities.

Typically, a complex sensing network connects home appliances. Therefore, monitoring and controlling home appliances and active systems with high performance require the integration of IoT and big data technologies in order to deal with large amounts of non-structured data that are generated from objects in smart buildings, such as smart meters and other deployed sensors [16,17]. In fact, IoT refers to a network of interconnected devices capable of collecting and exchanging data. Big data refers to the vast and intricate datasets that originate from a variety of sources, including IoT devices. They are characterized by high volume, velocity, and variety, requiring advanced tools for processing and analysis [18–20]. The multiple sources of data that are generated from deployed sensors involve the integration of recent big data techniques such as artificial intelligence algorithms and distributed ledger technology. Additionally, data generated from buildings are mostly asynchronous, heterogeneous, and stored in different formats. Therefore, in order to manage building data clearly, big data approaches are required to be integrated for processing large amounts of both historical and stream data to extract appropriate knowledge for making data-driven instructions [21].

In summary, the large-volume and complex data generated from smart objects in buildings necessitates the implementation of effective IoT architecture to efficiently collect data and communicate between building appliances via the integration of recent big data technologies for the processing already-stored or real-time stream data. Accordingly, the main purpose of this study is to illustrate the right path and propose optimal architecture for managing building services with high performance. Furthermore, this study presents and discusses a high level of data-driven approaches for data exchange, building data management, and real-time stream data processing. Further, existing works related to IoT and big data technologies are provided and compared in terms of their characteristics and performances.

The remainder of this paper is organized as follows: Section 2 introduces the methodology followed for collecting and analyzing studies related to data sensing and processing approaches for smart building applications. Section 3 presents an overview of the sensing technologies used for smart building applications. Section 4 sheds more light on different big data technologies, including batch-based and stream-based techniques that can be integrated into building management systems. The discussion of and comparison criteria between stream processing techniques and the proposition of a global IoT and big data architecture platform are presented in Section 5.

2. Materials and Methods

This section provides the methodology used for collecting, selecting, and analyzing studies related to the sensing and data processing approaches that can be integrated for improving smart building services management. The main objective of this work is to share efficient and novel technologies in order to help developers and researchers design an optimal IoT architecture to enhance the infrastructure deployment of new building management systems. This section presents a detailed review methodology.

2.1. *Narrative Review Method*

The use of IoT and big data technologies in smart buildings can help improve the building management system in order to reduce the energy consumption and maintain the occupants' comfort. However, selecting the most relevant technologies for setting up an optimal architecture has still proven challenging in many studies related to smart buildings. This study follows a narrative literature review approach to provide a comprehensive overview of IoT and big data technologies that can be applied to smart building applications.

2.2. *Data Collection Method*

In order to perform this review, we collected relevant scientific articles, technical reports, and international standards from well-known databases such as IEEE Xplore, ScienceDirect, and Google Scholar. The selection criteria focused mostly on recent publications (from 2015 to 2024) related to energy performance, comfort indicators, and smart building technologies.

Several keywords have been used for this study, including "occupancy detection", "buildings active services", "buildings management systems", and "IoT and big data for smart buildings". The first search in Scopus yielded 1416 documents. Furthermore, another search was made using secondary keywords for extracting IoT and data processing approaches that can be integrated into building management systems.

2.3. Data Analysis Method

The data collected in this work were analyzed using qualitative research content analysis by systematically examining textual information in order to explore how IoT and big data technologies are applied to enable and improve smart building services. This method involved an in-depth examination of the selected literature to categorize the technological methods of enhancing key performance metrics and to highlight their relevance and applications across various smart buildings services (i.e., energy management, occupancy detection, and indoor air quality).

Qualitative content analysis was chosen due to its effectiveness in synthesizing diverse and interdisciplinary sources, especially in studies related to smart buildings. It allowed for the extraction of both explicit outcomes (e.g., energy savings, latency, and data accuracy) and implicit patterns (e.g., integration challenges, privacy concerns, and scalability issues). By clustering findings into thematic categories (i.e., sensor networks, data analytics techniques, system interoperability, and real-time decision-making), the review introduces a comprehensive overview of the current landscape. Despite this approach's added value, it is important to identify numerous challenges. Like several literature-based studies, the analysis is constrained by the reporting quality of existing studies. Many innovations may have been excluded if they were not presented and published in an accessible academic source.

Based on most of the research work presented in this review, at the end of this work, an architecture is presented, as general as possible, with detailed descriptions of different components that can be integrated in smart building applications. Furthermore, a detailed prototype platform of one of the important services in smart buildings, namely, occupancy detection, is presented by illustrating significant hardware/software settings in order to facilitate the deployment design of the IoT infrastructure for improving smart building services.

3. Sensing Technologies Used in Smart Building Services

New generations of buildings, or smart buildings, aim at integrating various methos of ensuring the comfort of building occupants while also ensuring that the energy is well exploited. In fact, large amounts of data are generated daily from smart building-connected devices. Therefore, IoT, big data analytics, machine learning, and sensing techniques are cost-effective technologies that can be integrated into smart buildings with minimum infrastructure in order to facilitate the use of large amounts of data generated from building objects and to create an optimal decision to well manage building services [5]. Different types of sensors can be integrated for analyzing occupants' behavior in buildings and, therefore, for studying the occupants' preferences (i.e., visual and thermal) in order to introduce building automation and establish better control of active building systems [22]. Various studies related to energy management in buildings have investigated the integration of IoT and big data technologies with the aim of managing the trade-off between energy consumption and occupants' comfort. This section presents an overview of IoT integration into smart buildings, evaluating it based on diverse characteristics.

Indoor air quality is an important feature related to the comfort and well-being of building occupants. For example, the authors of [23] proposed a general architecture prototype for real-time monitoring and processing; the aim of this work is to monitor and analyze the influence of CO_2 on and the SPO2 of building occupants. In fact, IoT platforms can help gather environmental data and act on building equipment in order to make building occupants more comfortable. The authors of [24] present the usefulness of integrating IoT technology for regulating HVAC and determine the opening/closing patterns for windows based on indoor air quality data in order to improve the performance of students learning in a classroom. Moreover, advanced indoor air quality monitoring systems are presented in [25] by integrating an open-source IoT web server platform named Emoncms for long-term monitoring and the storing of useful environmental data. Furthermore, the authors of [26] integrated IoT techniques for monitoring indoor air quality data, including CO, NH_3, CH_4, NO_2, CO_2, and machine learning for predicting different pollutant concentrations. In setting up a low-cost solution for monitoring indoor air quality data, CO_2 is easy to measure and can be influenced by the building occupants. The authors of [27] propose an IoT-based solution for monitoring real-time CO_2 concentration in the closed area. An ESP8266 microcontroller, which comprises a Wi-Fi module, was integrated as a communication unit. Furthermore, the work presented in [28] proposes a cloud-based IoT platform using a Lora gateway for gathering indoor air quality data, such as $PM_{2.5}$/PM10, CO, SO_2, NO_x, and O_3 pollutants. Support vector machine regression (SVMR) and neural network algorithms were integrated and evaluated for predictive analysis. Moreover, different algorithms can be proposed for controlling building active systems based on different parameters such as occupancy information, air quality, and thermal parameters. For example, the authors of [29] present a fuzzy control for adjusting the working interval of the exhaust fan with the use of an IoT platform for monitoring the CO_2 level and conducting the control remotely. The work presented in [30] concerns a simple approach for real-time controlling HVAC systems and achieving building occupants' comfort. IoT technologies have been integrated to collect environmental thermal data. The authors of [31] introduce an advanced approach in order to choose a suitable control strategy for ventilation systems. IoT devices and big data techniques were integrated for real-time data monitoring and processing. Moreover, the work presented in [32] presents a smart home application for controlling HVAC and lighting equipment using the ESP8266 microcontroller and Blynk platform.

Several works related to energy management systems have demonstrated the importance of integrating IoT to enhance many building services, such as occupancy detection [33,34]. Table 1 presents the studies showing the importance of IoT and sensing technologies for improving building services by introducing different communication technologies that can be integrated into building equipment control. For example, the authors of [35] present an occupancy detection algorithm using an IoT platform for collecting four environmental sensors (i.e., temperature, humidity, and CO_2) to determine wasted spaces in the building. An IoT-based prototype has been proposed by the authors of [36] for gathering occupancy information in the framework of energy building management. Lattice iCE40-HX1K stick FPGA boards and Raspberry Pi are the main elements integrated for counting people walking through the door. Further, the authors of [37] present an IoT platform for collecting and processing environmental data to create a low-cost and accurate occupancy detection model. Moreover, the authors of [38] propose a mobile air quality monitoring system (i.e., named e-nose) of different environmental parameters such as CO, CO_2, NO_2, PM10, humidity, and temperature in order to evaluate the impact of human activities on the indoor environment with the integration of different sensing techniques and cloud services.

Smart buildings are considered as sociotechnical systems including various heterogeneous entities (e.g., lighting, HVAC, RES) to balance between energy use and user comfort [39]. Making living in buildings more comfortable requires the integration of advanced techniques allowing for the interaction of different building entities in order to perform appropriate actions (e.g., well-controlling HVAC and lighting), as depicted in Figure 1. For example, occupancy detection and information can help control the building's active systems [40–42]. Different studies have investigated the development of advanced and intelligent strategies for making occupants' lives more comfortable and therefore efficiently managing energy consumption. The main purpose of smart or intelligent buildings is the responsiveness to users' preference needs. Integrating recent technologies is the main challenge of developing advanced building services in terms of reliability and performance [10]. Several definitions are proposed for smart buildings due to the evolution of technology. The integration of IoT and big data become important in order to evolve the smart approaches of building services [39,43]. The main purpose of building an efficient institute is to provide a comfortable environment and optimize energy use and gas emission. As depicted in Figure 1, different services (i.e., such as lighting control, air conditioning, solar energy generators, temperature/humidity monitoring, energy use monitoring, and cameras) can be integrated into smart buildings in order to establish the trade-off between occupants' comfort and energy consumption [16,44,45]. Intelligent approaches and smart control strategies (e.g., automotive control, machine learning, and q-learning) of such building appliances (i.e., such as lighting, HVAC, and windows opening) can assist in ensuring efficient energy use [4,46].

Table 1. Overview of different building services using IoT technologies.

Building Services	Sensing Parameters	Communication Techniques	The Aim
Air quality control (HVAC and window opening)	CO_2, VOC, temperature, humidity	Restful API	Regulating HVAC and determining the opening/closing patterns for windows based on indoor air quality data [24]
Air quality monitoring	CO_2, CO, SO_2, NO_2, O_3, Cl_2, temperature, and humidity	IoT web server platform (Emoncms)	Long-term monitoring and storing environmental data [25]
Air quality predicting	NH_3, CO, NO_2, CH_4, CO_2	GSM/Wi-Fi	Predicting different indoor air quality pollutants [26]
Real-time CO_2 monitoring	CO_2	Wi-Fi (http requests)	Developing an IoT platform for monitoring CO_2 concentration in real-time [27]
Air quality monitoring and predicting	$PM_{2.5}$/PM10, CO, SO_2, NO_x, and O_3	Lora gateway	Collecting and predicting air pollutant data using artificial intelligent algorithms [28]
Air quality monitoring	CO, CO_2, NO_2, PM10, humidity, and temperature	Wi-Fi (http requests)	Evaluate the impact of human activities on the indoor environment [38]
HVAC control	Temperature	Wi-Fi (http requests)	Optimal control of HVAC system using IoT framework for collecting thermal data [30]
Ventilation control	CO_2, temperature, and humidity	Wi-Fi (MQTT protocol)	Comparing different ventilation-based control strategies in terms of reducing energy and indoor comfort [31]
HVAC and lighting control	Temperature	IoT web server platform (Blynk)	Controlling remotely HVAC and lighting systems [32]
Occupancy detection	temperature, humidity, PIR, and CO_2	Wi-Fi based mesh network topology	Detecting occupancy with high accuracy in a floor area using environmental data [35]
Occupancy counting	iCE40-HX1K stick FPGA boards	Wi-Fi	Counting people using 2 iCE40-HX1K stick FPGA boards deployed in the door [36]

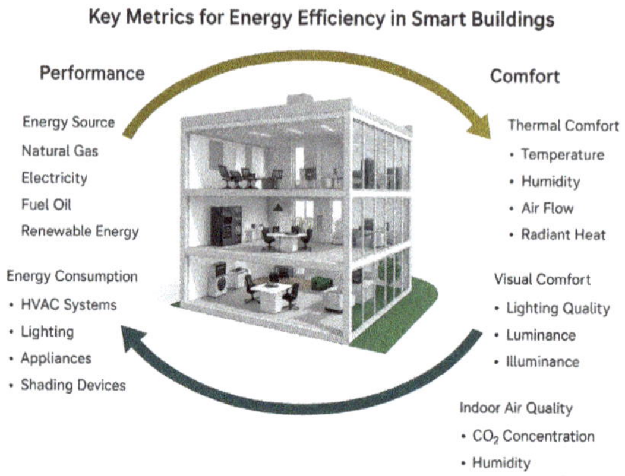

Figure 1. Smart building metrics.

IoT technology is the main concept employed to enable building entities to collect data, run commands, and control equipment. Therefore, the introduction of IoT can be considered an opportunity for developing products that save energy costs and increase user comfort and work efficiency. This subsection presents, as generally as possible, the architecture and detailed layers adopted by industries and researchers. Generally, there is no standard IoT architecture. Diverse architectures have been proposed by different studies. The most adopted IoT architecture consists of four main layers, namely, the perception, network, data processing, and application layers, as shown in Figure 2.

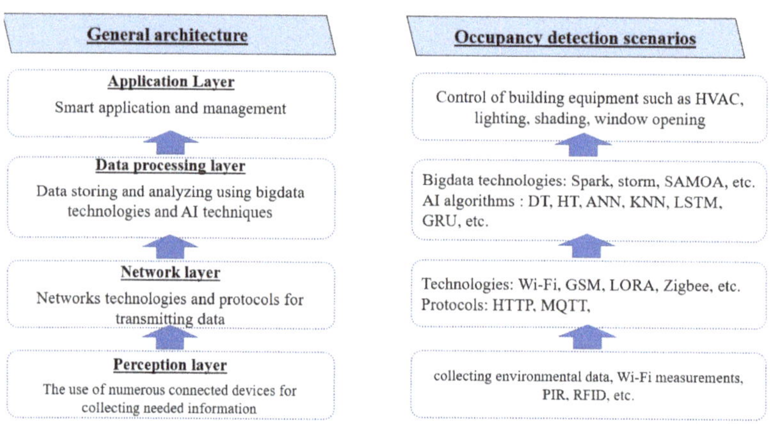

Figure 2. Mapping IoT layers for the occupancy detection scenario.

Deploying recent occupancy detection techniques requires the integration of advanced IoT architecture in order to deal with various numbers of connected devices deployed in distributed areas (e.g., offices, conference rooms, etc.) [37,41,47]. As illustrated in Figure 2, four main layers must be implemented in order to facilitate the deployment of advanced occupancy detection scenarios.

Perception layer: The perception layer (i.e., physical layer) consists mainly of sensing techniques. Various types of sensors can be deployed simply or are already installed in the building. Recent studies have revealed an interest in using environmental sensors (i.e., such as CO_2, temperature, humidity, light intensity, etc.) in order to fit the user privacy requirement contrary to camera sensors.

Network layer: The network layer is responsible for connecting all smart things and servers. As presented in Figure 2, different protocols (i.e., such as HTTP, MQTT, etc.) and techniques can be integrated for transmitting and processing sensor data.

Data processing layer: To set up smart building services using big data, traditional methods are not sufficient (i.e., due to latency and fault tolerance). The integration of big data techniques becomes a necessity for collecting and analyzing the numerous building data generated. Sophisticated IoT platforms such as Kaa and Thingsboard can provide advanced services such as the managing and integrating of various IoT devices, and they are horizontally scalable and fault tolerant. Big data tools such as Spark and Apache Storm can process real-time big data streams in a distributed manner [48].

Application layer: In order to efficiently use building systems such as HVAC, lighting, and ventilation, different control strategies have been developed for managing the use of these systems. For instance, building platforms could integrate MPC, GPC, and PID to efficiently control active equipment using environmental data and occupancy information [44].

In order to set up and integrate occupancy detection systems in the building, integrating IoT technologies becomes important to facilitate the collection and processing of data and therefore to make action efficient. Moreover, some works add more layers to their architecture due to a lack of certain features, such as security and business layers [49]. The aim is to facilitate the process of collecting and analyzing data in order to enable more efficient actions with the aim of reducing energy consumption and increasing the users' comfort.

4. Data Processing Technologies for Smart Building Services

Most modern buildings include a diversity of sensors, actuators, and advanced connectivity techniques. Further, these buildings generate a huge amount of data every second. In fact, it is difficult to organize, analyze, and store such a large amount of data. Therefore, large-scale data processing techniques have become increasingly important for efficiently analyzing building data and generating decisions in order to better control active equipment. The aim is to process a myriad of generated data in an intelligent manner with high-performance computing [39]. In the past, data storage was limited to spreadsheets and roles in database management systems. The information that could be structured in different row–column formats was disregarded [50]. Big data technologies are designed for analyzing different types of data and predicting accurate results in a short period of time. Big data techniques include five main characteristics, named the five Vs (i.e., volume, variety, velocity, value, veracity), in order to deal efficiently with different kinds of generated data.

Volume: Storing and analyzing a large amount of data is the main challenge of big data technologies. In fact, developing big data solutions means no longer dealing only with gigabytes; the current challenge is dealing with zettabytes.

Variety: The data can be categorized into different types (i.e., structured, unstructured, or structured). In fact, traditional database management tools are not sufficient to store and process a large amount of data of different types, and the data can be formed as documents, images, audio, logs, text, emails, or others.

Generally, processing big data consumes substantial resources and time. The aim is to ensure the processing of data streams in a short time with high performance and continuity. In fact, most research work related to big data development refers to the velocity of data streaming collected from websites, sensors, and video streams.

Value: The value refers to how the data can be beneficial in decision-making. For example, the work related to energy efficiency in buildings requires the implementation of big data technologies for managing energy use and increasing human comfort [31,51].

Veracity: This refers to data (i.e., accuracy and noise) and availability. This factor can affect the analysis methods, the time needed, and data dependency.

According to several studies and norms related to big data development, the above-mentioned characteristics do not comprise definitive rule. Several other characteristics can be taken into consideration such as variability (changeability in data flow), viscosity (degree of correlation or complexity), viability (ability to be active), volatility (how long-time data are valid and how long they should be stored (i.e., durability)), and validity (understandable in order to find the hidden relationships) [52].

Combining wireless communication technologies and electronic devices (i.e., smartphones, tablets, sensors, and house appliances) led to the emergence of the concept of the IoT. While the development of IoT infrastructure is growing, the need for platforms for managing data generated by large numbers of devices has become more important. This phenomenon has resulted in the generation of copious amounts of data from diverse sources.

Big data technology is the best solution to the problem of valuing and exploiting the huge data generated from IoT. Furthermore, industries and organizations should take advantage of the massive amount of data collected. The quick growth of information technologies, such as smartphones, social applications, and machine-to-machine communication, has led to the development of big data technologies [53]. Several studies have investigated the integration of IoT and big data for ensuring the development and implementation of different services such as energy building management, healthcare, and electrical vehicles. For example, in the healthcare field, some solutions have been proposed using IoT for remotely monitoring patients and analyzing the data collected from the embedded connected devices such as glycosometers and tension meters [54]. Further, in the building sector, for the successful management of the active equipment, the integration of IoT and data analytics techniques is greatly needed due to the big data generated from sensors deployed in different areas in residentials or industrial buildings [31,47]. Recently, car companies have invested in the development of new strategies in order to manage the reduction in energy consumption of vehicle components (i.e., HVAC) and the deployment of recharge stations using the IoT to collect data and to employ big data tools for analyzing the behavior of vehicles and drivers in order to provide efficient action and control [55,56].

In fact, the IoT and big data development necessitate the integration of different technologies and frameworks in order to transfer and gather the generated data to be processed. The following technologies have been presented in many research projects and are used by different companies to deploy their own solutions.

(1) **Fog computing techniques:** Fog computing is an extension of cloud computing. It can be considered a mini data center integrated between cloud computing and the edge of the network. Moreover, it is a highly virtualized platform that offers storage, processing, and networking services between edge and traditional cloud servers. Furthermore, depending on much research work, fog computing is described as a bringer that makes cloud computing closer to IoT devices. Any computing device with data processing and storage ability can be considered a fog node [57]. Among the advantages of fog nodes is their ability to reduce the amount of data transferred to the cloud server, as presented in Figure 3.

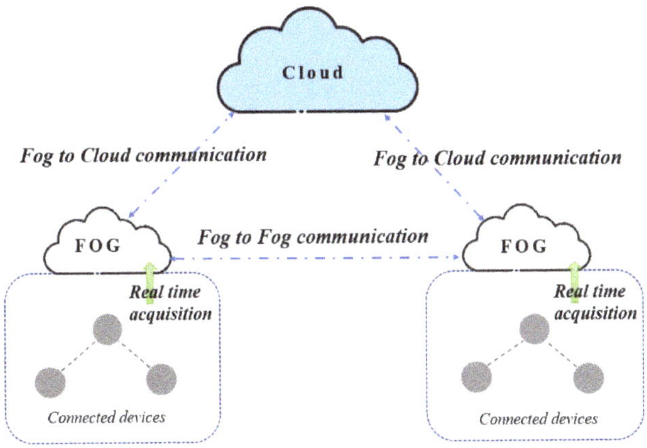

Figure 3. Fog-based IoT architecture [57].

(2) **Data aggregation and ingestion**. In order to link IoT devices to the processing unit (i.e., such as Spark and Apache Storm), different linker techniques can be deployed and implemented in order to enable the connection between big data processing elements and the data collected from different sources such as IoT devices and SaaS apps, ending up in different target environments. Data ingestion is an important technology that can help make sense of the increasing volume and data complexity. In fact, two main methods can be adopted for data ingestion, namely, batches and real-time. Real-time data ingestion refers to a process of collecting and transferring data to be processed directly from sources (i.e., see Figure 4). Real-time ingesting is an important technique for applications that are time-sensitive, i.e., when organizations need to rapidly react to new information (i.e., such as stock market trading, power grid monitoring, etc.). Also, the high-performance data pipeline refers to a crucial technique employed in order to help improve real-time data processing when systems are used to make rapid operational decisions and act on new insights. However, batches ingestion is designed for transferring data as batches depending on the interval of time. The ingestion, in this case, is useful when the user needs to gather a specific set of data already stored on a daily basis and not for real-time decision-making.

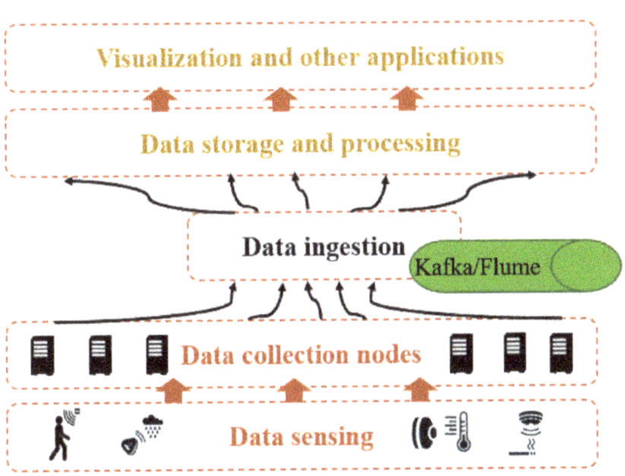

Figure 4. Data ingestion procedure [58].

Several techniques have been developed for data ingestion into different big data technologies (i.e., such as Spark, Storm, Samza, S4, etc.):

Apache Flume (https://flume.apache.org/): This is a reliable and distributed service for efficiently collecting and transferring large amounts of streaming data and log data into the big data technologies employed such as the Hadoop distributed file system (HDFS). It is robust, scales horizontally, and is fault tolerant, with tunable reliability mechanisms. It integrates an extensive data model to enable online data processing techniques.

Apache Kafka (https://kafka.apache.org/): This is component of a data stream platform used for collecting, processing, storing, and integrating data at scale. It has many use cases, including real-time pub/sub messaging, stream processing, and distributed streaming.

Different technologies have been developed in order to improve the management of a large amount of data needing to be stored or processed. Data ingestion is more important in order to ensure the transmission of well-structured data into a processing platform, whether stream (i.e., near real-time) or batch processing.

4.1. Batches-Based Processing Techniques

Batch setting, or batch-based processing, involves procedures and operations pertaining to a big and static dataset. Moreover, the results can be returned at a later time when processing is definitively completed. In fact, batch processing is when the set of data is already stored over a period of time in the server, and a big set of data can be totally analyzed and processed using different batch-based techniques such as Hadoop and Spark. For instance, weekly or monthly processing is required for billing systems and payrolls. Hadoop ecosystems and Apache Spark are the most popular technologies adopted to store, process, and analyze data in batches [59].

Hadoop MapReduce: This is the main programming model and execution engine for analyzing massive data in Hadoop framework is MapReduce. As a programming model, MapReduce includes two main functions, namely, the Map (Ki, Vi) and the reduce (Kj, list (Vj)). Each map function of input key-value (Ki, Vi) is used to output different intermediate key values (Kj, Vj). The intermediate values Vj should be grouped in order to generate the pairs Kj, list (Vj). The reduce function of the Kj list (Vj) summarizes the entire dataset into pairs (i.e., Kk, Vk) after the shuffling phase [59–61].

Apache Spark: This is popular for simplifying the development of applications that can be executed in parallel and in a distributed cluster in a fault-tolerant way. The most common programming support for Spark is Scala. The implementations of the tasks are based on the resilient distributed dataset (RDD) as a multiset of objects with the "read-only" feature propagating over a cluster with the fault-tolerant system [59]. In fact, the whole architecture components include Spark Core and a set of libraries. Most of the functionalities provided by Apache Spark are built on top of Spark Core. Moreover, it is considered to be the basis of the distributed and parallel processing of large amounts of data. It contains Spark Core at its top layer, including Spark SQL, Spark streaming, Mlib, and Spark GraphX, which are utilized for data monitoring and processing [62].

The above-mentioned techniques (i.e., the batches processing techniques) can be integrated for setting up different scenarios related to energy management in buildings. Using data already stored in the past can help create the best approaches and methods for increasing profit and decreasing costs. As an example, several works have investigated the integration of batch processing techniques in order to forecast the behavior of building occupants using historical data from previous years in order to develop the best planning and schedules of the occupancy in a specific area (i.e., such as offices, floor, buildings, and conference rooms). The main purpose of such studies is to efficiently control the active equipment of buildings (i.e., turn on/off HVAC, and lighting systems) in order to reduce the wasted energy in buildings.

Better control of building services requires real-time action. Several works investigate the use of environmental data in order to manage the use of most energy-consuming equipment [3,4]. Analyzing all generated data from IoT and acting directly on this equipment necessitate the integration of stream data processing techniques in order to analyze continuously generated data with high performance (i.e., latency and accuracy) [63].

4.2. Stream-Based Processing Techniques

Processing a myriad of generated data in batch is not sufficient when a new instance of data should be processed rapidly in order to ensure a fast reaction. Specifically, the stream processing paradigm online enables the analysis of continuous and dynamic data. Therefore, stream data processing techniques have become more important to industries and researchers who are working on real-time services such as building energy management systems [5]. The common streaming engines that have a large-scale adoption are Apache Spark Streaming (https://spark.apache.org/), Apache Flink (https://flink.apache.org/), and Apache Storm (https://storm.apache.org/). In fact, several applications necessitate the integration of these data stream processing tools. For example, the purchases of tasks for some applications, such as online video games that are downloaded from different sources, must analyze the users' actions.

Apache Storm: This is a distributed real-time processing system. It is easy to process and consumes a large amount of stream data in order to ensure real-time processing. It includes several processing techniques, such as API and SQL stream processing. The architecture of the Storm is composed of three essential elements: spouts as a source unit, and bolts and topologies as processing units. A spout is considered a source of streams. It consumes data from different sources, such as message queues and databases, and then transfers the tuples into the bolts in the topology without any processing tasks. Moreover, bolts consume the tuple from the spout, apply some tasks, such as aggregation, joining operations, or filtering, and then emit the new tuples to its output streams. The topology of a Storm application covers the multi-stage stream communication and computation of spouts and bolts, which are connected to stream grouping [64].

Apache Spark streaming: Spark streaming enables real-time data processing of data coming from different sources such as Flume, Kafka, and Amazon Kinesis. In fact, these processed data can be emitted to file systems and live dashboards. It is considered a micro batch generic-purpose data stream processing engine, an extension of the Spark Core API. The main data abstraction in Spark is called the resilient distributed dataset (RDD). Furthermore, Spark streaming offers the concept of discretized stream (DStream) in order to process a continuous stream of data, represented as a sequence of RDDs [65].

Apache Flink: This is designed for the distributed processing and stateful computation of continuous data streams (i.e., unbounded or bounded data streams). It is also considered a generic-purpose data stream processing engine. Each dataflow or instance in Flink starts with one or several sources and ends in multiple sinks in order to submit data to different outputs such as file systems or databases [66,67]. Apache Flink can be integrated into different cluster resource managers such as Hadoop and Kubernetes, and it can also be set up to be executed as a stand-alone cluster.

4.3. Big Data and Machine Leaning

Industry parts need to gather useful information from data streams (i.e., such as social media, credit card transactions, deployed sensors, etc.). Data collected from high-speed streams are more and more common in modern data ecosystems [68]. The real-time processing of a huge amount of data is not easy. For example, web companies must integrate systems that are able to process each instance of incoming data, demonstrating high performance in terms of time of execution and memory use [69]. Furthermore, Machine learning approaches have has an impact on many types of applications and systems, such as speech processing, natural language, computer vision, etc. Batch-based learning techniques require the training phase on the data already stored in file systems or databases. Different frameworks are designed and developed for applying machine learning algorithms on big data as one batch. Distributed machine learning frameworks (i.e., MLBase and Mahout) have demonstrated their effectiveness in terms of variety and volume [70]. However, different frameworks have been developed in order to address the issue of applying machine learning techniques on data coming from different sources (e.g., social media, building sensors, internet traffic data) as a data stream. In fact, streaming machine learning frameworks (i.e., such as massive online analysis (MOA)) also address the dimensions of variety and velocity [69,70].

Batch-based machine learning integration (Apache Mahout): Most of the work related to building energy management, such as via occupancy detection systems, has investigated the integration of batch machine learning techniques. This means that the data should be already available in a specific database or file system in order to build a model by training it on this data and building an efficient model. In this case, the model cannot be updated or improved using stream data. Further, Apache Mahout is an open-source framework that offers the free implementation of scalable and distributed machine learning algorithms, including several approaches such as classification, clustering, and recommender engine. The authors of [71] present the Hadoop ecosystems architecture, which includes several libraries, including the Mahout library, which is suitable for applications necessitating the scale of large amount of data executed on top of the Apache Hadoop. Figure 5 presents the global architecture of Hadoop ecosystems presented by [71], including the Mahout Library. In summary, the main purpose of Apache Mahout is to execute and apply machine learning algorithms on data already stored in HDFS.

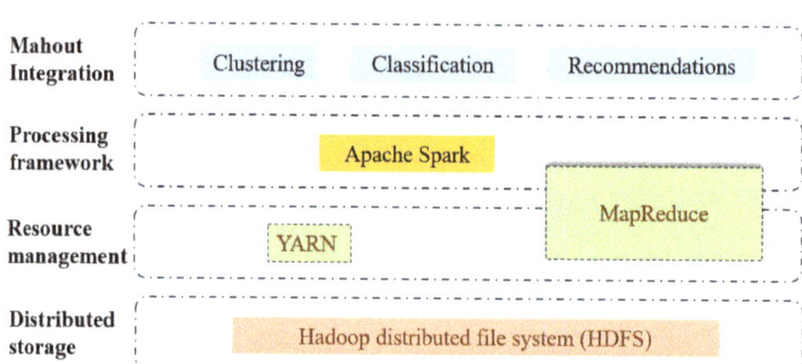

Figure 5. Global architecture of Hadoop ecosystems [71].

In fact, batch machine learning is extremely important in order to improve the different services that need to process historical data in order to make efficient decisions and to plan for the present or the future. However, many issues cannot deal with batch-based methods because of the environment change, and the static model will be unserviceable over time.

4.4. Online and Real-Time Machine Learning Techniques

Most developed solutions and frameworks related to machine learning techniques address just two out of three dimensions (i.e., variety and volume). In fact, some solutions have emerged for the three most popular dimensions (i.e., variety, volume, and velocity) in order to perform the stream mining of big data [72]. The main idea of stream machine learning or online machine learning is to update the model continuously and improve performance when each instance arrives at the processing unit in real or near real time. In fact, batch-based machine learning algorithms are static. These algorithms should exploit a large dataset to train and identify the model results before moving it to production. In contrast, online learning algorithms are processed incrementally when new data become available. Actually, stream learning is still developing its philosophy, and the documentation of its general properties is not yet comprehensive, as in batch learning. In this subsection, we discuss the two main frameworks that involve machine learning for the data stream: Apache Samoa, developed by Apache [69], and StreamDM, by Huawei [68].

Apache SAMOA (scalable advanced massive online analysis) is a distributed system for applying and programming streaming algorithms in order to develop new ML algorithms by avoiding the complication of streaming processing engines (SPEs), and developers can integrate new platforms using SAMOA APIs. The SAMOA project is divided into two main elements: SAMOA-Platform and SAMOA-API. The SAMOA-API can be integrated with different SPEs without any restriction. If new SPEs are being released, a new SAMOA platform module can be included. In order to perform ML algorithm development in SAMOA, five components are included in the global architecture: processing items, content event, stream, topology, and task. Processors are computational units in SAMOA use for executing algorithms separately on a defined SPE. In fact, these processing items (PIs) refer to the particular implementation of processors in different SPEs, such as Bolt in Storm. Two different PIs type are presented by the SAMOA community, namely, the entrance PI and the normal PI. The entrance PI consumes events from external sources or generates new events and transfers them to the normal PI to be processed and exploited for ML algorithms. The normal PI can send the processed event or new event to the outgoing streams. The SAMOA framework facilitates the implementation of ML algorithms in a parallel way by specifying the parallelism hint (i.e., the PI runtime during the task execution), as presented in Figure 6. Also, the stream can be considered as a communication tool between PIs (source PI and destination PI), as shown in Figure 6. The content event handles the data transferred from one PI to another via a stream. The source Pi sends an event to the destination PI via a stream, while the destination PI consumes the event coming from source PI. In fact, in order to consume the content event coming from source PI, the destination PI should integrate different "grouping" mechanisms, which define the stream route of content events. Three grouping mechanisms in SAMOA can be recognized, namely, shuffle grouping, all grouping, and key grouping. Moreover, the topology is a set of connected PIs and streams. The distributed streaming ML algorithms can be integrated into the SAMOA architecture. Tasks are related to ML integration. It is often used for evaluating the performance of algorithms. For example, the prequential evaluation task is the most popular task in the SAMOA architecture. It is used for testing the performance of models and for exploiting the current instance to train and update the model [69].

StreamDM is an open-source framework for applying machine learning algorithms on data streams. It has emerged on top of Spark streaming. The main goal is to connect the Spark framework using the Spark streaming library and online machine learning algorithms in a distributed system. The Spark environment already includes machine learning libraries such as MLlib and spark.ml. The aim of StreamDM is to develop advanced methods of data mining and machine learning in order to apply them to large-scale data streams. In fact, StreamDM has been developed for practical users, only using algorithms that are already presented by the framework, i.e., without the need to develop new algorithms, and researchers and developers can implement new algorithms in StreamDM. The Spark streaming is the main provider of data streams in StreamDM. In fact, it is not an instance-by-instance setting of streams. The idea is to process mini batches of data. Therefore, the latency of data needs more time. Further, StreamDM is mostly based on tasks that describe its functionality in mining data streams.

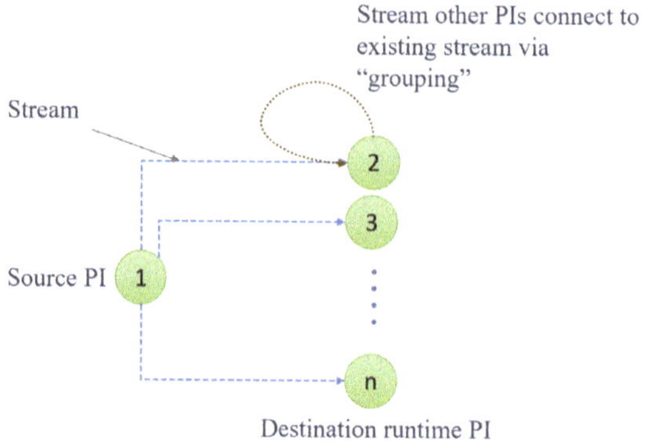

Figure 6. Parallelism hint and stream grouping in SAMOA.

Figure 7 describes as generally as possible the workflow of an example of a StreamDM task named "EvaluatePrequential". The task processes contain four main classes. First, the Reader class reads data streams from discretized streams. In addition, the testing and training data streams are consumed by the Learner class, which contains machine learning algorithms. Further, the predicted data are transmitted to the Evaluator class in order to evaluate the performance of the applied algorithm. Then, the results are sent by the Output Writer class into other data stream consumers such as file systems, consoles, or other tasks.

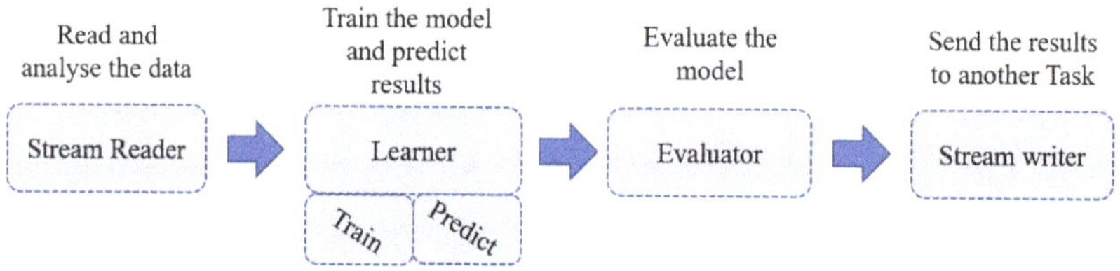

Figure 7. The workflow of "EvaluatePrequential" Task in StreamDM.

5. Discussion

Most digital solutions exploit a huge amount of data generated every millisecond. The main aim is to process these data quickly with high performance. Furthermore, many applications and systems require the near-real-time processing of large amounts of data streams and the real-time acting and controlling of a myriad of deployed devices simultaneously, such as building energy management systems. Numerous open-source frameworks and systems have been designed and developed for processing data streams. In fact, generating significant knowledge from unbounded data in time has become more difficult. The various number of developed frameworks and systems presents a substantial challenge for researchers and developers in choosing the best elements and determining which framework is suitable for their use cases. There are many features for processing data streams that should be taken into account in order to make the right design and develop an appropriate approach.

The benchmarking of some services is an important tool that can be used for evaluating the performance of big data technologies, especially stream processing frameworks [73]. Latency and fault tolerance are the main parameters that can help researchers and industries to choose an appropriate framework for their use cases. Several works have evaluated the performance of different stream processing frameworks. The aim is to choose a suitable system that can help facilitate the implementation of desirable solutions. Most works related to big data stream technologies focus on evaluating three popular data stream processing systems, namely, Apache Storm, Apache Spark, and Apache Flink. These systems can be integrated in order to process data streams with high performance. For example, the authors of [74] set up a benchmarking method for testing and evaluating the performance of Apache Storm, Spark streaming, and Flink in terms of throughput and latency. The results show that Apache Flink and Strom are close in terms of latency; however, Apache Spark has the biggest latency. Moreover, the work presented in [75] shows the potential of Apache Storm in terms of latency and node failure. The results of the experiment showed that Storm has better performance (i.e., by at least 50%) than Apache Flink and Apache Spark. Table 2 presents the comparison criteria of the two most popular big data stream processing systems.

Table 2. Comparison criteria of data stream processing technologies.

Technology	Processing Model	Streaming Type	Data Type	Latency	Machine Learning Support
Apache Storm	At least once	Native streaming	Tuple	Milliseconds	Apache SAMOA
Apache Spark	Exactly one	Mini batches	Discretized stream	Seconds	StreamDM + MLlib
Apache Flink	Exactly one	Native streaming	Data stream	Milliseconds	FlinkML

The main objective of this work is to design and choose the best elements in order to set up occupancy detection systems using environmental data generated from sensors deployed in buildings. Most building services require control of active equipment in time; this is why the integration of real-time monitoring and processing large amounts of environmental data have become more important in order to set up effective control of building active equipment. In fact, occupancy information is considered the main input for applying different control approaches such as PID, MPC, and GPC on HVAC, ventilation, or lighting systems in order to save energy and maintain the users' comfort.

Machine learning (ML) algorithms are the most popular solution that can help researchers and developers to use environmental data for detecting indoor building occupancy, i.e., either the presence of an occupant or the number thereof. Most of the works related to occupancy detection systems use the traditional ML algorithm. This means that the environmental data should already be collected and stored for exploitation in order to build the model using many ML algorithms (i.e., such as SVM, ANN, DT, etc.). However, dealing with environmental data is not easy because it is not stationary and changes over time. The main purpose of this paper is to design and develop an appropriate architecture platform using recent big data stream processing technologies, which can involve the integration of machine learning algorithms that can deal with data streams and non-stationary data. Based on the results presented in the literature review of works that address the evaluation of stream processing techniques, Storm is still the best framework for implementing occupancy detection systems in terms of latency and fault tolerance. In this work, the objective is to present a general architecture that can be integrated into a smart building for detecting occupancy based on environmental data using recent IoT and big data technologies. The idea is to use valuable tools in terms of performance in order to provide efficient building services. Figure 8 illustrates the adopted architecture that can improve the performance of occupancy detection using IoT and big data technologies. The architecture includes four main layers.

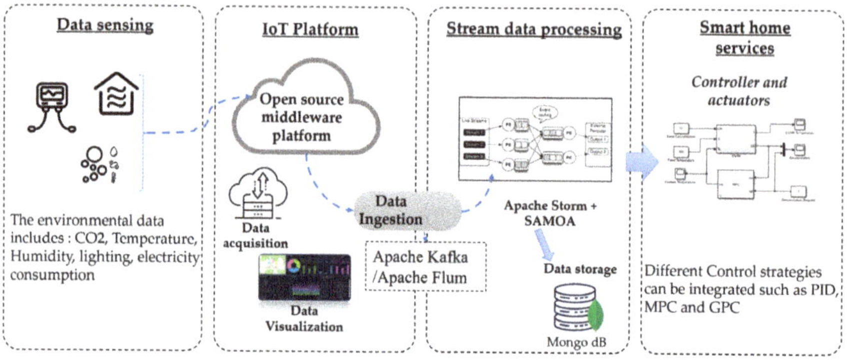

Figure 8. The global architecture describing IoT and big data layers for setting up smart home services based on occupancy detection systems.

Data sensing: Many types of sensors can be included in the proposed architecture. As an example, for collecting CO_2 data, there are different embedded technologies that can help in collecting data from sensors in the environment. The main types of environmental data that can be used for detecting occupancy include CO_2, temperature, humidity, lighting, and electricity consumption. Furthermore, infrared or camera sensors can be exploited to gather the true occupancy information, which can help to train the model (it is named the label).

IoT platform middleware: An IoT platform is a set of elements that enable developers to collect data from multiple sensors remotely while ensuring secure connectivity. The aim of each deployed IoT platform is fast communication between different connected devices and the cloud, using advanced networking protocols such as MQTT and HTTPs. Different IoT platforms provide a free service for data acquisition and visualization. In fact, these platforms can be implemented in the cluster. For example, Kaa and Thingsboard are the most popular open-source IoT platforms that include multiple services related to data management.

Stream data processing techniques: In the proposed architecture, Storm is the most suitable due to its effectiveness in terms of fault tolerance and latency. Moreover, it is a distributed system and can support an advanced Machine learning framework, such as Apache SAMOA.

Smart home services: Different applications can be developed for controlling active equipment in buildings. The aim is to manage building services such as HVAC and lighting systems in order to reduce energy consumption and maintain the occupants' comfort. Different approaches proposed in the literature aim at building control devices efficiently [29–31].

Furthermore, a prototype platform was deployed in the laboratory in order to set up the occupancy detection use case by integrating the above-mentioned technologies for ensuring the real-time stream processing of data coming from deployed sensors. Figure 9 illustrates different elements and components that are integrated into the platform in order to set up an occupancy detection system. In the proposed scenario, multiple types of sensors and microcontrollers have been exploited as hardware parts for collecting, gathering, and sending data to the remote server. In fact, the main hardware modules that have been deployed for data acquisition and monitoring in EEBLab are as follows.

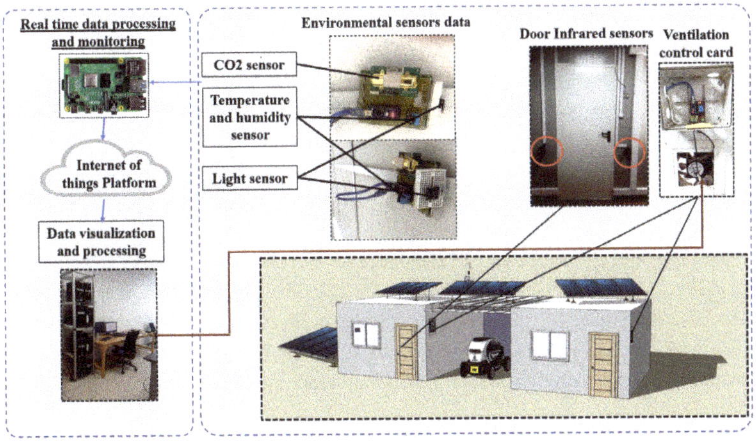

Figure 9. Prototype platform architecture of occupancy scenario deployment.

MH Z14A is an infrared CO_2 meter that could be integrated using either UART or I2C protocols. DHT22 is used in thermal comfort by measuring indoor/outdoor temperature and relative humidity in a given area. One wire communication protocol was used for gathering data. Further, Raspberry PI 3 is considered a bridge or edge of the deployed platform. The main task of Raspberry is to gather data using the UART protocol and to filter/reorganize sensors' data to be transmitted to the server.

NodeMCU is a low-cost Wi-Fi module with an 802.1 Wi-Fi direct protocol and a soft-AP-integrated TCP/IP protocol stack, Fan module (12 V) is used to control air quality in the room, and PZEM 004T current sensor is deployed for measuring the consumption of different building appliances and machines (e.g., ventilation, HVAC, lighting).

Regarding data transfer, communication techniques are required. The MQTT (message queue telemetry transport) protocol is deployed in the Ubuntu server. It is mainly used to transmit data from the monitoring prototype to the gateway actor. It is worth noting that MQTT protocol is a machine-to-machine connectivity protocol, and it was designed as an extremely lightweight publish/subscribe messaging transport. In fact, the sensing node starts sampling sensors' data and sends them via MQTT to the gateway, which transports them to the server using HTTP transport protocols. The deployed cluster includes the Apache Storm setup as a data stream engine for processing data coming directly from sensors without storing them and then resending the decision results for the ventilation system control.

6. Conclusions and Perspectives

The integration of IoT and big data is a crucial solution for improving building services. For example, managing building equipment requires the efficient manipulation of the large amount of data generated from connected devices. In this work, we provide an overview highlighting the transformative role of IoT and big data technologies in smart building applications in order to improve operational efficiency, energy management, and occupant comfort. Many technologies have been presented and discussed: both batch-based and stream-based techniques for developing the optimal architecture for processing building data and setting up active services. The key findings demonstrate that IoT devices, integrated with big data analytics, enable the real-time monitoring and optimization of building systems, leading to significant energy savings and improved sustainability. In fact, stream processing technologies are the most suitable for processing building environmental data.

For example, Storm is classified by various studies as the best framework that can be integrated for processing data streams in terms of latency and fault tolerance. The main objective of this study was to design an effective architecture, including the best tools for processing large amounts of environmental data, in order to implement efficient management building systems so as to improve building services with the aim of reducing energy and maintaining the occupants' comfort. The technologies and platforms studied are the main means for researchers to improve the deployment of next-generation approaches, such as digital twin, for critical environment monitoring and anomaly detection. Future research should focus on addressing these issues by developing robust data security frameworks, improving interoperability between IoT devices, and exploring advanced data processing techniques like edge computing. Overall, the continued development of IoT and big data technologies holds great promise for the evolution of smart buildings, fostering more intelligent, energy-efficient, and responsive environments.

Funding: This research received no external funding.

Conflicts of Interest: The authors declare no conflict of interest.

References

1. González-Torres, M.; Pérez-Lombard, L.; Coronel, J.F.; Maestre, I.R.; Yan, D. A review on buildings energy information: Trends end-uses, fuels and drivers. *Energy Rep.* **2022**, *8*, 626–637. [CrossRef]
2. El Asri, N.; Nouira, Y.; Maaroufi, I.; Marfak, A.; Saleh, N.; Mharzi, M. The policy of energy management in public buildings procurements through the study of the process of delegated project management-Case of Morocco. *Energy Policy* **2022**, *165*, 112944. [CrossRef]
3. Berouine, A.; Ouladsine, R.; Bakhouya, M.; Essaaidi, M. Towards a real-time predictive management approach of indoor air quality in energy-efficient buildings. *Energies* **2020**, *13*, 3246. [CrossRef]
4. Berouine, A.; Akssas, E.; Naitmalek, Y.; Lachhab, F.; Bakhouya, M.; Ouladsine, R.; Essaaidi, M. A fuzzy logic-based approach for HVAC systems control. In Proceedings of the 2019 6th International Conference on Control, Decision and Information Technologies (CoDIT), Paris, France, 23–26 April 2019. [CrossRef]
5. Hannan, M.A.; Faisal, M.; Ker, P.J.; Mun, L.H.; Parvin, K.; Mahlia, T.M.I.; Blaabjerg, F. A review of internet of energy based building energy management systems: Issues and recommendations. *IEEE Access* **2018**, *6*, 38997–39014. [CrossRef]
6. Michailidis, P.; Michailidis, I.; Gkelios, S.; Kosmatopoulos, E. Artificial Neural Network Applications for energy management in buildings: Current trends and future directions. *Energies* **2024**, *17*, 570. [CrossRef]
7. Salimi, S.; Hammad, A. Critical review and research roadmap of office building energy management based on occupancy monitoring. *Energy Build.* **2018**, *182*, 214–241. [CrossRef]
8. Im, P.; Bae, Y.; Cui, B.; Lee, S.; Bhattacharya, S.; Adetola, V.; Vrabie, D.; Zhang, L.; Leach, M. *Literature Review for Sensor Impact Evaluation and Verification Use Cases—Building Controls and Fault Detection and Diagnosis (FDD)*; US Department of Energy Washington, DC, USA, 2020. [CrossRef]
9. Smith, A.R.; Ghamari, M.; Velusamy, S.; Sundaram, S. Thin-Film technologies for sustainable Building-Integrated photovoltaics *Energies* **2024**, *17*, 6363. [CrossRef]
10. Bakhouya, M.; NaitMalek, Y.; Elmouatamid, A.; Lachhab, F.; Berouine, A.; Boulmrharj, S.; Ouladsine, R.; Felix, V.; Zine-dine K.; Khaidar, M.; et al. Towards a context-driven platform using IoT and big data technologies for energy efficient buildings In Proceedings of the 3rd International Conference of Cloud Computing Technologies and Applications (CloudTech), Rabat Morocco, 24–26 October 2017; pp. 1–5. [CrossRef]
11. Lachhab, F.; Bakhouya, M.; Ouladsine, R.; Essaaidi, M. Context-driven monitoring and control of buildings ventilation systems using big data and Internet of Things–based technologies. *Proc. Inst. Mech. Eng. Part I J. Syst. Control Eng.* **2018**, *233*, 276–288 [CrossRef]
12. Khraisat, A.; Gondal, I.; Vamplew, P.; Kamruzzaman, J. Survey of intrusion detection systems: Techniques, datasets and challenges *Cybersecurity* **2019**, *2*, 20. [CrossRef]
13. Dwivedi, Y.K.; Hughes, L.; Ismagilova, E.; Aarts, G.; Coombs, C.; Crick, T.; Duan, Y.; Dwivedi, R.; Edwards, J.; Eirug, A.; et al. Artificial Intelligence (AI): Multidisciplinary perspectives on emerging challenges, opportunities, and agenda for research, practice and policy. *Int. J. Inf. Manag.* **2019**, *57*, 101994. [CrossRef]
14. Miorandi, D.; Sicari, S.; De Pellegrini, F.; Chlamtac, I. Internet of things: Vision, applications and research challenges. *Ad Hoc Netw.* **2012**, *10*, 1497–1516. [CrossRef]

15. Minerva, R.; Lee, G.M.; Crespi, N. Digital Twin in the IoT context: A survey on technical features, scenarios, and architectural models. *Proc. IEEE* **2020**, *108*, 1785–1824. [CrossRef]
16. Alemazkoor, N.; Tootkaboni, M.; Nateghi, R.A. Louhghalam Smart-Meter Big Data for Load Forecasting: An Alternative Approach to Clustering. *IEEE Access* **2022**, *10*, 8377–8387. [CrossRef]
17. Berouine, A.; Lachhab, F.; Malek, Y.N.; Bakhouya, M.; Ouladsine, R. A smart metering platform using big data and IoT technologies. In Proceedings of the 3rd International Conference of Cloud Computing Technologies and Applications (CloudTech), Rabat, Morocco, 23–26 April 2017; pp. 1–6. [CrossRef]
18. Rizwan, B.M.; Gill, A.Q. Towards an IoT big data analytics framework: Smart buildings systems. In Proceedings of the IEEE 2016 IEEE 18th International Conference on High Performance Computing and Communications, Sydney, Australia, 12–14 December 2016; pp. 1325–1332. [CrossRef]
19. Plageras, A.P.; Psannis, K.E.; Stergiou, C.; Wang, H.; Gupta, B. Efficient IoT-based sensor BIG Data collection–processing and analysis in smart buildings. *Future Gener. Comput. Syst.* **2018**, *82*, 349–357. [CrossRef]
20. Bashir, M.R.; Gill, A.Q.; Beydoun, G.; Mccusker, B. Big data management and analytics metamodel for IoT-enabled smart buildings. *IEEE Access* **2020**, *8*, 169740–169758.
21. Priyanka, E.B.; Thangavel, S.; Meenakshipriya, B.; Prabu, D.V.; Sivakumar, N.S. Big Data Technologies with Computational Model Computing Using Hadoop with Scheduling Challenges. In *Studies in Computational Intelligence*; Springer: Cham, Switzerland, 2021.
22. Dong, B.; Prakash, V.; Feng, F.; O'Neill, Z. A review of smart building sensing system for better indoor environment control. *Energy Build.* **2019**, *199*, 29–46. [CrossRef]
23. Malek, Y.N.; Kharbouch, A.; El Khoukhi, H.; Bakhouya, M.; De Florio, V.; El Ouadghiri, D.; Latre, S.; Blondia, C. On the use of IoT and Big Data Technologies for Real-time Monitoring and Data Processing. *Procedia Comput. Sci.* **2017**, *113*, 429–434. [CrossRef]
24. Tagliabue, L.C.; Cecconi, F.R.; Rinaldi, S.; Ciribini, A.L.C. Data driven indoor air quality prediction in educational facilities based on IoT network. *Energy Build.* **2021**, *236*, 110782. [CrossRef]
25. Benammar, M.; Abdaoui, A.; Ahmad, S.H.M.; Touati, F.; Kadri, A. A modular IoT platform for real-time indoor air quality monitoring. *Sensors* **2018**, *18*, 581. [CrossRef]
26. Mumtaz, R.; Zaidi, S.M.H.; Shakir, M.Z.; Shafi, U.; Malik, M.M.; Haque, A.; Mumtaz, S.; Zaidi, S.A.R. Internet of things (Iot) based indoor air quality sensing and predictive analytic—A COVID-19 perspective. *Electron* **2021**, *10*, 184. [CrossRef]
27. Marques, G.; Ferreira, C.R.; Pitarma, R. Indoor Air Quality Assessment Using a CO_2 Monitoring System Based on Internet of Things. *J. Med. Syst.* **2019**, *43*, 67. [CrossRef] [PubMed]
28. Ramachandraarjunan, S.; Perumalsamy, V.; Narayanan, B. IoT based artificial intelligence indoor air quality monitoring system using enabled RNN algorithm techniques. *J. Intell. Fuzzy Syst.* **2022**, *43*, 2853–2868. [CrossRef]
29. Pradityo, F.; Surantha, N. Indoor air quality monitoring and controlling system based on IoT and fuzzy logic. In Proceedings of the 2019 7th International Conference on Information and Communication Technology (ICoICT), Kuala Lumpur, Malaysia, 26–29 July 2019. [CrossRef]
30. Rajith, A.; Soki, S.; Hiroshi, M. Real-time optimized HVAC control system on top of an IoT framework. In Proceedings of the Third International Conference on Fog and Mobile Edge Computing (FMEC), Malmö, Sweden, 23–26 April 2018. [CrossRef]
31. Lachhab, F.; Bakhouya, M.; Ouladsine, R.; Essaaidi, M. Towards an Intelligent Approach for Ventilation Systems Control Using IoT and Big Data Technologies. *Procedia Comput. Sci.* **2018**, *130*, 926–931. [CrossRef]
32. Taştan, M.; Gökozan, H. An Internet of Things Based Air Conditioning and Lighting Control System for Smart Home. *Am. Sci. Res. J. Eng. Technol. Sci.* **2018**, *50*, 181–189.
33. Elkhoukhi, H.; Bakhouya, M.; Elmouatamid, A.; El Ouadghiri, D.; Gaber, J. A review of occupancy sensing technologies and approaches in smart buildings. *Int. J. RF Technol.* **2024**, *15*, 19–48. [CrossRef]
34. de Bakker, C.; Aries, M.; Kort, H.; Rosemann, A. Occupancy-based lighting control in open-plan office spaces: A state-of-the-art review. *Build. Environ.* **2017**, *112*, 308–321. [CrossRef]
35. Ji, Y.; Ok, K.; Choi, W.S. Occupancy detection technology in the building based on IoT environment sensors. In Proceedings of the 8th International Conference on the Internet of Things, Santa Barbara, CA, USA, 15–18 October 2018. [CrossRef]
36. Huang, Q.; Rodriguez, K.; Whetstone, N.; Habel, S. Rapid internet of things (IoT) prototype for accurate people counting towards energy efficient buildings. *J. Inf. Technol. Constr.* **2019**, *24*, 1–13. [CrossRef]
37. Parise, A.; Manso-Callejo, M.A.; Cao, H.; Mendonca, M.; Kohli, H.; Wachowicz, M. Indoor Occupancy Prediction using an IoT Platform. In Proceedings of the 2019 Sixth International Conference on Internet of Things: Systems, Management and Security (IOTSMS), Granada, Spain, 22–25 October 2019; pp. 26–31. [CrossRef]
38. Taştan, M.; Gökozan, H. Real-time monitoring of indoor air quality with internet of things-based e-nose. *Appl. Sci.* **2019**, *9*, 3435. [CrossRef]
39. Daissaoui; Boulmakoul, A.; Karim, L.; Lbath, A. IoT and Big Data Analytics for Smart Buildings: A Survey. *Procedia Comput. Sci.* **2020**, *170*, 161–168. [CrossRef]

40. Kitzberger, T.; Kotik, J.; Pröll, T. Energy savings potential of occupancy-based HVAC control in laboratory buildings. *Energy Buil* **2022**, *263*, 112031. [CrossRef]
41. Elkhoukhi, H.; Naitmalek, Y.; Berouine, A.; Bakhouya, M.; Elouadghiri, D.; Essaaidi, M. Towards a Real-time Occupancy Detectic Approach for Smart Buildings. *Procedia Comput. Sci.* **2018**, *134*, 114–120. [CrossRef]
42. Rueda, L.; Agbossou, K.; Cardenas, A.; Henao, N.; Kelouwani, S. A comprehensive review of approaches to building occupanc detection. *Build. Environ.* **2020**, *180*, 106966. [CrossRef]
43. Qolomany, B.; Al-Fuqaha, A.; Gupta, A.; Benhaddou, D.; Alwajidi, S.; Qadir, J.; Fong, A.C. Leveraging Machine Learning and Bi Data for Smart Buildings: A Comprehensive Survey. *IEEE Access* **2019**, *7*, 90316–90356. [CrossRef]
44. Mariano-Hernández, D.; Hernández-Callejo, L.; Zorita-Lamadrid, A.; Duque-Pérez, O.; García, F.S. A review of strategies fc building energy management system: Model predictive control, demand side management, optimization, and fault detect diagnosis. *J. Build. Eng.* **2021**, *33*, 101692. [CrossRef]
45. Yu, L.; Qin, S.; Zhang, M.; Shen, C.; Jiang, T.; Guan, X. A Review of Deep Reinforcement Learning for Smart Building Energ Management. *IEEE Internet Things J.* **2021**, *8*, 12046–12063. [CrossRef]
46. Berouine, A.; Ouladsine, R.; Bakhouya, M.; Essaaidi, M. A Predictive Control Approach for Thermal Energy Management i Buildings. *SSRN Electron. J.* **2022**, *8*, 9127–9141. [CrossRef]
47. Elkhoukhi, H.; NaitMalek, Y.; Bakhouya, M.; Berouine, A.; Kharbouch, A.; Lachhab, F.; Hanifi, M.; El Ouadghiri, D.; Essaaidi, M A platform architecture for occupancy detection using stream processing and machine learning approaches. *Concurr. Comput. P Exp.* **2020**, *32*, e5651. [CrossRef]
48. Krishnamurthi, R.; Kumar, A.; Gopinathan, D.; Nayyar, A.; Qureshi, B. An overview of iot sensor data processing, fusion, an analysis techniques. *Sensors* **2020**, *20*, 6076. [CrossRef]
49. Antão, L.; Pinto, R.; Reis, J.; Gonçalves, G. Requirements for testing and validating the industrial internet of things. In Proceeding of the 2018 IEEE International Conference on Software Testing, Verification and Validation Workshops (ICSTW), Västerås, Swede 9–13 April 2018; pp. 110–115. [CrossRef]
50. Baig, M.I.; Shuib, L.; Yadegaridehkordi, E. Big data adoption: State of the art and research challenges. *Inf. Process. Manag.* **201** *56*, 102095. [CrossRef]
51. Farzaneh, H.; Malehmirchegini, L.; Bejan, A.; Afolabi, T.; Mulumba, A.; Daka, P.P. Artificial intelligence evolution in smar buildings for energy efficiency. *Appl. Sci.* **2021**, *11*, 763. [CrossRef]
52. Zanoon, N.; Al-Haj, A.; Khwaldeh, S.M. Cloud Computing and Big Data is there a Relation between the Two: A Study. *Int. Appl. Eng. Res.* **2017**, *12*, 6970–6982.
53. Alansari, Z.; Anuar, N.B.; Kamsin, A.; Soomro, S.; Belgaum, M.R.; Miraz, M.H.; Alshaer, J. Challenges of Internet of Thing and big data integration. In Proceedings of the Emerging Technologies in Computing: First International Conference (iCETiC London, UK, 24 August 2018; pp. 47–55. [CrossRef]
54. Kharbouch, A.; El Khoukhi, H.; Nait Malek, Y.; Bakhouya, M.; De Florio, V.; El Ouadghiri, D.; Latré, S.; Blondia, C. Towards a IoT and Big Data Analytics Platform for the Definition of Diabetes Telecare Services. *SSRN Electron. J.* **2018**. [CrossRef]
55. Lv, Z.; Qiao, L.; Cai, K.; Wang, Q. Big Data Analysis Technology for Electric Vehicle Networks in Smart Cities. *IEEE Trans. Intel Transp. Syst.* **2021**, *22*, 1807–1816. [CrossRef]
56. Malek, Y.N.; Najib, M.; Bakhouya, M.; Essaaidi, M. Multivariate deep learning approach for electric vehicle speed forecasting. *Bi Data Min. Anal.* **2021**, *4*, 56–64. [CrossRef]
57. Yu, T.; Wang, X.; Shami, A. A novel fog computing enabled temporal data reduction scheme in iot systems. In Proceedings of th 2017 IEEE Global Communications Conference, Singapore, 4–8 December 2017; pp. 1–5. [CrossRef]
58. Ullah, F.; Babar, M.A. Architectural Tactics for Big Data Cybersecurity Analytics Systems: A Review. *J. Syst. Softw.* **2019**, *151* 81–118. [CrossRef]
59. Herodotou, H.; Chen, Y.; Lu, J. A Survey on Automatic Parameter Tuning for Big Data Processing Systems. *ACM Comput. Surv* **2020**, *53*, 43. [CrossRef]
60. Herodotou, H.; Babu, S. Profiling, what-if analysis, and costbased optimization of mapreduce programs. *Proc. VLDB Endow.* **2011** *4*, 1111–1122. [CrossRef]
61. Herodotou, H.; Dong, F.; Babu, S. Mapreduce programming and costbased optimization? Crossing this chasm with starfish. *Proc VLDB Endow.* **2011**, *4*, 1446–1449. [CrossRef]
62. Santhanam, R. Comparative Study of XGBoost4j and Gradient Boosting for Linear Regression Comparative Study of XGBoost4 and Gradient Boosting for Linear Regression. *Int. J. Control Theory Appl.* **2016**, *9*, 40.
63. Elkhoukhi, H.; Bakhouya, M.; El Ouadghiri, D.; Hanifi, M. Using Stream Data Processing for Real-Time Occupancy Detection ir Smart Buildings. *Sensors* **2022**, *22*, 2371. [CrossRef]
64. Kharbouch, A.; Naitmalek, Y.; Elkhoukhi, H.; Bakhouya, M.; De Florio, V.; El Ouadghiri, M.D.; Umi; Latre, S.; Blondia, C. IoT and big data technologies for monitoring and processing real-time healthcare data. *Int. J. Distrib. Syst. Technol.* **2019**, *10*, 17–30 [CrossRef]

65. Apache, S. Spark Streaming Programming Guide. 2024. Available online: https://spark.apache.org/docs/latest/streaming-programming-guide.html (accessed on 20 December 2024).
66. Isah, H.; Abughofa, T.; Mahfuz, S.; Ajerla, D.; Zulkernine, F.; Khan, S. A survey of distributed data stream processing frameworks. *IEEE Access* **2019**, *7*, 154300–154316. [CrossRef]
67. Flink, A. Apache Flink Architecture. 2020. Available online: https://flink.apache.org/flink-architecture.html (accessed on 20 December 2024).
68. Bifet, A.; Maniu, S.; Qian, J.; Tian, G.; He, C.; Fan, W. StreamDM: Advanced Data Mining in Spark Streaming. In Proceedings of the IEEE International Conference on Data Mining Workshop (ICDMW), Atlantic City, NJ, USA, 14–17 November 2015. [CrossRef]
69. De Francisci Morales, G.; Bifet, A. SAMOA: Scalable advanced massive online analysis. *J. Mach. Learn. Res.* **2015**, *16*, 149–153.
70. Bifet, A.; Morales, G.D.F. Big data stream learning with SAMOA. In Proceedings of the IEEE International Conference on Data Mining Workshop (ICDMW), Shenzhen, China, 14 December 2014. [CrossRef]
71. Al-Barznji, K.; Atanassov, A. Collaborative Filtering Techniques for Generating Recommendations on Big Data. In Proceedings of the International Conference Automatics and Informatics, Sofia, Bulgaria, 4–6 October 2017.
72. Fernandes, L.F.D.O.; Bernardini, F.; Meza, E.M.; Miranda, L.; Viterbo, J. Energy Consumption Prediction using Data Stream Learning for Commercial Buildings. In Proceedings of the 2020 International Conference on Systems, Signals and Image Processing (IWSSIP), Niteroi, Brazil, 1–3 July 2020; pp. 441–446. [CrossRef]
73. Karimov, J.; Rabl, T.; Katsifodimos, A.; Samarev, R.; Heiskanen, H.; Markl, V. Benchmarking distributed stream data processing systems. In Proceedings of the 2018 IEEE 34th International Conference on Data Engineering (ICDE), Paris, France, 16–19 April 2018; pp. 1507–1518. [CrossRef]
74. Chintapalli, S.; Dagit, D.; Evans, B.; Farivar, R.; Graves, T.; Holderbaugh, M.; Liu, Z.; Nusbaum, K.; Patil, K.; Peng, B.J.; et al. Benchmarking streaming computation engines: Storm, flink and spark streaming. In Proceedings of the 2016 IEEE International Parallel and Distributed Processing Symposium Workshops (IPDPSW), Chicago, IL, USA, 23–27 May 2016; pp. 1789–1792. [CrossRef]
75. Lopez, M.A.; Lobato, A.G.P.; Duarte, O.C.M.B. Duarte A performance comparison of open-source stream processing platforms. In Proceedings of the IEEE Global Communications Conference (GLOBECOM), Washington, DC, USA, 4–8 December 2016. [CrossRef]

sustainability

MDPI

Article

A Workflow for a Building Information Modeling-Based Thermo-Hygrometric Digital Twin: An Experimentation in an Existing Building

Tullio De Rubeis [1], Annamaria Ciccozzi [2], Mattia Ragnoli [2], Vincenzo Stornelli [2], Stefano Brusaporci [1,*], Alessandra Tata [1] and Dario Ambrosini [2]

1 Department of Civil, Construction-Architectural and Environmental Engineering, University of L'Aquila, Piazzale Pontieri 1, Monteluco di Roio, 67100 L'Aquila, Italy; tullio.derubeis@univaq.it (T.D.R.); alessandra.tata@univaq.it (A.T.)

2 Department of Industrial and Information Engineering and Economics, University of L'Aquila, Piazzale Pontieri 1, Monteluco di Roio, 67100 L'Aquila, Italy; annamaria.ciccozzi@graduate.univaq.it (A.C.); mattia.ragnoli@graduate.univaq.it (M.R.); vincenzo.stornelli@univaq.it (V.S.); dario.ambrosini@univaq.it (D.A.)

* Correspondence: stefano.brusaporci@univaq.it

Abstract: Building Information Modeling (BIM)-based digital twin (DT) could play a fundamental role in overcoming the limitations of traditional monitoring methods by driving the digitalization of the construction sector. While existing studies on the topic have provided valuable insights, significant knowledge gaps remain, which continue to hinder the large-scale adoption of this approach. Moreover, to date, there is no standardized procedure available, able to guide the step-by-step creation of a DT. Another significant challenge concerns the choice of technologies able to integrate perfectly with each other throughout the process. This paper outlines a comprehensive workflow for creating a digital twin (DT) of an existing building and proposes various solutions to improve the integration of different technologies involved. These enhancements aim to address the limitations of current monitoring methods and leverage the advantages of BIM and DT for accessing and managing monitoring data, ultimately facilitating the implementation of energy-efficient interventions. This work examines the concept of "Living Lab" in an office building also used as an academic laboratory. The created DT allowed for real-time remote monitoring of four rooms, each with a different functional and occupational characteristic, useful also for future predictive analyses.

Keywords: Building Information Modeling (BIM); Digital Twin (DT); indoor environmental monitoring; energy optimization; Internet of Things (IoT); Wireless Sensor Networks (WSN)

1. Introduction

Building Information Modeling (BIM) began to establish itself about two decades ago and has since evolved into one of the most transformative technologies with the highest potential in the construction world. Originally conceived to improve the efficiency of design and construction processes, BIM has grown into a comprehensive approach to the lifecycle management of buildings and infrastructures. Over the years, BIM's role has expanded from being a tool primarily for improving project visualization and documentation to becoming a central process for decision-making throughout the entire building lifecycle, from conception through construction to operation and maintenance.

As BIM and related technology continue to evolve, there is a growing interest in real-time monitoring and control of buildings to enhance comfort, security, and overall quality of life through the control and management of real-time data from installed sensors. From a technical point of view, currently, there are various monitoring solutions with low impact on the building, low cost, and easy to install. One of the limits of traditional monitoring systems lies in the data visualization and processing phase, which, in most cases, is cumbersome, not very intuitive, and not accessible on a large scale [1]. Furthermore, the lack of seamless information exchange between physical and digital resources makes it difficult to quickly locate critical issues, not allowing efficient and intuitive real-time monitoring, especially in the case of multi-environmental monitoring. Promoting an intuitive interface to understand the relationships between internal comfort parameters and the structure itself would allow the involvement of users in energy issues by making them fully participants in the optimization process.

In this context BIM, and particularly digital twin technology, could be the key to overcoming some of the limitations of current monitoring methods, while supporting the integration between architecture and energy systems.

A digital twin is a dynamic digital representation of a physical asset, continuously updated with real-time data deriving from sensors or other sources. While BIM focuses on the creation of a "static", data-rich model for design, construction, and management, a digital twin extends this concept by incorporating live data and analytics during the operational phase of a building or infrastructure, allowing for real-time monitoring, predictive maintenance, and scenario simulation.

The synergy between BIM and digital twin technologies creates an even more powerful combination for asset management. By merging the static model with real-time operational data, digital twins can provide an in-depth understanding of how a building performs under different conditions, leading to more informed decision-making and enhancing the asset's efficiency, durability, and sustainability.

This work proposes a workflow for real-time indoor environment monitoring and visualization testing of the BIM-based DT approach on an office building, also used as an academic laboratory. The proposed DT allows partial monitoring of thermal comfort, being limited to the measurement of temperature and relative humidity only. In the future, the idea is to implement the system with the measurement of other parameters that together contribute to the determination of Indoor Environmental Quality (IEQ). The creation of the system was made possible thanks to the installation of a Wireless Sensor Network (WSN) directly connected to the BIM model via an online database used as a middleware tool. The aim of the contribution is to provide a guideline for the creation of an effective monitoring system, capable of overcoming the limits of traditional techniques, supporting the modernization and digitalization of the construction sector promoting the integration of DT platforms with modern information technologies.

2. State of the Art

2.1. A BIM-Based Digital Twin

The National Building Information Model Standard Project Committee in the United States defines BIM as "a digital representation of physical and functional characteristics of a facility" [2]. Through three-dimensional informative modeling of both new and existing buildings, BIM provides an exceptionally accurate virtual representation of all building components, serving as a reliable basis for decision-making throughout the entire life cycle of the facility [2–6]. In recent years, there has been growing interest in utilizing BIM for the development of green buildings, a trend increasingly reflected in the research community. This is further supported by evolving regulations, such as the Italian UNI 11337 standard [7], which incorporates sustainable building practices into the BIM dimensions it governs. However, the potential of this methodology has been explored mainly during the building design and construction phases, but not enough during the operational phase. For example, in the design phase, the potential interoperability of BIM with BEM (Building Energy Modeling) software makes it possible to immediately integrate the energy and architectural aspects of a project, allowing different design strategies to be evaluated simultaneously to identify the most efficient one from each point of view [8]. Furthermore, thanks to BIM it is possible to conduct very detailed solar studies, which have a great influence on the behavior of the building on the energy plan [8]. Ciccozzi et al. [9] explored the potential of a BIM software, specifically Revit, in the preliminary design of a photovoltaic system, conducting an in-depth study on shading and incident solar radiation. The integration of BIM with lifecycle assessment (LCA) applications offers significant benefits for improving the environmental performance of buildings [10]. Having access to a detailed and reliable model from the early stages of the design process facilitates conscious decision-making by allowing for the anticipation of choices early, thereby optimizing both sustainability and cost-effectiveness. This proactive approach not only supports the selection of materials with a low environmental impact during construction but also enhances waste management strategies for future demolition phases.

BIM was developed to manage the entire lifecycle of buildings within a single interoperable platform, focusing on data deriving from the design stages, and necessary to the construction and the managing phases. However, managing dynamic real-time data were not part of BIM's original design, resulting in a lack of such capabilities in BIM software [11]. Nevertheless, integrating monitoring systems into BIM models could revolutionize the way we interact with the building environment as it would combine the geometric and informative properties of BIM models with streaming environmental data (e.g., temperature, humidity, etc.) collected by the sensors allowing for immediate, intuitive, and open data visualization and processing [12].

In response to these limitations, the concept of the digital twin (DT) has emerged as a promising solution and is gaining significant attention [13]. A digital twin is defined as "the virtual representation of a physical entity, living or non-living, of a person or even a complex system connected to a physical part and with which it can exchange data and information, both in synchronous mode (in time real), than asynchronous (in subsequent times)" [14]. The digital twin could act as a bridge between the BIM model of the sensors within the building. A BIM-based DT would provide a detailed and reliable digital reproduction of the building while enabling real-time communication of monitoring data through its connection to the sensors installed in the building's physical environment. To date, several researchers have tested this approach. Marzouk et al. [15] connected wireless sensors to the BIM model of a subway to monitor its thermal conditions. Lin and Cheung [16] proposed a visually effective monitoring solution for environmental monitoring management. Wang et al. [17] supported real-time environmental monitoring through semantic enrichment of BIM with model language sensors for facility management purposes. In the study by Spudys et al. [18], DT technology was applied for the energy assessment of a case study building. Opoku et al. [19] integrated sensors with Building Information Modeling to constantly monitor the thermal, lighting, and air quality of a library. The study of Qian et al. [20] proposed an intelligent management approach for indoor air quality based on digital twin platforms. In another work, Qian et al. [21] used DT technology to fully investigate the different thermal environments of interior and semi-open spaces in Chinese courtyard dwellings.

Although digital twin technology is generating significant interest in the scientific research community, its use in the professional field is still in its early stages, hindered by knowledge gaps that currently limit its widespread adoption.

Filling the gaps currently present, the digital twin could represent the key to revolutionizing the Green Building sector. Furthermore, the possibility of integrating DT with artificial intelligence systems contains potential developments of the technology in the energy field, particularly in the area of predictive energy management. Indeed, effective planning and operation of modern energy systems depend largely on accurate medium- to long-term energy demand forecasts. In this context, deep learning algorithms have become increasingly valuable, as they are capable of modeling complex load patterns with considerable fluctuations and uncertainty. Furthermore, thanks to recent advances in the evaluation of different configurations of deep learning models, the forecast accuracy has been significantly optimized [22].

2.2. The Thermo-Hygrometric Monitoring

A fundamental characteristic for ensuring that an internal environment is healthy, safe, and comfortable for people is to ensure a good level of IEQ. IEQ is characterized by various parameters related to the quality of the indoor air (e.g., presence of carbon dioxide, volatile organic compounds, fine dust), lighting (e.g., light level), acoustics (e.g., sound level), and thermal comfort (e.g., temperature, relative humidity, and air flow rate) [23]. Among the aforementioned parameters, in particular, indoor thermo-hygrometric conditions play a fundamental role in occupants' health, work productivity, and learning performance [24]. In fact, several studies talk about how the temperature and relative humidity of the internal air are the main parameters that affect the conditions of comfort and well-being [25–34] and various works have specifically dealt with thermal or thermo-hygrometric monitoring [15,35–38]. Conditions that are too cool or too warm can cause discomfort and decrease work performance and well-being [25,26]. The impact of temperature on cognitive performance follows a U-shaped curve [27]. This indicates that rather than having a single ideal temperature, there is a range where cognitive performance remains stable regardless of temperature. Outside this range, cognitive performance declines. Additionally, poor sleep quality can also impair performance. Even moderate exposure to heat or cold can reduce sleep quality [28]. For instance, sleeping in humid heat has been shown to increase heart rate, sweat rate, thermal load, and wakefulness [29]. Furthermore, thermal comfort influences heart rate variability, which is widely recognized as an indicator of autonomic nervous system activity [30,31]. The ideal relative humidity (RH) for maintaining health and work performance indoors ranges from 40% to 60% [31]. Low indoor RH can lead to dry eyes and airways, as well as reduced effectiveness of mucociliary clearance. Additionally, indoor RH affects the perceived indoor air quality (IAQ), levels of house dust mite allergens, the infectivity of certain viruses, and skin dryness [32,33]. To ensure a comfortable environment from a thermo-hygrometric point of view, buildings often rely on heating, ventilation, and air conditioning (HVAC) systems, which can lead to increased energy consumption. In colder regions, the heating systems play an important role in providing comfortable indoor temperatures and helping minimize drafts from cold surfaces such as windows or poorly insulated walls. In warmer regions, heat stress from warm outside temperatures may be reduced by cooling systems applied indoors. The thermal environment is strongly linked to energy consumption via HVAC, with a direct impact on energy efficiency targets and related climate change considerations. The potential for energy savings in the thermal environment is related to the reduction in heating during winter or cooling during summer, achieved by improving the building envelope in terms of thermal resistance and leakage [34]. Designing high-performance buildings allows for maintaining a comfortable indoor environment while requiring moderate use of the HVAC systems and consequently limiting costs and consumption [39]. The Energy Performance of Building Directive (EPBD) aims to upgrade the building stock of the entire European Union (EU) [40]. Compared to older buildings, newer or renovated buildings tend to be more airtight and have better thermal insulation. Modernizing and renovating the building stock has, after the initial investment, the potential to reduce energy poverty by making indoor space heating more affordable and reducing the risk of damp and mold, improving both physical and mental health [34,41]. Before any energy optimization intervention, it is always necessary to carry out an accurate diagnosis, based on an in-depth study of the current state and an efficient monitoring campaign. Monitoring spaces during the operational phase represents a very effective approach to constantly keep environmental conditions under control and be able to highlight any critical issues affecting the building under investigation. Only by accurately analyzing the current state of the building is it possible to individuate targeted strategies that enhance thermo-hygrometric comfort while minimizing the increase in energy expenses [42].

3. Materials and Methods

Although digital twin technology has gained attention for indoor environment monitoring, its use in the industry remains limited due to unresolved challenges that prevent large-scale implementation. However, DT could play a crucial role in improving the communication and the processing of data deriving from sensors, offering several benefits in the energy field. First, the system's user-friendly design would improve interaction between the building, system, and user, supporting the increase in awareness of the user himself, who is mainly responsible for the correct management of living spaces. Involving end users in monitoring could stimulate them to improve their habits in terms of energy consumption [43]. Furthermore, the approach would bring significant benefits in the implementation of energy interventions, whether immediate or planned. Through the DT, in fact, the facility managers can constantly and simultaneously monitor the data coming from the various sensors placed in the building, instantly identifying the presence of anomalies and their exact spatial location, even remotely. This promotes proactive maintenance, allowing immediate interventions if necessary [44]. Moreover, the potential integration of the DT with automatic control systems would help to keep the measured environmental parameters within the imposed threshold values. On the other hand, in the context of the planned interventions, the historical data gathered during the monitoring period can be used to calibrate the building's thermal model, providing accurate and reliable energy analyses that can serve as the foundation for any energy optimization projects. Finally, in addition to the various advantages mentioned above, it should be highlighted that in the case of very long monitoring, which involves a large number of environments, the instantaneous and automatic processing of data, provided by the integrated BIM and DT platform, would help limit human errors related to data interpretation.

3.1. *Knowledge Gap and Significance of the Proposed Research*

The main purpose of this study is to promote the use of digital twin technology to overcome the current limitations of operational building assessment practices, supporting the digitalization of the construction sector. To do this, the research seeks to address the various challenges surrounding the application of this approach to encourage its adoption in professional settings.

Despite the importance of existing studies related to the realization of BIM-based digital twins, there are still three important knowledge gaps which are explained below.

First, in the cited literature works, the three-dimensional model of the building is treated more from the geometric side, than from the semantic one. This is because BIM, in most cases, is created through laser scanning technology, which aims to convert the point cloud data into a BIM model. Although this digital representation is compatible with BIM software, it contains limited information since point clouds mainly represent geometric data [45]. Therefore, information related to the semantics (e.g., material properties and construction details) of the objects in the point clouds is missing. In our case, the model was created from scratch on the BIM by inserting in the smallest detail all the data related to the stratigraphy of the various building components. This feature is essential to have a digital twin that is rich in information and that fully reflects the characteristics of the real building.

Secondly, the problem of interoperability between the various technologies involved in the development of DT has not been fully adequately addressed, although several studies have started to address this topic [15–21]. In fact, since there are many technologies available on the market, it is difficult to identify the most suitable ones for this purpose. Starting from the assumption that for the development of an efficient DT, the various technologies involved in the process must be well integrated, our article proposes solutions that can satisfy the interoperability requirement, explaining the technologies, protocols, and formats chosen for the workflow.

Thirdly, there is currently no standardized procedure capable of instructing and guiding the professional in the development of a DT. In the different studies analyzed, the system architecture for DT development has been presented generically, focusing specifically only on certain work phases, mainly related to the integration of sensors with the digital model and to the acquisition and visualization of monitored data. In our case, the developed workflow is explained in detail from the study phase of the physical building to the analysis and discussion phase of the monitored data. This approach was aimed at defining guidelines to facilitate the application of the workflow in a professional context, for each type of case study.

Finally, notwithstanding the efforts geared toward the integration of WSNs and BIM technologies, few studies have been directed toward the office buildings where laboratory activities also take place. For example, Opoku et al. [19] and Roda Sanchez et al. [46] chose to test the DT on academic buildings for their constant interaction with the occupants. The building we chose; however, in addition to having the requirement present in the aforementioned case studies [19,46], also represents an example of a multipurpose building and, as such, a building with a high dynamism and an enormous potential regarding the search for cutting-edge solutions for intelligent energy management.

The aim of the research is as follows: (i) To propose a monitoring technique that has a user-friendly interface accessible to everyone, optimizing the real-time data visualization and processing phase and overcoming the limits of traditional approaches. (ii) To explain in detail the proposed workflow and define a standardized procedure that can be applied for any circumstance and case study guiding the professional in choosing the most suitable technologies capable of communicating with each other fluidly and without interruptions. (iii) To experiment with the concept of "Living Lab", an office building is taken as a case study, which is also used as an academic laboratory.

3.2. Methodology

To achieve the aforementioned objectives, a real-time connection between the physical and virtual building was created, thanks to the use of a middleware tool that allowed a continuous exchange of information between the wireless sensors installed in the building and the BIM model of the building itself.

The proposed workflow can be substantially divided into three macro phases, which can be described as follows.

Phase 1—Physical Building study:

- study of the characteristics of the case study building through technical drawings and surveys.

Phase 2a—WSN configuration:

- installation of wireless sensors in the physical environment of the building to detect air temperature and humidity;
- network connection of the sensors via Wi-Fi router;
- sending the data detected by the sensors to a remote server.

Phase 2b—BIM modeling:

- three-dimensional informative BIM modeling of the chosen building;
- partition of the internal spaces of the building with an indication of the rooms and their intended uses.

Phase 3—Digital twin creation:

- importing the BIM model of the building onto an online platform to be used as a middleware tool;
- constant acquisition of data recorded by sensors and real-time processing of the trend of environmental parameters over time.

Figure 1 shows the architecture of the proposed system. The tools used for the application of the system will be explained in detail in Section 3.3.

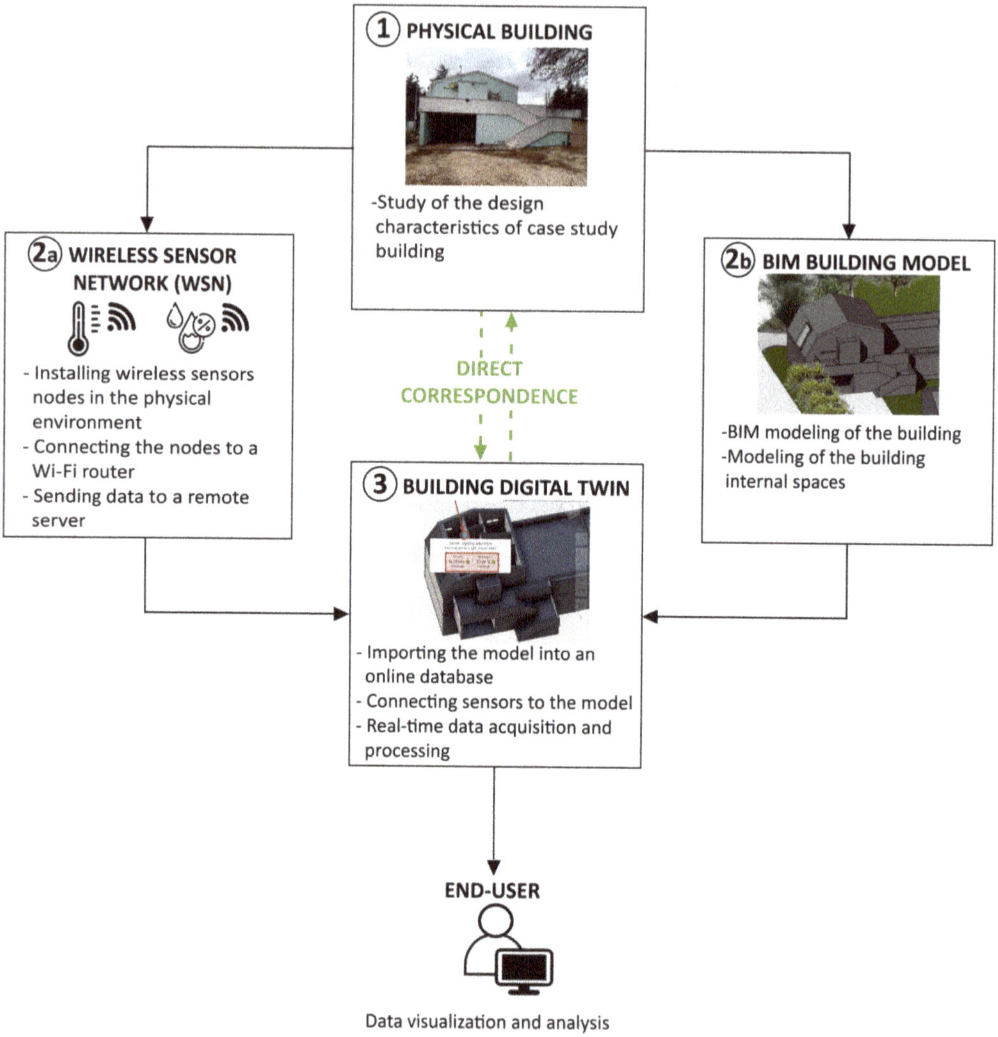

Figure 1. Flow chart representing the applied work methodology.

3.3. System Equipment

3.3.1. Wireless Sensors Nodes

In developing a DT, the foundation of the system is the data acquisition component, consisting of the nodes equipped with sensing elements. The sensor nodes have the possibility of being connected to computer networks and cloud infrastructures for analysis. Nowadays, the IoT paradigm represents the main methodology of connectivity structure for such devices [47]. Integrating IoT with BIM has demonstrated the potential to improving building management and operational efficiency. The IoT sensors offer real-time data on various environmental conditions, and when combined with BIM models, they enable more dynamic and responsive building management [48,49]. Different technologies can be used for data exchange, and those are evaluated with respect to the particular implementation under consideration. Indoor applications that rely on sensor networks can be developed using technologies with short-range coverage including Wi-Fi, Bluetooth, near-field communication (NFC), and Zigbee [19,50]. Other types of DT implementations can require a larger network coverage, so different technologies such as Low Power Wide Area Networks (LPWANs) are often employed [51]. In the case of this study, the sensor nodes allow the monitoring of internal environmental conditions by generating a stream of digitally encapsulated signals used for data collection. The network elements measure the following parameters: temperature and air humidity, which allow the behavior of the building to be scrupulously analyzed. The employed sensor network will be described in this paper.

3.3.2. BIM

Building Information Modeling is a methodology based on object-oriented modeling. The model consists of parametric objects that represent building components, each enriched with geometric and informative attributes that provide both qualitative and quantitative information. Utilizing a shared environment for storing and accessing all design and management information enhances collaboration among different professionals. This collaboration is further supported by interoperability, which ensures continuous data updates, reduces errors from outdated information, and optimizes the process in terms of time and cost efficiency. Therefore, BIM is a shared knowledge resource for information about a facility forming a reliable basis for decisions during its life-cycle; defined as existing from earliest conception to demolition [2]. Currently, BIM is used in Architecture Engineering Construction (AEC) for clear and detailed visualization of all building components, for detecting any design interferences, for estimating construction costs and times, for determining the exact quantity of material necessary for the construction of the building and to facilitate the exchange of information between the interested parties of a project (interoperability). Recently researchers began to explore the potential of this instrument in the energy field, discovering that BIM can constitute a useful support for reducing the environmental impact of a building during its entire lifecycle, starting from the design phase, up to the demolition/renovation phase [8]. Unfortunately, the advantages of BIM in the operational phase have not yet been adequately analyzed due to the tool's inability to acquire and manage data in real time. In this context, digital twin technology is designed to fill the gaps in BIM and allow its use even during the operation phase.

3.3.3. Digital Twin

BIM models can simulate and predict future conditions, but their capacity for self-learning, autonomy, and external data processing is limited [52]. In contrast, the digital twin is capable of acquiring, communicating, and processing data in real time. A DT is a dynamic virtual copy of the real-world physical object, user, or activity in a digital world that is continuously updated through data and information collected in real time via smart sensors while maintaining historical data to make any predictions or decisions through simulations, machine learning, and reasoning [47]. A DT essentially has three components: the physical system, the virtual model, and the data acquired by the sensors [53]. BIM-based DT allows the user to visualize sensor data in a simple and intuitive way via a three-dimensional model of the building, supporting visual communication of the system [54]. DT has the potential to encourage transparency and collaboration between project stakeholders, reduce energy demand during the operational phase, support proactive building maintenance through instant anomaly detection, and help designers develop well-thought-out energy retrofit interventions, based on real data.

According to the Verdantix maturity model [55], there are five stages of digital twin evolution (Table 1). Essentially, the more mature the twin is, the more valuable it becomes.

Table 1. Stages of digital twin evolution.

Stages	Description
1. Descriptive twin	The twin accurately represents the building structure in the real world.
2. Informative twin	The twin is able to connect to IoT and sensors to communicate real data to the user.
3. Predictive twin	The digital twin uses its historical data to run analytics and predict trends.
4. Comprehensive twin	The digital twin, as it evolves, allows real-time simulation for future "what-if" scenarios.
5. Autonomous twin	The twin has the data and processes established to automatically take action based on the insights it delivers.

4. Workflow Application: The "Solar House"

This section explains in detail the work phases presented in the flow diagram in Figure 1.

4.1. Phase 1—Physical Building Study

The case study chosen to develop the workflow is an office building, the so-called "Solar House", which is part of a complex of university buildings housing the Engineering Departments of the University of L'Aquila. The building is located in Monteluco di Roio, L'Aquila, near a wooded area characterized by a dense presence of trees (Figure 2).

Figure 2. Case study location, Monteluco di Roio, L'Aquila, Italy.

The building, dating back to the 1970s, is made of reinforced concrete. The name "Solar House" derives from the particular shape of the roof, designed to accommodate solar and photovoltaic panels which; however, are not currently present. The building consists of three floors, one basement, and two above grounds. Only the first floor, i.e., the floor containing the offices of the Las.E.R Laboratory (Laser and Electro-optical Research in engineering metrology) research group, was taken into consideration for the experimentation (Figure 3).

Figure 3. Case study front view with the plan under consideration boxed in red.

The floor consists of an entrance hall, four offices, a lighting laboratory, a hallway, and a bathroom (Figure 4a). The internal height of the floor is between 3.10 and 4.50 m (Figure 4b).

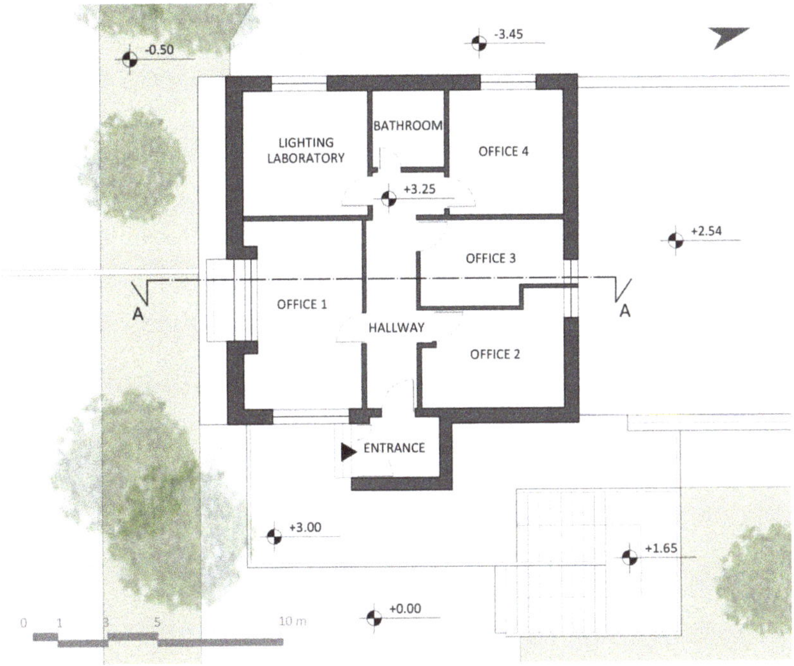

(a)

Figure 4. *Cont.*

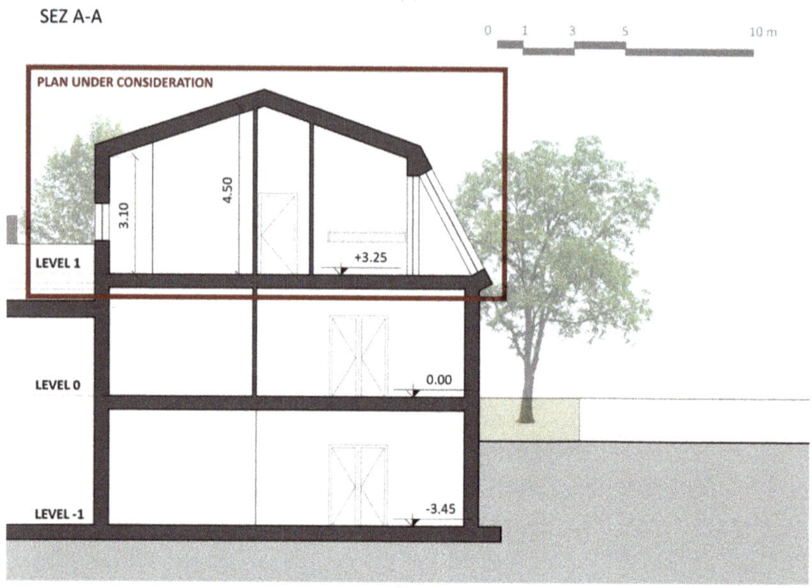

(b)

Figure 4. Project drawings of the "Solar House". (**a**) Plan with indication of the rooms. (**b**) Longitudinal section.

4.2. *Phase 2a—WSN Configuration*

The monitoring, which lasted two weeks, involved four rooms, namely the hallway, the lighting laboratory, and offices one and two, each with different functional and occupational characteristics. The choice of rooms is linked to specific reasons: the lighting laboratory is the room used for laboratory experiments, with non-constant occupancy, the hallway constitutes a passage space, where occupancy is variable, office one is a two-people cellular office, and office two is a one-person cellular office. Following the ISO 7726 standard [56], the sensors were positioned so that the thermal exchanges were more or less identical to those to which the room occupant was exposed. The location of the sensors is shown in Figure 5.

The block scheme of the monitoring system employed in this work is shown in Figure 6. The sensor nodes are equipped with wireless communication using Wi-Fi. These elements are powered directly from the electrical grid by means of a Direct Current (DC) power supply. In this case, there is no need for batteries as the grid supply is always available. The same principles apply to scenarios where energy harvesting can be exploited in remote or non-electrically served environments [57]. The Wi-Fi network access for internet connection is the same one used for normal operation in the offices. Using HTTPS requests, the nodes forward the sensor measurements to the Tandem platform that acts as a database. Then, the information can be accessed using the user front end.

The sensor nodes are structured as shown in Figure 7. A core ESP32 microcontroller [58], by Espressif, periodically retrieves information from the sensors, which are of the type AHT20 [59], from Asair. Data are encapsulated in JSON format and transmitted using POST requests. If the request is not successful, it is repeated after a guard time period. If the network connectivity is not available, the sensor nodes enter a sleep mode and the connection attempts loop.

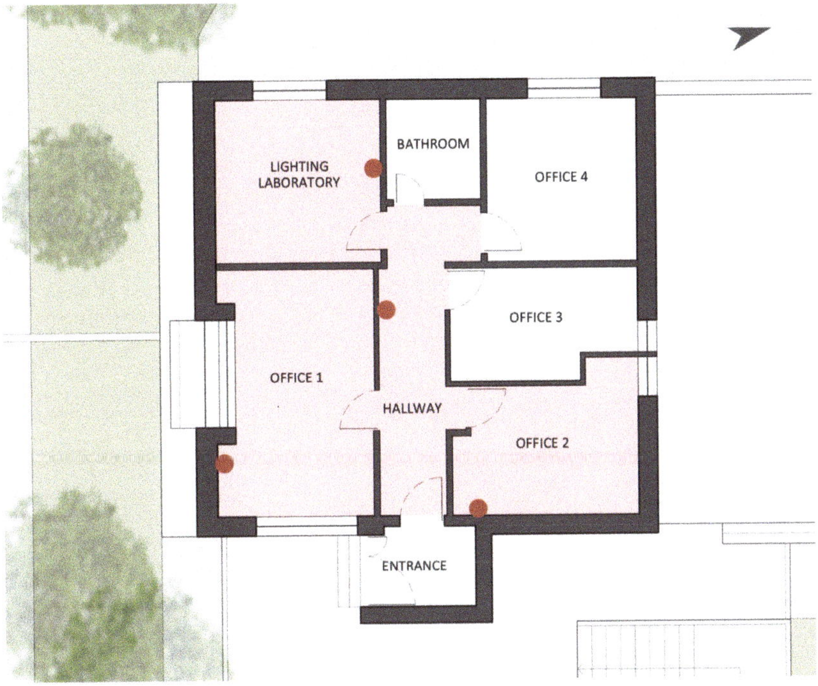

Figure 5. Areas affected by monitoring and indication of sensor nodes placement (red dots).

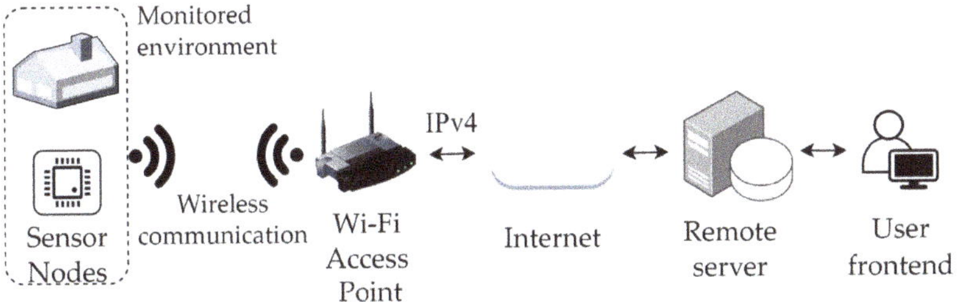

Figure 6. Block scheme of the IoT sensing system.

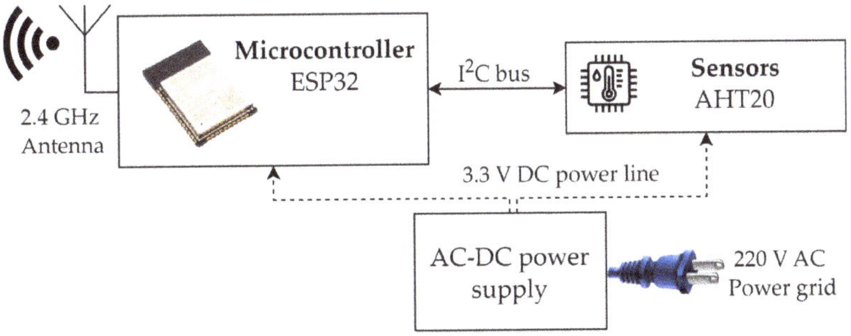

Figure 7. Block scheme of the sensor nodes.

The sensors characteristics are reported in Table 2.

Table 2. Sensor characteristics.

Characteristic	Description
Type	Integrated circuit, resistive–capacitive sensing technology
Measuring Range	0 to 100% RH, −40 to +85 °C
Recommended range	0 to 80% RH, −20 to +60 °C
Accuracy (Typical)	Humidity: 0.024% RH, Temperature: 0.01 °C

A diagram of the functionality of the system is shown in Figure 8.

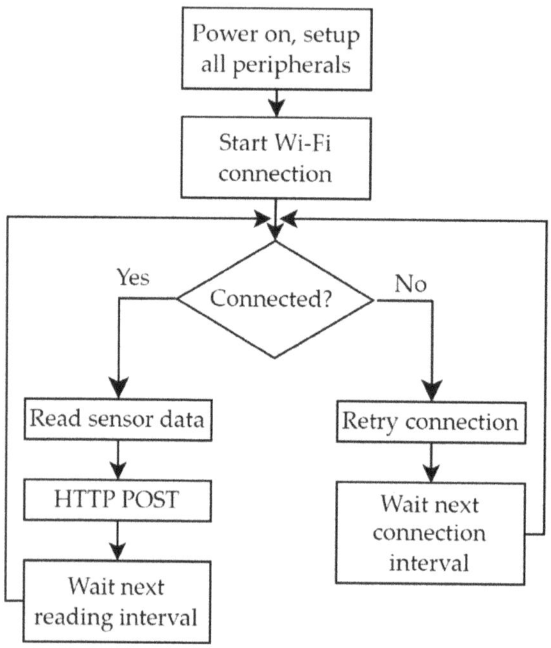

Figure 8. Functional diagram of the sensor nodes.

The sensor has a stable performance within the recommended working range, as the datasheet reports. Long-term exposure to conditions out of the normal range may cause a temporary signal drift. This is true especially when the humidity is >80%, where the datasheet reports a drift of +3% RH after 60 h. The sensor will slowly self-recover to the calibration state after returning to normal working conditions. This process can be executed in fast recovery by taking drying measures, as suggested by the sensor designers, by keeping it at 60–85 °C and <5%RH for 2–10 h.

Figure 9 shows a picture of the experimental setup.

The electronic components used for the implementation of the WSN nodes in this study have been chosen following two main criteria. The Espressif ESP series microcontrollers are well documented by the productor both in terms of hardware and firmware development [60]. The configuration of the nodes is made easier using wireless interfaces that can be integrated using open-source documentation [61], allowing for the reduction of the overall complexity of the deployment of a research-grade system. The availability of the mentioned circuits in the market is the one motivation, as the selected node elements are widespread in the free market at competitive prices. The use of a Wi-Fi connection enables the portability of the sensor system to different locations, as the only gateway needed to access the internet is a Wi-Fi access point with a 2.4 GHz band.

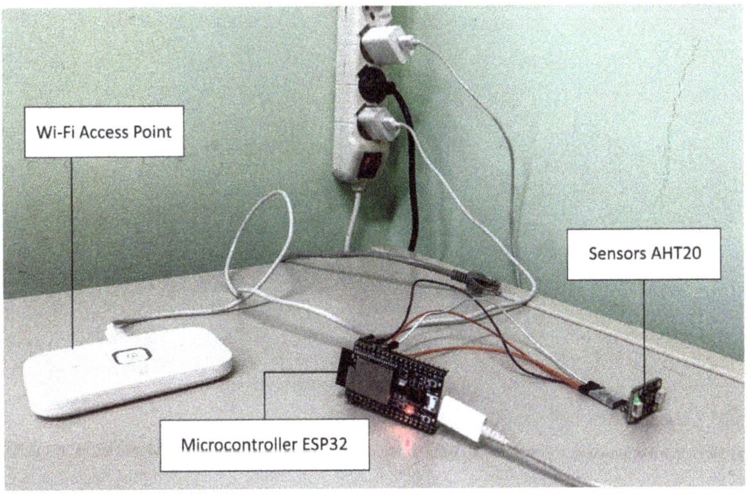

Figure 9. Experimental setup.

The integration of IoT–DT systems raises significant security and privacy concerns, especially regarding the protection of sensitive data transmitted and stored within these systems [62]. Huge efforts have been made in the search for solutions for the management of security data that provide accessibility, transparency, reliability, consistency, and traceability. Among these, a solution that seems to be very promising is the one that involves the use of blockchain technology for the reliable sharing of information between all the entities involved in a construction project [63].

Beyond the research currently underway, the basis for safeguarding data integrity and addressing privacy issues associated with real-time monitoring is certainly the implementation of robust security measures, such as encryption and access controls.

Encryption is fundamental for protecting data confidentiality in DT systems. Advanced encryption techniques ensure that data exchanged between the physical and virtual components of a DT remains secure from unauthorized access. Implementing end-to-end encryption protocols protects data both in transit and at rest. This approach ensures that even if data transmissions are intercepted, the information remains unintelligible without the proper decryption keys. In IoT-oriented systems, lightweight encryption algorithms are more suited as computational efficiency is critical. These algorithms provide a balance between security and performance, ensuring that encryption does not hinder the real-time capabilities of the DT system. Access control mechanisms ensure that only authorized personnel can access specific data and functionalities within DT systems. Role-Based Access Control (RBAC) models are widely used to assign permissions based on user roles and responsibilities, effectively limiting access to sensitive information. Access control is often coupled with Multi-Factor Authentication (MFA), that enhances security by requiring users to provide multiple forms of verification before accessing the system, including a password or a security token.

Real-time monitoring in DT systems involves continuous data collection and analysis, which may include sensitive operational and personal information. Addressing privacy concerns is crucial to complying with data protection regulations like the General Data Protection Regulation (GDPR). Collecting only the data necessary for specific purposes reduces the risk of unauthorized exposure. Implementing data minimization strategies ensures that unnecessary or redundant data are not collected or stored, thereby limiting the potential impact of data breaches. Techniques that anonymize or pseudonymize personal data can prevent the identification of individuals from datasets. This involves altering or removing personal identifiers from the data, making it difficult to trace back to any individual while still retaining the usefulness of the data for analysis. Developing clear and transparent data policies that outline data collection, usage, storage, and sharing practices helps build trust with stakeholders. Ensuring that all parties are informed about how their data are handled and protected is essential for compliance and accountability.

Using encryption standards, such as Secure Sockets Layer (SSL) for data in transit, ensures that data are unreadable to unauthorized parties. In this work, the authors implemented the use of Hypertext Transfer Protocol Secure (HTTPS) for data exchange as a first layer security enabling solution. Its widespread adoption and standardization make it a practical choice for securing web-based communications within these systems. However, relying solely on HTTPS has limitations. Performance impacts due to encryption overhead, the complexity of certificate management, and its inapplicability to non-HTTP protocols are notable drawbacks.

4.3. Phase 2b—BIM Modeling

The BIM model of the building was created with Autodesk Revit 2023 on the basis of available two-dimensional executive technical drawings and integrated with infrared thermography analysis. Revit was chosen due to its versatility in modeling various types of buildings and its adaptability to a wide range of studies [64]. Additionally, Revit offers a high level of interoperability compared to many other BIMs, providing greater flexibility when integrating with other software and other technologies for data exchange. The general overview of the model is presented in Figure 10.

Figure 10. BIM model of the "Solar House".

The BIM environment has the advantage of being able to recreate the building at the required level of detail and accuracy. In our case, defining the internal partitions was crucial, as monitoring was conducted at the level of individual rooms (Figure 11).

The selected building has regular and standardized characteristics, fully aligned with a BIM approach. Therefore, from a semantic perspective, the model was mostly developed using existing families from the software libraries and, only in specific cases, by creating new custom families, such as for the sensors, which were placed in the model in the same positions as in reality and linked to the designated rooms within the model. Therefore, for monitoring purposes, the rooms constituting the "Solar House" were created within the model, enabling the integration of data from environmental sensors assigned to each room (Figure 12).

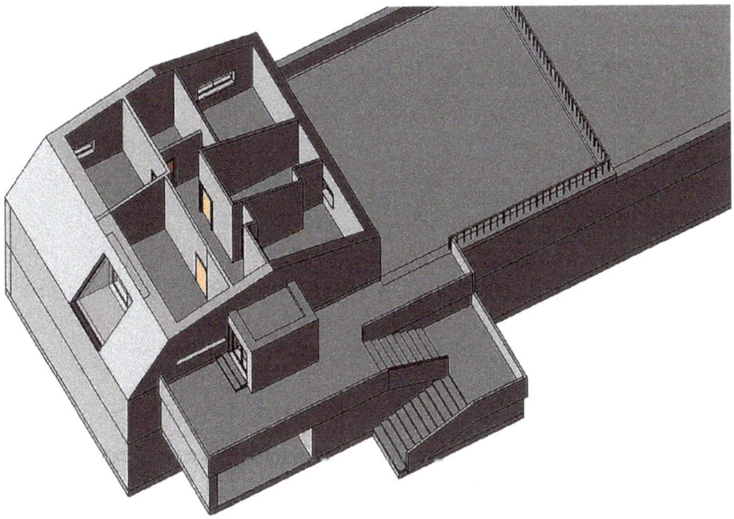

Figure 11. Distribution of the internal spaces of the floor in question.

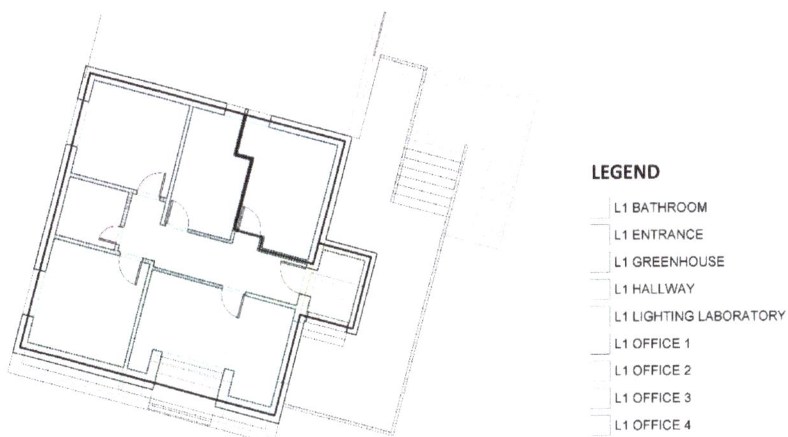

Figure 12. Creation of analytical spaces.

BIM allowed for the recreation of all the stratigraphies of the building components. In Revit, layered elements support the construction modeling process by allowing certain elements of the design intent model to be divided into distinct parts. In our case, the stratigraphies were generated by elements with layered structures, such as walls, floors, roofs, etc. The building envelope stratigraphy was reconstructed using executive technical drawings, supplemented by the infrared thermography (IRT) technique, useful in identifying any modification made during or after the initial construction. For instance, wall C1 shown in Table 3, originally indicated as concrete in the layer compositions, was actually constructed with hollow block masonry. Analyzing the building's design and identifying changes made to the original plans over the years was essential for understanding its history and initial goals, which aimed to reduce its environmental impact by creating a self-sustaining building. The DT was created based on the building's current characteristics with the aim of assessing its present features and performance to serve as a foundation for proposing future improvement interventions. Figure 13 shows all the types of opaque closures present in the building.

Table 3. Schematic representation of the opaque closures and the layers that compose them.

Type of Closure	Representation	Layers	Thermal Transmittance U $\left[\frac{W}{m^2K}\right]$
C1	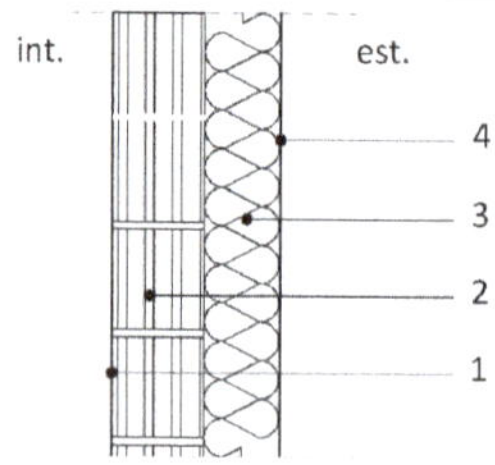	1. Internal plaster (0.5 cm) 2. Hollow blocks (20 cm) 3. Mineral wool insulation (16 cm) 4. External Plaster (0.5 cm)	0.23
C2	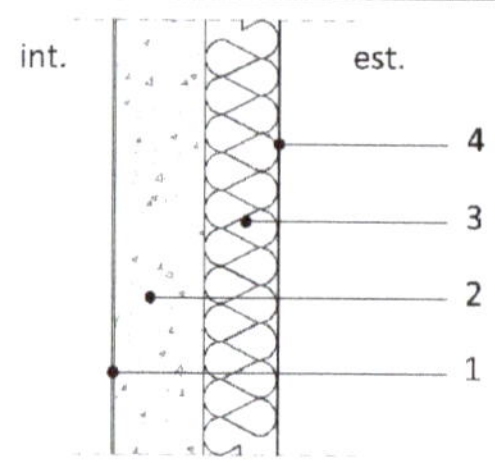	1. Internal plaster (0.5 cm) 2. Reinforced concrete (20 cm) 3. Mineral wool insulation (16 cm) 4. External plaster (0.5 cm)	0.26
C3	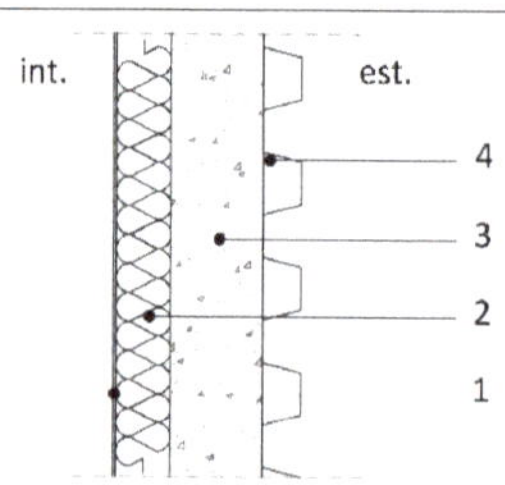	1. Internal plaster (0.5 cm) 2. Mineral wool insulation (12 cm) 3. Reinforced concrete (20 cm) 4. Corrugated sheet (8 cm)	0.34
C4	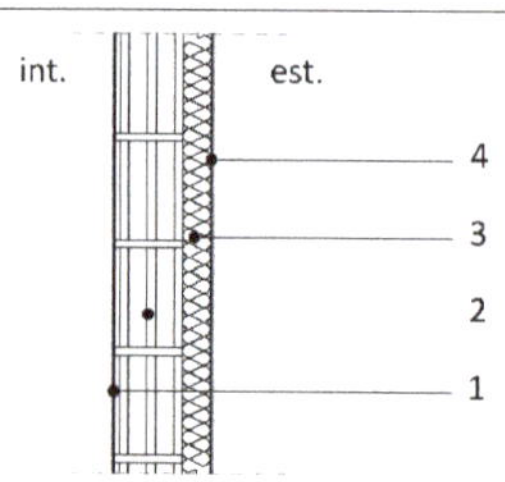	1. Internal plaster (0.5 cm) 2. Hollow blocks (15 cm) 3. Mineral wool insulation (6 cm) 4. External plaster (0.5 cm)	0.52
Roof	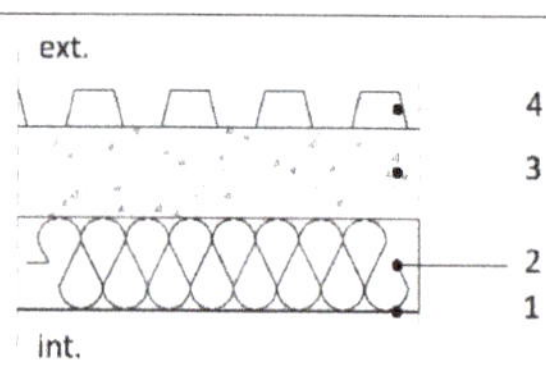	1. Internal plaster (0.5 cm) 2. Mineral wool insulation (20 cm) 3. Reinforced concrete (20 cm) 4. Corrugated sheet	0.21
Interfloor slab	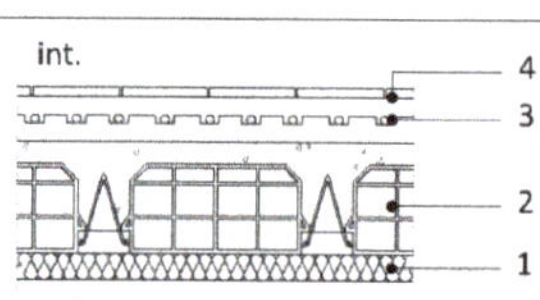	1. Expanded polystyrene (EPS) Insulation (6 cm) 2. Beam and block floor (20 + 5 cm) 3. Lightweight concrete (10 cm) 4. Porcelain stoneware (2 cm)	0.25

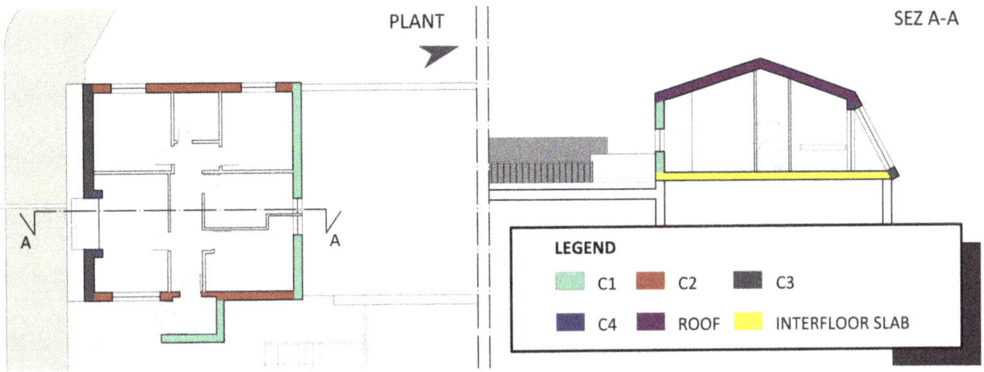

Figure 13. Types of opaque closures.

Table 3 summarizes the main characteristics of opaque closures modeled on BIM.

Assigning each architectural component its thermophysical characteristics is important in order to create a digital twin capable of offering a holistic view of the building. With all the data at hand, the user will be able to acquire an in-depth knowledge of the properties possessed by the building in question. Furthermore, a digital twin of the building rich in information can be subsequently imported into an energy simulation software in order to conduct more in-depth and accurate studies in view of any efficiency interventions.

4.4. Phase 3—Digital Twin Creation

Autodesk Tandem was used to create the digital twin of the building under investigation. The choice of middleware tool is linked to the fact that Tandem and Revit are released by the same software company, favoring the interoperability process and making Tandem able to read the BIM model correctly. Tandem is a cloud-based digital twin solution that enables collaboration in Architecture, Engineering, Construction and Operation (AECO). It consists of an online accessible platform that allows you to create an ecosystem in which all the resources, spaces, and systems of the building are related to each other. Therefore, Tandem can be seen as a middleware tool capable of connecting IoT systems to the BIM model of the building in a single environment, providing the end user with the key to control and interpret data easily and quickly from day zero. With the context of the robust 3D twin viewing and supporting charts and heatmaps, data insights are instantly visible. To use Tandem, you need to have an account. Since the twin is hosted in the cloud, the end users can access it from anywhere at any time, upon invitation from the account administrator. Based on their permission level, end users will be able to "Read", "Edit" or "Manage" the digital twin. Tandem gives the possibility to analyze both real-time and historical data, allowing an accurate analysis of the structure's performance.

The first step to creating a digital twin on Tandem involves "create a new facility". In this way, it is possible to import the BIM model of the building in ".rvt" or ".ifc" format onto the platform. To be able to send the data recorded by the sensors to Tandem, you need to create connections and assign them a classification. In Tandem, a connection is the digital representation of an IoT device or asset (for example thermostats, air handling units, or occupancy sensors). Once you have created all the necessary connections, you need to configure them. There are several ways to do this, in our case the used method is HTTPS requests, using the provided endpoint. Each node in the DT platform has a unique endpoint that can be used to access the database. Tandem gives the possibility to assign the connection to a host, which can be any element or space. In our case, the various connections were assigned to the corresponding sensors, previously modeled on Revit using a custom family. This feature is very interesting, as it makes it possible to view instant data from each sensor simply by clicking on its virtual model, making monitoring much more direct and intuitive (Figure 14).

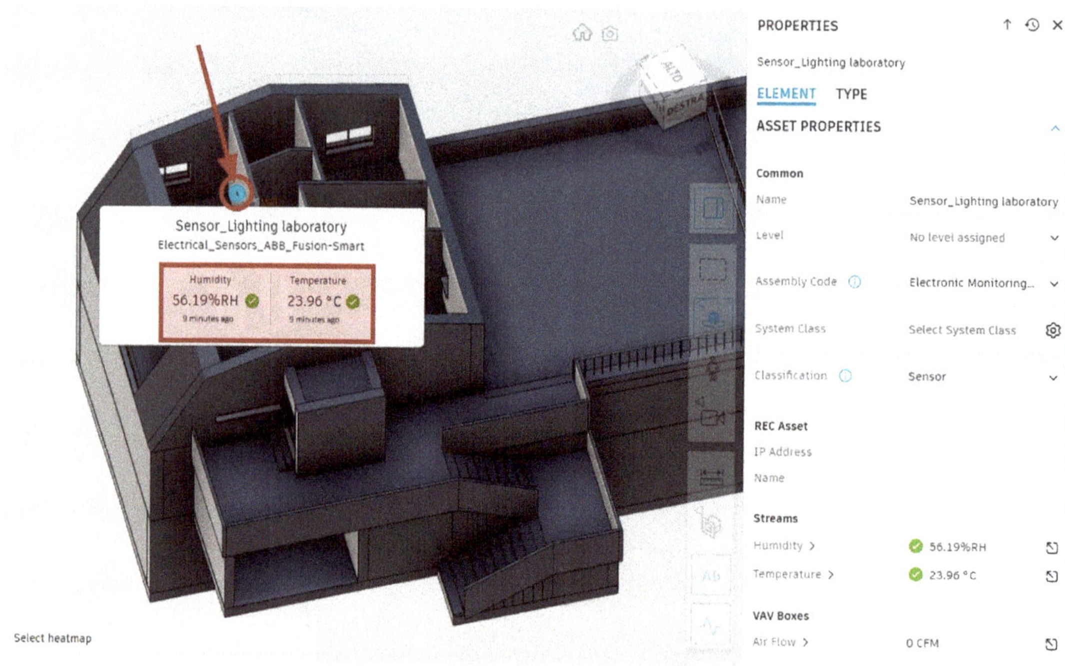

Figure 14. Example of instant data display for the lighting laboratory sensor.

In the "Streams" section it is possible to view the trend graphs of the various environmental parameters recorded over time, selecting the desired time interval. In Tandem it is possible to create thresholds, i.e., upper and lower limit values within which the environmental parameters should remain. When the values exceed the limits set by the user, the graphs signal it with red to facilitate the identification of critical issues. In our case, the limits were set in accordance with the criteria of the National Institute for Insurance of Accidents at Work (Inail), based on Legislative Decree 81/2008 [65]. According to the above criteria, in offices, it is advisable that the humidity remains between 40 and 60% and the temperature between 18 and 24 °C. Figure 15 shows the trends in humidity and temperature values for each room during three weeks of monitoring, from 30 May 2024 to 13 June 2024. The time intervals in which the parameters exceeded the imposed limits are highlighted in red.

Autodesk Tandem provides the ability to configure and customize maps to reflect connected streams (Figure 16). The Heatmap widget shows the last set of data received for the streams of data as they are associated with spaces or assets within the digital twin.

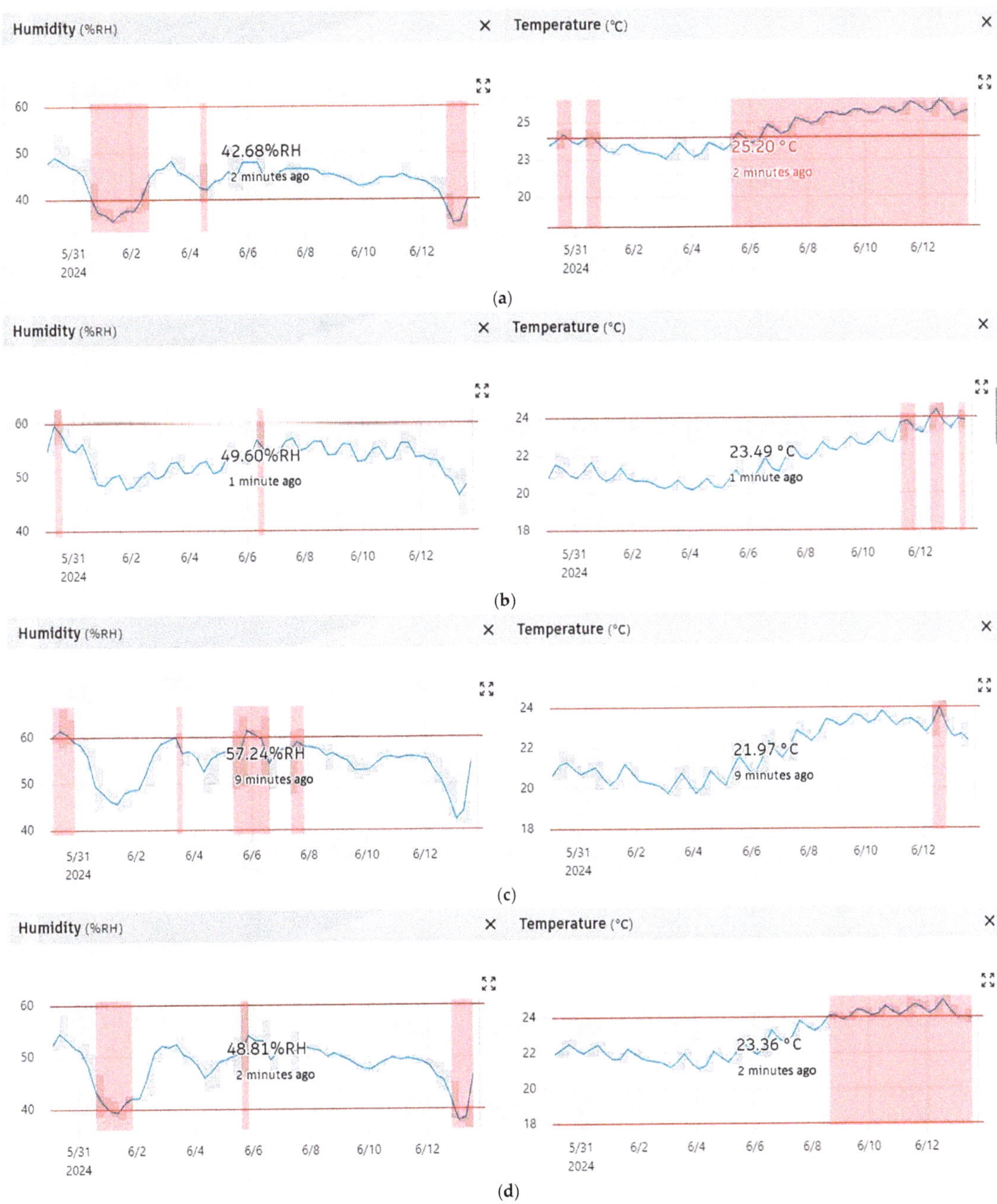

Figure 15. Humidity and temperature trend for each room. (**a**) Hallway. (**b**) Lighting laboratory. (**c**) Office 1. (**d**) Office 2. Horizontal axis: time; vertical axis: humidity or temperature value.

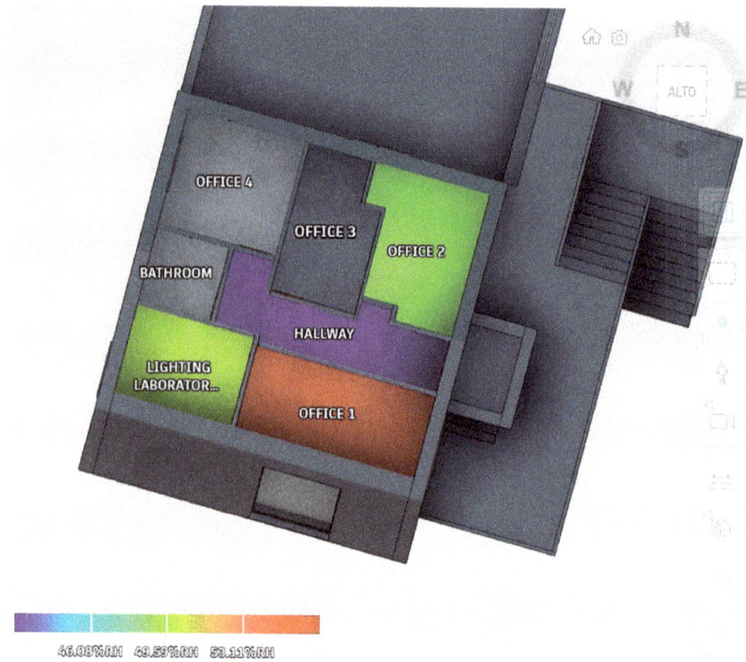

(**a**)

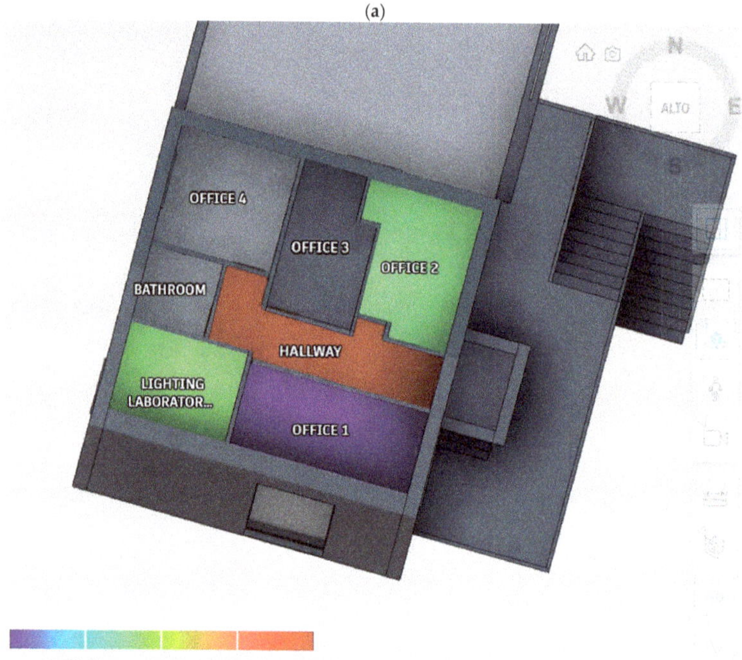

(**b**)

Figure 16. Mapping of average environmental parameters recorded during the monitoring period. (**a**) Humidity. (**b**) Temperature.

5. Results and Discussion

Figure 17 correlates the temperature and humidity values on the same graph, highlighting the peaks reached in the time intervals in which the values exceeded the limits imposed by the current decree.

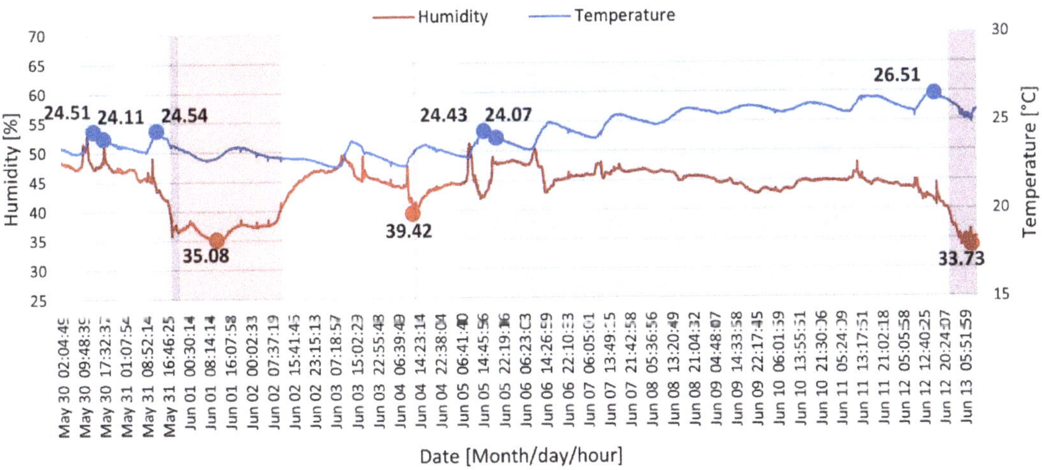

(a)

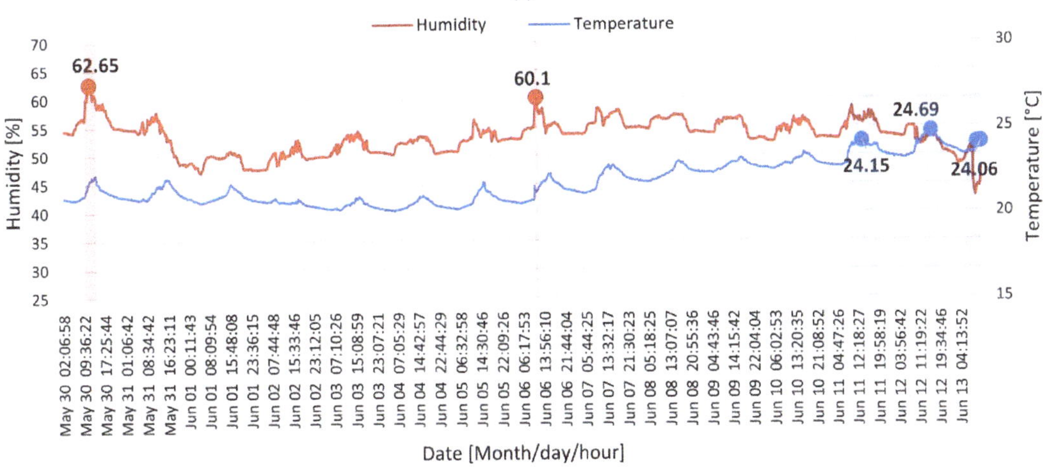

(b)

Figure 17. *Cont.*

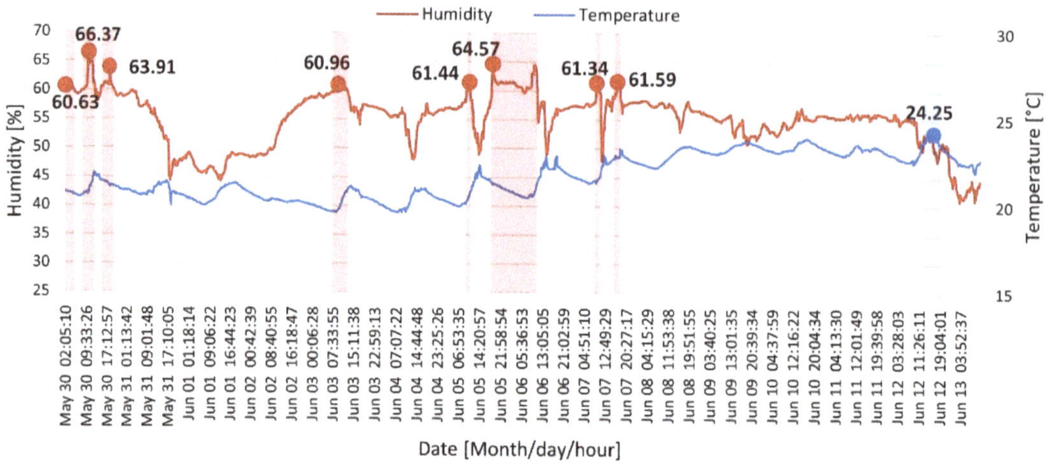

(**c**)

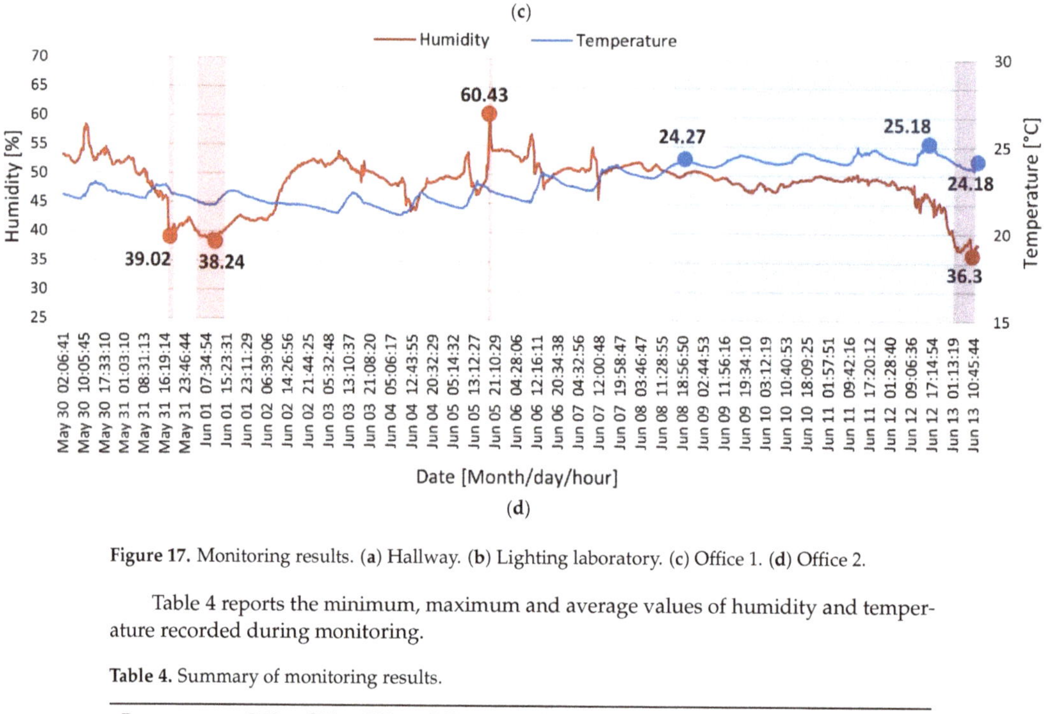

(**d**)

Figure 17. Monitoring results. (**a**) Hallway. (**b**) Lighting laboratory. (**c**) Office 1. (**d**) Office 2.

Table 4 reports the minimum, maximum and average values of humidity and temperature recorded during monitoring.

Table 4. Summary of monitoring results.

Room	RH_{min} [%]	RH_{max} [%]	RH_{ave} [°C]	T_{min} [°C]	T_{max} [°C]	T_{ave} [°C]
Hallway	33.73	54.21	43.95	22.47	26.51	24.38
Lighting lab.	42.63	62.65	53.27	20.09	24.69	21.72
Office 1	41.07	66.37	54.99	19.61	24.25	21.73
Office 2	36.30	60.43	48.55	20.97	25.18	22.87

Table 4 shows some initial assessments, where the temperature of the monitored rooms varies from 19.61 °C to 26.51 °C, while the humidity is from 33.73% to 66.37%. The hallway has the lowest humidity and the highest temperature values. The high temperature is probably due to the fact that the hallway occupies a central position inside the building and, for this reason, is not affected by heat losses through the envelope. The dryness of the environment is a direct consequence of the high temperature found there. On the contrary, office one presented the highest humidity and lowest temperature values. This is linked to the presence of large glazing in the room which favored heat dispersion in a period of the year when the external temperature was still relatively low. In any case, by averaging the humidity and temperature values recorded during the two weeks of monitoring for all the rooms under examination, it can be noted that the values recorded every 10 min fell within the limits imposed by the national decree in force, except in the case of the hallway in which the average value of temperature exceeds the threshold of 0.38 °C. To have an even broader vision of the situation, it would be advisable to repeat the monitoring during the summer and winter periods.

Figure 18 shows the correlation matrix of the monitored parameters for the different rooms. Each off-diagonal subplot shows a scatterplot of a pair of variables, including a least-squares reference line, where the slope represents the displayed correlation coefficient. The diagonal subplots contain the distribution of a variable as a histogram.

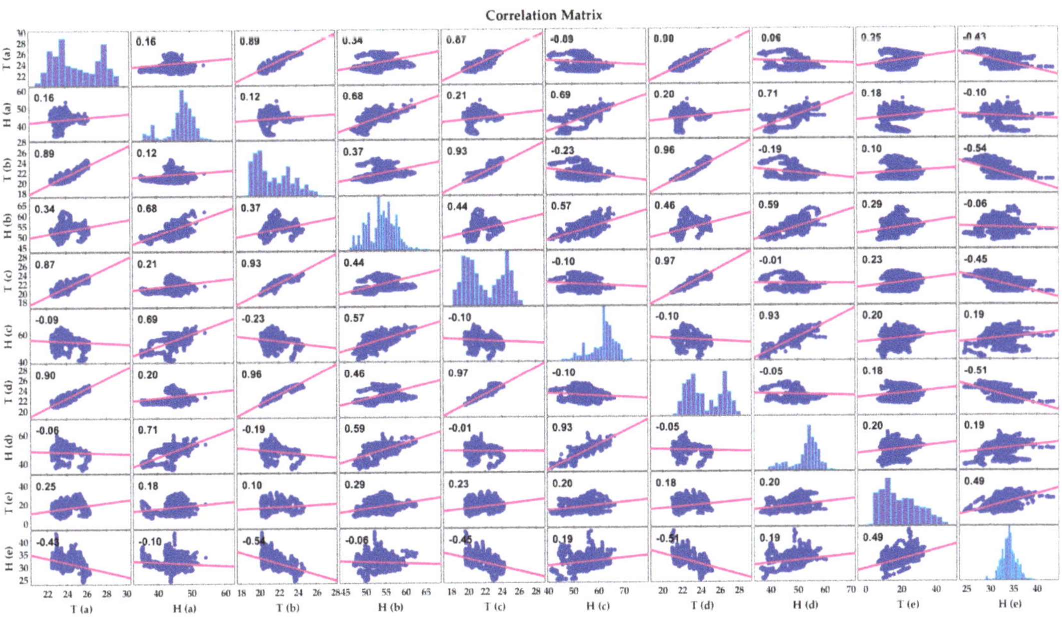

Figure 18. The correlation matrix of the monitored parameters for the rooms of interest. T stands for temperature, and H stands for relative humidity. (**a**) Hallway. (**b**) Lighting laboratory. (**c**) Office 1. (**d**) Office 2. (**e**) External environment.

It is possible to assess that the temperature values of the different zones can be considered very close to each other, as expected in a building that does not present a big volume and room-dedicated air exchange systems. This can be observed also for the values of zone relative humidity. The correlation of temperatures and relative humidity values can be considered low, as the factors that rule the variation in humidity in each monitored room are various and the interaction can be complex [66]. Furthermore, the relationships between internal temperature and humidity and the external climate are low; this can be due to a variety of factors, such as human activities, occupancy, and the behavior of the building envelope.

During monitoring, the DT helped in the constant display of the parameters in a simple and quick way, allowing continuous updating on the thermo-hygrometric condition of the different rooms at any time, even remotely.

The developed DT lays the foundation for future implementations of the system capable of promoting both proactive and corrective maintenance interventions, guiding facility managers in identifying the most appropriate energy efficiency strategies. On the proactive maintenance level, for example, the DT allows to instantly identify the presence of anomalies ensuring an immediate intervention by the facility managers in case of need. In the context of corrective maintenance, the historical data collected during the monitoring period could be used to calibrate the thermal model of the building in order to obtain accurate and reliable energy analyses on which to base any energy efficiency interventions, simulating different design alternatives. Furthermore, the ability to update the digital twin in the event of changes to the real building would mean having a virtual model capable of providing continuously evolving data, which constantly reflects the conditions of the physical entity.

Given that the DT would bring several advantages within the life cycle of a single building, recent studies are considering the possibility of exploiting this technology in the field of urban planning [67–70]. According to Deng et al. [69], digital twins, along with the Internet of Things, fifth-generation wireless systems, blockchain, collaborative computing, simulation, and artificial intelligence technologies, could be the key to the transformation of the current urban governance paradigm toward smart cities. Taking inspiration from the work of Halimi et al. [71] that offers a methodological framework for the evaluation of the impact of transportation infrastructure on community connectivity, it would be interesting, in the future, to analyze the socio-economic benefits of DT at the community level, testing the applicability of our research on an urban scale.

The main aim of the paper was to present a detailed workflow for the practical application of creating a DT, seeking to provide a more comprehensive view by highlighting the workflow and methodological aspects involved.

From the application of the developed workflow, it emerged that for the creation of an efficient and functioning DT it is important:

- to choose an appropriate BIM tool, which is capable of recreating the various building components in detail;
- to select suitable sensor devices capable of connecting to the appropriate network access point;
- to choose a reliable and efficient network connection that allows a fluid and constant flow of data;
- to identify a middleware tool capable of appropriately reading both the 3D model of the building and the data coming from the sensors;
- to ensure efficient synchronization between the BIM model and the IoT sensor devices by configuring the connection in the most appropriate way for the tools available.

Table 5 illustrates the technologies chosen for the creation of the DT, based on the needed interoperability requirements.

A problem that could arise during the transition of the model from one platform to another could be the incorrect export of geometric data or the loss of information. If these critical issues emerge, it is appropriate to make changes to the three-dimensional model in most cases by proceeding with its geometric and semantic simplification [8]. The data exchange among internet-based services can be considered less critical due to a large spectrum of possible exchange protocols and methods [72].

Figure 19 summarizes the technologies, protocols, and exchange formats used to achieve efficient interoperability between the various tools involved in the process.

Table 5. Technologies chosen for the creation of the DT.

Technology	Interoperability Requirement	Selected Solution
Sensor	- possibility of physically connecting to the data acquisition device	AHT20
Data acquisition and exchange device	- wireless connection capability - capability of data exchange in the format required by the DT platform	Espressif ESP Microcontroller
BIM software	- capability to export the model in the format accepted by the DT platform	Autodesk Revit
DT platform	- ability to be accessible via the Internet - ability to import the model in the format supported by BIM - ability to receive data in the storage format supported by the data acquisition device - capability to support storage of the received data	Autodesk Tandem

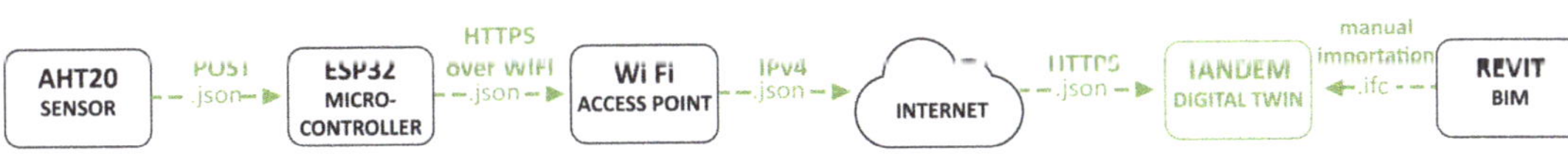

Figure 19. Schematic representation of interoperability between the tools involved in the creation of the digital twin.

Table 6 summarizes the main potentials and challenges in the application of BIM-based digital twin.

Table 6. BIM-based digital twin potentials and challenges.

Potentials	Challenges
• Digitalization of the construction sector	• Lack of a standardized procedure for applying the technology
• Integration between architecture and energy	• Use of technology still very limited
• Constant connection between physical and virtual entity	• Absence of BIM models of historical or ancient buildings in the archive
• Real-time remote monitoring	• Difficulty of interoperability between the different technologies and software involved in the application of DT.
• Optimizing the data visualization and processing phase	
• User involvement in energy issues	

6. Conclusions

For efficient energy management of buildings, it is crucial that energy use and the thermal environment are addressed simultaneously. Identifying a building's design criticalities can help choose the most appropriate energy optimization interventions, leading to a reduction in heat loss and, consequently, limiting the use of HVAC with a consequent reduction in costs related to its consumption. Accurate diagnoses through comprehensive internal comfort monitoring campaigns are essential to identify the most efficient energy optimization strategies. With the rapid progress of urban modernization and construction industrialization, it is necessary to overcome the limitations of current environmental monitoring systems to promote the digital transformation of the construction sector. In this context, BIM-based DT technology can play a fundamental role. Nonetheless, despite the unquestionable potential of DT, there are still challenges and unresolved issues due to its relatively recent development. This study presents a detailed workflow for creating a DT, defining standard guidelines applicable to various case studies, also offering different solutions to optimize the integration of the various technologies necessary for the creation of an efficient DT. The developed workflow was tested on a singular building, an office building also used as an academic laboratory, with the aim of experimenting with the "Living Lab" concept. The system integrates real-time data collected through a WNS with a BIM-based 3D model that automatically combines the monitored data. The study confirms that DT offers several benefits for the construction industry, such as facilitating the integration of architecture and energy systems, accelerating the digitalization of the construction sector, and enabling the direct and continuous connection between physical and virtual entities. Additionally, it optimizes the data visualization and processing phase of monitoring, providing a user-friendly interface that is accessible to all users, in real time and remotely.

This work represents an initial exploration of digital twin (DT) technology, focusing on the second stage of DT evolution as outlined by the Verdantix maturity model. Future developments should involve conducting repeated monitoring at different times of the year also including other parameters influencing the IEQ index, in order to gain a broader understanding of the building performance.

Further advancements could consist of exploiting the DT for predictive analyses and real-time simulations, enhancing the accuracy and effectiveness of the decision-making process in energy requalification interventions and the overall quality of life for users.

From a future perspective, integrating monitoring sensors with digital technologies, such as BIM, DT, Internet of Things (IoT), and Artificial Intelligence (AI) could significantly advance the building automation field by encouraging the application of automated control systems, preserving energy efficiency and ensuring a comfortable indoor environment throughout the building's lifespan [73].

Author Contributions: Conceptualization, T.D.R., A.C., A.T. and S.B.; methodology, T.D.R., A.C., M.R., V.S., D.A. and S.B.; software, A.C. and M.R.; validation, A.C., M.R. and A.T.; formal analysis, T.D.R., A.C. and M.R.; investigation, T.D.R., A.C., M.R. and A.T.; resources, T.D.R., A.C., M.R. and A.T.; data curation, T.D.R. and A.C.; M.R. and A.T.; writing—original draft preparation, T.D.R., A.C., M.R. and A.T.; writing—review and editing, T.D.R., V.S., S.B. and D.A.; visualization, T.D.R., A.C., M.R., V.S., S.B., A.T. and D.A.; supervision, T.D.R., V.S. and S.B. All authors have read and agreed to the published version of the manuscript.

Funding: This research received no external funding.

Data Availability Statement: The original contributions presented in the study are included in the article, further inquiries can be directed to the corresponding author.

Conflicts of Interest: The authors declare no conflicts of interest.

References

1. Calcagno, G.; Trombadore, A.; Pierucci, G.; Montoni, L. Untapping the Potential of the Digital Towards the Green Imperative The Interdisciplinary BeXLab Experience. In *Technological Imagination in the Green and Digital Transition*; Arbizzani, E., Cangell E., Clemente, C., Cumo, F., Giofrè, F., Giovenale, A.M., Palme, M., Paris, S., Eds.; CONF.ITECH 2022; The Urban Book Series Springer: Cham, Switzerland, 2022.
2. *ISO 29481-1*; 2016 Building Information Models—Information Delivery Manual—Part 1: Methodology and Format. Internationa Organization for Standardization: Geneva, Switzerland, 2016.
3. Eastman, C.; Teicholz, P.; Sacks, R.; Liston, K. *BIM Handbook—A Guide to Building Information Modeling for Owners, Managers Designers, Engineers, and Contractors*; John Wiley & Sons: Hoboken, NJ, USA, 2008.
4. Pavan, A.; Mirarchi, C.; Giani, M. *BIM: Metodi e Strumenti; Progettare, Costruire e Gestire Nell'era Digitale*; Tecniche Nuove: Milanc Italy, 2017.
5. Brusaporci, S.; Maiezza, P.; Tata, A. A Framework for Architectural Heritage HBIM Semantization and Development. *Int. Arch Photogramm. Remote Sens. Spat. Inf. Sci.* **2018**, *XLII-2*, 179–184. [CrossRef]
6. Fassi, F.; Achille, C.; Mandelli, A.; Rechichi, F.; Parri, S. A New Idea of BIM System for Visualization, Web Sharing and Using Huge Complex 3D Models for Facility Management. *Int. Arch. Photogramm. Remote Sens. Spat. Inf. Sci.* **2015**, *XL-5/W4*, 359–366 [CrossRef]
7. *UNI 11337:2017*; Gestione Digitale dei Processi Informativi Delle Costruzioni (BIM). Italian Standardization Body: Milar Italy, 2017.

8. Ciccozzi, A.; de Rubeis, T.; Paoletti, D.; Ambrosini, D. BIM to BEM for Building Energy Analysis: A Review of Interoperability Strategies. *Energies* **2023**, *16*, 7845. [CrossRef]
9. Ciccozzi, A.; de Rubeis, T.; Pasqualoni, G.; Paoletti, D.; Ambrosini, A. Approach to energy analysis using Building Information Modeling (BIM) software. In Proceedings of the XXIII CIRIAF National Congress-Sustainable Development, Protection of the Environment and Human Health, Perugia, Italy, 13–14 April 2023.
10. Antòn, L.A.; Dìaz, J. Integration of life cycle assessment in a BIM enviroment. *Procedia Eng.* **2014**, *85*, 26–32. [CrossRef]
11. Shahinmoghadam, M.; Natephra, W.; Motamedi, A. BIM- and IoT-based virtual reality tool for real-time thermal comfort assessment in building enclosures. *Build. Environ.* **2021**, *199*, 107905. [CrossRef]
12. Desogus, G.; Frau, C.; Quaquero, E.; Rubiu, G. From Building Information Model to Digital Twin: A Framework for Building Thermal Comfort Monitoring, Visualizing, and Assessment. *Buildings* **2023**, *13*, 1971. [CrossRef]
13. Shishehgarkhaneh, M.B.; Keivani, A.; Moehler, R.C.; Jelodari, N.; Laleh, S.R. Internet of Things (IoT), Building Information Modeling (BIM), and Digital Twin (DT) in construction industry: A review, bibliometric, and network analysis. *Buildings* **2022**, *12*, 1503. [CrossRef]
14. El Saddik, A. Digital Twins: The Convergence of Multimedia Technologies. *IEEE MultiMedia* **2018**, *25*, 87–92. [CrossRef]
15. Marzouk, M.; Abdelaty, A. Monitoring thermal comfort in subways using building information modeling. *Energy Build.* **2014**, *84*, 252–257. [CrossRef]
16. Lin, Y.C.; Cheung, W.F. Developing WSN/BIM-Based Environmental Monitoring Management System for Parking Garages in Smart Cities. *J. Manag. Eng.* **2020**, *36*, 04020012. [CrossRef]
17. Wang, T.; Gan, V.J.L.; Hu, D.; Liu, H. Digital twin-enabled built environment sensing and monitoring through semantic enrichment of BIM with SensorML. *Autom. Constr.* **2022**, *144*, 104625. [CrossRef]
18. Spudys, P.; Afxentiou, N.; Georgali, P.Z.; Klumbyte, E.; Jurelionis, A.; Fokaides, P. Classifying the operational energy performance of buildings with the use of digital twins. *Energy Build.* **2023**, *290*, 113106.
19. Opoku, D.G.J.; Perera, S.; Osei-Kyei, R.; Rashidi, M.; Bamdad, K.; Famakinwa, T. Digital twin for indoor condition monitoring in living labs: University library case study. *Autom. Constr.* **2024**, *157*, 105188. [CrossRef]
20. Qian, Y.; Leng, J.; Zhou, K.; Liu, Y. How to measure and control indoor air quality based on intelligent digital twin platforms: A case study in China. *Build. Environ.* **2024**, *253*, 111349. [CrossRef]
21. Qian, Y.; Leng, J.; Chun, Q.; Wang, H.; Zhou, K. A year-long field investigation on the spatio-temporal variations of occupant's thermal comfort in Chinese traditional courtyard dwellings. *Build. Environ.* **2023**, *228*, 109836. [CrossRef]
22. Morteza, A.; Yahyaeian, A.A.; Mirzaeibonehkhater, M.; Sadeghi, S.; Mohaimeni, A.; Taheri, S. Deep learning hyperparameter optimization: Application to electricity and heat demand prediction for buildings. *Energy Build.* **2023**, *208*, 113036. [CrossRef]
23. Indoor Environmental Quality (IEQ). Available online: https://www.efficienzaenergetica.enea.it/glossario-efficienza-energetica/lettera-q/qualita-ambientale-degli-interni.html (accessed on 30 October 2024).
24. Seppänen, O.A.; Fisk, W. Some Quantitative Relations between Indoor Environmental Quality and Work Performance or Health. *HVAC&R Res.* **2006**, *12*, 957–973.
25. Rupp, R.F.; Vàsquez, N.G.; Lamberts, R. A review of human thermal comfort in the built environment. *Energy Build.* **2015**, *105*, 178–205. [CrossRef]
26. Seppänen, O.; Fisk, W.J.; Faulkner, D. Cost benefit analysis of the night-time ventilative cooling in office building. In Proceedings of the Healthy Buildings Conference, Singapore, 7–11 December 2003.
27. Zhang, F.; de Dear, R.; Hancock, P. Effects of moderate thermal environments on cognitive performance: A multidisciplinary review. *Appl. Energy* **2019**, *236*, 760–777. [CrossRef]
28. Lan, L.; Tsuzuki, K.; Liu, Y.F.; Lian, Z.W. Thermal environment and sleep quality: A review. *Energy Build.* **2017**, *149*, 101–113. [CrossRef]
29. Lan, L.; Lian, Z. Ten questions concerning thermal environment and sleep quality. *Build. Environ.* **2016**, *99*, 252–259. [CrossRef]
30. Liu, W.; Lian, Z.; Liu, Y. Heart rate variability at different thermal comfort levels. *Eur. J. Appl. Physiol.* **2008**, *103*, 361–366. [CrossRef] [PubMed]
31. Yao, Y.; Lian, Z.; Liu, W.; Jiang, C.; Liu, Y.; Lu, H. Heart rate variation and electroencephalograph—The potential physiological factors for thermal comfort study. *Indoor Air* **2009**, *19*, 93–101. [CrossRef] [PubMed]
32. Wolkoff, P.; Azuma, K.; Carrer, P. Health, work performance, and risk of infection in office-like environments: The role of indoor temperature, air humidity, and ventilation. *Int. J. Hyg. Environ. Health* **2021**, *233*, 113709. [CrossRef] [PubMed]
33. Derby, M.M.; Hamehkasi, M.; Eckels, S.; Hwang, G.M.; Jones, B.; Maghirang, R.; Shulan, D. Update of the scientific evidence for specifying lower limit relative humidity levels for comfort, health, and indoor environmental quality in occupied spaces (RP-1630). *Sci. Technol. Built Environ.* **2017**, *23*, 30–45. [CrossRef]
34. Dimitroulopoulou, S.; Dudzińska, M.R.; Gunnarsen, L.; Hägerhed, L.; Maula, H.; Singh, R.; Toyinbo, O.; Haverinen-Shaughnessy, U. Indoor air quality guidelines from across the world: An appraisal considering energy saving, health, productivity, and comfort. *Environ. Int.* **2023**, *178*, 108127. [CrossRef]
35. de Rubeis, T.; Nardi, I.; Ambrosini, D.; Paoletti, D. Is a self-sufficient building energy efficient? Lesson learned from a case study in Mediterranean climate. *Appl. Energy* **2018**, *218*, 131–145. [CrossRef]
36. Smarra, F.; Jain, A.; de Rubeis, T.; Ambrosini, D.; D'Innocenzo, A.; Mangharam, R. Data-driven model predictive control using random forests for building energy optimization and climate control. *Appl. Energy* **2018**, *226*, 1252–1272. [CrossRef]

37. de Rubeis, T.; Nardi, I.; Ambrosini, D.; Paoletti, D. The restoration of severely damaged churches—Implications and opportunitie on cultural heritage conservation, thermal comfort and energy efficiency. *J. Cult. Herit.* **2020**, *43*, 186–203. [CrossRef]
38. ElArwady, Z.; Kandil, A.; Mohanad Afiffy, M.; Marzouk, M. Modeling indoor thermal comfort in buildings using digital twin an machine learning. *DIBE* **2024**, *19*, 100480. [CrossRef]
39. Rosenfeld, A.H.; Akbari, H.; Bretz, S.; Fishman, B.L.; Kurn, D.M.; Sailor, D.; Taha, H. Mitigation of urban heat islands: Material utility programs, updates. *Energy Build.* **1995**, *22*, 255–265. [CrossRef]
40. Directive (Eu) 2024/1275 of the European Parliament and of the Council of 24 April 2024 on the Energy Performance of Building Available online: https://www.europarl.europa.eu/news/en/press-room/20230310IPR77228/meps-back-plans-for-a-climate neutral-building-sector-by-2050 (accessed on 22 November 2024).
41. Kaewunruen, S.; Rungskunroch, P.; Welsh, J. A Digital-Twin Evaluation of Net Zero Energy Building for Existing Building *Sustainability* **2019**, *11*, 159. [CrossRef]
42. Hong, S.; Jang, E.; Cho, J.; Lee, J.; Rhee, J.H.; Lee, H.; Lee, M.; Cha, S.H.; Koo, C.; Baik, O.M.; et al. A living lab to develop smar home services for the residential welfare of older adults. *Technol. Soc.* **2024**, *77*, 102577. [CrossRef]
43. Tagliabue, L.C.; Cecconi, F.R.; Maltese, S.; Rinaldi, S.; Ciribini, A.L.C.; Flammini, A. Leveraging Digital Twin for Sustainabilit Assessment of an Educational Building. *Sustainability* **2021**, *13*, 480. [CrossRef]
44. Ma, X.; Chen, F.; Wang, Z.; Li, K.; Tian, C. Digital twin model for chiller fault diagnosis based on SSAE and transfer learning *Build. Environ.* **2023**, *243*, 110718. [CrossRef]
45. Weinmann, M.; Jutzi, B.; Hinz, S.; Mallet, C. Semantic point cloud interpretation based on optimal neighborhoods, relevan features and efficient classifiers. *ISPRS J. Photogrammetry Remote Sens.* **2015**, *105*, 286–304. [CrossRef]
46. Roda-Sanchez, L.; Cirillo, F.; Solmaz, G.; Jacobs, T.; Garrido-Hidalgo, C.; Olivares, T. Building a Smart Campus Digital Twir System, Analytics, and Lessons Learned From a Real-World Project. *IEEE Internet Things J.* **2024**, *11*, 4614–4627. [CrossRef]
47. Asif, M.; Naeem, G.; Khalid, M. Digitalization for sustainable buildings: Technologies, applications, potential, and challenges. J *Clean. Prod.* **2024**, 141814. [CrossRef]
48. Cascone, S.; Parisi, G.; Caponetto, R. BIM-Based Strategies for the Revitalization and Automated Management of Buildings: A Case Study. *Sustainability* **2024**, *16*, 6720. [CrossRef]
49. Desogus, G.; Quaquero, E.; Rubiu, G.; Gatto, G.; Perra, C. BIM and IoT Sensors Integration: A Framework for Consumption anc Indoor Conditions Data Monitoring of Existing Buildings. *Sustainability* **2021**, *13*, 4496. [CrossRef]
50. de Rubeis, T.; Ragnoli, M.; Leoni, A.; Ambrosini, D.; Stornelli, V. A Proposal for A Human-in-the-Loop Daylight Contro System—Preliminary Experimental Results. *Energies* **2024**, *17*, 544. [CrossRef]
51. Ragnoli, M.; Esposito, P.; Stornelli, V.; Barile, G.; de Santis, E.; Sciarra, N. A LoRa-based Wireless Sensor Network monitoring system for urban areas subjected to landslide. In Proceedings of the 2023 8th International Conference on Cloud Computing and Internet of Things, Okinawa, Japan, 22–24 September 2023.
52. Nguyen, T.D.; Adhikari, S. The Role of BIM in Integrating Digital Twin in Building Construction: A Literature Review. *Sustainability* **2023**, *15*, 10462. [CrossRef]
53. Cai, J.; Chen, J.; Hu, Y.; Li, S.; He, Q. Digital twin for healthy indoor environment: A vision for the post-pandemic era. *Front. Eng Manag.* **2023**, *10*, 300–318. [CrossRef]
54. Lee, K.S.; Lee, J.J.; Aucremanne, C.; Shah, I.; Ghahramani, A. Towards democratization of digital twins: Design principles for transformation into a human-building interface. *Build. Environ.* **2023**, *244*, 110771. [CrossRef]
55. Learn How Digital Twins Mature over Time. Available online: https://intandem.autodesk.com/about-digital-twins/ (accessed on 11 June 2024).
56. *ISO 7726:1998*; Ergonomics of the Thermal Environment—Instruments for Measuring Physical Quantities. International Organization for Standardization: Geneva, Switzerland, 1998.
57. Ragnoli, M.; Stornelli, V.; Tosto, D.D.; Barile, G.; Leoni, A.; Ferri, G. Flood monitoring: A LoRa based case-study in the city of L'Aquila. In Proceedings of the 17th Conference on Ph.D Research in Microelectronics and Electronics (PRIME), Villasimius, SU Italy, 12–16 June 2022.
58. ESP32. Available online: https://www.espressif.com/en/products/socs/esp32#:~:text=ESP32%20is%20highly-integrated%20with,Hybrid%20Wi-Fi%20&%20Bluetooth%20Chip (accessed on 14 June 2024).
59. AHT20 Integrated Temperature and Humidity Sensor. Available online: https://asairsensors.com/product/aht20-integrated-temperature-and-humidity-sensor/ (accessed on 14 June 2024).
60. Espressif ESP-IDF Github Repository. Available online: https://github.com/espressif/esp-idf (accessed on 20 September 2024).
61. WiFi Connection Manager with Web Captive Portal Github Repository. Available online: https://github.com/tzapu/WiFiManager (accessed on 20 September 2024).
62. Alcaraz, C.; Lopez, J. Digital Twin: A Comprehensive Survey of Security Threats. *IEEE Commun. Surv. Tutor.* **2022**, *24*, 1475–1503. [CrossRef]
63. Morteza, A.; Ilbeigi, M.; Schwed, J. A Blockchain Information Management Framework for Construction Safety. In Proceedings of the ASCE International Conference on Computing in Civil Engineering, Orlando, FL, USA, 12–14 September 2021.
64. Di Biccari, C.; Calcerano, F.; D'Uffizi, F.; Esposito, A.; Campari, M.; Gigliarelli, E. Building information modeling and building performance simulation interoperability: State-of-the-art and trends in current literature. *Adv. Eng. Inform.* **2022**, *54*, 101753. [CrossRef]

5. DECRETO LEGISLATIVO 9 Aprile 2008, n. 81. Available online: https://www.bosettiegatti.eu/info/norme/statali/2008_0081.htm (accessed on 11 June 2024).
6. Lechner, N. *Heating, Cooling, Lighting. Sustainable Design Methods for Architects*, 4th ed.; John Wiley & Sons: Hoboken, NJ, USA, 2015; pp. 67–69.
7. Schrotter, G.; Hürzeler, C. The Digital Twin of the City of Zurich for Urban Planning. *PFG* **2020**, *88*, 99–112. [CrossRef]
8. Caprari, G.; Castelli, G.; Montuori, M.; Camardelli, M.; Malvezzi, R. Digital Twin for Urban Planning in the Green Deal Era: A State of the Art and Future Perspectives. *Sustainability* **2022**, *14*, 6263. [CrossRef]
9. Deng, T.; Zhang, K.; Shen, Z.J. A systematic review of a digital twin city: A new pattern of urban governance toward smart cities. *JMSE* **2021**, *6*, 125–134. [CrossRef]
10. Jiang, F.; Ma, L.; Broyd, T.; Chen, W.; Luo, H. Digital twin enabled sustainable urban road planning. *SCS* **2022**, *78*, 103645. [CrossRef]
11. Halimi, Z.; Bavafa, A.; Cui, Q. Barriers to Community Connectivity: An Assessment of Reconnecting Communities Pilot Program. In Proceedings of the ICTD, Atlanta, Georgia, 15–18 June 2024.
12. Afanasieva, I.; Golian, N.; Hnatenko, O.; Daniiel, Y.; Onyshchenko, K. Data exchange model in the Internet of Things concept. *Telecommun. Radio Eng.* **2019**, *10*, 869–878. [CrossRef]
13. Qiang, G.; Tang, S.; Hao, J.; Di Sarno, L.; Wu, G.; Ren, S. Building automation systems for energy and comfort management in green buildings: A critical review and future directions. *Renew. Sustain. Energy Rev.* **2023**, *179*, 113301. [CrossRef]

sustainability

MDPI

Article

Adaptive Grey-Box Modelling for Energy-Efficient Building Retrofits: Case Studies in Denmark

Yujie Yang and Muhyiddine Jradi *

Center for Energy Informatics, The Maersk Mc-Kinney Moller Institute, University of Southern Denmark, Campusvej 55, 5230 Odense, Denmark
* Correspondence: mjr@mmmi.sdu.dk

Abstract: Optimizing energy efficiency in existing buildings can yield substantial savings, though collecting the necessary data for energy modelling often poses challenges. This study developed a flexible, room-level framework for evaluating retrofit strategies using simplified energy models. The approach, based on the RC model, estimated parameters from readily available data such as solar radiation, indoor and outdoor temperatures, and heating system characteristics. The model was validated through case studies of an office and a daycare room in Denmark, guiding energy retrofit decisions. Results showed that adding roof insulation provided greater energy savings compared to wall insulation. A multi-objective optimization was employed to balance energy efficiency and thermal comfort, achieving a 6.58% reduction in energy demand during January while maintaining occupant comfort for 744 h. This framework not only facilitates building–energy retrofitting but also supports the development of digital twins and operational optimization, improving both energy performance and indoor environmental quality.

Keywords: grey-box model; model identification; building–energy optimization; thermal comfort; digital twin

1. Introduction

Growing concerns about the use of fossil energy and its environmental implications are intensifying due to the increasing threat of global warming and climate change. This has increased awareness of the relationship between economic growth, energy consumption, and environmental pollution [1]. Therefore, efforts to reduce the use of fossil energy are essential. In the energy sector, buildings account for about 30% of global total energy consumption, and in Europe, 30% of energy is used in buildings, which contribute to over 37% of CO_2 emissions [2]. Therefore, addressing building–energy issues and improving energy efficiency represents a promising strategy to alleviate the worsening energy challenges.

A digital twin for buildings is a virtual replica of a physical building, created using real-time data and advanced simulation techniques [3]. This digital model mirrors the building's structure, systems, and environment, allowing for continuous monitoring, analysis, and optimization [4]. It enables stakeholders to simulate scenarios, predict outcomes, and make data-driven decisions to improve building performance, energy efficiency, and occupant comfort [5]. This initiative not only helps reduce operational and maintenance costs by lowering energy consumption but also aids in reducing CO_2 emissions from the building sector [6].

To implement digital twins for buildings, building–energy models (BEM) are essential. These models, with the aid of historical and real-time data, are able to simulate predicted

https://doi.org/10.3390/su1704170

loads and estimate potential cost savings, enabling continuous monitoring, analysis, and optimization of building performance [7]. Generally, BEM can be categorized into three types. The first type is the building physical energy model, also known as the white-box model. This model relies on detailed building parameters and heat balance equations. Common tools used for simulating building physical energy include EnergyPlus [8] and TRNSYS [9]. However, constructing a white-box model requires obtaining parameters for the building's structure, systems, and equipment from their physical characteristics [10]. These data are typically sourced from design plans, manufacturer catalogues, or on-site measurements, which can sometimes be difficult to acquire. To overcome this difficulty, the second model, known as the 'black box', utilizes statistical methods to directly capture the correlation between building–energy consumption and operational data [11]. This model predicts the building's energy performance using on-site measurements gathered over a specific period, thus reducing reliance on data from design plans and manufacturer catalogues [12]. However, black-box models lack transparency and provide limited physical insight, and they often require a long training period and are bound to the operating conditions for which they were trained, which can sometimes lead to significant forecasting errors when the training data do not cover the entire forecasting range [13]. The third type of BEM is the grey-box model, which serves as an intermediate approach between the white-box and black-box models. Grey-box models integrate physical principles with stochastic elements to account for unexplained variations in data. These models maintain essential physical meanings while also benefiting from the strengths of purely data-driven approaches. Parameters within grey-box models can be estimated and monitored over time, allowing for the detection of systematic changes that directly correlate with operational status [14]. Consequently, grey-box models generally show better generalization properties compared to black-box and white-box models and offer greater reliability for practical implementations [15].

A large amount of research has been conducted on building grey-box models using building performance data. Brastein et al. [16] constructed a grey-box model using measured data from an experimental wood house, which included indoor air temperature, solar radiation, ambient air temperature, and electric heat. Jacopo et al. [17] evaluates two simplified building models for their accuracy in predicting energy needs and dynamic thermal behaviour. Vivian [18] compared the grey-box model and the black-box model, showing that the grey-box model can significantly reduce the required training data and offer higher reliability under normal operating conditions. Wang et al. [19] proposed an effective method to obtain a reliable grey-box model for houses. Arroyo et al. [20] compared grey-box models of different orders and used them for predictive controllers.

In recent decades, numerous grey-box models with varying structures and complexities have been developed for different applications [21]. This study advances the field by introducing a streamlined yet robust grey-box energy model, identified using readily measurable indoor, outdoor, and energy system data, to support digital twin-based energy retrofits. The originality of this work lies in the development of a flexible and scalable grey-box modelling framework that effectively balances physical principles with empirical data, ensuring practical applicability while maintaining predictive accuracy.

A key contribution of this study is its focus on real-world feasibility, as the proposed model relies on commonly available environmental and heating system data. Unlike traditional white-box models, which require extensive input data and computational resources [22], this grey-box approach offers a more accessible and scalable solution for evaluating energy retrofits. Additionally, the integration of the grey-box model with digital twins enables a dynamic and continuously adaptable framework [14], ensuring that the model remains relevant as new data becomes available.

Beyond energy modelling, this study introduces a multi-objective optimization process that systematically balances energy efficiency and thermal comfort—a critical aspect often overlooked in energy retrofitting studies. By applying this approach to real-world cases, including measured data from Aalborg University and a validated EnergyPlus model of a daycare facility in Aarhus, the model demonstrates its adaptability across different building types and operational conditions. Results highlight the impact of envelope optimization, with findings aligning with previous research showing that roof-insulation improvements can achieve energy savings of up to 43.4% in Danish conditions.

The added value of this work extends beyond conventional energy modelling by bridging grey-box methodologies with digital twin technology, supporting faster and more efficient retrofitting strategies. This integration contributes to the advancement of flexible digital twins in practical engineering, offering a scalable and data-driven approach to long-term building–energy management [23]. By simplifying retrofit evaluations while ensuring high accuracy, this study addresses critical challenges in building performance optimization, enhancing operational efficiency, and accelerating the adoption of digital twin-based solutions in the built environment.

This paper is organized as follows: Section 2 introduces the models, identification methods, and details of the case studies, a Danish office building, and a daycare centre. Section 3 focuses on validating the models for these buildings. In Section 4, both negative and positive retrofit strategies are examined for their effects on energy consumption and thermal comfort, using the identified building–energy models. Section 5 presents the main conclusions and discusses the key limitations of this research.

2. Methodology

2.1. *Building the Grey-Box Model*

2.1.1. Basic Structure of Grey-Box Model

Overall, when using a grey-box model for modelling a certain system, some parameters are fully understood and known, while others are unknown. The basic structure of the grey-box model can be written as follows:

$$dx_t = f(x_t, u_t, t, \theta)dt + \sigma_t(\theta)d\omega_t \quad (1)$$

The first part of the model, $f(x_t, u_t, t, \theta)dt$, is usually built by parameters with certain physical meanings. The diffusion term, $\sigma_t(\theta)d\omega_t$, differentiates the stochastic calculus from white-box model which is used to predict the error of the system that is not measured. $\sigma_t(\theta)$ represents the associated standard deviation and $\{\omega_t\}$ refers to n-dimensional Standard Weiner processes.

2.1.2. Basic RC Model of a Room

The energy flow for a single room with a façade connected to the outdoor environment is illustrated in Figure 1a. By drawing an analogy with thermal resistors and capacitors in electric circuits, the building's thermal dynamics can be simplified to a thermal RC model. Several assumptions are made to simplify the BEM:

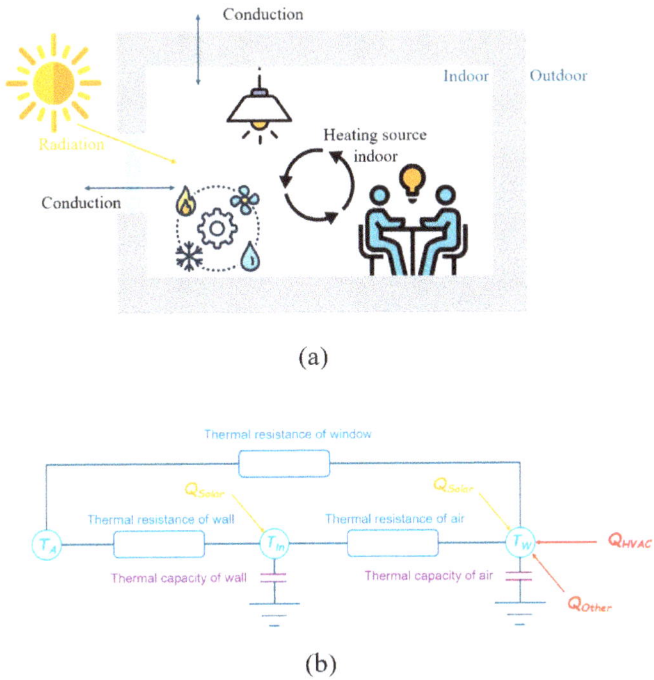

Figure 1. (**a**) The Schematic of Energy flow of the room and (**b**) Schematic of 2R2C model.

1. All rooms are assumed to be at a similar temperature level, meaning the room exchanges energy only with the exterior environment rather than adjoining rooms;
2. The thermal–physical properties of the envelope are considered uniform, with only one temperature;
3. The indoor air is also considered uniform, with only one temperature.

Based on these assumptions, the indoor air, the wall, and the outdoor air are simplified as temperature nodes. Consequently, the thermal performance of the single room can be represented by the 2R2C model, as shown in Figure 1b. In this model, the resistances and capacitances in the network are referred to as equivalent thermal parameters.

Considering the 2R2C scheme of modelling, the energy balance for the room can be represented by the following equations:

$$C_{in.air}\frac{dT_{in.air}}{d\tau} = \frac{T_{out.air} - T_{in.air}}{R_{window}} + \frac{T_{wall} - T_{in.air}}{R_{in.air}} + q_{heating} + q_{other} + A_{window}I_{solar} \tag{2}$$

$$C_{wall}\frac{dT_{wall}}{d\tau} = \frac{T_{out.air} - T_{wall}}{R_{wall}} + \frac{T_{in.air} - T_{wall}}{R_{in.air}} + A_{wall}I_{solar} \tag{3}$$

where $C_{in.air}$ (J/°C) is the thermal capacity of the indoor environment, mainly consisting of an air zone; C_{wall} (J/°C) is the thermal capacity of the external wall; R_{window} (°C/W), $R_{in.air}$ (°C/W), and R_{wall} (°C/W) are the thermal resistances of the window, air, and wall, respectively; $T_{in.air}$ (°C), $T_{out.air}$ (°C), and T_{wall} (°C) are the temperatures of the indoor air, outdoor air, and wall, respectively; I_{solar} (J/m^2) is the solar radiation; A_{window} (m^2) and A_{wall} (m^2) are the effective areas of the window and wall, respectively; $q_{heating}$ (J/m^2) is the energy delivered by the heating system, and q_{other} (J/m^2) is the unmeasured energy in the room.

Therefore, for the building, if the diffusion terms are used to represent the unmeasured energy flow, the grey-box model for a single room can be then written as follows:

$$C_{in.air}\frac{dT_{in.air}}{d\tau} = \frac{T_{out.air} - T_{in.air}}{R_{window}} + \frac{T_{wall} - T_{in.air}}{R_{in.air}} + q_{heating} + A_{window}I_{solar} + \sigma_{in}d\omega_{in} \quad (4)$$

$$C_{wall}\frac{dT_{wall}}{d\tau} = \frac{T_{out.air} - T_{wall}}{R_{wall}} + \frac{T_{in.air} - T_{wall}}{R_{in.air}} + A_{wall}I_{solar} + \sigma_{wall}d\omega_{wall} \quad (5)$$

However, in practical measurements of the building, for most of the cases, the heating energy data from a single heating radiator are not measured. Other available and more commonly employed measurements include temperature, water flow, or valve opening, which is closely related to the flow. Therefore, in this study, a space heater model is also used to represent the heating energy delivered and connect the system with the space model.

Based on the energy balance of the space heater, its respective energy model can be written as follows:

$$C_{heater}\frac{dT_r}{d\tau} = c_{p.water}\dot{m}_{water}(T_{water.in} - T_{water.out}) - UA(T_r - T_{in.air}) \quad (6)$$

where C_{heater} (J/°C) is the thermal capacity of the heater; T_r (°C) is the effective temperature of the heater; $T_{water.in}$ (°C) and $T_{water.out}$ (°C) refer to the inlet and outlet temperatures of the heater, respectively; $c_{p.water}$ (J/kg °C) is the specific heat capacity of water; $\dot{m}_{water}$ (kg/s) is the mass flow rate of water; and UA (W/°C) is the heat transfer coefficient. For the heater, the energy delivered to the indoor space environment can be represented by the following:

$$q_{heating} = UA(T_r - T_{in.air}) \quad (7)$$

Similarly, the diffusion term is added to Equation (6) to obtain the following representation:

$$C_{heater}\frac{dT_r}{d\tau} = c_{p.water}\dot{m}_{water}(T_{water.in} - T_{water.out}) - UA(T_r - T_{in.air}) + \sigma_r d\omega_r \quad (8)$$

Overall, for the presented energy model of the room, the known parameters are indoor and outdoor temperature, solar radiation, inlet temperature, and the mass flow of the heater. The parameters to be identified are the thermal resistance and the thermal capacity.

2.1.3. Parameter Estimation Method

In this study, the maximum likelihood estimation (MLE) is adopted to estimate the unknown parameters in the model. For the observed data sequence $Y_n = (y_1, y_0, \ldots, y_{n-1}, y_n)$, the MLE first builds the likelihood function related to the observed data sequence Y_n and the unknown parameter θ from a particular model structure as follows:

$$L(\theta; Y_n) = p(y_0|\theta)\left(\prod_{k=1}^{n} p(y_k|Y_{k-1}, \theta)\right) \quad (9)$$

where p refers to the probability density function.

The log-likelihood function is commonly used to facilitate numerical calculations for the likelihood function, as shown below:

$$\updownarrow(\theta; Y_n) = \log L(\theta, Y_n) \quad (10)$$

The parameter estimates can be determined by solving the following optimization problem:

$$\hat{\theta} = \underset{\theta}{argmin} - \ell(\theta; Y_n) \tag{11}$$

The detailed mathematical formulation of the algorithms is presented in [24]. In this study, the grey-box model for the single room consists of Equations (4), (5), and (8), where the observation data are measured data from the room and the estimated parameters are the corresponding thermal properties of the considered equations. The parameter estimation process is carried out in R [25], employing the CTSM package [26], a tool designed for estimating embedded parameters in continuous-time stochastic state–space models. This package has been developed over many years, with its accuracy validated in various fields including building systems, refrigeration, and heat pumps [14,27,28].

2.1.4. Case Study Description

In this study, two real case study buildings are considered to aid the investigation. First, the proposed energy model is developed and validated, considering an office building case study. Then, the validated model is implemented, and the energy retrofit measures are evaluated, considering a daycare centre.

The first case considered is a university office building in Denmark. For this case, Melgaard et al. [29] collected operational and indoor comfort data from six office rooms in the building shown in Figure 2a. The dataset consists of measurements of occupancy, indoor environmental quality, room-level, system-level heating, ventilation, and lighting, with a resolution of 5 min. A detailed description of the building can be found in the literature [30]. For the model development and validation process, one room in the case study building is chosen, and its floor plan can be seen in Figure 2b.

(a) (b)

Figure 2. (**a**) View of the case study office building and (**b**) floor plan of the considered room [29].

This building is designed according to the Danish building regulation BR class 2015 [30], the U value of the walls and windows is designed to be less than 0.15 W/(m^2·K) and 1.2 W/(m^2·K) [31].

The second case study considered in this investigation is a daycare centre which is also located in Denmark. For this case, data from a validated white-box full-scale building–energy model is considered [32]. The building's geometric shape is shown in Figure 3, where the room marked by the blue box is the room used for model implementation and evaluation.

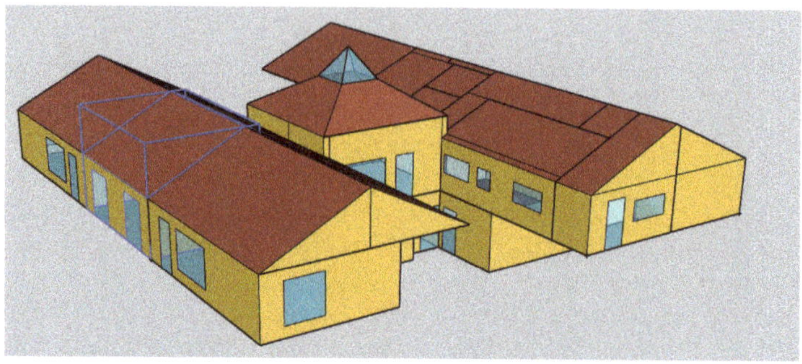

Figure 3. The geometric modelling of daycare in Denmark [32].

2.1.5. Energy Efficiency Optimization of Buildings

Energy efficiency in buildings primarily depends on a building's thermal properties [31] and the efficiency of its internal equipment [33]. Therefore, to optimize the energy efficiency of existing buildings, this study discusses passive strategies aimed at enhancing the thermal properties of the building, as well as active strategies focused on optimizing its systems, particularly the heating system in this work.

It is noted that the energy retrofit analysis and operational strategies implementation will be demonstrated in the daycare centre case, where the building was built in the 1970s, with an outdated building physical envelope and conventional energy systems operation mode.

2.1.6. Optimization of Building Envelope

The structure and properties of the external walls and roof of the daycare room are detailed in Tables 1 and 2. The external walls are composed of four material layers, while the roof consists of three layers. Additionally, the U-value of the roof is slightly smaller than that of the external wall.

Table 1. The structure of the external wall and its properties.

	Thermal Conductivity (W/(m·K))	Thickness (m)
Brick	1.02	0.1016
Wall insulation	0.049	0.08
Mortar	0.44	0.12
Gypsum	0.25	0.013

Table 2. The structure of the roof and its properties.

	Thermal Conductivity (W/(m·K))	Thickness (m)
Slate	1.59	0.0127
Roof insulation	0.048	0.09
Mortar	0.44	0.12

The most basic method to alter the thermal properties of the wall and roof is by adding insulation. For this study, insulation material with a thermal conductivity of $0.034\ \mathrm{W/(m^2{\cdot}K)}$ [34] is applied to the external surface of the building to enhance its insulation performance. As a result, the thermal resistance of the building envelope R_{env} can be calculated as follows:

$$R_{env} = \sum(R_i) \tag{12}$$

where R_i (°C/W) is the thermal resistance of the layer i of the building envelope. Therefore, the increased thermal resistance resulting from the addition of insulation, with a thickness ranging from 0 to 200 mm in this study [35], can be directly incorporated into the grey-box model presented in the previous section to observe its impact on the building's energy performance.

Windows and doors also are surfaces that connect the indoor environment to the outdoor environment. To enhance the thermal properties of these elements, the general approach is to replace them with components that have better thermal properties. For the daycare room, double-glazed structures are already in use. Therefore, we aim to optimize their thermal properties by replacing them with triple-glazed structures.

2.1.7. Optimization of Heating Systems

While adding insulation to the building envelope and replacing windows/doors is seen as a passive retrofitting approach, this study also considers the implementation of an active retrofitting measure, primarily considering the heating system. The aim is to reduce the heating energy demand by optimizing the operation of its heating system, specifically its respective control strategy. In the research of Baghoolizadeh et al. [36], the setpoints are set at 17 °C for non-working hours and 20 °C for working hours. To enhance energy-saving efficiency, the non-working hour setpoint is lowered by 1 °C to 16 °C. Regarding the setpoint during working hours, the standard suggests maintaining indoor temperatures in accordance with Category II of DS/EN 16798-1 [37]. While the exact temperature range is not specified in the provided excerpts, it is known to the authors that, in response to rising energy costs, the university has adjusted the setpoint to 20 °C. Therefore, 20 °C is chosen based on this practical consideration. Figure 4 presents the currently implemented setpoints for the heating system at the daycare centre case. The setpoints for this building are divided into two categories: working days and days off. During working days, as indicated by the blue pattern in Figure 4, the setpoint is set to 21 °C for working hours and 18 °C for off-hours. On off days, shown in the pink pattern in Figure 4, the setpoint is 16 °C.

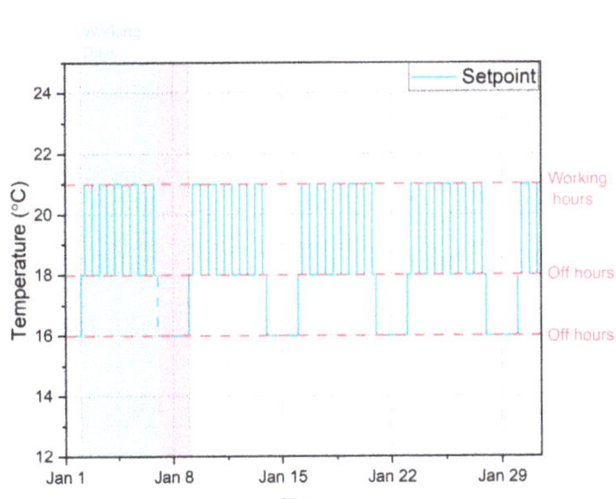

Figure 4. The set point of the heating system for the daycare in January.

Therefore, to reduce the energy required by the building, the basic method is to lower the setpoint. However, a building is not merely a structure that is meant only to consume energy; it primarily serves to provide comfort to the people inside. Therefore, when adjusting the setpoint, the thermal comfort of the indoor environment must be taken into account.

The Predicted Mean Vote (PMV) and the Predicted Percentage Dissatisfied (PPD) are common indices used to measure indoor thermal comfort [38]. The PMV directly shows how people feel about the indoor environment, and the relationship between the PMV and thermal sensation can be seen in Table 3. This parameter can be calculated as follows:

$$\begin{aligned} PMV = {} & \left(0.303e^{-0.036M} + 0.028\right) \\ & \times \big((M - W) - 3.05 \times 10^{-3}(5733 - 6.99(M - W) - P_a) \\ & - 0.42((M - W) - 58.15) - 1.7 \times 10^{-5}M(5867 - P_a) \\ & - 1.4 \times 10^{-3}M(34 - t_a) \\ & - 3.96 \times 10^{-8} f_{cl}\left((t_{cl} + 237)^4 - (t_r + 273)^4\right) \\ & - f_{cl}h_c(t_{cl} - t_a)\big) \end{aligned} \tag{13}$$

where M (W/m^2) is the metabolic rate, with a value of 60 W/m^2; W (W/m^2) is the effective mechanical power, with a value of 0 W/m^2; f_{cl} (−) is the ratio of the surface area of the body with clothes to the surface area of the body without clothes, with a value of 1.1; h_c ($W/m^2{\cdot}K$) is the convective heat transfer coefficient, with a value of 3 $W/m^2{\cdot}K$; t_{cl} (°C) is the clothing surface temperature, with a value of 25 °C; t_a (°C) is the air temperature; t_r (°C) is the mean radiant temperature; and P_a (Pa) is the water vapour partial pressure. More details of the parameters calculation can be found in [39].

Table 3. The qualitative and quantitative statements of thermal sense.

PMV	−3	−2	−1	0	1	2	3
Thermal sense	Cold	Coo1	Slightly cool	Neutral	Slightly warm	Warm	Hot

PPD (%) is an index related to the PWV and can be calculated as follows:

$$PPD = 100 - 95exp\left(-0.03353PMV^4 - 0.2179PMV^2\right) \tag{14}$$

when −0.5 < PMV < 0.5, in other words when the PPD < 10%, the indoor environment is deemed to be comfortable. For the hour that satisfies this condition, it is regarded as the comfortable hour for the room.

To reduce energy consumption, one possible approach is to lower the setpoint during different periods. However, lowering the indoor temperature may decrease thermal comfort. Therefore, the challenge becomes how to minimize energy consumption while maximizing thermal comfort in indoor environments, resulting in a multi-objective problem. This study aims to find the optimal combination of setpoints for working hours, off-hours, and holiday hours to minimize energy consumption and maximize indoor thermal comfort. To achieve this, the multi-objective optimization package, pymoo, is utilized [40], and the NSGA-II is applied for the optimization process.

In addition, since people are the primary focus of building services, another method to reduce energy consumption while maintaining indoor thermal comfort is to implement a control based on the occupancy sensing system (PIR sensors). Unlike rule-based control, where indoor temperatures are regulated to a scheduled setpoint [41], the control based on a sensing system ensures that the indoor temperature is maintained at a specific level based on the presence of people indoors [42]. Although this control method requires additional indoor sensors to monitor the presence of people, it has the potential to minimize the building's energy demand while keeping occupants comfortable.

3. Model Validation

3.1. Unconditional Predictions

After model development and establishment, an unconditional prediction simulation was conducted to validate the model and to ensure that it can predict the actual performance with an acceptable margin of uncertainty. Figure 5 shows a comparison between the measured indoor temperatures and those predicted unconditionally for the two case study buildings, the office building, and the daycare centre. The area with a pink background indicates the training period of the model. For the room of the office building, 6 days of measured data were used for model identification, while for the daycare room, data for the full month of January were used. For the room in an office building, the actual data collected on-site is used for comparison, while for the daycare centre room, data from the validated EnergyPlus model is used. Based on the comparison, it is noted that the model predicts the indoor temperature with an acceptable margin of uncertainty, with RMSE of 0.61 °C and 1.45 °C for the office room and daycare room. Figure 6 illustrates the correlation between the observed and simulated indoor temperatures using the grey-box model for the rooms in the office and the daycare centre. The Pearson correlation coefficients for the two buildings are 0.53 and 0.91, respectively. Considering that the data from the office are measured data, there is significantly more uncertainty in the dataset; therefore, a lower correlation coefficient is acceptable in this study.

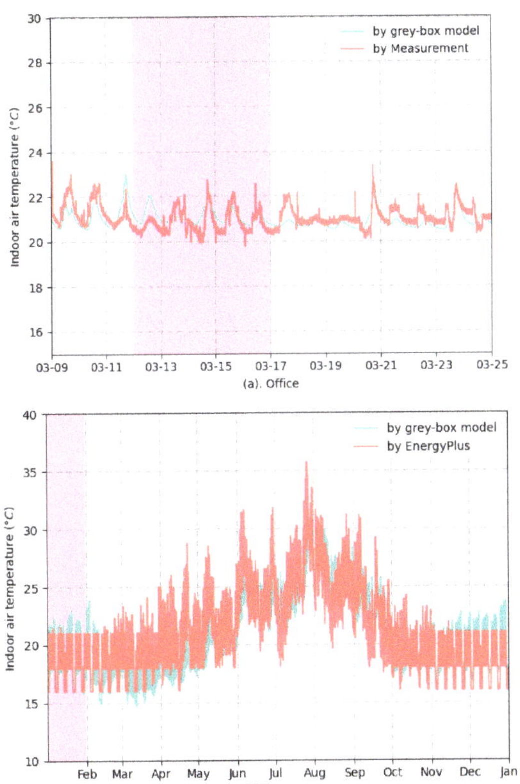

Figure 5. Comparisons between the measured indoor temperatures and those predicted unconditionally are shown for (**a**) the office room and (**b**) the daycare center room, with a pink area indicating that data from the period were used for model identification.

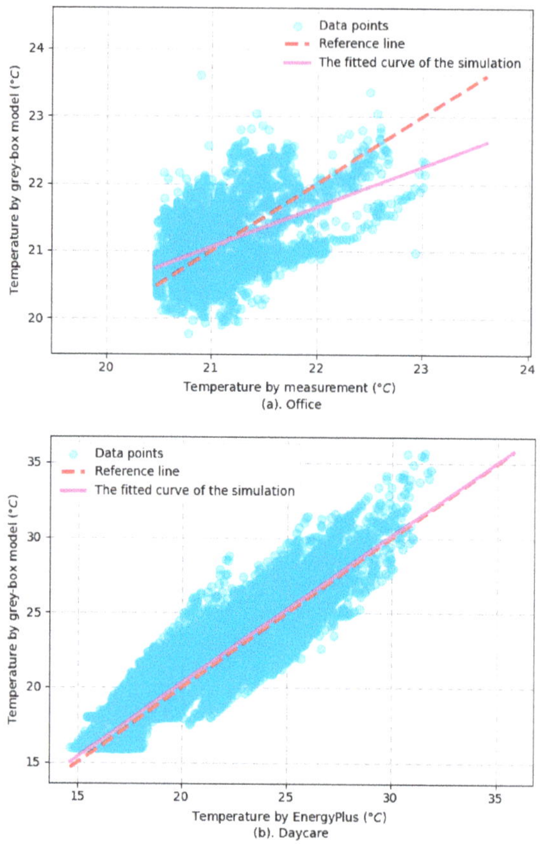

Figure 6. The correlation between the observed indoor temperatures and simulated indoor temperatures in (**a**) office room and (**b**) daycare centre room.

3.2. Residual Analysis

The autocorrelation function (ACF) quantifies the relationship between data points in a time series and their historical values [43]. In the context of stochastic differential equations, the residuals of the model estimation should exhibit no correlation, confirming that the model accounts for all system autocorrelations [14]. Figure 7 illustrates the ACF and the cumulative periodogram for the grey-box model at the office. As depicted in Figure 7, the majority of autocorrelations fall within the 5% significance limits, and the cumulative periodogram remains within the confidence interval. Thus, the residuals from the office model can be classified as white noise, demonstrating that the model effectively captures the building's heat dynamics.

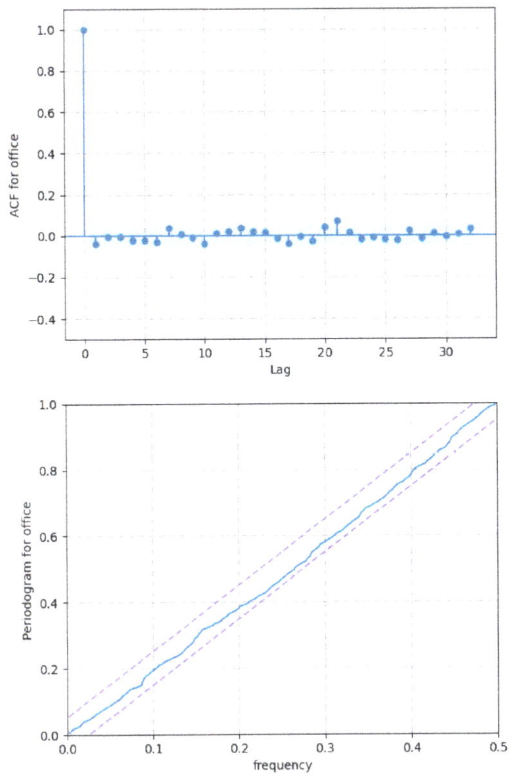

Figure 7. ACF for the office room and its cumulated periodogram.

Furthermore, Figure 8 illustrates the ACF and cumulative periodogram for the residuals of the daycare model. While there are some minor deviations in the ACF, the overall trend within the significance boundaries and the apparent periodicity in the cumulative periodogram suggest that these residuals can be considered white noise. Consequently, the residuals can also be considered as white noise.

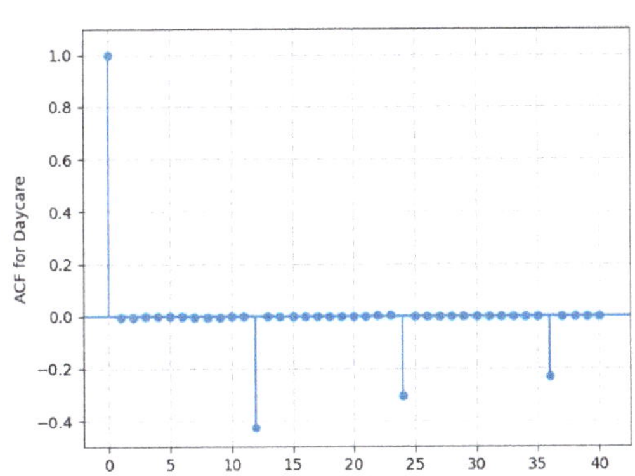

Figure 8. *Cont.*

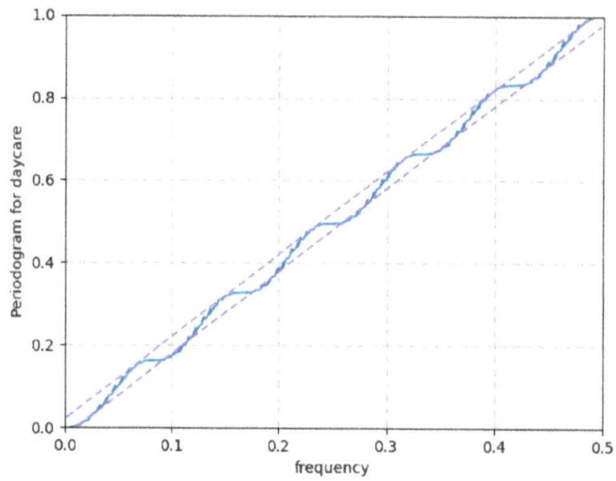

Figure 8. ACF for the daycare centre room and its cumulated periodogram.

3.3. Comparison of the Heating Energy Demand

In addition to predicting the indoor temperature in space, the model is designed to predict the respective heating energy demand. Therefore, the energy supply by the heating radiators to maintain the heating setpoint is compared between the grey-box model and observed data to verify the model's accuracy. In this section, the heating energy demand predicted by the validated white-box model of the daycare centre is used as a baseline to compare the predictions of the grey-box model developed.

Figure 9a shows the monthly energy demand as calculated by both EnergyPlus and the developed grey-box model in this work. It is observed that the heating energy demands calculated by both models are comparable on a monthly basis. On the annual scale, EnergyPlus estimates the heating energy demand at 2679.81 kWh, whereas it is 2638.36 kWh as predicted by the grey-box model, resulting in a deviation of 1.55%. Additionally, the heating demand for the daycare room in January by these two models can also be seen in Figure 9b. It can be observed that the hourly energy demand predicted by our model and EnergyPlus are at a similar level. Consequently, given the validation results discussed in this section, the grey-box model will be employed for subsequent energy retrofitting and optimization measures implementation in the following parts.

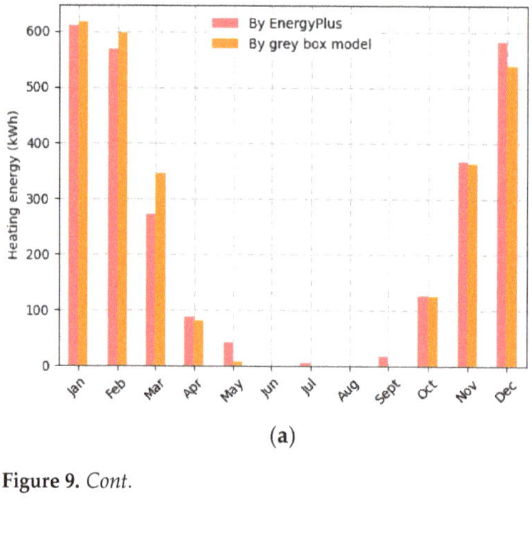

(a)

Figure 9. *Cont.*

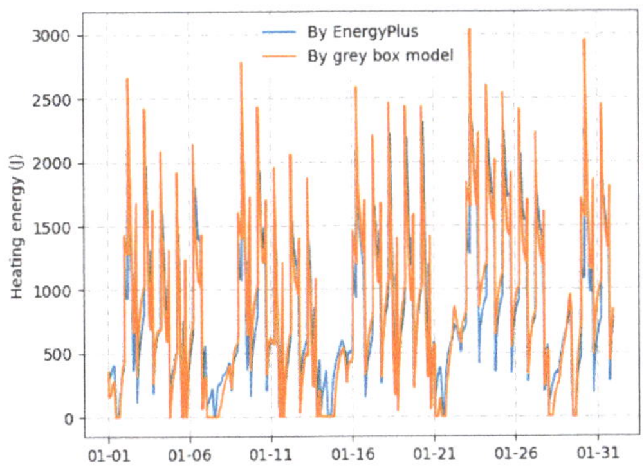

(b)

Figure 9. (**a**) Monthly heating energy demand for the daycare centre and (**b**) heating energy demand for the daycare room over January.

4. Results and Discussion

4.1. Retrofitting of Building Envelope of the Daycare Centre

The room which is the object of this investigation in the daycare centre interfaces with the outdoor environment via three distinct envelop components: the external wall, the roof, and the glass windows/doors. Consequently, methods for retrofitting and enhancing the thermal properties of these surfaces will be discussed in this section with the aid of the developed grey-box model, aiming to improve the building's energy performance.

For the external surfaces, enhancing thermal properties can be most effectively achieved by adding insulation. Figure 10 presents the relationship between the thickness of added insulation on the external surface and the annual energy demand of the room. As expected, the annual heat demand of the building decreases with an increase in insulation thickness; and the rate of reduction diminishes as the insulation becomes thicker. This trend can be attributed to the suboptimal thermal properties of the external walls and roof of the daycare room, which have U-values of 0.49 W/(m^2·K) and 0.38 W/(m^2·K), respectively. Additionally, given that the U-value is inversely proportional to thermal resistance and that thermal resistances combine in series according to the principle of accumulation and the proposed concept of grey-box model development, the increase in insulation thickness results in a decrease in the rate of annual heating energy demand.

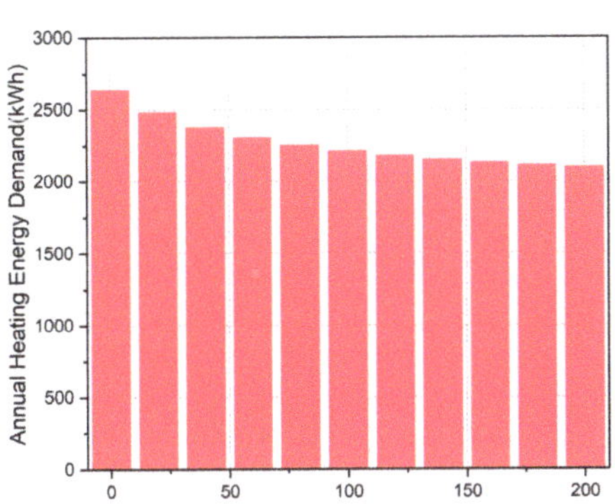

(a)

Figure 10. *Cont.*

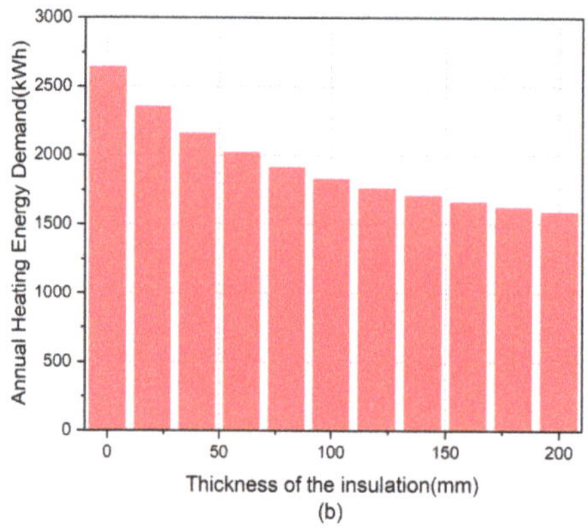

Figure 10. Relationship between the (**a**) wall insulation thickness and (**b**) roof-insulation thickness and the annual heating energy demand of the daycare room.

Additionally, since the roof area is larger than that of the external walls, applying insulation of the same thickness results in greater energy savings for the roof compared to the walls. As shown in Figure 10, increasing the insulation thickness from 0 to 200 mm leads to heating energy savings of 20.4% in the wall insulation scenario and 39.6% in the roof-insulation scenario. This aligns with previous research conducted by Zhou et al. [44], which reported that adding insulation to the building envelope can achieve energy savings of up to 43.4%. Therefore, the energy savings estimated by the grey-box model can be considered reasonable.

Changing the glass windows to those with better thermal performance is another way to enhance a building's thermal efficiency. Table 4 displays the U-values for the current doors and windows (double-glazed) compared to the newly fitted proposed doors and windows (triple-glazed) for the considered space. The U-value of the triple-glazed windows and doors is almost half that of the original, which also meets the Danish building standards [31]. After applying the triple-glazed structure, the heating energy demand of the room is reduced to 2414.75 kWh.

Table 4. The U value for double-glazed and triple-glazed and the corresponding heating energy demand.

	U Value (W/(m^2·K))	Annual Heating Energy Demand (kWh)
Double-glazed	3.48	2638.36
Triple-glazed	1.76	2414.75

4.2. Heating System Management

4.2.1. Heating Setpoint Control

Since lowering the setpoint results in a direct reduction in the building's heating demand, Figure 11 illustrates the monthly heating energy demand by decreasing the setpoint during working and off-hours on working days. It can be observed that the monthly heating energy demand decreases in the month when the setpoint is reduced during these periods. Throughout the year, adjusting the setpoint during working hours results in slightly more energy savings compared to adjustments during off-hours on working days. By reducing the setpoint by 1 °C, the annual energy demand for working hours is reduced by 5.80%, while for the same temperature reduction during off-hours, the annual energy demand is reduced by 5.27%.

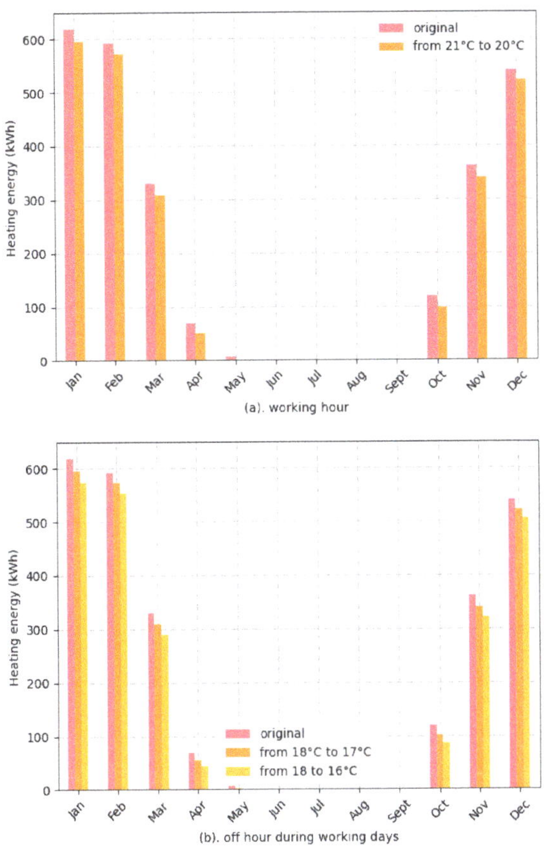

Figure 11. The monthly heating energy demand for the original case and the cases lower the setpoint in (**a**) working hours and (**b**) off-hours during the working days.

To figure out why this difference appears, Figure 12 shows the heating energy demand of the daycare room for the first day of January at different setpoints. The optimized results are compared to the results from the validated model [32]. The peak energy demand for this room occurs at 7:00, due to the change in the setpoint from off-hour to working-hour. After that, due to the increase in outdoor air temperature and solar radiation, the heating energy demand decreases, even though the setpoint during working hours is higher than during off-hours. By reducing the heating setpoint during working hours, the heating energy demand for the room is lowered. However, it should be noticed that even though the heating setpoint for off-hours does not change, the heating energy demand increases slightly during off-hours. In addition, when changing the heating setpoint during off-hours, decreasing the setpoint decreases the heating energy demand during off-hours. However, to improve the indoor air temperature to the working-hour setpoint, the energy demand is greater when the heating setpoint changes, and after that, the heating energy demand is at a similar level to the original heating setpoint case.

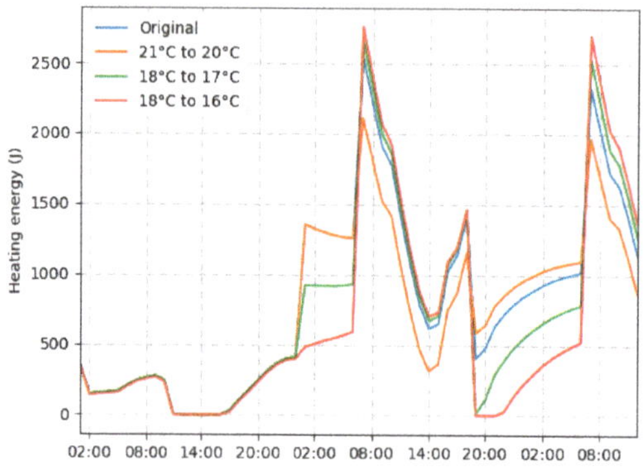

Figure 12. The heating energy demand of the daycare room for the first day of January by different setpoints.

Figure 13 shows the distribution of the resulting PMV for each month when changing the setpoint during working hours and off-hours on working days. As observed in Figure 13a,b, PMV increases and then decreases over the course of a year, with the PMV during July, keeping an average value of 0.38, indicating that the building remains quite comfortable during the summer hours. In addition, altering the setpoint during working days does not affect the PMV during June, July, August, and September in a significant way. The boxplots for the PMV during these four months remain unchanged, which is attributed to the absence of heating demand shown in Figure 11. However, lowering the setpoints will decrease indoor thermal comfort during the heating season, as evidenced by the reduced average values of the PMV. Furthermore, it is noticeable that changing the setpoint during working hours primarily affects the upper quartile, while changes during off-hours impact the lower quartile—narrowing the box for the former and widening it for the latter.

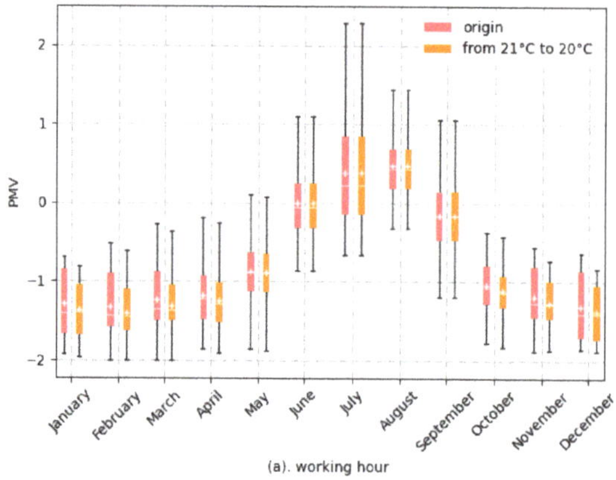

Figure 13. *Cont.*

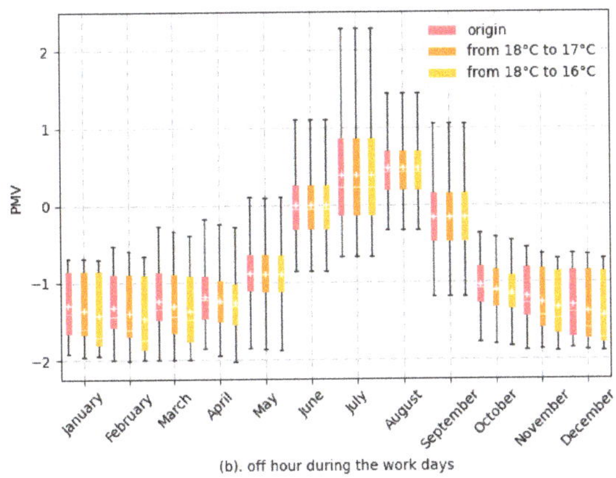

Figure 13. The monthly statistical PMV for the adjustments of the setpoint in (**a**) working hours and (**b**) off-hours during the working days.

Figure 14 is also shown here to explain this. It illustrates the PMV in the daycare room over January and July, representing the winter and summer months, respectively. For winter, changing the heating setpoint from 21 °C to 20 °C during working hours decreases the PMV by 0.2 during working hours, leading to the thermal comfort in the room being slightly cool according to Table 3. Additionally, this change has a minimal impact on thermal comfort during off-hours. Conversely, changing the heating setpoint during off-hours has little impact on the PMV during working hours but decreases the PMV during off-hours. However, the thermal comfort during the off-hours remains slightly cool and cool, according to Table 3. Overall, what is shown in Figure 14a leads to the change in the boxplot in Figure 10 for January. However, it can be noted from Figure 14b that the PMV remains at a similar level during the summer. Therefore, the boxplot in Figure 13 for July remains almost the same.

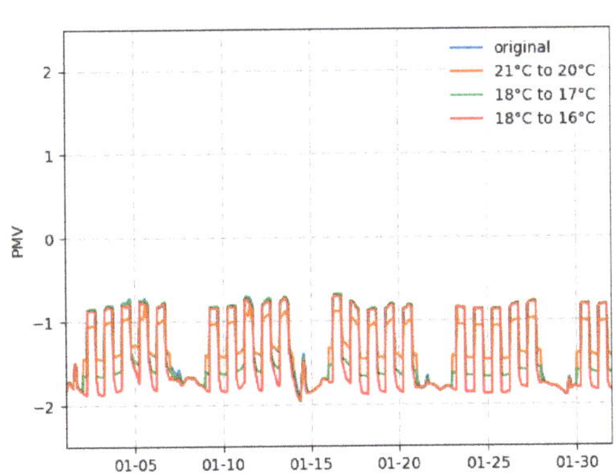

Figure 14. *Cont.*

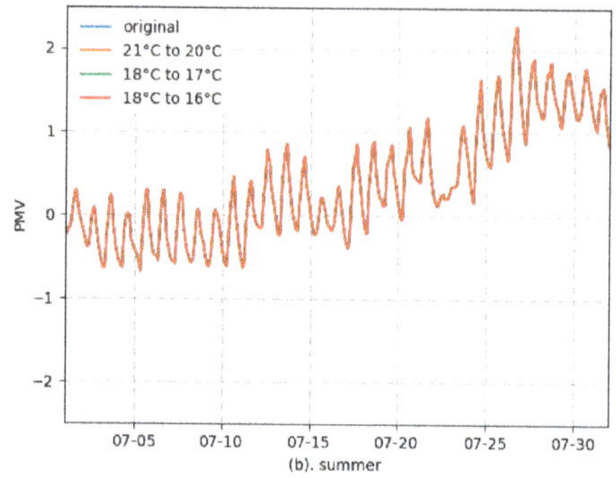

Figure 14. The PMV in the daycare room over one month: (**a**) winter and (**b**) summer.

The PPD for the two types of setpoint change schedules can be seen in Figure 15. An indoor environment with a PPD lower than 10% is considered comfortable [45]. For this building, the room will only be comfortable during the summer. Therefore, in this study, we slightly broaden the definition of thermal comfort by setting the absolute value of PMV smaller than 1 as the threshold for thermal comfort, which means that the PPD value should be lower than 26%. Consequently, reducing the setpoint during working hours raises the lower quartile, thereby increasing the amount of uncomfortable time by reducing the comfortable hours. Similarly, lowering the setpoint during off-hours on working days raises the upper quartile, further increasing discomfort by making already uncomfortable hours more uncomfortable.

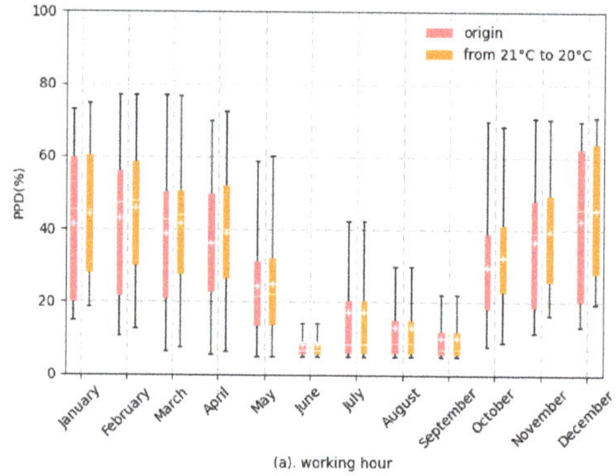

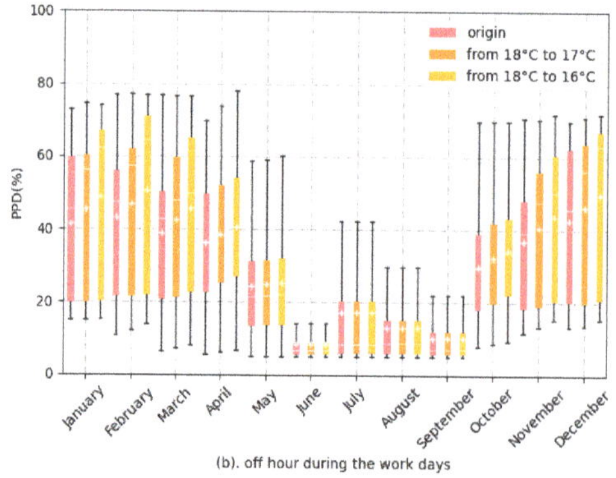

Figure 15. The monthly statistical PPD for the adjustments of the setpoint in (**a**) working hours and (**b**) off-hours during the working days.

4.2.2. Heating System Optimization

Based on Energy Demand and Thermal Comfort

While it is important to reduce the heating energy demand of the building, the implemented optimization measure should ensure proper indoor thermal comfort, leading to a multi-objective optimization problem. To carry out such an optimization process on an annual basis, substantial computing resources are required. To simplify the complexity, data from January is used to conduct the investigation in this section. Implementing the multi-objective optimization framework described in Section 2.1.7. Unlike single-objective optimization, which focuses on optimizing a single criterion, multi-objective optimization involves finding trade-offs among two or more conflicting objectives. This approach results in a set of solutions known as the Pareto Set, where no single solution is universally superior, but each solution represents an optimal balance among the different objectives. The optimal setpoints for minimizing heating energy demand while maximizing indoor thermal comfort are presented in Figure 16. For off-hours on both off days and working days, the setpoint solutions fall within a relatively narrow range, from 16.03 °C to 16.16 °C and from 17.01 °C to 17.06 °C, respectively. The solution for the setpoint during working hours spans a wider range, from 20.00 °C to 21.16 °C.

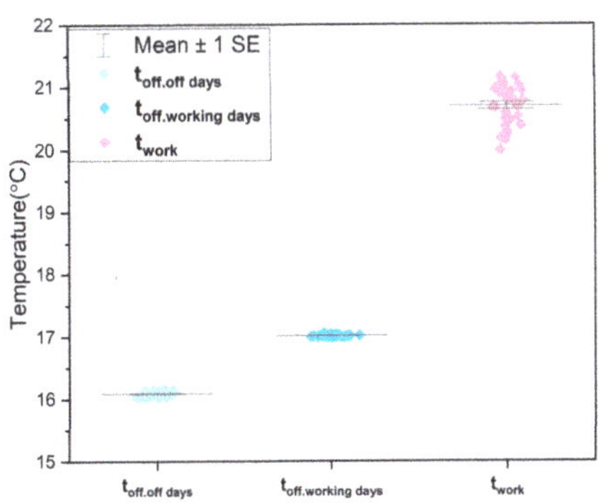

Figure 16. The solutions for the multi-objective optimization for energy consumption and thermal comfort in January.

To make decisions in this multi-objective optimization problem, weights are applied to convert it into a single-objective optimization problem. In this study, equal weights of 0.5 and 0.5 are assigned to energy consumption and thermal comfort. Table 5 displays the heating energy consumption and count of comfortable hours for January, comparing the original case with the optimized case. It can be seen that, in January, optimizing the setpoint saves the heating energy consumption by 4.49% while maintaining an unchanged indoor comfort level. On this basis, optimizing the heating setpoint effectively saves energy and preserves thermal comfort.

Table 5. The heating energy consumption and comfortable hours count of January of the day-care room.

	Heating Energy Demand (kWh)	Comfortable Hours
Original case	617.85	744
Optimization case	590.12	744

Based on an Occupancy Sensing System

Figure 17 compares the indoor air temperatures controlled by the standard rule-based control method implemented currently in the building against the proposed indoor temperature based on an occupancy sensing system over one week in January. It can be observed that for the control method based on the sensing system, the heating system is only activated when the sensing system detects people. Therefore, a slight increase can be observed during the off-hours, as people start arriving at the building at 7:00. As shown in Table 6, the heating energy demand decreased by 6.58%. Similar to the results in the Section Based on Energy Demand and Thermal Comfort, the comfortable hours do not change compared to the original case.

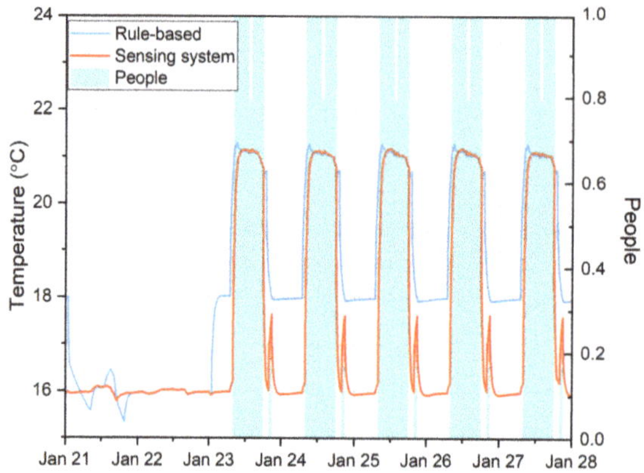

Figure 17. The indoor air temperature based on the rule-based control and occupancy sensing system methods.

Table 6. The heating energy consumption and comfortable hours of January of the daycare room by rule-based control and building–energy sensing system.

	Heating Energy Demand (kWh)	Comfortable Hours
Original case	617.85	744
Optimization case	577.16	744

5. Limitations of the Research

This study presents a grey-box modelling approach for evaluating building–energy retrofits, leveraging readily available environmental and heating system data. While the approach provides a practical and scalable framework for optimization, certain limitations must be acknowledged. The primary limitation lies in the model's scope, which focuses exclusively on the heating system, omitting other critical building systems such as ventilation, cooling, lighting, and occupancy-driven dynamics. By simplifying the representation of energy interactions, the model may not fully capture the complex dependencies between different building components. More comprehensive models, integrating additional building systems, could offer deeper insights into overall energy performance.

Moreover, while the grey-box approach balances physical modelling with empirical data, its reliance on simplified parameterization may lead to reduced accuracy in cases where highly dynamic conditions or intricate heat transfer mechanisms are present [46]. The model assumes uniform thermal properties across surfaces and does not account for detailed spatial variations in heat distribution, which could be significant in larger or more heterogeneous buildings. Another limitation pertains to the data sources used for model validation. The study utilizes measured data from a specific office and daycare building in Denmark, which may not fully generalize to other climate zones, construction types, or occupancy patterns. Expanding the dataset to include diverse building typologies, climatic conditions, and operational scenarios would enhance the robustness and applicability of the framework.

Furthermore, the study does not explicitly incorporate real-time adaptive control strategies. While the model supports digital twin development, its current implementation remains primarily static, relying on historical or periodic data inputs. The integration of real-time sensor networks and advanced machine learning techniques could improve predictive capabilities, allowing for more responsive and adaptive optimization strategies.

Future research should address these limitations by extending the model to encompass a broader range of building systems and operational parameters. Incorporating HVAC dynamics, lighting control, and occupancy-driven behavioural models would enable a more holistic assessment of energy efficiency measures. Additionally, refining the parameter identification process using advanced optimization techniques and diverse datasets could further enhance model accuracy. Exploring reinforcement learning or hybrid AI-driven approaches may also facilitate automated decision-making in digital twin applications, paving the way for continuous performance optimization in real-world buildings. By addressing these challenges, future work can enhance the flexibility, accuracy, and real-time applicability of grey-box models, contributing to more effective and scalable solutions for building–energy retrofits and digital twin integration.

6. Conclusions

Developing an accurate and reliable building–energy model requires comprehensive data to capture both the physical characteristics of the building and its internal activities. However, obtaining detailed and extensive datasets in real-world settings is often challenging due to measurement constraints and data availability issues. To overcome this limitation, the present study employed a grey-box modelling approach that relies on easily measurable environmental and system-related parameters. These include solar radiation, indoor and outdoor air temperatures, as well as critical heating system variables such as inlet and outlet water temperatures and water flow rates. By leveraging such data, the grey-box model offers a balance between computational efficiency and predictive accuracy, making it particularly suitable for energy performance analysis and digital twin applications.

The model was validated using empirical data from a real building at Aalborg University and simulated data from a validated EnergyPlus model of a daycare facility in Aarhus. The results demonstrated that the grey-box model effectively captures key thermal dynamics while remaining adaptable for future refinements. A major advantage of this approach is its modularity—new physical models can be integrated as additional data becomes available, enhancing the model's accuracy and relevance over time. This adaptability is particularly valuable for digital twin frameworks, where continuous updates and real-time data integration are essential for long-term performance monitoring and optimization. The key findings and implications of this study are as follows:

1. Passive Optimization Strategies: Among the evaluated energy efficiency measures, improving the thermal insulation of the building envelope proved to be the most effective passive strategy. Increasing the roof-insulation thickness from 0 to 200 mm led to a 39.6% reduction in heating energy consumption, highlighting the significant impact of envelope improvements in cold climates. This finding underscores the importance of targeted retrofitting strategies that maximize energy savings with minimal operational complexity;
2. Heating System Optimization: Optimizing heating system operation was also explored as a means of enhancing energy efficiency. However, modifications in heating control parameters can directly influence indoor thermal comfort. The study found

that reducing temperature setpoints during working hours led to a decline in comfort hours, whereas adjusting setpoints during unoccupied periods intensified discomfort. These insights emphasize the need for a balanced approach when implementing control-based energy-saving measures;

3. Multi-Objective Optimization for Energy and Comfort Trade-offs: To address the trade-off between energy efficiency and thermal comfort, a multi-objective optimization framework was developed. This approach successfully minimized heating energy demand while maintaining indoor comfort levels at an acceptable threshold. The findings demonstrate that optimization-based decision support tools can play a crucial role in developing energy-efficient yet occupant-friendly buildings.

The insights gained from this study reinforce the potential of grey-box modelling as a practical and scalable solution for building–energy analysis, optimization, and digital twin applications. Future work should focus on further refining model granularity by incorporating additional real-time data sources, such as occupancy patterns and dynamic shading control. Additionally, integrating machine learning techniques for automated parameter calibration could enhance model adaptability and predictive accuracy. Moreover, there is a need to combine the technical assessment and analysis of retrofit solutions presented in this paper with a thorough economic analysis for more effective decision-making. Overall, this study contributes to advancing data-driven energy modelling methodologies by demonstrating how simplified grey-box models can provide actionable insights for building performance optimization, ultimately supporting more sustainable and efficient built environments.

Author Contributions: Conceptualization, M.J. and Y.Y.; methodology, M.J. and Y.Y.; software, Y.Y.; validation, Y.Y.; formal analysis, M.J. and Y.Y.; investigation, M.J. and Y.Y.; resources, M.J.; data curation, Y.Y.; writing—original draft preparation, M.J. and Y.Y.; writing—review and editing, M.J.; visualization, Y.Y.; supervision, M.J.; project administration, M.J.; funding acquisition, M.J. All authors have read and agreed to the published version of the manuscript.

Funding: This research was carried out under the "DanRETwin: A Digital Twin solution for optimal energy retrofit decision-making and decarbonization of the Danish building stock" project, funded by the Danish Energy Agency (Energistyrelsen) under the Energy Technology Development and Demonstration Program (EUDP), ID number: 640222-496754.

Institutional Review Board Statement: Not applicable.

Informed Consent Statement: Not applicable.

Data Availability Statement: The data presented in this study are available on request from the corresponding author.

Conflicts of Interest: The authors declare no conflicts of interest. The funders had no role in the design of the study; in the collection, analyses, or interpretation of data; in the writing of the manuscript, or in the decision to publish the results.

References

1. Hooshmand, S.M.; Zhang, H.; Javidanfar, H.; Zhai, Y.; Wagner, A. A Review of Local Radiant Heating Systems and Their Effects on Thermal Comfort and Sensation. *Energy Build.* **2023**, *296*, 113331. [CrossRef]
2. Zhang, L.; Wen, J.; Li, Y.; Chen, J.; Ye, Y.; Fu, Y.; Livingood, W. A Review of Machine Learning in Building Load Prediction. *Appl. Energy* **2021**, *285*, 116452. [CrossRef]
3. Cespedes-Cubides, A.S.; Jradi, M. A Review of Building Digital Twins to Improve Energy Efficiency in the Building Operational Stage. *Energy Inf.* **2024**, *7*, 11. [CrossRef]
4. de Wilde, P. Building Performance Simulation in the Brave New World of Artificial Intelligence and Digital Twins: A Systematic Review. *Energy Build.* **2023**, *292*, 113171. [CrossRef]

Pan, Y.; Zhu, M.; Lv, Y.; Yang, Y.; Liang, Y.; Yin, R.; Yang, Y.; Jia, X.; Wang, X.; Zeng, F.; et al. Building Energy Simulation and Its Application for Building Performance Optimization: A Review of Methods, Tools, and Case Studies. *Adv. Appl. Energy* **2023**, *10*, 100135. [CrossRef]

Galiano-Garrigós, A.; García-Figueroa, A.; Rizo-Maestre, C.; González-Avilés, Á. Evaluation of BIM Energy Performance and CO_2 Emissions Assessment Tools: A Case Study in Warm Weather. *Build. Res. Inf.* **2019**, *47*, 787–812. [CrossRef]

Hafez, F.S.; Sa'di, B.; Safa-Gamal, M.; Taufiq-Yap, Y.H.; Alrifaey, M.; Seyedmahmoudian, M.; Stojcevski, A.; Horan, B.; Mekhilef, S. Energy Efficiency in Sustainable Buildings: A Systematic Review with Taxonomy, Challenges, Motivations, Methodological Aspects, Recommendations, and Pathways for Future Research. *Energy Strategy Rev.* **2023**, *45*, 101013. [CrossRef]

Yang, Y.; Zhang, G.; Rong, L. Assessing Energy Saving Potential of Wavelength-Dependent Passive Daytime Radiative Cooler Implemented with EnergyPlus by a Roof Model. *Heliyon* **2024**, *10*, e26428. [CrossRef] [PubMed]

TRNSYS: Transient System Simulation Tool. Available online: https://www.trnsys.com/ (accessed on 28 April 2024).

10. Yang, Y.; Zhang, G.; Rong, L. Temperature Adaptive Passive Daytime Radiative Cooling: The Impact of Their Essential Properties on Energy Performance of Different Building Types. *Energy Build.* **2024**, *325*, 114955. [CrossRef]
11. Herbinger, F.; Vandenhof, C.; Kummert, M. Building Energy Model Calibration Using a Surrogate Neural Network. *Energy Build.* **2023**, *289*, 113057. [CrossRef]
12. Chen, Y.; Guo, M.; Chen, Z.; Chen, Z.; Ji, Y. Physical Energy and Data-Driven Models in Building Energy Prediction: A Review. *Energy Rep.* **2022**, *8*, 2656–2671. [CrossRef]
13. Bourdeau, M.; Zhai, X.Q.; Nefzaoui, E.; Guo, X.; Chatellier, P. Modeling and Forecasting Building Energy Consumption: A Review of Data-Driven Techniques. *Sustain. Cities Soc.* **2019**, *48*, 101533. [CrossRef]
14. Huang, T.; Bacher, P.; Møller, J.K.; D'Ettorre, F.; Markussen, W.B. A Step towards Digital Operations—A Novel Grey-Box Approach for Modelling the Heat Dynamics of Ultra-Low Temperature Freezing Chambers. *Appl. Energy* **2023**, *349*, 121630. [CrossRef]
15. Li, Y.; O'Neill, Z.; Zhang, L.; Chen, J.; Im, P.; DeGraw, J. Grey-Box Modeling and Application for Building Energy Simulations—A Critical Review. *Renew. Sustain. Energy Rev.* **2021**, *146*, 111174. [CrossRef]
16. Brastein, O.M.; Perera, D.W.U.; Pfeifer, C.; Skeie, N.-O. Parameter Estimation for Grey-Box Models of Building Thermal Behaviour. *Energy Build.* **2018**, *169*, 58–68. [CrossRef]
17. Vivian, J.; Zarrella, A.; Emmi, G.; De Carli, M. An Evaluation of the Suitability of Lumped-Capacitance Models in Calculating Energy Needs and Thermal Behaviour of Buildings. *Energy Build.* **2017**, *150*, 447–465. [CrossRef]
18. Vivian, J.; Prataviera, E.; Gastaldello, N.; Zarrella, A. A Comparison between Grey-Box Models and Neural Networks for Indoor Air Temperature Prediction in Buildings. *J. Build. Eng.* **2024**, *84*, 108583. [CrossRef]
19. Wang, Z.; Chen, Y.; Li, Y. Development of RC Model for Thermal Dynamic Analysis of Buildings through Model Structure Simplification. *Energy Build.* **2019**, *195*, 51–67. [CrossRef]
20. Arroyo, J.; Spiessens, F.; Helsen, L. Identification of Multi-Zone Grey-Box Building Models for Use in Model Predictive Control. *J. Build. Perform. Simul.* **2020**, *13*, 472–486. [CrossRef]
21. Zhang, X.; Rasmussen, C.; Saelens, D.; Roels, S. Time-Dependent Solar Aperture Estimation of a Building: Comparing Grey-Box and White-Box Approaches. *Renew. Sustain. Energy Rev.* **2022**, *161*, 112337. [CrossRef]
22. Klanatsky, P.; Veynandt, F.; Heschl, C. Grey-Box Model for Model Predictive Control of Buildings. *Energy Build.* **2023**, *300*, 113624. [CrossRef]
23. Sun, Y.; Haghighat, F.; Fung, B.C.M. A Review of The-State-of-the-Art in Data-Driven Approaches for Building Energy Prediction. *Energy Build.* **2020**, *221*, 110022. [CrossRef]
24. Thilker, C.A.; Bacher, P.; Cali, D.; Madsen, H. Identification of Non-Linear Autoregressive Models with Exogenous Inputs for Room Air Temperature Modelling. *Energy AI* **2022**, *9*, 100165. [CrossRef]
25. R: The R Project for Statistical Computing [WWW Document]. Available online: https://www.r-project.org/ (accessed on 11 November 2024).
26. ctsmr—Continuous Time Stochastic Modelling for R. Available online: https://ctsm.info/index.html (accessed on 11 November 2024).
27. Leerbeck, K.; Bacher, P.; Heerup, C.; Madsen, H. Grey Box Modeling of Supermarket Refrigeration Cabinets. *Energy AI* **2023**, *11*, 100211. [CrossRef]
28. Leerbeck, K.; Bacher, P.; Junker, R.G.; Tveit, A.; Corradi, O.; Madsen, H.; Ebrahimy, R. Control of Heat Pumps with CO_2 Emission Intensity Forecasts. *Energies* **2020**, *13*, 2851. [CrossRef]
29. Melgaard, S.P.; Johra, H.; Nyborg, V.Ø.; Marszal-Pomianowska, A.; Jensen, R.L.; Kantas, C.; Larsen, O.K.; Hu, Y.; Frandsen, K.M.; Larsen, T.S.; et al. A Danish High-Resolution Dataset for Six Office Rooms with Occupancy, Indoor Environment, Heating, Ventilation, Lighting and Room Control Monitoring. 2024. Available online: https://zenodo.org/records/10673763 (accessed on 11 November 2024).
30. Johra, H.; Aalborg University. *General Study Case Description of TMV 23: A Multi-Storey Office Building and Living Lab in Denmark*; Department of the Built Environment: Aalborg, Denmark, 2023. [CrossRef]

31. Hansen, M.L. Danish Building Regulations 2015. Available online: https://www.google.com/url?sa=t&source=web rct=j&opi=89978449&url=https://historisk.bygningsreglementet.dk/file/591081/br15_english.pdf&ved=2ahUKEwj5jcvl2 smLAxUWgVYBHerCEJ0QFnoECBYQAQ&usg=AOvVaw2RQ8-CcfWoFBDucScZkNFx (accessed on 11 November 2024).
32. Jradi, M.; Lecuelle, P.; Høj Madsen, K.M.; Veje, C.; Jørgensen, B.N. Dynamic Model-Driven Energy Retrofit of Bøgevangen an Runevej Daycare Centers in Aarhus. *Energy Procedia* **2017**, *132*, 975–981. [CrossRef]
33. Stoffel, P.; Maier, L.; Kümpel, A.; Schreiber, T.; Müller, D. Evaluation of Advanced Control Strategies for Building Energy System *Energy Build.* **2023**, *280*, 112709. [CrossRef]
34. Gaarder, J.E.; Friis, N.K.; Larsen, I.S.; Time, B.; Møller, E.B.; Kvande, T. Optimization of Thermal Insulation Thickness Pertainin to Embodied and Operational GHG Emissions in Cold Climates—Future and Present Cases. *Build. Environ.* **2023**, *234*, 11018 [CrossRef]
35. Kaynakli, O. A Review of the Economical and Optimum Thermal Insulation Thickness for Building Applications. *Renew. Sustai Energy Rev.* **2012**, *16*, 415–425. [CrossRef]
36. Baghoolizadeh, M.; Rostamzadeh-Renani, M.; Rostamzadeh-Renani, R.; Toghraie, D. Multi-Objective Optimization of Venetia Blinds in Office Buildings to Reduce Electricity Consumption and Improve Visual and Thermal Comfort by NSGA-II. *Energ Build.* **2023**, *278*, 112639. [CrossRef]
37. DS/EN 16798-1:2019Webshop Dansk Standard [WWW Document]. Available online: https://webshop.ds.dk/standard/M2974 9/ds-en-16798-1-2019 (accessed on 11 November 2024).
38. Dong, Z.; Boyi, Q.; Pengfei, L.; Zhoujian, A. Comprehensive Evaluation and Optimization of Rural Space Heating Modes in Col Areas Based on PMV-PPD. *Energy Build.* **2021**, *246*, 111120. [CrossRef]
39. Cheung, T.; Schiavon, S.; Parkinson, T.; Li, P.; Brager, G. Analysis of the Accuracy on PMV—PPD Model Using the ASHRA Global Thermal Comfort Database II. *Build. Environ.* **2019**, *153*, 205–217. [CrossRef]
40. Blank, J.; Deb, K. Pymoo: Multi-Objective Optimization in Python. *IEEE Access* **2020**, *8*, 89497–89509. [CrossRef]
41. Arendt, K.; Clausen, A.; Mattera, C.G.; Jradi, M.; Johansen, A.; Veje, C.T.; Kjærgaard, M.B.; Jørgensen, B.N. Multi-Objective Mode Predictive Control Framework for Buildings. In Proceedings of the 16th IBPSA Conference, Rome, Italy, 2–4 September 201 pp. 2779–2786. [CrossRef]
42. Dong, B.; Prakash, V.; Feng, F.; O'Neill, Z. A Review of Smart Building Sensing System for Better Indoor Environment Contro *Energy Build.* **2019**, *199*, 29–46. [CrossRef]
43. Box, G.E.P.; Jenkins, G.M.; Reinsel, G.C.; Ljung, G.M. *Time Series Analysis: Forecasting and Control*; John Wiley & Sons: Hoboker NJ, USA, 2015.
44. Zhou, S.; Zhao, J. Optimum Combinations of Building Envelop Energy-Saving Technologies for Office Buildings in Differen Climatic Regions of China. *Energy Build.* **2013**, *57*, 103–109. [CrossRef]
45. Omidvar, A.; Kim, J. Modification of Sweat Evaporative Heat Loss in the PMV/PPD Model to Improve Thermal Comfor Prediction in Warm Climates. *Build. Environ.* **2020**, *176*, 106868. [CrossRef]
46. Serasinghe, R.; Long, N.; Clark, J.D. Parameter Identification Methods for Low-Order Gray Box Building Energy Models: A Critical Review. *Energy Build.* **2024**, *311*, 114123. [CrossRef]

MDPI AG
Grosspeteranlage 5
4052 Basel
Switzerland
Tel.: +41 61 683 77 34

Sustainability Editorial Office
E-mail: sustainability@mdpi.com
www.mdpi.com/journal/sustainability

www.ingramcontent.com/pod-product-compliance
Lightning Source LLC
Chambersburg PA
CBHW061558220926
51850CB00005B/365
9783725878413